Y0-BTA-819

# Number Theory

# Number Theory

Proceedings of the First Conference of the
Canadian Number Theory Association
held at the Banff Center, Banff, Alberta, April 17-27, 1988

Edited by

Richard A. Mollin

Walter de Gruyter
Berlin · New York 1990

Editor

Richard A. Mollin
Mathematics Department
The University of Calgary
Calgary, Alberta, Canada T2N 1N4

1980 Mathematics Subject Classification (1985 Revision): 11-06; 05A15, 11A25, 11B27, 11E12, 11L05, 11N35, 11K65, 11R29, 11R42, 11S40, 11T21, 13A20, 14K07, 20C05, 33A25, 65D20, 68Q25.

*Library of Congress Cataloging-in-Publication Data*

Canadian Number Theory Association. Conference (1st : 1988 : Banff Center)
Number theory.

1. Number theory--Congresses. I. Mollin, Richard A., 1947- . II. Title.
QA241.C275 1988 512'.7 89-25900
ISBN 0-89925-570-1 (alk. paper)

*Deutsche Bibliothek Cataloging-in-Publication Data*

**Number theory** : held at the Banff Center, Banff, Alberta, April 17-27, 1988 / ed. by Richard A. Mollin. - Berlin ; New York : de Gruyter, 1990
(Proceedings of the ... conference of the Canadian Number Theory Associaton ; 1)
ISBN 3-11-011723-1
NE: Mollin, Richard A. [Hrsg.]; Canadian Number Theory Association: Proceedings of the ...

∞ Printed on acid-free paper.



Printed in Germany
Printing: Ratzlow-Druck, Berlin; Binding: Lüderitz & Bauer, Berlin; Cover design: Thomas Bonnie, Hamburg

# Preface

The Canadian Number Theory Association (CNTA) is an informal organization of Canadian number theorists. It essentially began at the International Number Theory Meeting held at the University of Laval in Quebec during the summer of 1987, when I presented my idea for such an organization to my Canadian colleagues. Claude Levesque graciously volunteered his house for the meeting, where my idea for holding a Number Theory Conference on Canadian soil on a regular basis was well received. However, there was relatively uniform opinion that the organization not be formalized, and so it remains.

The first CNTA conference held at the Banff Center on April 17–27, 1988 was an enormous success due in large part to the enthusiasm and involvement of the participants. There were 114 registered participants from 15 different countries. The meeting had 46 special session talks and 25 one–hour talks given by invited speakers. Moreover, there were numerous social events which met with high approval.

Since the organization was a solo effort on my part, I do not have an organizing committee to thank per se. However, there were certain individuals who helped me out a great deal before and during the conference and deserve my deepest gratitude. Among them are J. Buchmann, K.E. Kallstrom, P.G. Walsh, H.C. Williams and R. Woodrow. I also owe thanks to all of my Canadian number theory colleagues who gave me valued advice during the preparation of the conference. In particular, however, I must cite Claude Levesque and J.M. DeKoninck, the organizers of the aforementioned Quebec number theory meeting, for passing on their advice (well learned from their own organizational experience). Thanks must also go to the typists, Ms. B.L. Kennedy and Ms. K.E. Kallstrom, for the onerous task of retyping *all* manuscripts. Finally, since virtually all papers had two referees, then thanks must go to them for what is otherwise a thankless task.

At the time of this writing, the second CNTA conference has just successfully concluded at the University of British Columbia in Vancouver. The meeting was organized by David Boyd who secured roughly 120 registered participants. The CNTA committee met during the Vancouver meeting to decide on the third meeting, and we agreed that it would be held in 1991 at Queen's University at Kingston.

With this solid foundation for the CNTA meetings, I am confident that we will be hosts to the world number theory community for some time to come, and that much valued research will follow from the contacts made, as has already occurred with the first two conferences.

August 31, 1989 Richard Mollin

# List of Participants

M. Baica (U.S.A.)
M. Beattie (Canada)
P. Borwein (Canada)
A. Bremner (U.S.A.)
J. Buchmann (F.R.G.)
J. Chahal (U.S.A.)
C. Corrales Rodrigañez (Spain)
M.J. Deleon (U.S.A.)
H. Diamond (U.S.A.)
A.G. Earnest (U.S.A.)
R.J. Evans (U.S.A.)
L.J. Federer (U.S.A.)
A.R. Freedman (Canada)
D. Goldfeld (U.S.A.)
A. Granville (Canada)
R. Gupta (Canada)
J. Hafner (U.S.A.)
D. Hayes (U.S.A.)
R. Hudson (U.S.A.)
E. Jacobson (U.S.A.)
M. Jutila (Finland)
P. Kaplan (France)
H. Kisilevsky (Canada)
M. Knopp (U.S.A.)
M. Langevin (France)
C. Levesque (Canada)
L. Lipschitz (U.S.A.)
Y.V. Matijasevic (U.S.S.R.)
A. Mercier (Canada)
H. Montgomery (U.S.A.)
R. Murty (Canada)
K. Nagasaka (U.S.A.)
M. Nathanson (U.S.A.)
A.E. Ozluk (U.S.A.)

P. Bao (Canada)
B.C. Berndt (U.S.A.)
D. Boyd (Canada)
T. Brown (Canada)
P. Bundschuh (F.R.G.)
E.L. Cohen (Canada)
J.M. DeKoninck (Canada)
F. DeMeyer (U.S.A.)
J. Dixon (Canada)
P. Erdös (Hungary)
J. Fabrykowski (Canada)
W. Forsythe (U.S.A.)
J. Friedlander (Canada)
D. Grant (U.S.A.)
G. Greenfield (U.S.A.)
R.K. Guy (Canada)
K. Hardy (Canada)
A. Hildebrand (U.S.A.)
H. Ito (U.S.A.)
J.P. Jones (Canada)
E. Kani (Canada)
I. Katai (Hungary)
M.S. Klamkin (Canada)
C. Lacampagne (U.S.A.)
A.J. Lazarus (U.S.A.)
H. Levitz (U.S.A.)
E. Liverance (U.S.A.)
M. Mays (U.S.A.)
R. Mollin (Canada)
W. Müller (Austria)
G. Myerson (Australia)
T. Nakahara (Japan)
W.G. Nowak (Austria)
A. Pethö (Hungary)

A. Pfister (F.R.G.)
C. Pomerance (U.S.A.)
D. Richard (France)
J. Riddell (Canada)
L. Roberts (Canada)
H.G. Rück (F.R.G.)
R. Sasaki (Japan)
W.M. Schmidt (U.S.A.)
R. Sczech (U.S.A.)
J. Selfridge (U.S.A.)
P. Shiue (U.S.A.)
V. Skarda (U.S.A.)
M. Stone (Canada)
K. Takeuchi (Japan)
P.E. Thomas (U.S.A.)
R. Tijdeman (The Netherlands)
A. Van der Poorten (Australia)
P. Vojta (U.S.A.)
L. Washington (U.S.A.)
H. Williams (Canada)
R.E. Woodrow (Canada)
L. Yip (Canada)
H.J. Zassenhaus (U.S.A.)

A. Pollington (U.S.A.)
J. Propp (U.S.A.)
B. Richmond (Canada)
C. Riehm (Canada)
M. Rosen (U.S.A.)
J. Sands (U.S.A.)
A. Schinzel (Poland)
J. Schönheim (Israel)
S. Sehgal (Canada)
Z. Shan (China)
B. Silverman (U.S.A.)
C. Stewart (Canada)
M.V. Subbarao (Canada)
F. Thaine (Brazil)
P. Thurnheer (France)
L. Van den Dries (U.S.A.)
K. Varadarajan (Canada)
G. Walsh (Canada)
W. Webb (U.S.A.)
K.S. Williams (Canada)
S. Yates (U.S.A.)
H. Yokoi (Japan)
H. Zimmer (F.R.G.)

**1** C. Levesque, Canada
**2** A. van der Poorten, Australia
**3** A. Pollington, USA
**4** H. Williams, Canada
**5** V. Skarda, USA
**6** M. Klamkin, Canada
**7** J. Selfridge, USA
**8** J. Propp, USA
**9** K.S. Williams, Canada
**10** H. Abbot, Canada
**11** F. Thaine, Brazil
**12** J. Chahal, USA
**13** C. Lacampagne, USA
**14** H. G. Ruck, F.R.G.
**15** P. Vojta, USA
**16** R. Guy, Canada
**17** R. A. Mollin, Canada
**18** G. Walsh, Canada
**19** H. Zimmer, F.R.G.
**20** A. Pfister, F.R.G.
**21** J. Riddell, Canada
**22** E. Kani, Canada
**23** G. Myerson, Australia
**24** H. Yokoi, Japan
**25** S. Yates, USA
**26** A. Pethö, Hungary
**27** C. Stewart, Canada
**28** P. Kaplan, France
**29** H. Montgomery, USA
**30** J. Sands, USA
**31** J. Friedlander, Canada
**32** D. Boyd, Canada
**33** R. Silverman, USA
**34** R. Gupta, Canada
**35** W. Nowak, Austria
**36** Mrs. Nowak, Austria
**37** W. Muller, Austria
**38** K. Takeuchi, Japan
**39** D. Grant, USA
**40** A. Granville, Canada
**41** B. Berndt, USA
**42** A. Lazarus, USA
**43** P. Bundschuh, F.R.G.
**44** M. Mays, USA
**45** W. Forsythe, USA
**46** A. Özlük, USA
**47** M. Jutila, Finland
**48** D. Richard, France
**49** J. Jones, Canada
**50** Y.V. Matijasevic, Russia
**51** J. Buchmann, F.R.G.
**52** R. Sasaki, Japan
**53** H. Levitz, USA
**54** A. Freedman, Canada
**55** G. Greenfield, USA
**56** R. Woodrow, Canada
**57** R. Tijdeman, Netherlands
**58** J. Dixon, Canada
**59** E. Liverance, USA
**60** M. Langevin, France
**61** Z. Shan, China
**62** M. Deleon, USA
**63** T. Nakahara, Japan
**64** J. Fabrykowski, Canada

# Table of Contents

# An Integral Functional Equation of Ramanujan Related to the Dilogarithm

*Bruce C. Berndt* [1] *and Ronald J. Evans*

Ramanujan's third notebook encompasses only 33 pages of unorganized material. In this notebook, on page 373 of volume 2 of the Tata Institute's edition [3], Ramanujan considers a certain logarithmic integral. Let $n \geq 0$, put $v = u^n - u^{n-1}$, and define

$$\phi(n) = \int_0^1 \frac{\log u}{v} dv .$$

Ramanujan offers the values

$$\phi(0) = \frac{\pi^2}{6}, \tag{1}$$

$$\phi(1) = \frac{\pi^2}{12}, \tag{2}$$

and

$$\phi(2) = \frac{\pi^2}{15}, \tag{3}$$

and then asserts the following beautiful, and perhaps surprising, functional equation.

**Theorem 1.** *For* $n > 0$,

$$\phi(n) + \phi\left(\frac{1}{n}\right) = \frac{\pi^2}{6}. \tag{4}$$

---

[1]Research partially supported by a grant from the Vaughn Foundation.

As we shall see in the sequel, $\phi(n)$ is closely related to the dilogarithm $Li_2(z)$ defined by

$$Li_2(z) = -\int_0^z \frac{\log(1-t)}{t}\,dt, \tag{5}$$

where $z$ belongs to the cut complex $z$-plane $C - (1,\infty)$. By expanding $\log(1-t)$ in its Maclaurin series and integrating termwise, we easily find that

$$Li_2(z) = \sum_{n=1}^{\infty} \frac{z^n}{n^2}, \qquad |z| \le 1. \tag{6}$$

Theorem 1, in fact, is a special case of the following much more general theorem.

**Theorem 2.** *Let $g$ be a strictly increasing, differentiable function on $[0,\infty)$ with $g(0) = 1$ and $g(\infty) = \infty$. For $n > 0$ and $t \ge 0$, define*

$$v(t) = \frac{g(t)^n}{g(t^{-1})}. \tag{7}$$

*Suppose that*

$$\phi(n) := \int_0^1 \log g(t) \frac{dv}{v} \tag{8}$$

*converges. Then*

$$\phi(n) + \phi\left(\frac{1}{n}\right) = 2\phi(1). \tag{9}$$

Let $g(t) = 1 + t$. Observing that $v(t) = t(1+t)^{n-1}$, setting $u = 1 + t$, and using (2), which is proved below, we find that Theorem 2 reduces to Theorem 1.

Further examples for which Theorem 2 is valid can be obtained from setting $g(t) = \log(t + e)$ and $g(t) = 1 + a_1 t + a_2 t^2 + \ldots + a_k t^k$, where $k \ge 1$ and $a_j > 0$, $1 \le j \le k$.

**Proof of Theorem 2.** Since $g(\infty) = \infty$, it follows from (7) that $v(0) = 0$. Also, $v(w) = 1$, where $w$ is defined by

$$n = \frac{\log g(w^{-1})}{\log g(w)}. \tag{10}$$

Since $g(0) = 1$ and $g$ is increasing, we see from (10) that each $n > 0$ determines a unique $w > 0$, and conversely each $w > 0$ determines a unique $n > 0$. We also see from (10) that interchanging $n$ and $n^{-1}$ corresponds to interchanging $w$ and $w^{-1}$. Define a real valued function $F$ on $[0,\infty)$ by

$$F(w) = \phi(n). \tag{11}$$

Then (9) is equivalent to

$$F(w) + F(w^{-1}) = 2F(1). \tag{12}$$

Differentiating (12), we find that

$$F'(w) = w^{-2}F'(w^{-1}). \tag{13}$$

Since (12) is clearly valid for $w = 1$, it then suffices to prove (13).

By (7),

$$\frac{dv}{v} = nd \log g(t) - d \log g(t^{-1}). \tag{14}$$

Therefore, by (11), (8), (10), and (14),

$$F(w) = \frac{\log g(w^{-1})}{\log g(w)} \int_{t=0}^{w} \log g(t)\, d \log g(t) - \int_{t=0}^{w} \log g(t)\, d \log g(t^{-1})$$

$$= \frac{1}{2} \log g(w^{-1}) \log g(w) - \int_{t=0}^{w} \log g(t)\, d \log g(t^{-1}). \tag{15}$$

Thus,

$$F'(w) = \frac{1}{2}\left\{ \log g(w^{-1})\frac{d}{dw}\log g(w) + \log g(w)\frac{d}{dw}\log g(w^{-1})\right\}$$

$$- \log g(w)\frac{d}{dw}\log g(w^{-1})$$

$$= \frac{1}{2}\left\{ \log g(w^{-1})\frac{d}{dw}\log g(w) - \log g(w)\frac{d}{dw}\log g(w^{-1})\right\}.$$

The equality in (13) now easily follows, and so the proof of Theorem 2 is complete.

Setting $g(t) = 1 + t$ in (15), we find that

$$F(w) = \frac{1}{2}\{\log(1+w) - \log w\}\log(1+w) - \int_0^w \log(1+t)\left\{\frac{1}{1+t} - \frac{1}{t}\right\}dt$$

$$= -\frac{1}{2}\log w\ \log(1+w) + \int_0^w \frac{\log(1+t)}{t}\,dt. \tag{16}$$

Expanding $\log(1+t)$ in its Maclaurin series and integrating termwise, we deduce that

$$F(1) = \sum_{n=1}^{\infty}\frac{(-1)^{n-1}}{n^2} = \frac{\pi^2}{12}. \tag{17}$$

This proves (2).

By (5), we may rewrite (16) in the form

$$F(w) = -\frac{1}{2}\log w\ \log(1+w) - Li_2(-w). \tag{18}$$

Hence, we easily deduce from Theorem 1 (or Theorem 2) that

$$\frac{\pi^2}{6} = F(w) + F(w^{-1}) = -\frac{1}{2}\log^2 w - Li_2(-w) - Li_2(-w^{-1}),$$

which is a well-known functional equation for the dilogarithm [2], p. 4. Thus, Theorem 2 can be regarded as a vast generalization of this classical functional equation. One might then ask how much of the general theory of the dilogarithm can be generalized for the more general function $\phi(n)$ defined by (8). Ramanujan, in fact, considered some different generalizations of $Li_2(z)$ [3], Chapter 9, [1], pp. 249—260.

Observe that if we let $n$ tend to $\infty$ in (4), we deduce (1), since $\phi(\infty) = 0$. Of course, it is not difficult to prove (1) directly from the definition of $\phi(0)$.

Finally, we prove (3). By (18) and (11) with $n = 2$, $w = \dfrac{\sqrt{5}-1}{2}$, we have

$$\phi(2) = \frac{1}{2}\log^2 w - Li_2(-w) = \frac{\pi^2}{15},$$

where we have used a well-known value for $Li_2(-w)$ [2], p. 7.

Added in proof: The problem of Ramanujan that we addressed in this paper was also submitted by Ramanujan as a problem in The Journal of the Indian Mathematical Society 10 (1918), 397–399. (Also see Ramanujan's Collected Papers, p. 334.) One solution, somewhat lengthier than that of this paper, was received and published. No generalization was indicated.

## References

[1] *B.C. Berndt*, Ramanujan's Notebooks, Part I. Springer–Verlag, New York, 1985.

[2] *L. Lewin*, Polylogarithms and Associated Functions. North–Holland, New York, 1981.

[3] *S. Ramanujan*, Notebooks (2 volumes). Tata Institute of Fundamental Research, Bombay, 1957.

---

Dept. of Mathematics, University of Illinois, Urbana, IL 61801, USA
Dept. of Mathematics, University of California, La Jolla, CA 92093, USA

# Cyclotomic Partitions

*David W. Boyd* [1] *and Hugh L. Montgomery* [2]

## 1. Introduction

Let

$$\phi_q(z) = \prod_{\substack{a=1 \\ (a,q)=1}}^{q} (z - e(a/q))$$

denote the $q$–th cyclotomic polynomial. This polynomial is monic, irreducible, of degree $\varphi(q)$, and its roots are the primitive $q$–th roots of unity. More generally, we call a monic polynomial *cyclotomic* if it has integral coefficients and its roots lie on the unit circle $|z| = 1$. In view of a classical theorem of Kronecker, it follows that the roots of such a polynomial are roots of unity. Consequently, a cyclotomic polynomial is simply a product of the irreducible polynomials $\phi_q$. Our object is to estimate the number $c(n)$ of such polynomials of degree $n$, and also the number $c_d(n)$ of such polynomials of degree $n$ in which the factors $\phi_q$ are distinct. The question of the size of these functions arises naturally when one considers the problem of listing all monic polynomials $P \in \mathbb{Z}[x]$ of degree $n$ whose Mahler measure lies below a given bound. Since the cyclotomic polynomials will fall in such a class, the number $c(n)$ provides a trivial lower bound for the computational complexity of the task.

It is convenient to adopt the convention that $c(0) = c_d(0) = 1$. For positive $n$ we see that $c(n)$ is the number of solutions of the equation

---

1 Supported in part by an NSERC operating grant.
2 Research supported in part by National Science Foundation grant NSF–DMS–85–02804.

$$\sum_q k_q\,\varphi(q) = n \tag{1}$$

in non–negative integers $k_q$. Thus we have a partition problem in which the summands are the numbers $\varphi(q)$. Similarly, $c_d(n)$ is the number of solutions of (1) in which the $k_q \in \{0,1\}$. Thus $c_d(n)$ counts the number of square–free cyclotomic polynomials of degree $n$.

After deriving suitable estimates for the underlying power series generating functions, we use Cauchy's formula to express $c(n)$ in terms of the power series, and then employ the saddle point method to complete the estimation. Our analysis amounts to a crude application of the Hardy–Littlewood circle method, in which we have only one major arc and one minor arc. An asymptotic formula for $c(n)$ could be derived from the major arc estimate alone, by appealing to the Tauberian theorem for partitions of Ingham [5], but then one would have only a qualitative estimate. To apply Ingham's theorem, one must establish a major arc estimate (our Lemmas 2 and 3 suffice), and also that the coefficients $c(n)$ and $c_d(n)$ are monotonically increasing. These latter facts are easily derived from (31) and (50) below. Alternatively, one may show that $c(n)$ is strictly increasing by appealing to the criterion of Bateman and Erdös [2]. The asymptotic behaviour of $c(n)$ and $c_d(n)$ cannot be determined by means of the simpler Tauberian theorem of Meinardus [6], as the underlying Dirichlet series is not regular at $s = 0$ (see Lemma 1).

**Theorem 1.** *Let $c(n)$ and $c_d(n)$ be defined as above. Then*

$$c(n) = A(\log n)^{-1/2}\, n^{-1} \exp(B\sqrt{n})\left(1 + O\left(\frac{\log\log n}{\log n}\right)\right) \tag{2}$$

*as $n \to \infty$, where*

$$\begin{aligned} A &= (2\pi)^{-1} \exp(-C_0/2)\,\zeta(2)\,\zeta(3)^{1/2}\,\zeta(6)^{-1/2} \\ &= \frac{1}{4\pi^2}\left(105\,\zeta(3)/e^{C_0}\right)^{1/2} = 0.213234\ldots, \end{aligned} \tag{3}$$

$$B = 2\,\zeta(2)\,\zeta(3)^{1/2}\,\zeta(6)^{-1/2} = \frac{1}{\pi}\left(105\,\zeta(3)\right)^{1/2} = 3.57608\ldots, \tag{4}$$

*and $C_0 = 0.5772156\ldots$ denotes Euler's constant. Moreover,*

$$c_d(n) = A_d n^{-3/4} \exp(B_d \sqrt{n}) \left(1 + O\left(\frac{1}{\log n}\right)\right) \tag{5}$$

*where*

$$A_d = 2^{-7/4} \pi^{-1/2} \zeta(2)^{1/2} \zeta(3)^{1/4} \zeta(6)^{-1/4} = \frac{1}{4\pi} (105\,\zeta(3)/2)^{1/4} = 0.22429\ldots$$

*and*

$$B_d = 2^{1/2} \zeta(2) \zeta(3)^{1/2} \zeta(6)^{-1/2} = \frac{1}{\pi} (105\,\zeta(3)/2)^{1/2} = 2.52867\ldots$$

The error terms here are best possible, in the sense that if one wanted a smaller error term then it would be necessary to introduce a more precise main term. The form of the expression for $c(n)$ is different from that for the partition function $p(n)$ to the extent that we have a factor $(\log n)^{-1/2}$ which does not arise in the classical case. The constant $B$ is larger than its counterpart $\pi\sqrt{2/3} = 2.56509\ldots$ in the case of $p(n)$, as is to be expected, since we have replaced the classical summands $q$ by the smaller summands $\varphi(q)$. One may note that $B_d = B/\sqrt{2}$, which corresponds to the relation between the constants arising in the classical situation. It is to be expected that numerical values will be seen to converge only slowly to the asymptotic behaviour. For example, when $n = 5000$, the main term in (2) is $9.6275 \times 10^{104}$, which is about 36% higher than the actual value,

$$\begin{aligned} c(5000) = {} & 705600973140112635289309723191853991132386116 4961 \\ & 1563571865172379358873064300578247330339523112477811247 0. \end{aligned}$$

Similarly, when $n = 5000$, the main term in (5) is $1.6990 \times 10^{74}$, which is about 6% higher than the actual value,

$$\begin{aligned} c(5000) = {} & 16052872050208292629647322031344666 77 \\ & 6053783193474490950500380231993236377 2. \end{aligned}$$

The main difficulty in the proof is that the summands $\varphi(q)$ are not very evenly distributed. For example, Erdös [4] and Dressler [3] showed that

$$\operatorname{card}\{q : \varphi(q) \le x\} = \frac{\zeta(2)\zeta(3)}{\zeta(6)} x + o(x), \tag{6}$$

and Bateman [1] gave quantitative estimates for the error term, but it is conjectured that this error term is not $O(x^{1-\varepsilon})$. The underlying Dirichlet series generating function

$$F(s) = \sum_{q=1}^{\infty} \varphi(q)^{-s} \tag{7}$$

is initially defined in the half–plane $\sigma > 1$. Apart from a pole at $s = 1$, it can be continued to the strip $0 < \sigma < 1$, but it takes on large values as $|\text{Im } s|$ grows. For our present purposes we need consider only a power series formed from the $\varphi(q)$, so the behaviour when $|\text{Im } s|$ is large is less important. However, to obtain the necessary precision we must continue $F$ into the half–plane $\sigma < 0$. We find that this can be done, but that $F$ has a transcendental singularity at $s = 0$, which forces us to slit the plane along the negative real axis. It is this singularity at $s = 0$ which gives rise to the unusual factor $(\log n)^{-1/2}$ in (2).

In the subsequent discussion, $c$ denotes a small positive constant, not necessarily the same at each occurrence, while $C$ similarly denotes a large positive constant, but not always the same one.

## 2. Analytic Continuation of the Generating Dirichlet Series

The function $F(s)$ defined in (7) is an ordinary Dirichlet series, which is to say that it can be written in the form $\sum b_n n^{-s}$, where $b_n$ is the number of $q$ for which $\varphi(q) = n$. However, the $b_n$ are not multiplicative, and are irregularly distributed. Erdös [4] proved that there is a $\delta > 0$ such that $b_n > n^{\delta}$ for infinitely many $n$, and Pomerance [7] has shown that one may take $\delta = 3/5$. Nevertheless, for $\sigma > 1$ we see that

$$\begin{aligned} F(s) &= \prod_p (1 + (p-1)^{-s} (1 - p^{-s})^{-1}) \\ &= \zeta(s) \prod_p (1 - p^{-s} + (p-1)^{-s}) \\ &= \zeta(s) Q(s), \end{aligned} \tag{8}$$

say. Since

$$(p-1)^{-s} - p^{-s} = s \int_{p-1}^{p} u^{-s-1} du \ll |s| p^{-\sigma-1}, \tag{9}$$

the product $Q(s)$ is absolutely and uniformly convergent in compact portions of the half–plane $\sigma > 0$. Indeed, we may estimate $Q(s)$ as follows: Write $Q(s) =$

$\Pi_{p\le\tau}\cdot\Pi_{p>\tau}$ where $\tau=|t|+4$ and $s=\sigma+it$. We estimate the first product trivially:

$$\Pi_{p\le\tau} \ll \exp\left(C\frac{\tau^{1-\alpha}}{\log\tau}\right)$$

uniformly for $\sigma\ge\alpha$, where $\alpha\le 1-\delta$. This estimate is achieved by dividing the range in question into dyadic blocks, and using the Chebyshev estimate for $\pi(x)$. To bound the second product we use (9), which gives

$$\Pi_{p>\tau} \ll \exp\left(C\frac{\tau^{1-\alpha}}{\log\tau}\right)$$

uniformly for $\sigma\ge\alpha\ge\delta>0$. Thus

$$Q(s) \ll \exp\left(C\frac{\tau^{1-\alpha}}{\log\tau}\right) \tag{10}$$

uniformly for $\sigma\ge\alpha$, where $0<\delta\le\alpha\le 1-\delta$. It is quite probable that this estimate is far from the truth. One might conjecture that

$$Q(s) \ll \exp\left(\exp\left((\log\tau)^{1-\sigma+\varepsilon}\right)\right)$$

for $0<\sigma<1$, and that this is best possible in the sense that if $0<\sigma<1$ then there exist arbitrarily large $t$ for which

$$|Q(s)| > \exp\left(\exp\left((\log\tau)^{1-\sigma-\varepsilon}\right)\right).$$

When $\sigma<0$, the product $Q(s)$ fails to converge. Nevertheless, the function it defines can be continued to the region $\mathcal{D}=\{s\in\mathbb{C}:\sigma>-c/\log\tau,\ t=0\Rightarrow\sigma>0\}$.

**Lemma 1.** *The function $Q(s)$ is regular in the domain $\mathcal{D}$. If $s\in\mathcal{D}$ and $|s|\le 1/3$, then*

$$Q(s)=1-s\log s+\frac{1}{2}s^2(\log s)^2+c_1s^2+O(|s|^3|\log s|^3). \tag{11}$$

*Moreover,*

$$Q(s)\ll\exp\left(\frac{C\tau}{\log\tau}\right) \tag{12}$$

*uniformly for $s\in\mathcal{D}$.*

**Proof.** Let $\delta_p=\delta_p(s)$ denote the left hand side of (9). From the integral representation in (9) we see that if $|s|<1/3$ then $|\delta_p|\le|s|(p-1)^{-\sigma-1}<1/3$ for all primes $p$. Moreover, we see that $\sum|\delta_p|^2 \ll \sum p^{-4/3} < \infty$. For $\sigma > 0$, $|s| \le 1/3$ we write

$$Q(s) = \prod(1+\delta_p) = s^{-s} R(s)S(s)T(s)U(s) \tag{13}$$

where

$$R(s) = (s\,\zeta(s+1))^s = 1 + c_2 s^2 + O(|s|^3), \tag{14}$$

$$S(s) = \exp\left(-s \sum_{p,k>1} \frac{1}{k} p^{-k(s+1)}\right) \tag{15}$$

$$= 1 - \left(\sum_{p,k>1} \frac{1}{k} p^{-k}\right)s + c_3 s^2 + O(|s|^3),$$

$$T(s) = \exp\left(\sum_p (\delta_p - s/p^{s+1})\right) \tag{16}$$

$$= 1 + \left(\sum_p (\log(1-1/p)^{-1} - 1/p)\right)s + c_4 s^2 + O(|s|^3),$$

and

$$U(s) = \prod(1+\delta_p)e^{-\delta_p} = 1 + c_5 s^2 + O(|s|^3). \tag{17}$$

Each of these functions is regular in the disc $|s| \le 1/3$, and the initial power series coefficients are as indicated. The only difficult verification is in the case of $T(s)$. By integrating by parts in the integral in (9), it is easy to show that the summand in (16) is

$$s(s+1)\int_{p-1}^{p} u^{-s-2}\{u\}\,du \ll p^{-5/3}. \tag{18}$$

Thus the formula (13) serves to continue $Q(s)$ to the slit disc $|s| \le 1/3$, $t = 0 \Rightarrow \sigma > 0$. In this slit disc we have

$$s^{-s} = 1 - s\log s + \frac{1}{2}s^2(\log s)^2 + O(|s\log s|^3),$$

while $R(s)S(s)T(s)U(s) = 1 + c_1 s^2 + O(|s|^3)$, so we have (11).

We now modify the above method to establish (12). If $\tau$ is large and $p = o(\tau)$ then the estimate (7) is worse than the trivial estimate

$$\delta_p \ll p^{-\sigma}.$$

That is, the two terms which comprise $\delta_p$ may have quite different arguments. Thus we split $Q$ into two parts, and treat the small primes trivially, as we did in proving (10). Rather than switch abruptly between the two methods, we pass smoothly from one to the other. Let $X$ be a parameter to be chosen later, put

$$w_1(n) = \begin{cases} 1 & 1 \le n \le X \ , \\ (2X - n)/X & X < n \le 2X \ , \\ 0 & n > 2X \ , \end{cases}$$

and put $w_2(n) = 1 - w_1(n)$. If $s \in \mathcal{D}$, $\sigma \le 1$, then by (9) we see that

$$|\delta_p| \le |s|(p-1)^{-\sigma-1} \le \frac{|s|}{p-1} \exp\left(\frac{c \log p}{\log \tau}\right).$$

If $p \ge 4\tau$ then the above is $\le 1/2$. That is, for these larger primes, we may pass from a factor $(1 + \delta_p)$ to $\exp(\delta_p)$ as before. We suppose now that

$$4\tau \le X \le 8\tau, \tag{19}$$

and we write

$$Q(s) = \prod (1 + \delta_p)^{w_1(p)} \prod (1 + \delta_p)^{w_2(p)} = Q_1(s) Q_2(s), \tag{20}$$

say. If $s \in \mathcal{D}$ then $|1 + \delta_p| \le 3 \exp\left(\frac{c \log p}{\log \tau}\right)$, so that

$$|Q_1(s)| \le \exp\left((\log 3)\pi(2X) + \frac{c}{\log \tau}\vartheta(2X)\right) \le \exp\left(\frac{CX}{\log \tau}\right) \tag{21}$$

by the Chebyshev estimates. For small primes $p$ the factor $1 + \delta_p$ may have zeros in the region under consideration, but $w_1(p) = 1$ for such primes, so that $Q_1(s)$ is regular. If $w_2(p) \ne 0$ then $|\delta_p| \le 1/2$, so that for $\sigma > 0$ we may write

$$Q_2(s) = R_2(s) S_2(s) T_2(s) U_2(s) \tag{22}$$

where

$$R_2(s) = \exp\left(s \sum \frac{\Lambda(n)}{\log n} w_2(n) n^{-s-1}\right), \tag{23}$$

$$S_2(s) = \exp\left(-s \sum_{p,\, k > 1} \frac{1}{k} w_2(p^k) p^{-k(s+1)}\right), \tag{24}$$

$$T_2(s)=\exp\left(\sum_p w_2(p)(\delta_p - s/p^{s+1})\right), \tag{25}$$

and

$$U_2(s)=\prod\left((1+\delta_p)^{w_2(p)}\exp(w_2(p)\delta_p)\right). \tag{26}$$

From (9) we see that

$$\sum_{p>X}|\delta_p|^2 \ll |s|^2\sum_{p>X}p^{-2\sigma-2} \ll \tau^2 X^{-2\sigma-1}/\log X$$

uniformly for $-1/2 < \alpha \le \sigma \le 1$. Thus $U_2(s)$ is regular in this strip, and for $s \in \mathcal{D}$ the above is $\ll \tau/\log\tau$, so that

$$|U_2(s)| \le \exp\left(\frac{C\tau}{\log\tau}\right) \tag{27}$$

From (18) we know that $\delta_p - sp^{-s-1} \ll \tau^2 p^{-\sigma-2}$ for $-1 \le \sigma \le 1$. Thus the sum in (25) is

$$\ll \tau^2 X^{-\sigma-1}/\log X$$

uniformly for $-1 < \alpha \le \sigma \le 1$. Hence $T_2(s)$ is regular in this strip. If $s \in \mathcal{D}$ then the above estimate is $\ll \tau/\log\tau$, so that

$$|T_2(s)| \le \exp\left(\frac{C\tau}{\log\tau}\right) \tag{28}$$

uniformly for $s \in \mathcal{D}$. The sum in (24) is absolutely convergent for $\sigma > -1/2$, and for $-1/2 < \alpha < \sigma \le 1$ it is $\ll X^{-\sigma-1/2}/\log X$. Thus $S_2(s)$ is regular in $\mathcal{D}$. and satisfies the inequality

$$|S_2(s)| \le \exp\left(\frac{C\tau^{1/2}}{\log\tau}\right). \tag{29}$$

As for $R_2(s)$, suppose first that $\sigma > 0$. We note that

$$\sum_n \frac{\Lambda(n)}{\log n}w_1(n)n^{-s-1} = \frac{1}{2\pi i X}\int_{a-i\infty}^{a+i\infty}\log\zeta(s+1+w)\frac{(2X)^{w+1}-X^{w+1}}{w(w+1)}\,dw\,.$$

At this point we invoke the Vinogradov zero–free region for this Riemann zeta function, as found in Titchmarsh [8], §6.15. This permits us to replace the contour of integration by the contour $\mathcal{C}$ which consists of the piecewise linear path from $1/\log X - i\infty$ to $1/\log X - it/2$, to $-1/(\log\tau)^{\alpha} - it/2$, to $-1/(\log\tau)^{\alpha} + it/2$, to $-1/\log X + it/2$, to $1/\log X + i\infty$. Here $\alpha < 1$. In moving the contour we gain a residue at $w = 0$. Thus the above is

$$= \log\zeta(s+1) + \frac{1}{2\pi i X}\int_{\mathcal{C}} \cdots .$$

We express $\log\zeta(s+1)$ by its Dirichlet series, and subtract, to find that

$$\sum_n \frac{\Lambda(n)}{\log n} w_2(n) n^{-s-1} = \frac{-1}{2\pi i X}\int_{\mathcal{C}} \log\zeta(s+1+w)\frac{(2X)^{w+1} - X^{w+1}}{w(w+1)}\, dw .$$

The advantage of the weighting is that this integral is absolutely convergent. As defined, the contour $\mathcal{C}$ depends on $s$, but if we leave $\mathcal{C}$ fixed and make a small change in $s$ we find that the right hand side is regular, and hence the sum on the left is analytically continued to $\mathcal{D}$.

For $s \in \mathcal{D}$, the integral is $\ll \exp(-c(\log\tau)^{1-\alpha})$, so that

$$|R_2(s)| \le \exp(C\tau\exp(-c\log\tau)^{1-\alpha}). \tag{30}$$

At this point we have achieved the analytic continuation of $Q(s)$. Moreover, we may combine (27)–(30) in (22), and combine (21) and (22) in (20) to deduce the desired estimate (12).

## 3. The Power Series Generating Functions

For $\operatorname{Re} z > 0$ we write

$$P(z) = \sum_{n=0}^{\infty} c(n) e^{-nz} = \prod_{q=1}^{\infty} (1 - e^{-\varphi(q)z})^{-1}. \tag{31}$$

On appealing to the power series for $\log(1-w)^{-1}$, we find that this is

$$\exp\left(\sum_{q=1}^{\infty}\sum_{m=1}^{\infty}\frac{1}{m}e^{-m\varphi(q)z}\right)=\exp\left(\sum_{n=1}^{\infty}g(n)e^{-nz}\right) \tag{32}$$

where

$$g(n)=\sum_{m\varphi(q)=n}\frac{1}{m}.$$

This latter power series may be expressed as a Mellin transformation of the corresponding Dirichlet series; for $\sigma>1$ we put

$$G(s)=\sum_{n=1}^{\infty}g(n)n^{-s} \tag{33}$$

and note that this is $\zeta(s+1)F(s)$ where $F(s)$ is the Dirichlet series (7) already considered. On combining (31), (32), (33), (7), and (8), we see that

$$\log P(z)=\frac{1}{2\pi i}\int_{\alpha-i\infty}^{\alpha+i\infty}\zeta(s)\zeta(s+1)Q(s)\Gamma(s)z^{-s}\,ds \tag{34}$$

where $\alpha>1$. Using this representation, we now derive an asymptotic estimate for $P(z)$ when $z$ is near 0.

**Lemma 2.** *Let* $\mathcal{R}=\left\{z\in\mathbb{C}:\ \operatorname{Re}z>0, |\arg z|\le\frac{\pi}{2}-\frac{C}{\log\log 1/|z|}\right\}$. *As $z$ tends to 0 in $\mathcal{R}$, we have*

$$P(z)=a(\log 1/z)^{-1/2}z^{1/2}\exp(b/z)E(z) \tag{35}$$

*where*

$$E(z)=1-\frac{1}{2}(\log\log 1/z)/\log 1/z+(\log a)/\log 1/z$$
$$+O\left((\log\log 1/|z|)^{3}(\log 1/|z|)^{2}\right),$$

$a=(2\pi)^{-1/2}\exp(-\frac{1}{2}C_0)$, *$C_0$ is Euler's constant, and* $b=\zeta(2)^2\zeta(3)/\zeta(6)$.

**Proof.** We replace the contour of integration in (34) by a new curve $\mathcal{C}$ which is defined as follows: Put $r=1/\log 1/|z|$. For $-\infty<t\le -r$, follow the path $s=-c/\log\tau+it$. Then proceed by a line segment to the point $-ir$, then by a semicircle in the positive sense, centred at the origin, to the point $ir$, then by a line

segment to the point $-c/\log(r+4)+ir$, then along the curve $s=-c/\log\tau+it$ for $r\le t<\infty$. In moving the contour we encounter a residue at $s=1$. Thus

$$\log P(z)=\zeta(2)Q(1)\frac{1}{z}+\frac{1}{2\pi i}\int_{\mathcal{C}}\zeta(s)\zeta(s+1)Q(s)\Gamma(s)z^{-s}\,ds. \tag{36}$$

Let $\mathcal{C}_1$ denote that part of $\mathcal{C}$ which loops around the origin, while $\mathcal{C}_2$ is the part which runs roughly parallel to the imaginary axis.

From (2.1.16), (2.4.3), and (2.4.5) of Titchmarsh [8] we are able to determine the initial terms of the Laurent expansions of the first two factors of the above integrand about $s=0$. We estimate the third factor by means of (11) of Lemma 1. From Weierstrass's definition of the gamma function as given by Whittaker and Watson [9], p. 236, we recall the initial terms of the Laurent expansion $\Gamma(s)$ about $s=0$. Thus we find that for $s\in\mathcal{C}_1$, the first four factors in the integrand are

$$\begin{aligned}&\left(-\frac{1}{2}-\frac{1}{2}(\log 2\pi)s+\ldots\right)\cdot\left(\frac{1}{s}+C_0+\ldots\right)\\ &\quad\cdot\left(1-s\log s+\frac{1}{2}s^2(\log s)^2+c_1s^2+O(|s\log s|^3)\right)\cdot\left(\frac{1}{s}-C_0+\ldots\right)\\ &=-\frac{1}{2s^2}+\frac{1}{2s}\log s-\frac{\log 2\pi}{2s}-\frac{1}{4}(\log s)^2+\frac{1}{2}(\log 2\pi)\log s+c_6\\ &\quad+O(|s(\log s)^3|.\end{aligned}$$

The contribution of the error term is

$$\ll\int_{\mathcal{C}_1}|s(\log s)^3|\,|z^{-s}|\,|ds|\ll(\log\log 1/|z|)^3/(\log 1/|z|)^2.$$

To determine the contribution of the main terms, we extend $\mathcal{C}_1$ to $-\infty$, both above and below the real axis. As a consequence, the value of the integral is altered by an amount

$$\ll\exp(-c\log 1/|z|).$$

To evaluate the integrals that the main terms give rise to, we recall from Whittaker and Watson [9], p. 245, that for any $w$,

$$\frac{1}{\Gamma(w)}=\frac{1}{2\pi i}\int_{\mathcal{H}}e^s s^{-w}\,ds \tag{37}$$

Where $\mathcal{H}$ is a Hankel contour which loops around the negative real axis in the positive sense. Taking $w = 2$, we see that

$$\frac{1}{2\pi i}\int_{\mathcal{H}} s^{-2} z^{-s}\, ds = -\log z$$

provided that $z$ lies in the slit unit disc $\mathbb{S} = \{z \in \mathbb{C} : |z| < 1,\ \operatorname{Im} z = 0 \Rightarrow \operatorname{Re} z > 0\}$. Taking $w = 1$ gives

$$\frac{1}{2\pi i}\int_{\mathcal{H}} s^{-1} z^{-s}\, ds = 1$$

for $z \in \mathbb{S}$. We now show that if $z \in \mathbb{S}$ then

$$\frac{1}{2\pi i}\int_{\mathcal{H}} s^{-1}(\log s) z^{-s}\, ds = -C_0 - \log\log 1/z\,. \tag{38}$$

for $z \in \mathbb{S}$, both sides of the above represent analytic functions of $z$. Hence it suffices to demonstrate the identity when $z$ is real, $0 < z < 1$. For such $z$ we replace $-(\log z)s$ by $s$, and find that the integral above is

$$\frac{1}{2\pi i}\int_{\mathcal{H}} s^{-1}(\log s) e^{s}\, ds - (\log\log 1/z)\frac{1}{2\pi i}\int_{\mathcal{H}} s^{-1} e^{s}\, ds\,. \tag{39}$$

By (37), the second integral is $2\pi i$. As for the first, we differentiate both sides of (37) to see that

$$\frac{\Gamma'(w)}{\Gamma(w)^2} = \frac{1}{2\pi i}\int_{\mathcal{H}} e^{s}(\log s) s^{-w}\, ds\,. \tag{40}$$

On setting $w = 1$, we find that the first term in (39) is $-C_0$, and thus we have (38). By the same method, we find that

$$\frac{1}{2\pi i}\int_{\mathcal{H}} z^{-s}\, ds = 0,$$

since the integrand is regular. To evaluate the remaining integrals we write the left hand side of (37) in the form $w/\Gamma(w+1)$, differentiate twice, and evaluate the first and second derivatives when $w = 0$. Proceeding as in the derivation of (38), we deduce that

$$\frac{1}{2\pi i}\int_{\mathcal{H}} (\log s) z^{-s}\, ds = -1/\log 1/z ,$$

and that

$$\frac{1}{2\pi i}\int_{\mathcal{H}} (\log s)^2 z^{-s}\, ds = (2(\log\log 1/z) + 2C_0)/\log 1/z .$$

On combining these estimates, we find that

$$\begin{aligned}\frac{1}{2\pi i}\int_{\mathcal{C}_1} &\zeta(s)\zeta(s+1)Q(s)\Gamma(s)z^{-s}\, ds \qquad (41)\\ &= \frac{1}{2}\log z - \frac{1}{2}\log\log 1/z + \log a\\ &-\frac{1}{2}(\log\log 1/z)/\log 1/z + (\log a)/\log 1/z\\ &+O((\log\log 1/|z|)^3 (\log 1/|z|)^2)\end{aligned}$$

uniformly for $\operatorname{Re} z > 0$, $|z| \le 1/4$.

From standard estimates (obtained by combining the functional equation (2.1.1) with Theorem 3.5 of Titchmarsh [8]) we see that for $s \in \mathcal{C}_2$, $\zeta(s) \ll \tau^{1/2}\log\tau$ and $\zeta(s+1) \ll \log\tau$. By Stirling's formula (Whittaker and Watson [9], p. 252) we find that $\Gamma(s) \ll \tau^{-1/2}\exp(-\frac{\pi}{2}\tau)$ for $s \in \mathcal{C}_2$. By combining these estimates with the estimate (12) of Lemma 1, we see that if $|\arg z| \le \frac{\pi}{2} - \delta$ then

$$\begin{aligned}\frac{1}{2\pi i}\int_{\mathcal{C}_2} &\zeta(s)\zeta(s+1)Q(s)\Gamma(s)z^{-s}\, ds\\ &\ll \int_0^\infty (\log\tau)^2 \exp\left(\frac{C\tau}{\log\tau} - c\frac{\log 1/|z|}{\log\tau} - \delta\tau\right) dt .\end{aligned}$$

Over the restricted interval $0 \le \tau \le \frac{c}{4C}\log 1/|z|$, this integral is

$$\ll \exp\left(-\frac{c\log 1/|z|}{4\log\log 1/|z|}\right).$$

If $\delta \ge C_1/\log\log 1/|z|$ then the integral over $\tau \ge \frac{c}{4C}\log 1/|z|$ is

$$\ll \exp\left(-\frac{\log 1/|z|}{\log\log 1/|z|}\right).$$

Thus

$$\frac{1}{2\pi i}\int_{C_2} \ldots \ll \exp\left(-c\,\frac{\log 1/|z|}{\log\log 1/|z|}\right)$$

for $|z| \le 1/4$, $|\arg z| \le \frac{\pi}{2} - C/\log\log 1/|z|$. On combining this with (41) in (36), we obtain the stated result.

To treat $c_d(n)$, we first note that

$$P_d(z) = \sum_{n=0}^{\infty} c_d(n) e^{-nz} = \prod_{q=1}^{\infty} (1 + e^{-\varphi(q)z}) \tag{42}$$

for $\operatorname{Re} z > 0$. Since $1 + w = (1 - w^2)/(1 - w)$, we see that

$$P_d(z) = P(z)/P(2z). \tag{43}$$

Thus we are able to derive an asymptotic formula for $P_d(z)$ directly from Lemma 2.

**Lemma 3.** *Let $\mathcal{R}$ be as in Lemma 2. As $z$ tends to $0$ in $\mathcal{R}$,*

$$P_d(z) = a_d \exp(b_d/z) E_d(z) \tag{44}$$

*where*

$$E_d(z) = 1 - \frac{1}{2}(\log 2)/\log 1/|z| + O((\log\log 1/|z|)^3/(\log 1/|z|)^2),$$

$a_d = \dfrac{1}{\sqrt{2}}$, *and* $b_d = \zeta(2)^2\zeta(3)/(2\zeta(6))$.

## 4. Minor Arc Estimates

As a complement to Lemma 2, we establish the following auxiliary estimate.

**Lemma 4.** *If $0 < x \le 1/4$ then*

$$\int_x^{\pi} |P(x + iy)|\, dy \ll x^2(\log 1/x)P(x). \tag{45}$$

By taking more care, it could be shown that this integral is $\asymp x^{7/2}P(x)$. That no better estimate is possible is easily seen from the observation that $|P(x + iy)| \asymp x^2 P(x)$ when $|y - \pi| \le x^{3/2}$.

**Proof.** We show that if $0 < x \le 1/4$, $x < y \le \pi$, then

$$P(z) \ll P(x)\exp(-c/y), \tag{46}$$

and that also

$$P(z) \ll x^2y^{-2}P(x). \tag{47}$$

Once these inequalities are established, the stated estimate (45) follows by using (46) for $x \le y \le c/(2\log x)$, and (47) for $c/(2\log x) \le y \le \pi$.

Let $y_0$ be a small positive constant. Since $|P(z)| \le P(x)$ for all $y$, (46) is trivial when $y_0 \le y \le \pi$. To establish (46) for $0 < y < y_0$, we note that by (32),

$$|P(z)|/P(x) = \exp\left(-\sum_{n=1}^{\infty} g(n)e^{-nx}(1-\cos ny)\right).$$

Using only the term $m = 1$ in the definition of $g(n)$, we see that

$$\sum_{\pi/(2y) \le n \le \pi/y} g(n) \ge \text{card}\,\{n:\ \pi/(2y) \le \varphi(n) \le \pi/y\}.$$

By (6) we see that this latter quantity is $\gg 1/y$ when $y < y_0$. Since $y \ge x$, we see that if $n \le \pi/y$ then $e^{-nx} \ge e^{-ny} \ge e^{-\pi}$, and thus we deduce that the sum in the exponential is $\ge c/y$. This gives (46).

As for (47), we see by (31) that

$$P(z) = (1-e^{-z})^{-2}\prod_{q>2}(1-e^{-\varphi(q)z})^{-1} = (1-e^{-z})^{-2}P_e(z),$$

say. (Here the subscript reflects the fact that the product is over those $q$ for which $\varphi(q)$ is even.) Now $|P_e(z)| \le P_e(x) \asymp x^2P(x)$, and $|1-e^{-z}| \asymp y$ for $x \le y \le \pi$, so we have (47).

We now establish a corresponding estimate for $P_d(z)$; this is done in a similar manner, although one new idea is required.

**Lemma 5.** *If* $0 < x < 1/4$, *then*

$$\int_x^{\pi} |P_d(x+iy)|\,dy \ll x^2(\log 1/x)P_d(x).$$

**Proof.** We show that if $0 < x \le 1/4$, $x \le y \le \pi$, then

$$P_d(z) \ll P_d(x)\exp(-c/y), \tag{48}$$

and that also

$$P_d(z) \ll x^2 y^{-2} P_d(x). \tag{49}$$

Once these inequalities are established, the stated estimate follows as in the preceding proof.

In the classical theory of partitions, it is a familiar fact that the number of partitions into distinct parts is equal to the number of partitions into odd parts. When expressed using generating functions, this reflects the identity

$$\prod_{k=0}^{\infty}(1 + w^{2^k}) = \frac{1}{1-w},$$

valid for $|w| < 1$. Write $q = 2^k r$ where $r$ is odd. Then $\varphi(q) = \varphi(2^k)\varphi(r)$, and $\varphi(2^k) = 2^{k-1}$ when $k > 0$, so by the identity above we see that

$$P_d(z) = \prod_{\substack{r=1 \\ 2\nmid r}}^{\infty} \frac{1 + e^{-\varphi(r)z}}{1 - e^{-\varphi(r)z}}. \tag{50}$$

Using the power series for $\log \frac{1+w}{1-w}$, we see that the above is

$$\exp\left(2\sum_{\substack{r=1 \\ 2\nmid r}}^{\infty} \sum_{\substack{m=1 \\ 2\nmid m}}^{\infty} \frac{1}{m} e^{-m\varphi(r)z}\right) = \exp\left(\sum_{n=1}^{\infty} h(n)\, e^{-nz}\right)$$

where

$$h(n) = 2 \sum_{\substack{m\,\varphi(r)=n \\ 2\nmid mr}} 1/m .$$

Thus

$$|P_d(z)|/P_d(x) = \exp\left(-\sum_{n=1}^{\infty} h(n) e^{-nx}(1 - \cos ny)\right).$$

Let $A(u)$ denote the number of $q$ such that $\varphi(q) \le u$, and let $B(u)$ denote the number of odd $r$ for which $\varphi(r) \le u$. If $k > 0$ then the number of $q$ such that $2^k \| q$ and $\varphi(q) \le u$ is $B(u/2^{k-1})$. On summing over $k$ we deduce that

$$A(u) = 2B(u) + \sum_{k=1}^{\infty} B(u/2^k).$$

This may be inverted, and we find that

$$B(u) = \frac{1}{2}A(u) - \sum_{k=1}^{\infty} A(u/2^k)/2^{k+1}.$$

Then from (6) we deduce that

$$B(u) \sim \frac{\zeta(2)\zeta(3)}{3\zeta(6)} u$$

as $u \to \infty$. Thus if $y_0$ is a small positive constant and $y \le y_0$, then the number of odd $r$ for which $\pi/(2y) \le \varphi(r) \le \pi/y$ is $\gg 1/y$, and hence

$$\sum_{\pi/(2y) < n \le \pi/y} h(n) \gg 1/y.$$

Now (48) proceeds as in the preceding proof.

To prove (49) we note that the combined contribution of $q = 1$ and $q = 3$ in (50) is $(1 + e^{-2z})/(1 - e^{-z})^2$. This is $\ll y^{-2}$ for $x \le y \le \pi$. Let $P_0(z)$ denote the product (50) restricted to odd $r > 3$. For any real $y$, $|P_0(z)| \le P_0(x) \approx x^2 P_d(x)$. These estimates give (49).

## 5. Completion of the Proof of the Theorem

Let $P(z)$ be as in section 3. For $x > 0$ we see by Cauchy's formula for the coefficient of a power series that

$$c(n) = \frac{1}{2\pi}\int_{-\pi}^{\pi} P(z)e^{nz}\,dy = \frac{1}{\pi}\int_{0}^{\pi} P(z)e^{nz}\,dy.$$

Noting that $\exp(b/x + nx)$ is minimized when $x = \sqrt{b/n}$, we take this value for $x$. Let $I_0 = [0, x^{7/5}]$. We see by Lemma 2 that if $y \in I_0$ then

$$P(z)e^{nz} = a(\log 1/x)^{-1/2} x^{1/2} \exp(b/x + nx) \cdot \exp(-by^2/x^3)\left(1 + O\left(\frac{\log\log 1/x}{\log 1/x}\right)\right).$$

Hence

$$\frac{1}{\pi}\int_{I_0} P(z)e^{nz}\,dy = \frac{1}{2\pi} a(\log 1/x)^{-1/2} x^{1/2} \exp(b/x + nx) \cdot (\pi x^3/b)^{1/2}\left(1 + O\left(\frac{\log\log 1/x}{\log 1/x}\right)\right),$$

which gives the main term in (2). Suppose that $I_1 = [x^{7/5}, x]$. By Lemma 2 we see that if $y \in I_1$ then $P(z) \ll P(x)\exp(-cx^{-1/5})$, so that

$$\frac{1}{\pi}\int_{I_1} P(z)e^{nz}\,dy \ll \exp(B\sqrt{n})/\exp(-cn^{-1/10}).$$

Finally, let $I_2 = [x, \pi]$. By Lemma 4 we see that

$$\frac{1}{\pi}\int_{I_2} P(z)e^{nz}\,dy \ll x^2(\log x)P(x)e^{nx}$$

$$\ll (\log n)^{1/2} n^{-5/4} \exp(B\sqrt{n}).$$

Thus we have (2).

In deriving (2) we did not make use of the full precision of Lemma 2. By taking more care one could show that (2) can be replaced by a more precise estimate in which the last factor on the right is replaced by

$$1 - \frac{\log\log n}{\log n} + \frac{c_7}{\log n} + O\left(\frac{(\log\log n)^3}{(\log n)^2}\right)$$

where $c_7 = -\log\pi - C_0 + \frac{1}{2}\log b = -1.140825\ldots$ When evaluated at $n = 5000$, this more precise estimate is $5.9166 \times 10^{104}$, which is about 16% below that actual value.

The proof of (5) is similar, except that we take $x = \sqrt{b_d/n}$, and employ Lemmas 3 and 5.

## References

[1] *P.T. Bateman*, The distribution of values of the Euler function. Acta Arith., **21** (1972), 329—345.

[2] *P.T. Bateman* and *P. Erdös*, Monotonicity of partition functions. Mathematika, **3** (1956), 1—14.

[3] *R.E. Dressler*, A density which counts multiplicity. Pacific J. Math., **34** (1970), 371—378.

[4] *P. Erdös*, On the normal number of prime factors of $p$ - 1 and some other related problems concerning Euler's $\varphi$–function. Quart. J. Math. (Oxford Series), **6** (1935), 205—213.

[5] *A.E. Ingham*, A Tauberian theorem for partitions. Ann. Math., **42** (1941), 1075—1090.

[6] *G. Meinardus*, Asymptotische Aussagen über Partitionen. Math. Zeit., **59** (1954), 388—398.

[7] *C. Pomerance*, Popular values of Euler's function. Mathematika, **27** (1980), 84—89.

[8] *E.C. Titchmarsh*, The Theory of the Riemann Zeta–Function. Oxford University Press (Oxford), 1951, 346 pp.

---

Department of Mathematics, University of British Columbia, Vancouver, BC, CANADA V6T 1Y4.

Department of Mathematics, University of Michigan, Ann Arbor, MI, USA 48109.

# Limit Points of the Salem Numbers

*David W. Boyd and Walter Parry*

## 1. Introduction

As usual, $S$ and $T$ will denote the sets of Pisot–Vijayaraghavan numbers and Salem numbers, respectively. In [8] (see also [9], page 30), Salem showed that each element of $S$ is a limit point of $T$ and in [9], page 62, raised the question whether there are other limit points. This paper continues the study of this question along the lines of [1]—[5].

The main purpose here is to point out an error which occurs in the proofs of Theorem 6.1 of [2] and Lemma 4 of [3]. We describe some results which correct these errors, to some extent. Only special cases will be proved here since a complete treatment requires a new technique which will be described in detail in a subsequent paper.

## 2. Definitions and Basic Results

The basic facts about $S$ and $T$, recalled here, may be found in [9]. The elements of $S$ ($T$) are those algebraic integers $\theta > 1$ for which all conjugates $\theta_i \neq \theta$ of $\theta$ satisfy $|\theta_i| < 1$ ($|\theta_i| \leq 1$, with at least one case of equality).

If $\theta$ is in $T$ then $\theta$ is *reciprocal*: if $R(x)$ is the minimal polynomial of $\theta$, and if $R^*(x)$ is defined to be $x^{\deg(R)} R(x^{-1})$, then $R^*(x) = R(x)$. If $\theta$ is in $T$ then $\deg(\theta)$ is even and $\geq 4$; all conjugates of $\theta_i$ but $\theta$ and $\theta^{-1}$ satisfy $|\theta_i| = 1$. The reciprocal elements of $S$ are of degree 2 and may be regarded as degenerate elements of $T$. The union of $T$ and these quadratic integers is denoted $T_+$. The set of non-reciprocal elements of $S$ is denoted $S_-$.

## 3. The Set $\mathbb{C}$ of Rational Functions

Details missing from this section may be found in [5]. The basic tool for studying $S$ is a set of rational functions $\mathbb{C}$ first studied by Dufresnoy and Pisot [7]. Let $\theta$ in $S$ have minimal polynomial $P(x)$ and write $Q(x) = P^*(x)$. We associate with $\theta$ the set of rational functions $f(x) = A(x)/Q(x)$ such that $A(x)$ has integer coefficients, $A(0) > 0$, $A \neq Q$ and $|A(x)| \leq |Q(x)|$ for $|x| = 1$. The union of these sets over all $\theta$ in $S$ is denoted $\mathbb{C}$.

Thus, each $f$ in $\mathbb{C}$ is meromorphic in $|x| < 1$, has a single pole $x = \theta^{-1}$ in $|x| < 1$, satisfies $|f(x)| \leq 1$ for $|x| = 1$, and the Maclaurin expansion $f(x) = u_0 + u_1 x + \ldots$ has integer coefficients, with $u_0 > 0$. These properties characterize the elements of $\mathbb{C}$.

The expansions $\{u_n\}$ of elements of $\mathbb{C}$ can be characterized by a system of inequalities determined by Schur's algorithm. This embeds $\mathbb{C}$ naturally into an infinite tree described in [3] and [5].

If $f_n$ is a sequence in $\mathbb{C}$ which converges uniformly to $f$ in a neighbourhood of $x = 0$ then $f$ is in $\mathbb{C}$. The *isolated points* of $\mathbb{C}$ in this topology are those $f$ with $|f(x)| = 1$ for all $|x| = 1$. Thus they are of the form $f(x) = \pm P(x)/Q(x)$ (with $P^*(x) = Q(x)$). The *limit points* are those $f$ with $|f(x)| < 1$ for some $|x| = 1$ (and hence for all but a finite subset of $|x| = 1$). We write $\mathbb{C}'$ for the set of limit points of $\mathbb{C}$.

Dufresnoy and Pisot [6] showed that $\theta$ is a limit point of $S$ if and only if there is an $f$ in $\mathbb{C}'$ with a pole at $\theta^{-1}$. It is easy to see that, if $f = A/Q$ is in $\mathbb{C}'$ then the following special sequences, (called the *principal branches* of $f$ in [5]), converge to $f$:

$$f_n^{\pm}(x) = \frac{A(x) \pm x^{n+r} Q(x^{-1})}{Q(x) \pm x^{n+r} A(x^{-1})}. \tag{1}$$

For reasons explained in [5]. the integer $r > 0$ is chosen to be the degree of the Laurent polynomial $Q(x)Q(x^{-1}) - A(x)A(x^{-1})$, so that

$$\Omega(x) := x^r (Q(x)Q(x^{-1}) - A(x)A(x^{-1})) \tag{2}$$

is a polynomial with $\Omega(0) \neq 0$. Thus $f_n^{\pm}(x) - f(x) = \pm\,\Omega(0)x^n + O(x^{n+1})$ as $x \to 0$.

There may, of course, be other essentially different sequences $g_n$ of isolated points of $\mathfrak{C}$ which converge to $f$. For example, if $f$ is in the second derived set of $\mathfrak{C}$, this is always possible.

## 4. Salem's Construction and Limit Points of $T$.

By Rouché's theorem, if $f = A/Q$ is in $\mathfrak{C}$ then, for any integer $m > 0$, and $\epsilon = \pm 1$, the equation $1 + \epsilon\, x^m f(x) = 0$ has at most one zero in $|x| < 1$. We denote this zero by $1/\theta_m^{\epsilon}$, adopting the convention that $\theta_m^{\epsilon} = 1$ if no such zero exists, (in this case $1 + \epsilon f(1) = 0$). Since $\theta_m^{\epsilon}$ is an algebraic integer, it follows that $\theta_m^{\epsilon}$ is in $\{1\} \cup S \cup T$.

If $f$ is isolated, so $A = \pm P$, then $\theta_m^{\epsilon}$ is reciprocal so $\theta_m^{\epsilon}$ is in $\{1\} \cup T_+$. As $m \to \infty$, $\theta_m^{\epsilon} \to \theta$, the pole of $f$. This is, in fact, Salem's proof that the elements of $S_-$ are limit points of $T$.

There is another way to use this construction to generate limit points of $T$. Suppose that $f = A/Q$ is in $\mathfrak{C}'$ with $f_n \to f$. Fixing $m$, and $\epsilon$, let $\sigma_n^{-1} \leq 1$ solve $1 + \epsilon x^m f_n(x) = 0$ and $\sigma^{-1} \leq 1$ solve $1 + \epsilon x^m f(x) = 0$. Then $\sigma_n$ is a sequence of elements of $\{1\} \cup T_+$ with $\sigma_n \to \sigma$. If the $\sigma_n$ are distinct then $\sigma$ is a limit point of $T$.

It is asserted in [2], page 323, that if $f = A/Q$ is in $\mathfrak{C}'$ then the zero of $1 + \epsilon x^m f(x) = 0$ cannot be in $T$. This used in the proof of Theorem 6.1 of [2] and Lemma 4 of [3]. However, this is false, as the following example shows: let

$$f(x) = A(x)/Q(x) = \frac{1 - x^2}{1 - 2x - x^2 + x^4}. \tag{3}$$

Then $1 + xf(x) = (1 - x - x^2 - x^3 + x^4)/Q(x)$, where $Q(x) + xA(x) = 1 - x - x^2 - x^3 + x^4$ is the minimal polynomial of a Salem number.

Indeed, our Theorem 1 shows that *every* Salem number $\sigma$ occurs as the zero of (infinitely many) equations $1 + xf(x) = 0$ with $f$ in $\mathfrak{C}'$.

Since the roots of $1+xf(x)=0$ with $f$ in $\mathbb{C}'$ can thus include all $\sigma$ in $\{1\}\cup T$, the observation made above opens up the possibility that such $\sigma$ may be limit points of $T$. After all, given $f$ in $\mathbb{C}'$, one always has the sequences $f_n^{\pm}$ of (1) with $f_n^{\pm}\to f$. If $\sigma_n^{\pm}$ solves $1+xf_n^{\pm}(x)=0$ then indeed $\sigma_n^{\pm}$ are in $\{1\}\cup T$ and $\sigma_n^{\pm}\to\sigma$. Fortunately, this does not show that $\sigma$ is a limit point of $T$ since $\sigma_n^{\pm}=\sigma$ for all $n$ in case $\sigma\in\{1\}\cup T_+$.

This is easy to see: if $\sigma$ is in $\{1\}\cup T_+$ the polynomial $Q(x)+xA(x)$ is reciprocal, so the numerator of $1+xf_n^{\pm}(x)$ is

$$Q(x)\pm x^{n+r}A(x^{-1})+x\,(A(x)\pm x^{n+r}Q(x^{-1}))$$
$$=(Q(x)+xA(x))(1\pm x^{n+r+1-s}),$$

where $s=\deg(Q(x)+xA(x))$. Thus $\sigma_n^{\pm}=\sigma$ for all $n$.

It is still conceivable that this construction yields a limit point of $T$ in the set $\{1\}\cup T_+$, but only if there is a sequence $g_n\to f$ which is essentially distinct from $f_n^{\pm}$. This possibility is the subject of Theorem 3 of the next section.

## 5. The Main Results

In this section, we state three main results, Theorems 1 to 3. We sketch complete proofs of Theorem 1 and 3 and prove only a special case of Theorem 2. Further details and discussion will appear in a forthcoming joint paper.

**Theorem 1.** *Given any $\sigma$ in $T$ there are infinitely many $f$ in $\mathbb{C}'$ such that $\sigma^{-1}$ solves $1+xf(x)=0$.*

**Theorem 2.** *If $f$ is in $\mathbb{C}'$, $m\geq 2$ and $\epsilon=\pm 1$, and if $\sigma^{-1}<1$ solves $1+\epsilon x^m f(x)=0$ then $\sigma$ is in $S$.*

**Theorem 3.** *Suppose $f=A/Q$ is in $\mathbb{C}'$ and $1+\epsilon xf(x)=0$ has a zero $\sigma^{-1}$ with $\sigma$ in $T$. If $1+\deg A<\deg Q$ and if $g_n$ in $\mathbb{C}$ has $g_n-f=cx^n+O(x^{n+1})$, with $c\neq 0$, then for sufficiently large $n$, $g_n=f_n^{\pm}$, as given by* (1).

**Proof of Theorem 1.** This is based on the ideas of [4], section 6. Let $R(x)$ be the minimal polynomial of $\sigma$, so $R^*(x)=R(x)$. We seek an $A(x)$ with integer coefficients, $\deg(A) = \deg(R) - 2$, and $A^*(x) = -A(x)$ such that $Q(x) = R(x) - xA(x)$ has a single zero in $|x| < 1$.

We examine the zeros of $Q(x,t) = R(x) - txA(x)$ as $t$ varies from 0 to 1. The conditions $R^* = R$ and $A^* = -A$ imply that

$$|Q(x,t)|^2 = |R(x)|^2 + t^2|A(x)|^2, \quad \text{for } |x| = 1. \tag{3}$$

Thus $Q(x,t) = 0$ has no zeros on $|x| = 1$ if $t > 0$. If $x = z(t)$ denotes a typical zero of $Q(x,t) = 0$ then $z(0)$ is a zero of $R(x) = 0$. For $z(0) = \sigma^{-1} < 1$, we must have $|z(1)| < 1$ since $z(t)$ never crosses $|x| = 1$. If $z(0) = \alpha$ when $|\alpha| = 1$, then $|z(1)| > 1$ provided $z^{\cdot}(0)/z(0) > 0$.

A computation shows that this condition is equivalent to $A(\alpha)/R'(\alpha) > 0$. This is a homogeneous linear inequality in the coefficients of $A$. Since there is one such inequality for each pair of complex conjugate zeros $\alpha$ of $R$ on $|\alpha| = 1$, one finds as in [2], or [4], that there are infinitely many $A$ satisfying all these conditions.

The corresponding $Q(x) = R(x) - xA(x)$ has a single root in $|x| < 1$, and (3) shows that $|A(x)| \le |Q(x)|$ on $|x| = 1$ so $f = A/Q$ is in $\mathbf{C}$. Since $1 + xf(x) = R(x)/Q(x)$ has zero $\sigma^{-1}$, this completes the proof of Theorem 1.

**Remark 1.** An alternate approach to the proof that the inequalities $A(\alpha)/R'(\alpha) > 0$ have infinitely many solutions can be based on the ideas of [1]. Since $R^* = R$ and $A^* = -A$, we can write $R(x) = x^k r(x + x^{-1})$ and $A(x) = (x^2 - 1)x^{k-2}a(x + x^{-1})$, where $r$ and $a$ are polynomials with integer coefficients. If $R(\alpha) = 0$ and $\rho = \alpha + \alpha^{-1}$, then $A(\alpha) = (\alpha^2 - 1)\alpha^{k-2}a(\rho)$ and $R'(\alpha) = (1 - \alpha^{-2})\alpha^k r'(\rho)$ so $A(\alpha)/R'(\alpha) = a(\rho)/r'(\rho)$. Let the zeros of $r$ be $\rho_1 > 2 > \rho_2 > \ldots > \rho_k > -2$. The conditions $a(\rho_i)/r'(\rho_i) > 0$, for $i \ge 2$, and the alternating signs of $r'(\rho_i)$ show that $a$ must have zeros $\gamma_2 > \ldots > \gamma_{k-1}$ interlacing those of $r$ in the sense that $\rho_2 > \gamma_2 > \rho_3 > \ldots > \gamma_{k-1} > \rho_k$. It is easy to construct $a(x)$ with integer coefficients with these properties. For example one can take the $\gamma_j$ to be rational numbers.

Returning to $A(x)$, we see that it must have zeros $1, e^{\pm i\phi_2}, \ldots, e^{\pm i\phi_{k-1}}, -1$ interlacing the zeros of $R(x)$ on $|x| = 1$, where $2\cos\phi_j = \gamma_j$. The results of [1] show that for 41 of the 43 known Salem numbers $\sigma < 1.3$, it is possible to choose $A(x)$ to be cyclotomic.

**Remark 2.** In the proof of Theorem 1, $R(x)$ may be taken to be the product of the minimal polynomial of $\sigma$ and a cyclotomic polynomial with simple roots which does not vanish at $\pm 1$.

One can also treat the case $1 - xf(x)$ in exactly the same way.

**Lemma 1.** *Suppose that* $A/Q$ *is in* $\mathcal{C}'$ *and for some* $m > 0$ *and* $\epsilon = \pm 1$, *that* $R(x) = Q(x) + \epsilon x^m A(x)$ *has a root* $\sigma > 1$ *which is a Salem number. Let* $\Omega$ *be given by* (2).

*Then*

(i) $R^2$ *divides* $\Omega$

(ii) $m + \deg A \leq \deg Q$

(iii) *if* $m + \deg A < \deg Q$ *then* $\Omega = \pm R^2$.

**Proof.** Clearly $R(x)$ is divisible by the minimal polynomial of $\sigma$. Furthermore $\sigma^{-1}$ is the unique zero of $R$ with $|x| < 1$. The product of all the conjugates of $\sigma$ is 1 and the product of all roots of $R$ is $\pm R(0) = \pm 1$, hence $\sigma$ is the only root of $R$ with $|x| > 1$.

Thus, all roots $\alpha$ of $R$ other than $\sigma$ and $\sigma^{-1}$ satisfy $|\alpha| = 1$. For such a root, since $Q(\alpha) = -\epsilon\alpha^m A(\alpha)$, one has $|Q(\alpha)| = |A(\alpha)|$. But then

$$\begin{aligned}\Omega(\alpha) &= \alpha^r (Q(\alpha)Q(\alpha^{-1}) - A(\alpha)A(\alpha^{-1})) \\ &= \alpha^r (|Q(\alpha)|^2 - |A(\alpha)|^2) = 0.\end{aligned}$$

Since the roots of $R$ are simple, this shows $R \mid \Omega$. But all zeros of $\Omega$ with $|x| = 1$ are of even multiplicity [5], page 23, so in fact $R^2 \mid \Omega$.

(ii) Comparing degrees, $2 \deg R \leq \deg \Omega = 2r$. If $m + \deg A > \deg Q$ then $\deg R = m + \deg A$. Clearly $r \leq \max(\deg Q, \deg A)$. Thus $\deg R \leq r$ implies

$$\deg Q < m + \deg A \leq r \leq \max(\deg Q, \deg A),$$

Which contradicts $m > 0$. This proves (ii).

(iii) If $m + \deg A < \deg Q$, then $\deg(A) < \deg(R) = \deg(Q) = r$, and comparing the leading coefficients of $R^2$ and $\Omega$ shows that $\Omega = \pm R^2$.

**Partial Proof of Theorem 2.** We treat only the case where $m + \deg A < \deg Q$. Then, by Lemma 1 (iii), on $|x| = 1$,

$$|R(x)|^2 = |\Omega(x)| = |Q(x)|^2 - |A(x)|^2.$$

So $|A(x)|^2 + |R(x)|^2 = |Q(x)|^2$, showing that both $A/Q$ and $R/Q$ are in $\mathbb{C}'$. But $R/Q = 1 + \epsilon x^m (A/Q)$ so, if $m \geq 2$, the Maclaurin expansion of $R/Q$ begins $1 + 0 \bullet x + O(x^2)$. However, this violates Schur's inequalities [5], page 21, hence $m \geq 2$ is not possible.

**Remark 3.** Theorem 2 is also valid in case $m + \deg A = \deg Q$ but the proof is entirely different.

**Proof of Theorem 3.** By (iii) of Lemma 1, $\Omega = \pm R^2$. This implies that $\Omega$ is *minimal* in the sense of [5], page 35. That is, there are no polynomials $S(x)$ with integer coefficients and $S(0) \neq 0$ satisfying $|S(x)| < |\Omega(x)|$ for all but a finite subset of $|x| = 1$. This is clear since such an $S$ would have double zeros at all the zeros of $\Omega$ on $|x| = 1$ and hence we would have $S = \pm\Omega$. Furthermore, $\Omega$ has no proper factors which do not vanish on $|x| = 1$. Applying Theorem 8.5 of [5] proves Theorem 3.

## 6. Conclusions

In this section, we describe the extent to which Theorem 6.1 of [2] and Lemma 4 of [3] need to be modified in light of the results of this paper. We begin by quoting from [2]:

"**Theorem 6.1.** *Let* $\sigma_k$ *be a sequence of distinct numbers of* $T$ *and suppose* $\sigma_k$ *converges to* $\sigma > 1$ *and that* $\sigma$ *is not in* $S$. *If* $\sigma_k$ *corresponds to* $\theta_k \in S$ *(as in Theorem 4.1 )* then $\theta_k \to \infty$".

The simplest way to convert this result is to replace "$\sigma$ is not in $S$ " by "$\sigma$ is not in $S \cup T$." The proof, as given in [2], is then valid. It is likely that Theorem 6.1 is correct as originally stated. This would follow if Theorem 3 could be proved in the remaining case $1 + \deg A = \deg Q$. However, it is possible to construct examples with $1 + \deg A = \deg Q$ for which $\Omega$ is not minimal, so the proof of Theorem 3 cannot be carried over to this case.

In our forthcoming paper we will show that the assumption "$\sigma > 1$" in Theorem 6.1 is unnecessary.

Next we quote Lemma 4 of [3]. For unfamiliar terminology, the reader should consult [3].

"**Lemma 4.** *Suppose* $[\alpha, \beta]$ *does not intersect the set* $\{1\} \cup S$. *Let* $\mathbf{P}$ *be the tree of Section 2 restricted by the further conditions*

$$s_n(\alpha) \leq u_n \leq s_n^*(\beta) \quad \text{for all } n \geq 0.$$

*If* $m \geq 2$, *then* $\mathbf{P}$ *is a finite tree. If* $m = 1$, *and* $u_0$ *is fixed, then the tree defined by the remaining conditions is finite.*"

In partial explanation, the tree $\mathbf{P}$ referred to here is defined by Schur's inequalities for the coefficient sequences $\{u_n\}$ of the elements of $\mathbf{C}$. The supplementary inequalities $s_n(\alpha) \leq u_n \leq s_n^*(\beta)$ restrict attention to those $f$ for which the solution $\theta_m^{-1} \leq 1$ of $1 + x^m f(x) = 0$ satisfies $\alpha \leq \theta_m \leq \beta$. The proof uses the assertion that if $f \in \mathbf{C}'$ then $\theta_m = 1$ or $\theta_m \in S$. Theorem 2 shows that this is correct if $m \geq 2$ but Theorem 1 shows that it is incorrect if $m = 1$. Thus to correct Lemma 4, one must omit the sentence "If $m = 1$, ... finite."

It should be emphasized that the error in Lemma 4 does not affect the correctness of the computations reported in [3]. Indeed, the algorithm described there has the property that it terminates if and only if the corresponding tree is finite. Thus, independently of Theorem 2, the results of [3] showed that no $\sigma$ in $T \cap [1.125, 1.3]$ solves an equation $1 + x^m f(x) = 0$ with $m \geq 2$ and $f$ in $\mathbf{C}'$. Theorem 2 proves this in general.

Note that no computational results were reported in [3] for the case $m = 1$, since they appeared to be too time–consuming to complete. Theorem 1 provides a theoretical basis for this: in fact the computation time would be infinite!

**Acknowledgement.** The work of the first author was supported in part by an NSERC operating grant.

## References

[1] *M–J. Bertin* and *D.W. Boyd*, A characterization of two related classes of Salem numbers. J. Number Theory, (to appear).

[2] *D.W. Boyd*, Small Salem numbers. Duke Math. J. **44** (1977), 315—328.

[3] *D.W. Boyd*, Pisot and Salem numbers in intervals of the real line. Math. Comp. **32** (1978), 1244—1260.

[4] *D.W. Boyd*, Families of Pisot and Salem numbers. Sém. Théorie des Nombres, Paris 1980–81, Birkauser (1982), 19—33.

[5] *D.W. Boyd*, Pisot numbers in the neighbourhood of a limit point, I. J. Number Theory **21** (1985), 17—43.

[6] *J. Dufresnoy* and *Ch. Pisot*, Sur un ensemble fermé d'entiers algébriques. Ann. Sci. École Norm. Sup. (3) **70**(1953), 105—133.

[7] *J. Dufresnoy* and *Ch. Pisot*, Études de certaines fonctions méromorphes bornées sur le cercle unité. Application a un ensemble fermé d'entiers algébriques, Ann. Sci. École Norm. Sup. (3) **82** (1955). 69—92.

[8] *R. Salem*, Power series with integral coefficients. Duke Math. J. **12** (1945), 153—172.

[9] *R. Salem*, Algebraic Numbers and Fourier Analysis.D.C. Heath & Co., 1963.

---

Dept. of Mathematics, University of British Columbia, Vancouver, B.C. V6T 1Y4, CANADA.

Department of Mathematics, Eastern Michigan University, Ypsilanti, Michigan 48197, USA.

# Complexity of Algorithms in Algebraic Number Theory

*Johannes Buchmann*

## 1. Introduction

Algorithms in algebraic number theory have recently received a lot of attention. This is not only because of the numerous applications of these algorithms, e.g., in cryptography, primality testing and factoring, but also because numerical results and tables allow for new insights into the old problems of number theory and help formulate new conjectures and theorems.

Many new algorithms have been proposed during the last years and it seems now appropriate to apply complexity theory in order to get an idea how well those algorithms will perform asymptotically. In this paper we give a brief overview of the complexity theoretic results known so far in this area.

## 2. The Problems

Let $f(x) = x^n + a_1 x^{n-1} + \ldots + a_{n-1}x + a_n \in \mathbb{Z}[x]$ be irreducible and let $\mathcal{F} = \mathbb{Q}[x]/f\,\mathbb{Q}[x]$ be the corresponding number field. Then $\mathcal{F} = \mathbb{Q}[\alpha]$ with $\alpha = x + f\,\mathbb{Q}[x]$. Important problems of algorithmic algebraic number theory are the following:

### 2.1 Maximal orders

Compute the maximal order of $\mathcal{O}$ of $\mathcal{F}$, more precisely, compute the *standard* $\mathbb{Z}$*-basis* $\omega_1, \ldots, \omega_n$ of $\mathcal{O}$ which is defined by the following properties:

$$\omega_j = \frac{1}{d} \sum_{i=1}^{n} w_{i,j}\, \alpha^{i-1}$$

with $d \in \mathbb{Z}_{\geq 1}$, $(w_{i,j}) \in \mathbb{Z}^{n \times n}$ in Hermite normal form, $\gcd(d, w_{1,1} w_{1,2}, \ldots, w_{n,n}) = 1$. So, the problem is the determination of the denominator $d$ and the matrix $(w_{i,j})$.

## 2.2 Unit group

### 2.2.1 *Basis*

Compute a basis $\epsilon_0, \epsilon_1, \ldots, \epsilon_r$ of the unit group $\mathcal{O}^\times$ of $\mathcal{O}$:

$$\mathcal{O}^\times = <\epsilon_0> \times <\epsilon_1> \times \ldots \times <\epsilon_r>$$

where $<\epsilon_0>$ is the torsion subgroup of $\mathcal{O}^\times$ and $<\epsilon_j>$ $(1 \leq i \geq r)$ are infinite cyclic groups. Each $\epsilon_j$ is given as

$$\epsilon_j = \sum_{k=1}^{n} e_{k,j} \, \omega_k .$$

So, the problem is to compute the integer matrix $E = (e_{k,j})$.

### 2.2.2 *Regulator*

Compute an approximation $\hat{R}$ to the regulator

$$R = |\det(\log |\epsilon_j|_i)_{1 \leq i, j \leq r}|,$$

$|\ |_i$ denoting the archimedian valuations on $\mathcal{F}$.

## 2.3 Class group

### 2.3.1 *Basis*

Compute a basis $\mathbf{a}_1, \ldots, \mathbf{a}_l$ of the class group $Cl = I / H$, $I$ denoting the group of fractional ideals of $\mathcal{F}$ and $\mathcal{H}$ denoting the subgroup of fractional principal ideals. The class group has a presentation

$$Cl = <\mathbf{a}_1 H> \times \ldots \times <\mathbf{a}_m H>$$

where $\mathbf{a}_i$ is an ideal of $O$ and $m_i \mid m_{i+1}$, $m_i$ being the order of the cyclic subgroup $< \mathbf{a}_i H >$. Each $\mathbf{a}_i$, in turn, is given a $\mathbb{Z}$-basis $\alpha_1(i), \ldots, \alpha_n(i)$, each $\alpha_j(i)$ represented as

$$\alpha_j(i) = \sum_{k=1}^{n} a_{j,k}(i)\,\omega_k .$$

So, the problem is, to compute the integer matrices $(a_{j,k}(i))$ and the orders $m_i$.

### 2.3.2 Type

Compute the type of the class group $Cl$ i.e., only the invariants $m_1, \ldots, m_m$ but no basis.

### 2.3.3 Class number

Compute the class number $h = |Cl|$.

## 3 What is Known and What is Desired

We would like to have polynomial time algorithms for solving all the problems mentioned in section 2. **Polynomial time** means that the running time is a polynomial in the binary length of the input or output data, whichever is larger. The **running time** is computed using the assumption that the number of bit operations necessary to perform elementary operations on integers (addition, subtraction, multiplication, division with remainder, comparison) is a polynomial in the binary length of the operands. Two problems will be called **polynomial time equivalent** if the existence of a polynomial time algorithm for the one problem is equivalent to the existence of a polynomial time algorithm of the other problem.

Since the problem of computing fundamental units and class groups of number fields is very closely related to the problem of computing shortest vectors in lattices and because the latter problem is conjecturally $NP$-complete, we analyze the algorithms for class group and unit computation for fixed degree and varying discriminant.

## 3.1 Maximal orders

The input data of an algorithm for computing the maximal order $O$ as described in 2.1 is a vector containing the $n$ coefficients of the defining polynomial $f$ of $\mathcal{F}$. The output is the denominator $d$ and the integer matrix $(w_{i,j})$ describing the standard $\mathbb{Z}$–basis of $O$. Since

$$|w_{i,j}| \leq d \leq \sqrt{|D_f|} \quad \text{for } 1 \leq i, j \leq n,$$

where $D_f$ denotes the absolute value of the discriminant of $f$, and since $D_f$ is bounded by a polynomial in the $|a_i|$, the output data are bounded by a polynomial in the input data. We therefore expect to be able to compute the maximal order in time polynomial in $n$ and $\log(|a_i|+1)$, $(1 \leq i \leq n)$. Analyzing an algorithm of Zassenhaus and Ford [11], Lenstra [4] was able to prove

**Theorem 1.** *Computing maximal orders is polynomial time equivalent to computing the largest square dividing a positive integer.*

We will briefly sketch the proof of this theorem. First, assume that one is able to compute maximal orders in polynomial time. Take any positive integer $k$. Let $k_0$ be the largest square dividing $k$. In case $k$ is not a square compute the standard basis of the maximal order of $\mathcal{F} = \mathbb{Q}(\sqrt{k})$ which is $\omega_1 = 1, \omega_2$ with $\omega_2 = \dfrac{\sqrt{k}}{k_0}$ if $k \equiv 2,3 \bmod 4$ and $\omega_2 = \dfrac{1+\sqrt{k}}{2k_0}$ if $k \equiv 1 \bmod 4$. So, in both cases we can read off the largest square divisor $k_0$ of $k$ from the denominator $d$ of the standard basis.

Now assume that we can compute largest square divisors in polynomial time. We want to compute the standard basis of the maximal order of the algebraic number field of $\mathcal{F}$.

The basic procedure of the algorithm does the following: Given an order $R \supset \mathbb{Z}[\alpha]$ in terms of its standard basis and a square free $p \in \mathbb{Z}_{\geq 1}$ it either finds a nontrivial factorization $p = p_1 p_2$ in $\mathbb{Z}$ of $p$ or it determines the standard basis of an order $R' \supset R$ with the property that for $R = R'$ the order is maximal at $p$ which means that $\gcd([O,R],p) = 1$. The running time of this algorithm is polynomial in $\log p$, $n$ and $\log(|a_i|+1)$, $(1 \leq i \leq n)$. We initialize $p$ to the product of all primes which divide $D_f$ at least in power 2. Then $p$ can be computed in polynomial time by applying the algorithm for finding largest square divisors repeatedly. We use the algorithm to compute a sequence of orders

$$\mathbb{Z}[\alpha] = R_0 \subset R_1 \subset \ldots$$

Either we find in this sequence two subsequent orders which are equal, then we have found the maximal order of $\mathcal{F}$, or at a certain point we find a nontrivial factorization of $p$ in which case we have to maximize the order first with respect to $p_1$ and then with respect to $p_2$. Iterating this process we get a polynomial time algorithm for computing maximal orders.

It remains to explain the enlarging procedure. So let $R \supset \mathbb{Z}[\alpha]$ be an order of $\mathcal{F}$ and let $p \in \mathbb{Z}_{\geq 1}$ be square free. The prime factors in $p$ which are less than $n$ are removed by trial division so that we may assume that $p$ is either prime or all the prime factors of $p$ exceed $n$. We now compute the pseudo radical $\bar{I_p}$ of $R/pR$ which for prime numbers $p$ is the kernel of the map

$$R/pR \to R/pR, \ \bar{\beta} \mapsto \bar{\beta}^{p^j}$$

where $j$ has to be chosen such that $p^j \geq n$. For square free $p$ with all prime factors exceeding $n$ the pseudo radical of $R/pR$ is the kernel of the linear map

$$R/pR \to \mathrm{Hom}(R/pR \to \mathbb{Z}/p\mathbb{Z}), \ \bar{\beta} \to (\bar{\lambda} \to \mathrm{Tr}(\overline{\beta\lambda}))$$

where $\mathrm{Tr}(\bar{\delta})$ is the trace map of the linear map

$$R/pR \to R/pR, \ \bar{\beta} \mapsto \delta\beta.$$

So, in both cases the radical can be computed by means of linear algebra algorithms over $\mathbb{Z}/p\mathbb{Z}$ which are the same as the linear algebra algorithms over finite fields except that a division might fail in which case we find a nontrivial factorization of $p$ and this is also considered a legal output of the algorithm. Of course, those procedures are polynomial time.

The enlarged order $R'$ is the ring of multipliers of the inverse image $I_p$ of $\bar{I_p}$ under the natural map $R \to R/pR$. $R'$ is computed as follows: First determine the kernel $\bar{U}_p$ of the map

$$R/pR \to \mathrm{End}(I_p/pI_p), \ \bar{\beta} \mapsto (\bar{\lambda} \mapsto \beta\lambda).$$

Then

$$R' = \frac{1}{p} U_p$$

where $U_p$ is the inverse image of $\bar{U}_p$ under the natural map $R \to R/pR$. So again, $R'$ can be computed by linear algebra and it can be shown that $R'$ satisfies the above requirements.

## 3.2 Unit group

### 3.2.1 *Fundamental units*

The input for the unit algorithms will be an integral basis $\omega_1, \ldots, \omega_n$ of $\mathcal{O}$, more precisely rational approximations to the conjugates $\omega_j^{(i)}$ of the basis elements and a multiplication table for the basis. We will assume that the entries of the multiplication table as well as the $|\omega_j^{(i)}|$ are of order of magnitude $O(\log \mathcal{D})$ where $\mathcal{D}$ is the discriminant of $\mathcal{F}$. Such a basis can be computed by means of LLL–reduction. It can be shown that the number of digits in the binary representation of the $\omega_j^{(i)}$ which one has to know in order to be able to carry out the computations below correctly is of order of magnitude $O(\log \mathcal{D})$. Therefore the input data are of length polynomial in $n$ and $\log \mathcal{D}$.

The binary length of the coefficients $O(\log \mathcal{D})$. in the representation of a system of fundamental units (see 2.2.1) can be much larger, however. The only upper bound we know can be derived from a theorem of Siegel [21] which says that $hR = O(\mathcal{D}^{1/2+\epsilon})$ implying $R = O(\mathcal{D}^{1/2+\epsilon})$. It is, however, not clear whether there is an infinite family of fields $\mathcal{F}$ for which the regulators satisfy $R \geq C\sqrt{\mathcal{D}}$ for some constant $C$ which is independent of $\mathcal{D}$. Proving this would mean proving a weak form of the famous Gauss conjecture.

Nevertheless, the output data in this case are much longer than the input data and it is therefore sensible to measure the complexity of unit algorithms in terms of the regulator. This was done in the following theorem which is due to Buchmann [1]

**Theorem 2.** *There is an algorithm which computes a system of fundamental units in* $O(R^{1/2}\mathcal{D}^{\epsilon})$ *bit operations.*

Again we sketch the proof of this theorem.

We start by introducing some basic notions and facts. A number $\mu$ in a fractional ideal **a** of $\mathcal{O}$ is called a **minimum** of **a** if there is no *nonzero* $\beta \in \mathbf{a}$ such that $|\beta|_i < |\mu|_i$

for $1 \leq i \leq m$, $m$ being the number of archimedian valuations on $\mathcal{F}$. The norm of such a minimum is bounded:

$$|N(\mu)| \leq \sqrt{\mathcal{D}}\, N(\mathbf{a})$$

The minimum $\upsilon$ of **a** is called a **neighbour** of the minimum $\mu$ of **a** if the set $\{\beta \in \mathbf{a} : |\beta|_i < \max\{|\upsilon|_i, |\mu|_i\}$ for $1 \leq i \leq r+1\}$ only contains the element 0.

The minima of **a** together with this neighbour relation form a graph $G$. The unit group $\mathcal{O}^{\times}$ of $\mathcal{F}$ acts on $G$ and $G / \mathcal{O}^{\times}$ is finite. More precisely, it was shown in Buchmann [2] that $|G / \mathcal{O}^{\times}| = O(R)$ and that the number of neighbours of each minimum is $O(\mathcal{D}^{\epsilon})$. A fractional ideal **a** of $\mathcal{O}$ is called **reduced** if 1 is a minimum of **a**. If **a** is a reduced ideal and if $\gamma$ is a nieghbour of 1 in **a** then the ideal $\mathbf{a}' = \frac{1}{\gamma_i}\mathbf{a}$ is reduced and this reduced ideal is called the **neighbour** of **a**. Hence the reduced ideals together wiht this neighbour relation form again a directed graph which is isomorphic to $G / \mathcal{O}^{\times}$. Each reduced ideal **a** of $\mathcal{O}$ can be written in the form $\mathbf{a} = \frac{1}{\mu}\mathcal{O}$ with a minimum $\mu$ of $\mathcal{O}$ and two reduced ideals $\frac{1}{\mu}\mathcal{O}$ and $\frac{1}{\upsilon}\mathcal{O}$ are equal if and only if $\mu/\upsilon$ is a unit in $\mathcal{O}$.

It is, however, not advisable to store a reduced ideal **a** in terms of the corresponding minimum $\mu$ since those numbers become extremely large. We rather use the fact that there is precisely one positive integer $d$ and one integer matrix $(a_{i,j})$ in Hermite normal form such that the gcd of its entries is coprime to $d$ and the numbers

$$\beta_j = \frac{1}{d}\sum_{k=1}^{n} a_{j,k}\,\omega_k \quad \text{for } 1 \leq j \leq n$$

form a $\mathbb{Z}$-basis of **a**. $d$ is called the **denominator** of **a** and $(a_{i,j})$ is called the **HNF-matrix** of **a**. Hence reduced ideals are stored and easily compared in terms of their denominators and their HNF-matrices. The storage requirement for each reduced ideal is $O(\mathcal{D}^{\epsilon})$.

Before computing a unit $\epsilon$ explicitly, we compute its **logarithm vector**

$$\mathbf{Log}\,\epsilon = (\log|\epsilon|_1, \ldots, \log|\epsilon|_m).$$

Only if we know that the unit $\epsilon$ enlarges the unit group generated by the units which we have found previously do we compute this unit explicitly.

Now we can present the algorithm:

**Algorithm 1**

1. *(Initialization)*

   $p \leftarrow 1, k \leftarrow 1, \mathbf{a}_1 \leftarrow O, \mathbf{v}_1 \leftarrow \mathbf{0}, r \leftarrow 0, U \leftarrow \varnothing.$

2. *(Computation of the neighbours)*

   *Compute the set N of all neighbours of* 1 *in the reduced ideal* $\mathbf{a}_k$.

3. *(comparison of the new reduced ideals with the old ones)*

   *For each* $\eta \in N$ *execute the following steps:*
   *Compute the reduced ideal* $\mathbf{b} = \frac{1}{\eta}\mathbf{a}_k$. *Compare* **b** *with all reduced ideals* $\mathbf{a}_l$, $1 \le l \le p$, *which have already been computed. If* **b** *is distinct from all those reduced ideals then set* $p \leftarrow p + 1$, $\mathbf{a}_p \leftarrow \mathbf{b}$ *and* $\eta_p \leftarrow \eta$. *Otherwise a unit* $\in$ *of* $O$ *has been found which can be computed by multiplying appropriate* $\eta_j$ *'s. If this uint enlarges the unit group found so far, update U to a basis.*

4. *Set* $k \leftarrow k + 1$. *If* $k > p$ *terminate, else go to step 2.*

It can be shown that there exist number geometric algorithms for computing neighbours in $O(\mathcal{D}^{\epsilon})$ bit operations and for updating the basis $U$ in $O(R\ \mathcal{D}^{\epsilon})$ bit operations. Moreover, it can be shown that the number of neighbours of reduced ideals in each ideal class is $O(R\ )$, that the number of neighbours of each reduced ideal is $O(\mathcal{D}^{\epsilon})$ and that the number of updates of $U$ is $O(\mathcal{D}^{\epsilon})$. Those facts are quite complicated to prove but too technical to be explained here. For details see [1], [2], [3]. Taking this for granted, it follows that the algorithm computes a basis for the unit group in $O(R\ \mathcal{D}^{\epsilon})$ bit operations, as asserted.

Another important application of this theory is principal ideal testing in $\mathcal{F}$. Algorithm 1 computes computes all the reduced ideals $\mathbf{a}_1, \ldots, \mathbf{a}_p$. So suppose that **a** is an ideal of $\mathcal{F}$ then in order to test **a** for principality, one only needs to determine a reduced ideal in the class of **a** and to compare this ideal to the reduced principal ideals. Reduction of **a** is performed by computing a minimum $\mu$ in **a**: then $\frac{1}{\mu}\mathbf{a}$ is reduced. We can for example take a number $\mu \in \mathbf{a}$ whose image under the Minkowski map

$$\beta \rightarrow (\beta^{(1)}, \ldots, \beta^{(s)}, \mathrm{Re}(\beta^{(s+1)}), \ldots, \mathrm{Re}(\beta^{(m)}), \mathrm{Im}(\beta^{(s+1)}), \ldots, \mathrm{Im}(\beta^{(m)}))^{t}$$

is a shortest nonzero vector in the lattice $\underline{\mathbf{a}}$. For fixed degree $n$ the reduction can be performed in polynomial time in **Log** $\mathcal{D}$ and $a$ if **a** is given by $a$ bits, as was pointed out in [6]. Since all the reduced principal ideals can be computed and lexicographically ordered by means of Algorithm 1 in $O(\mathcal{D}^{1/2+\epsilon})$ bit operations we get

**Theorem 3** *After precomputing all the reduced principal ideals of* $\mathcal{F}$ *in* $O(\mathcal{D}^{1/2+\epsilon})$ *bit operations, each ideal* **a** *of* $\mathcal{F}$ *which is given by* $a$ *bits can be tested for principality in polynomial time in* **Log** $\mathcal{D}$ *and* $a$.

### *3.2.2 Regulator*

Although the algorithm of the previous section is in some sense optimal for unit computation because the size of the matrix $(e_{i,j})$ from 2.2.1 this is no longer the case when one is only interested in the regulator. This, for example, is the case if one computes class numbers using the analytic class number formula.

Shanks [20] has presented a speed up for the continued fraction algorithm which computes the regulator of a real quadratic field. According to the analysis of Lenstra [14] this improved algorithm evaluates the regulator in time $O(R^{1/2}\mathcal{D}^{\epsilon})$. In [3] the author was able to generalize this result to arbitrary number fields.

In order to compute the regulator of a real quadratic number field by reduction theory one has to compute the graph of all the reduced principal ideals which in this case is a sequence.

$$O = \mathbf{a}_1, \mathbf{a}_2, \ldots, \mathbf{a}_p$$

and the corresponding sequence of logarithm vectors of minima

$$\mathbf{Log}\ \alpha_1, \ldots \mathbf{Log}\ \alpha_2, \ldots, \mathbf{Log}\ \alpha_p = R\,.$$

Rather than compute the whole sequence Shanks' idea is to compute only the reduced ideals up to a certain point (**babysteps**). The last ideal $\mathbf{a}_k$ calculated this way is then squared and reduced:

$$\mathbf{b}_1 = \mathrm{red}(\mathbf{a}_k\, \mathbf{a}_k) = \frac{1}{\gamma_1}\mathbf{a}_k\, \mathbf{a}_k$$

Then

$$\mathbf{Log}\ \beta_1 = 2\mathbf{Log}\ \alpha_k + \mathbf{Log}\ \gamma_1$$

is the logarithm vector of a minimum whose inverse generates $\mathbf{b}_1$. Fortunately, $|\mathbf{Log}\gamma_1|$ is very small, namely of order of magnitude $O(\mathcal{D}^{\epsilon})$. This procedure is called **giantstep**. In general, executing a giant step means computing

$$\mathbf{b}_{j+1} = \mathrm{red}(\mathbf{b}_j \mathbf{a}_k) = \frac{1}{\gamma_j} \mathbf{b}_j \mathbf{a}_k$$

and

$$\mathbf{Log}\,\beta_{j+1} = \mathbf{Log}\beta_j + \mathbf{Log}\,\alpha_k + \mathbf{Log}\gamma_j .$$

Eventually, one will find a repetition, namely

$$\mathbf{b}_q = \mathbf{a}_l$$

with $1 \leq l \leq k$ and then the regulator is

$$R = \mathbf{Log}\,\beta_q - \mathbf{Log}\,\alpha_l .$$

If one chooses the right number of baby steps then the regulator of the real quadratic field $\mathcal{F}$ can be computed by this method in $O(\mathcal{D}^{1/2+\epsilon})$ bit operations.

Shanks' method was generalized by Buchmann and Williams [7] to number fields of unit rank 1.

The generalization of this method to arbitrary fields is quite complicated since the fundamental parallelepiped of a lattice of dimension greater than one has no longer a canonical form. The use of old and recent results of the geometry of numbers is required and those methods are beyond the scope of this paper. For details see [3].

We do, however, state the following:

**Theorem 4.** *The regulator of* $\mathcal{F}$ *can be computed in* $O(\mathcal{D}^{1/2+\epsilon})$ *bit operations and an ideal* $\mathcal{F}$ *which is given by* $a$ *bits can be tested in time* $O(\mathcal{D}^{1/2+\epsilon})$*+polynomial in* $a$ *and* $\mathbf{Log}\ \mathcal{D}$ *for principality.*

## 3.3 Class group and class numbers

### 3.3.1 Class group

If $P$ is the set of all nonzero prime ideals of $O$ whose norm is below the Minkowski bound

$$M = \frac{n!}{n^n}\left(\frac{4}{\pi}\right)^t \sqrt{\mathcal{D}}$$

then

$$\{\mathbf{p}H : \mathbf{p} \in P\}$$

is a system of generators of the class group $Cl$. The prime ideals $p$ in $P$ can be computed by determining the decomposition

$$pO = \prod_{i=1}^{g} \mathbf{p}_i^{e_i}$$

into prime ideals for each prime number $p$ below the Minkowski bound.

Let $p$ be such a prime number. If $p$ does not divide the index $[O : \mathbb{Z}[\alpha]]$ then this decomposition can be computed by decomposing the defining polynomial $f$ into irreducibles:

$$f(x) \equiv \prod_{i=1}^{g} f_i(x)^{e_i} \bmod p$$

Then the corresponding prime ideals are

$$\mathbf{p}_i = pO + f_i(\alpha)O$$

and the norm of $\mathbf{p}_i$ is $p^{\deg f_i}$.

But if $p$ divides the index $[O : \mathbb{Z}[\alpha]]$, then one has to apply the following strategy (see [4]): Compute the $p$-radical $I_p$ as the inverse image of the kernel of

$$O/pO \to O/pO, \quad \bar{\beta} \to \bar{\beta}^{p^j}$$

under the natural map $O/pO$ by means of linear algebra. Then

$$I_p = \prod_{i=1}^{g} \mathbf{p}_i .$$

Also, $I_p$ is nilpotent in the sense that $I_p^k \subset p\mathcal{O}$ for some $k \in \mathbb{Z}_{\geq 1}$. Now one has to compute the (finitely many) integral ideals

$$I_i = I_p^i + p\mathcal{O}$$
$$J_i = I_{i-1}^{-1} I_i^2 I_{i+1}^{-1}.$$

Then it is easy to verify that $J_i$ is the product of all those prime ideals which divide $p\mathcal{O}$ precisely in the power $i$. So those ideals are integral and coprime in pairs and we have

$$p\mathcal{O} = \prod_{i=1}^{e} J_i^i.$$

Using the well known primitive element construction, we now compute polynomials $g_i \in \mathbb{Z}[x]$ $(1 \leq i \leq e)$ such that

$$\mathcal{O}/J_i \cong F_p[x]/\bar{g}_i F_p[x]$$

and those polynomials are decomposed mod $p$ into irreducibles

$$g_i(x) \equiv \prod_{j=1}^{t_i} g_{i,j}(x) \bmod p.$$

Finally, if $\beta_i \in \mathcal{F}$ with $g_i(x) \equiv 0 \bmod p$, then

$$J_i = \prod_{j=1}^{t_i} \mathfrak{p}_{i,j}$$

with prime ideals

$$\mathfrak{p}_{i,j} = J_i + g_{i,j}(\beta_i)\mathcal{O}.$$

The main ingredients of this method for computing a system of generators for the class group are linear algebra algorithms over $F_p$ which clearly are polynomial time both in $n$ and in $\log p$ and decomposition of polynomials over $F_p$. The best known factoring algorithm over $F_p$ has running time $O(\sqrt{p})$ (see Lenstra [15]) where the $O$-constant polynomially depends on the degree. We therefore have

**Theorem 4.** *A system of generators for the class group can be computed in $O(\mathcal{D}^{3/4+\epsilon})$ bit operations.*

The generating prime ideals **p** are stored in terms of their HNF–matrices and their denominators. This representation can easily be computed from the 2–element representation by applying the HNF–algorithm of Bachem and Kannan [9] or by using the modular HNF–algorithm of Domich, Kannan and Trotter [8] (with module $p$ ).

Once we have computed $P$ we can determine the structure of $Cl$. For this purpose, it is very important to be able to find out whether two fractional ideals **a** and **b** are equivalent modulo $H$, i.e., whether $\mathbf{ab}^{-1}$ is a principal ideal. This decision can be made by means of the reduction theory described above.

Applying Theorem 3 and the results of Pohst and Zassenhaus [17] the structure of $Cl$ can be determined by

**Algorithm 2**

- **Input** *The set* $P = \{\mathbf{p}_1, \ldots, \mathbf{p}_l\}$ *of prime ideals.*
- **Output** *The orders* $N_i$ *of* $C_i$ *and reduced ideals* $\mathbf{a}_i$ *such that* $\langle \mathbf{a}_i H \rangle = C_i \quad (1 \le i \le k)$.
  *The (lexicographically ordered) set* $G$ *of all reduced ideals of* $\mathcal{O}$ *which also contains for each element* **a** *the representation of* $\mathbf{a}H$ *in terms of the basis elements* $\mathbf{a}_i H$.

1. *(Initialization) Set* $k \leftarrow 0, i \leftarrow 0, G \leftarrow \{\mathcal{O}\}$.
2. *(Increase i) Set* $i \leftarrow i+1$. *For* $i > l$ *terminate otherwise set* $a \leftarrow \mathrm{red}(\mathbf{p}_i)$
3. *If* $\mathbf{a} \in G$ *go to* 2.
4. 
   - *Compute the order* $\mu$ *of* $\mathbf{a}H$ *in* $Cl$ *and factorize* $\mu$.
   - *Determine the least positive integer* $\lambda$ *such that the* $\mathrm{red}(\mathbf{a}^\lambda)$ *belongs to* $G$, *i.e.,*

$$(\mathbf{a}H)^\lambda = \prod_{i=1}^{k} (a_i H)^{\lambda_i} .$$

- *Compute the Smith normal form* $S = \mathrm{diag}(\tilde{N}_0, \ldots, \tilde{N}_k)$ *of the matrix*

$$M = \begin{pmatrix} \lambda & 0 & 0 & \ldots & 0 \\ \lambda_1 & N_1 & 0 & \ldots & 0 \\ \vdots & \vdots & \ddots & & \vdots \\ \lambda_k & 0 & 0 & \ldots & N_k \end{pmatrix},$$

  $S = UMV$ *where* $U, V \in Gl_{k+1}(\mathbb{Z})$. *Also compute the matrix* $U^{-1} = (\bar{u}_{i,j})$. *Set* $A_0 \leftarrow \mathbf{a}$.

- *For* $\tilde{N}_0 = 1$ *compute the new generators*

$$\mathbf{a}_i \leftarrow \mathrm{red}\left(\prod_{j=0}^{k} \mathbf{a}_j^{\bar{u}_{i\,j}}\right) \quad (1 \le i \le k)$$

  *and set* $N_i \leftarrow \tilde{N}_i$ *for* $1 \le i \le k$.

- For $\tilde{N}_0 \neq 1$ *compute the new generators*

$$\mathbf{a}_i \leftarrow \mathrm{red}\left(\prod_{j=0}^{k} \mathbf{a}_j^{\bar{u}_{i\,j}}\right) \quad (1 \le i \le k).$$

  *and set* $N_{i+1} \leftarrow \tilde{N}_i$ *for* $1 \le i \le k+1$, $k \leftarrow k+1$.

- *Update G and go to* 2.

Once we know the decomposition of $Cl$ into its cyclic factors $C_i$, it is very easy to determine the $p-$part of $Cl$ for each prime number $p$ dividing the class number. We simply have to figure out the $p-$part of each $C_i$.

The precomputation of all the reduced principal ideals takes $O(\mathcal{D}^{1/2+\epsilon})$ operations. In Algorithm 2 the full system $G$ of representatives for the part of the class group which we know already is stored in lexicographical order. In view of Theorem 3, step 3 of Algorithm 2 can be executed in $O(\mathcal{D}^{\epsilon})$ binary operations. Moreover, this step is executed at most $\mathcal{D}^{1/2+\epsilon}$ times since the list $P$ of prime ideals contains $\mathcal{D}^{1/2+\epsilon}$ elements.

Now we know from a theorem of Siegel [21] that the order of the class group is $\mathcal{D}^{1/2+\epsilon}$. In step 4 of Algorithm 2 the index of the part of the class group whose structure we have determined already in the full class group is diminished by a factor at least 2. Therefore this step is executed $O(\mathcal{D}^{\epsilon})$ times.

Now we analyze the complexity of the computations which have to be carried out in step 4. Again, by Siegel's Theorem the order $\mu$ can be computed in $\mathcal{D}^{1/2+\epsilon}$ binary operations and this order can be factored in $\mathcal{D}^{1/2+\epsilon}$ binary operations. Since the exponent $\lambda$ of **a** is necessarily a divisor of the order $\mu$ it can be determined in $O(\mathcal{D}^{\epsilon})$ bunary operations. The matrix $M$ has $O(\mathcal{D}^{\epsilon})$ columns and rows and its entries are of size $\mathcal{D}^{1/2+\epsilon}$; thus, by means of the algorithm of Bachem and Kannan [9] the Smith normal form of this matrix can be computed in $O(\mathcal{D}^{\epsilon})$ binary operations. Finally, by using the sorting techniques described in Knuth [12] the set $G$ can be updated in $\mathcal{D}^{1/2+\epsilon}$ binary operations.

**Theorem 5.** *The structure of the class group of an algebraic number field can be computed from the generating set P in* $\mathcal{D}^{1/2+\epsilon}$ *binary operations. The storage requirement is* $\mathcal{D}^{1/2+\epsilon}$ *bits.*

### 3.3.2 *Class number*

In order to compute the class number only, one can utilize the analytic class number formula

$$\lim_{s \to 1} (s - 1)\zeta_{\mathcal{F}}(s) = \frac{2^{r_1}(2\pi)^{r_2} Rh}{\omega\sqrt{\mathcal{D}}}$$

Here $\zeta_{\mathcal{F}}$ is the Dedekind zeta function of $\mathcal{F}$ $w$ is the number of roots of unity contained in $\mathcal{F}$, $r_1$ is the number of real roots of the generating polynomial $f$ and $2r_1$ is the number of its complex roots. If the regulator is known (for example by the algorithm from section 3.2.2) then a sufficient approximation of the residue of this zeta function at $s = 1$ will yield the class number. There are two methods known for approximating this residue. The one is due to Eckhardt [10]. It is based on a suitable series expansion of the zeta function. Using the method from above for determining the splitting behaviour of the prime numbers into prime ideals it can be shown that this method is of complexity $O(\mathcal{D}^{1/2+\epsilon})$.

The second method uses the Euler product expansion for the residue of the zeta function. Its correctness depends on the validity of the generalized Riemann hypothesis, more precisely on the effective version of the Chebotarev density theorem of Lagarias and Odlyzko [13] with the constants of Oesterlé [16].

As in the real quadratic case (see Lenstra [14] and Schoof [19]) it seems to be possible to prove

**Conjecture 1.** *Assuming the truth of the GRH for its normal closure, the class number of $\mathcal{F}$ can be computed in time* $O(\mathcal{D}^{1/5+\epsilon})$.

First results in this direction have been recently obtained by the author and H.C. Williams.

## References


[1] *J. Buchmann,* On the computation of units and class numbers by a generalization of Lagrange's algorithm. J. Number Theory **26** (1987), 8—30.

[2] *J. Buchmann,* On the period length of the generalized Lagrange algorithm. J. Number Theory **26** (1987), 31—37.

[3] *J. Buchmann,* Zur Komplexität der Berechnung von Einheiten und Klassenzahlen algebraischer Zahlkörper. Habilitationsschrift, Düsseldorf, 1988.

[4] *J. Buchmann* and *H.W. Lenstra,* To appear.

[5] *J. Buchmann* and *M. Pohst,* On the complexity of computing class groups of algebraic number fields. To appear.

[6] *J. Buchmann* and *H.C. Williams,* On principal ideal testing in algebraic number fields. J. Symbolic Computation **4** (1987), 11—19.

[7] *J. Buchmann* and *H.C. Williams,* On the infrastructure of the principal ideal class of an algebraic number field of unit rank one. Math. Comp.

[8] *P.D. Domich, R. Kannan* and *L.E. Trotter Jr.,* Hermite normal form computation using modulo determinant arithmetic. Math. Oper. Research **12** (1987), 50—59.

[9] *R. Kannan* and *A. Bachem,* Polynomial algorithms for computing Smith and Hermite normal forms. Siam J. Comput. **8** (1979),. 499—507.

[10] *C. Eckhardt,* Computation of class numbers by an analytic method. J. Symbolic Computation **4** (1987), 41—52.

[11] *C. Ford,* Computation of maximal orders over a Dedekind domain. Thesis, OSU 178.

[12] *D.E. Knuth,* The Art of Computer Programming, Vol. 3: Sorting and Searching. Addison–Wesley, Reading, Mass., 1973.

[13] *J.C. Lagarias* and *A.M. Odlyzko,* Effective versions of the Chebotarev density theorem. In Algebraic Number Fields, Academic Press, New York 1977, 409—464.

[14] *H.W. Lenstra Jr.,* On the computation of regulators and class numbers of quadratic fields. Lond. Math. Soc. Lect. Notes Ser. **56** (1982), 123—150.

[15] *H.W. Lenstra,* Private communication.

[16] *J. Oesterlé,* Versions effectives du théorème de Chebotarev sous l'hypothése de Riemann généralisée. Asterisque **61** (1979), 165—167.

[17] *M. Pohst* and *H. Zassenhaus,* Über die Berechnung von Klassenzahlen und Klassengruppen algebraischer Zahlkörper, J. Reine angew. Math. **361** (1985), 50—72.

[18] *M. Pohst* and *H. Zassenhaus,* Algorithmic Algebraic Number Theory. Cambridge University Press, to appear 1988.

[19] *R.J Schoof,* Quadratic fields and factorization. In Computational Methods in Number Theory, H.W. Lenstra Jr. and R. Tijdeman eds., Math Centrum tracts **155** Part II (1983), 235—286.

[20] *D. Shanks,* The infrastructure of a real quadratic field and its applications. Proc. 1972 Number Theory Conference, Boulder 1972), 217—224.

[21] *C.L. Siegel,* Abschätzung von Einheiten. Ges. Abh. IV, Berlin, Heidelberg, New York 1979, 66—81.

[22] *H.C. Williams,* Continued fractions and number theoretic computations. Rocky Mountain J. Math. **15** (1985),621—655.

[22] *H.C. Williams, G.W. Dueck* and *B.K. Schmidt,* A rapid method for evaluating the regulator and class number of a pure cubic field. Mat. Comp. **41** (1983), 235—286.

---

Mathematisches Institut, Universität Düsseldorf, 4000 Düsseldorf, FRG

# Perturbations and Complementary Sequences of Integers

*John R. Burke*
*William A. Webb*

## 1. Introduction

Throughout this article $\mathbb{Z}_0$ will denote the sequence of nonnegative integers and $A = \{a_i\}$, $B = \{b_i\}$, will be subsequences of $\mathbb{Z}_0$.

As usual, $A + B$ is defined as the sequence of all $n = a_i + b_j$ for some $a_i \in A$, $b_j \in B$ ; and $hA = (h-1)A + A$ for $h \geq 2$. $A$ is said to be a basis of order $h$ if $hA = \mathbb{Z}_0$ and an asymptotic basis of order $h$ if $hA$ contains all but finitely many elements of $\mathbb{Z}_0$. $A$ is a sub–basis of order $h$ if $hA$ contains an infinite arithmetic progression. Finally, $A$ and $B$ are complementary sequences if $A + B = \mathbb{Z}_0$.

Let $f$ be a function from $\mathbb{Z}_0$ to the nonnegative reals. A sequence $B^* = \{b_i^*\}$ is said to be a perturbation of $B$ of order $f$ if $\left|b_j^* - b_j\right| = O(f(b_j))$ for all $j$. (The constant implied by the big $O$ notation is absolute.) $B^*$ is a finite perturbation of $B$ if $\left|b_j^* - b_j\right| = O(1)$, and is a $k$–perturbation if $0 \leq b_j^* - b_j \leq k$ for all $j$.

Burr [3] and Burr and Erdös [4] have discussed the relation of perturbations to completeness properties of sums of sequences. Recently Burke and Webb [2] have discussed basis properties of perturbed sequences. It is proved the every 1–perturbation of $\mathbb{Z}_0$ is a sub–basis of order 2, but for every sequence $A$ there exists a 2–perturbation of $A$ which is not a sub–basis of order 2. Basis properties of perturbed sequences of primes and the relation to the Goldbach conjecture are also discussed.

In this article we will discuss the problem of perturbing a sequence $B$ into a complementary sequence to a given sequence $A$. An application will be given to the problem of finding a perturbation of $S$, the set of squares, which is an asymptotic basis of order 3.

## 2. Perturbing to a Complementary Sequence

Let $A$ and $B$ be increasing sequences of nonnegative integers. For $\alpha \in \mathbb{Z}_0$, let $I_\alpha = [2^\alpha, 2^{\alpha+1})$ and for $\alpha \geq 1$, let $B_\alpha = I_{\alpha-1} \cap B$ and $c(\alpha) = \max_{a_i \leq 3 \cdot 2^{\alpha-1}} (a_i - a_{i-1})$.

We will construct a $c(\alpha)$–perturbation of the elements of $B_\alpha$ to a set $B_\alpha^*$ (thus forming a $c(\alpha)$–perturbation of $B$ into a sequence $B^*$) such that $I_\alpha \subseteq B_\alpha^* + A \subseteq B^* + A$. (This, of course, puts a restriction on the structure of $B$ which will become apparent later.) Thus the sequence $B^*$ will be complementary to $A$.

Let $B_\alpha = \{b_1, \ldots, b_{t(\alpha)}\}$. Consider the set $\{b_i + j : 1 \leq i \leq t(\alpha), 0 \leq j \leq c(\alpha)\}$. Define $b_1^*$ to be the least element of this set such that $|(A + b_i + j) \cap I_\alpha|$ is maximal. Thus $b_1^*$ is a $c(\alpha)$–perturbation of some $b_i \in B_\alpha$, say $b_1$. Now choose $b_2^*$ from the set $\{b_i + j : 2 \leq i \leq t(\alpha), 0 \leq j \leq c(\alpha)\}$ such that $|(A + (b_i + j)) \cap I_\alpha \setminus (b_1^* + A)|$ is maximal.

Continue in this fashion until each of the terms in $B_\alpha$ whose perturbation will cover at least two new elements in $I_\alpha$ have been perturbed. (It may be that a term of $B_\alpha$ has been perturbed outside of $I_{\alpha-1}$.) Let $k$ denote the number of perturbed terms so that $0 \leq k \leq t(\alpha)$. We next find an upper bound for $k$. To do this, for an arbitrary but fixed $\alpha$ define

$$K_j = \{b_i^* : A + b_i^* \text{ covered } j \text{ new elements of } I_\alpha \text{ when it was chosen}\}.$$

$$k_j = |K_j|$$

$$R_s = \{n \in I_\alpha : n \text{ is not covered by } A + b^* \text{ for any } b^* \in K_j,\ j > s\}.$$

$$r_s = |R_s|$$

In particular, this leads to

$$r_s - r_{s-1} = sk_s.$$

Next, let $b \in B_\alpha$, $b$ is perturbed to $b^*$, and $b^* \notin K_j$, $j > s$. That is, when $b$ was perturbed to $b^*$, the number of new terms covered in $I_\alpha$ was no more than $s$. Let $z$ be the natural number of $n \in R_s$ (including repetitions) in at least one of the sets $A + b + i$, $0 \le i \le c(\alpha)$. Since the gap size in $A$ is no more than $c(\alpha)$ every $n \in R_s$ is in one of the sets $A + b + i$, $0 \le i \le c(\alpha)$ so that $z \ge r_s$.

On the other hand, none of the sets $A + b + i$, $0 \le i \le c(\alpha)$ contains more than $s$ new elements in $I_\alpha$ and there are $c(\alpha) + 1$ such sets so $z \le s(c+1)$. Thus $r_s \le s(c+1)$.

We have, letting $c = c(\alpha)$ and $M$ the largest $j$ such that $k_j \ne 0$,

$$k = \sum_{j=2}^{M} k_j = \sum_{j=2}^{M} \frac{r_j - r_{j-1}}{j} = \sum_{j=2}^{M} \frac{r_j}{j} - \sum_{j=1}^{M-1} \frac{r_j}{j+1}$$

$$= \frac{r_M}{M} + \sum_{j=2}^{M-1} \frac{r_j}{j(j+1)} - \frac{r_1}{2} \le \frac{r_M}{M} + \sum_{j=2}^{M-1} \frac{j(c+1)}{j(j+1)}$$

$$\le (c+1) + (c+1) \sum_{j=2}^{M-1} \frac{1}{j+1} \le \left(\frac{1}{2} + \log(M)\right)(c+1).$$

At this point the only unperturbed terms of $B_\alpha$ correspond to those in $K_1$ and if $B_\alpha^*$ is to be a complementary sequence, there must be no more elements in $R_1$ than there are in $K_1$. That is, each element of $K_1$ is perturbed to cover one element in $R_1$. This leads to

$$t(\alpha) - k = k_1 \ge r_1 \text{ or}$$

$t(\alpha) \ge k + r_1$. Thus it is sufficient that

$$t(\alpha) \ge \left(\frac{1}{2} + \log(M)\right)(c+1) + (c+1) = \left(\frac{3}{2} + \log(M)\right)(c+1) \tag{1}$$

In most applications $A$ will be a relatively dense set. Thus using the trivial estimate $M \le 2^\alpha$ the following theorem is an immediate corollary of inequality (1).

**Theorem.** *Let $A$ and $B$ be increasing sequences of nonnegative integers and*

$$c(\alpha) = \max_{a_i \leq 3 \cdot 2^{\alpha-1}} (a_i - a_{i-1}).$$

*Then if the number of elements $b_i \in B$, $2^{\alpha-1} \leq b_i < 2^{\alpha}$ is at least $(\frac{3}{2} + \alpha \log 2) \times (c(\alpha) + 1)$ for all sufficiently large $\alpha$, then there exists a perturbation of $B$ such that*

i) *no element $b_i \in B$, $b_i < 2^{\alpha}$, is perturbed by more than $c(\alpha)$*

ii) *$A + B$ contains all sufficiently large integers.*

## 3. Constructing a Basis Using Squares

One of the first articles to deal with perturbation of sequences was by Atkin [1], where it was shown there exists a perturbation of order $\log n$ of the sequence $S$ of perfect squares such that $2S$ has positive lower asymptotic density. Although any sequence can be perturbed into any other sequence if the size of the perturbation is sufficiently large, simple density considerations make it clear that any "small" perturbation of $S$ cannot be a basis of order 2. However, $S$ is dense enough to be a basis of order 3 but simple congruence properties prevent this.

This raises the question of whether a small perturbation of $S$ is a basis of order 3. What should be considered a small perturbation of $S$ ? Since a perturbation of order $\sqrt{n}$ would allow a given square $k^2$ to be perturbed into any element between $k^2$ and $(k+1)^2$, we desire a perturbation of order smaller that $\sqrt{n}$ . Although it seems quite possible that a finite perturbation of $S$ would be a basis of order 3, we have not yet been able to show the existence of such a sequence. Using the results of the preceding section, we can prove that a perturbation of order $n^{1/4}$ is sufficient.

If $S$ denotes the sequence of squares, let $A = 2S_E$ where $S_E$ is the sequence of even squares and let $B$ equal the sequence of odd squares. Clearly $3S \supseteq A + B$ and it is sufficient to find a perturbation of $B$ which is a complementary sequence to $A$.

For any $k$, consider the elements $(2k)^2, (2k)^2 + 4, \ldots, (2k)^2 + (2t)^2, (2k+2)^2$ where $(2k)^2 + (2t)^2 < (2k+2)^2 \leq (2k)^2 + (2t+2)^2$. All of these elements are in $A$ and the largest gap is at most $8t + 4$. Thus, in (1) $c \leq 8(3 \cdot 2^{\alpha-1})^{1/4} + 5$.

Since $\log M \leq \alpha \log 2$ and $t(\alpha) \geq (\sqrt{2}-1)2^{(\alpha-3)/2}-1$, the inequality (1) will be satisfied for $\alpha$ sufficiently large. Thus there exists a perturbation of $S$ of order $n^{1/4}$ which is an asymptotic basis of order 3.

## References


[1] *A.O.L. Atkin,* On pseudo–squares. Proc. Lond. Math. Soc. **14A** (1965), 22—27.

[2] *J.R. Burke* and *W.A. Webb,* On sums of perturbed integers. J. of Number Theory **29** (1988) no. 3, 264—270.

[3] *S.A. Burr,* On the completeness of sequences of perturbed polynomial values. Pacific J. Math. **86** (1979) no. 2, 355—360.

[4] *S.A. Burr* and *P. Erdös,* Completeness properties of perturbed sequences. J. of Number Theory, ser A **28** (1980), 150—155.



Department of Mathematics and Computer Science, Gonzaga University, Spokane, WA 99258

Department of Pure and Applied Mathematics, Washington State University, Pullman, WA 99164–2930

# On the Number of Sets of Integers With Various Properties

*P.J. Cameron and P. Erdös*

## Introduction

A number of questions in combinatorial number theory concern the maximum number of integers in $[1,n]$ satisfying some constraint, for example, containing no solution to a specified equation. It is our purpose here to examine a related question: what is the number of subsets satisfying the same constraint?

The two problems are closely related. For, suppose that a family $\mathscr{A}$ of subsets of $[1,n]$ is closed under taking subsets and that the largest cardinality of a set in $\mathscr{A}$ is $m$. Then

$$2^m \leq |\mathscr{A}| \leq \sum_{i=0}^{m} \binom{n}{i} \leq n^m + 1. \qquad (*)$$

In particular, suppose that we have a family $\mathscr{A}(n)$ for each $n$, and that the largest set in $\mathscr{A}(n)$ has size $m(n)$. Then

(i) if $m(n) \sim cn$ $(c > 0)$, then $2^{cn} \lesssim |\mathscr{A}(n)| \lesssim 2^{c'n}$;

(ii) if $m(n) = o(n)$, then $|\mathscr{A}(n)| = 2^{o(n)}$.

However, better estimates for $|\mathscr{A}|$ are closely related to the structure of $\mathscr{A}$. For example, if $|\mathscr{A}|$ is close to the lower bound in $(*)$, then we expect that there are a few large sets in $\mathscr{A}$ which contain most members of $\mathscr{A}$.

In almost all of our applications, the sets in $\mathcal{A}$ are those satisfying some condition such as that of being sum–free. (This of course guarantees the closure of $\mathcal{A}$ under taking subsets.) If $\mathcal{B}$ denotes the set of all subsets of $\mathbb{N}$ satisfying this condition, then

$$\mathcal{A}(n) = \{B \cap [1,n] : B \in \mathcal{B}\}.$$

In this case, other methods are available for "measuring the size" of $\mathcal{B}$. One such measure is the Hausdorff dimension $\dim(\mathcal{B})$ (see (1), (16), or, for a form adapted for our purposes here, (4)). For completeness, we give the definition.

In any metric space $(X,d)$, for $s$, $\delta > 0$ and $A \subseteq X$, define

$$\mu_\delta^s(A) = \inf\{\textstyle\sum(\operatorname{diam} U)^s : U \in \mathcal{U}\},$$

where $\mathcal{U}$ is a covering of $A$ by sets $U$ with diameter less than $\delta$, and the infimum is taken over all such coverings. Then $\mu_\delta^s(A)$ increases as $\delta$ decreases (the infimum is taken over a smaller set); so $\mu^s(A) = \lim_{\delta \to 0} \mu_\delta^s(A)$ exists. Then $\mu^s$ is an outermeasure, and defines in the dimension of $A$ is the unique $s_0 \in (0,\infty)$ such that

$$\mu^s(A) = \begin{cases} \infty & \text{if } s < s_0 \\ 0 & \text{if } s > s_0 \ . \end{cases}$$

We always apply this to the metric space on the power set of $\mathbb{N}$ in which $d(A,B) = 2^{-k+1}$, where

$$k = \min\{n : A \cap (1,n) \neq B \cap (1,n)\}.$$

Thus, for example, the fact that a set $\mathcal{B}$ can be covered by $|\mathcal{A}(n)|$ sets of diameter $2^{-n}$ shows that

$$\dim(\mathcal{B}) \leq \liminf\{\log |\mathcal{A}(n)| \,/\, n \log 2\,). \qquad (**)$$

Also (see (1)), if every member of $\mathcal{B}$ has lower density 0, then $\dim(\mathcal{B}) = 0$. On the other hand, if there is a set $B \in \mathcal{B}$ with lower density $d$, then $\dim(\mathcal{B}) \geq d$.

(These bounds are strict, in general. For example, the set of all finite sequences is countable and so has dimension 0, but $|\mathscr{A}(n)| = 2^n$ for all $n$. For a less trivial example, see Section 3.2.)

The remainder of the paper consists of a sequence of examples, illustrating the kind of behaviour which can occur. We begin with two well-behaved examples.

(i) The number of sequences $a_1 < \dots < a_t \leq n$ with $a_i$ odd for all $i$ is $2^{\lceil \frac{1}{2}n \rceil}$; and the set of infinite sequences with this contraint has Hausdorff dimension $\frac{1}{2}$ (and $\frac{1}{2}$-dimensional measure 1).

(ii) The number of sequences in $(1,n)$ with $a_{i+1} \neq a_i + 1$ (i.e. no two members consecutive) is the Fibonacci number $F_{n+2} \sim c \cdot \tau^n$, where $\tau = \frac{1}{2}(\sqrt{5}+1)$; the set of infinite sequences with this property has Hausdorff dimension

$$\log \tau / \log 2 = 0.69424 \dots .$$

## 1. Additive conditions

### 1.1. Sum-free sequences ($a_i + a_j \neq a_k$)

Let $A$ be a sum-free subset of $[1,n]$, with largest element $k$. Then $A$ contains at most one of $i$ and $k-i$, for each $i < k$ (and, if $k$ is even, then $\frac{1}{2}k \notin A$). Thus $|A| \leq \lceil \frac{1}{2}k \rceil \leq \lceil \frac{1}{2}n \rceil$. Furthermore, the only sum-free sets of size $\lceil \frac{1}{2}n \rceil$ are

(i) the set consisting of all odd numbers in $[1,n]$;
(ii) if $n$ is odd, $[\frac{1}{2}(n+1), n]$; if $n$ is even, $[\frac{1}{2}n, n-1]$ and $[\frac{1}{2}n+1, n]$.

Thus, if $f(n)$ denotes the number of sum-free subsets of $[1,n]$, then

$$f(n)/2^{\frac{1}{2}n} > 1.$$

(In fact, the sets displayed show that this ratio is at least $2\sqrt{2} - \epsilon$ if $n$ is odd, and at least $5/2 - \epsilon$ if $n$ is even.)

The second author and Andrew Granville, and independently Neil Calkin, have shown that

$$f(n) = 2^{(\frac{1}{2}+o(1))n} .$$

The proofs will appear elsewhere. Curiously, both use non-trivial results of Szemerédi – the first, a graph-theoretic result; the second, his celebrated theorem on arithmetic progressions. But we *conjecture* that much more is true, and that $f(n)/2^{\frac{1}{2}n}$ is bounded. Indeed, we conjecture that this ratio tends to limits $c$ and $c'$ as $n$ tends to infinity through odd and even values respectively, where $c$ and $c'$ are approximately 6.8 and 6.0 respectively. (The meaning of these numbers will be explained shortly.)

So far as Hausdorff dimension goes, the lower and upper bounds for $f(n)$ show that the dimension is $\frac{1}{2}$. Our conjecture would imply that the $\frac{1}{2}$-dimensional measure is finite. But note that, of the large sets we recorded above, sets of odd numbers have $\frac{1}{2}$-dimensional measure 1 (see Introduction), while the other type contribute nothing to the measure. We *conjecture* that, with respect to $\frac{1}{2}$-dimensional measure, almsot all sum-free sets consist of odd numbers. (See (4) for another "natural" measure on sum-free sets for which the measure of the set of sets of odd numbers lies strictly between 0 and 1; see also (3).)

Relatively many sum-free sets can be obtained by allowing a few numbers just smaller than $\frac{1}{2}n$. Our main result is that this does not produce enough sets to violate the conjecture.

**Theorem 1.1.** *There is an absolute constant $c$ such that the number of sum-free subsets of $[1,n]$ whose least element is greater than $n/3$ does not exceed $c\cdot 2^{\frac{1}{2}n}$.*

The proof will give the following extra information.

(i) The ratio of the number of sets in the theorem to $2^{\frac{1}{2}n}$ tends to limits $d$, $d'$ as $n \to \infty$ through odd, respectively even, values. (The numbers $c$, $c'$ of the conjecture are obtained by adding in the numbers of sets of odd numbers; that is, $c = d + \sqrt{2}$, $c' = d' + 1$.)

(ii) Of the sets in the statement of the theorem, almost all have least element exceeding $\frac{1}{2}n - w(n)$, where $w(n)$ is any function tending to $\infty$, however slowly.

The proof requires a couple of lemmas.

**Lemma 1.2.** *Let $k(n)$ denote the sum $\sum \frac{1}{2}^{|S|}$, summed over all independent sets in a path on $n$ vertices. Then $k(n) \leq c_1((\sqrt{3}+1)/2)^n$, for some constant $c_1$.*

**Proof.** Write $k(n) = k_0(n) + k_1(n)$, where $k_0$ and $k_1$ are the sums over independent sets not containing, respectively containing, a fixed end vertex of the path. Then $k_0(n) = k(n-1)$ and $k_1(n) = \frac{1}{2}k_0(n-1)$. So the recurrence for $(k_0, k_1)$ has matrix

$$\begin{pmatrix} 1 & 1 \\ \frac{1}{2} & 0 \end{pmatrix}$$

with greatest eigenvalue $\frac{1}{2}(\sqrt{3}+1)$, and the result follows.

**Lemma 1.3.** *Let $g(k) = (\frac{1}{2}^k) \sum t_k(S)$, where the sum is over all subsets $S$ of $[0, 2k-1]$, and*

$$t_k(S) = \frac{1}{2}^{|(S+S) \cap [0,2k-1]|}.$$

*Then $g(k)$ tends to a finite limit as $k \to \infty$.*

**Proof.** First, write $g(k) = g_0(k) + g_1(k)$, where $g_0$ and $g_1$ involve sums over sums over subsets $S$ with $0 \notin S$, respectively $0 \in S$.

We claim that $g_0(k) = g(k-1)$. To each subset $S$ of $[0, 2k-3]$ corresponds two subsets $S_1 = S + 1$ and $S_2 = S_1 \cup \{2k-1\}$ of $[1, 2k-1]$; and

$$\begin{aligned}(S_1 + S_1) \cap [0, 2k-1] &= (S_2 + S_2) \cap [0, 2k-1] \\ &= ((S+S) \cap [0, 2k-3]) + 2.\end{aligned}$$

Thus $t_k(S_1) = t_k(S_2) = t_{k-1}(S)$.

Since $S_1$ and $S_2$ run over all subsets of $[1, 2k-1]$, we have

$$g_0(k) = g(k-1).$$

Put $h(k) = g_1(k)$. We have to prove that $\sum h(k)$ converges. Write

$$h(k) = h_{00}(k) + h_{01}(k) + h_{10}(k) + h_{11}(k),$$

where $h_{ij}(k)$ involves sets $S$ with $0 \in S$ and

$$2k \in S+S \Longleftrightarrow i = 1, \quad 2k + 1 \in S+S \Longleftrightarrow j = 1.$$

Each set $S$ gives rise to four sets in the sum for $h(k+1)$, namely $S$, $S \cup \{2k\}$, $S \cup \{2k+1\}$, $S \cup \{2k, 2k+1\}$. Since $0 \in S$, we have $S \subseteq S+S$. Thus

$$\begin{aligned} h(k+1) &= \tfrac{1}{2}((1+\tfrac{1}{2}+\tfrac{1}{2}+\tfrac{1}{4})h_{00}(k) + (\tfrac{1}{2}+\tfrac{1}{2}+\tfrac{1}{4}+\tfrac{1}{4})(h_{01}(k) + h_{10}(k)) \\ &\qquad + (\tfrac{1}{4}+\tfrac{1}{4}+\tfrac{1}{4}+\tfrac{1}{4})h_{11}(k)) \\ &= (9/8)h_{00}(k) + (3/4)(h_{01}(k)+h_{10}(k)) + \tfrac{1}{2}h_{11}(k) \\ &\leq (3/4)h(k) + (3/8)h_{00}(k). \end{aligned}$$

A set $S$ contributes to $h_{00}(k)$ if and only if $2k, 2k+1 \notin S+S$. Thus it contains at most one of each of the pairs $\{1, 2k-1\}$, $\{2k-1, 2\}$, $\{2, 2k-2\}, \ldots, \{k-1, k+1\}$; and it contains 0 but not $k$. So it consists of 0 together with an independent set in the path $\{1, 2k-1, 2, \ldots\}$. By Lemma 1.2,

$$h_{00}(k) \leq (\tfrac{1}{2}^k) \cdot \tfrac{1}{2} \cdot c_1((\sqrt{3}+1)/2)^{2k-2}$$

$$\leq c_2((\sqrt{3}+1)/2\sqrt{2})^{2k}.$$

Thus we have

$$h(k+1) \leq \alpha h(k) + c_3 \beta^k,$$

with $\alpha, \beta < 1$. From this the convergence of $\sum h(k)$ follows.

**Proof of the Theorem.** Fix $k \leq n/3$, and count sum–free subsets of $[1, 2n-1]$ which contain no element smaller than $n-k$. Such a set is described by

(a) a subset of $[n-k, n+k-1]$ of the form $S + (n-k)$, where $S \subseteq [0, 2k-1]$ is arbitrary;

(b) an arbitrary subset of $[n+k, 2n-2k-1]$;

(c) a subset of $[2n-2k, 2n-1]$ of the form $T + (2n-2k)$, where $T \subseteq [0, 2k-1]$ and $T \cap (S+S) = \emptyset$.

So the number of such sets is

$$2^{n-3k} \cdot \sum_{S \subseteq [0\ ,\ 2k-1]} 2^{2k-|(S+S) \cap [0,2k-1)|}$$

which is at most $c \cdot 2^{n}$, by Lemma 1.3.

The case where $n$ is even is similar.

Convergents to the limits $d$ and $d'$ mentioned in the remark after the statement of the theorem have been obtained by calculation. These give lower bounds of 6.457... and 5.773... for the limits $c$ and $c'$ of $f(n)/2^{\frac{1}{2}n}$ for odd and even $n$ respectively.

The function $f(n)$ itself has been computed for $n \leq 45$. The parity effect is clearly visible. Allowing for this, $f(n)/2^{\frac{1}{2}n}$ increases to roughly 14.5 and 13.5 at $n = 39, 40$, and then begins to decrease. We *conjecture* that the maxima of $f(n)/2^{\frac{1}{2}n}$ for odd and even $n$ are attained for $n = 39, 40$.

### 1.2. Sidon sequences $(a_i + a_j \neq a_k + a_\ell$ for $\{i,j\} \neq \{k,\ell\})$

A Sidon sequence has its pairwise sums all distinct. It is well–known (see (12) that the maximum number of terms in a Sidon sequence is $(1+o(1))\sqrt{n}$; so, by (*), there are at least $2^{(1+o(1))\sqrt{n}}$ such sequences, and at most $n^{(1+o(1))\sqrt{n}}$ (and the set of infinite Sidon sequences has dimension 0).

We cannot decide which extreme is closer to the truth. We have only succeeded in making a slight improvement to the lower bound. If $f(n)$ is the number of Sidon sequences in $(1,n)$, and $s(n)$ the size of the longest such sequence, then

$$\limsup f(n)/2^{s(n)} = \infty .$$

This is immediate from the next result.

**Proposition 1.4.** *Let $S$ be a class of sequences closed under taking subsequences and under translation. Suppose that the maximum length $s(n)$ of an $S$-sequence in $[1,n]$ is $o(n)$. Then the number $f(n)$ of $S$-sequences in $[1,n]$ satisfies*

$$\limsup f(n)/2^{s(n)} = \infty .$$

**Proof.** For any $d$, there are infinitely many values of $n$ for which $s(n) = s(n-d)$; this gives $\frac{1}{2}\cdot 2^{s(n)}$ sequences whose (fixed) greatest element does not exceed $n-d$. By translation we obtain

$$f(n) \geq \tfrac{1}{2}(d+1)\cdot 2^{s(n)}.$$

The following related result is of some interest. Any Sidon sequence has the following property:

(P) The number of pairs $(a_i,a_j)$ with $i < j$ and $a_j - a_i < x$ is less than $x$, for any $x$.

**Proposition 1.5.** *There are $n^{c\sqrt{n}}$ sequences*

$$a_1 < \dots < a_t \leq n$$

*having property (P).*

**Proof.** Take sequences satisfying

$$a_i \in (\, i\sqrt{n},\ (i+1/10)\sqrt{n})$$

for $i = 1,\dots, \lfloor (9/10)\sqrt{n} \rfloor = t$.

### 1.3. All partial sums distinct

Let $f(n)$ be the number of sequences

$$a_1 < \ldots < a_t \leq n$$

for which all sums $\sum \epsilon_i a_i$ $(\epsilon_i = 0,1)$ are distinct.

**Proposition 1.6.**

$$n^{(1+o(1))\log n/\log 3} \leq f(n) \leq n^{(1+o(1))\log n/\log 2}.$$

**Proof.** If $t$ is the greatest length of such a sequence, then the $2^t - 1$ non-zero sums are all at most $tn$; so $2^t - 1 \leq tn$, and $t \leq (1+o(1))\log n /\log 2$. Then the upper bound follows from (*).

If we choose the terms successively (but not necessarily in increasing order), $a_{i+1}$ cannot be equal to any expression $\sum_{j<i} \delta_j a_j (\delta_j = -1, 0 \text{ or } 1)$, so at most $(3^i - 1)/2$ numbers are forbidden; thus we can make at least $t = \lfloor \log n/\log 3 \rfloor$ choices, and obtain $n^{(1+o(1))\log n/\log 3}$ sequences. We still have to divide by the number of orderings of the sequence, but this is $t! < n^{\log\log n}$, and does not affect the estimate.

We note in passing two subclasses for which a more precise estimate can be obtained. The number of sequences which satisfy

(a) $a_i \geq 2a_{i-1}$,

and the numbers which satisfy

(b) $a_i > \sum_{j<i} a_j$,

are both

$$n^{(1+o(1))\log n/(2\log 2)}.$$

For these numbers are related to the binary partition function $b(n)$, the number of ways of writing $n$ as a sum of powers of 2: there are $b(2n)$

sequences satisfying (a), and $\frac{1}{2}\left(1 + \sum_{i=0}^{n} b(2i)\right)$ satisfying (b). For the asymptotics of $b(n)$, see Mahler (15).

## 2. Multiplicative conditions

### 2.1. Product–free sequences ($a_i a_j \neq a_k$)

Trivially, the set of all numbers between $\lfloor\sqrt{n}\rfloor + 1$ and $n$ is product–free; so there are at least $2^{n-\sqrt{n}}$ product–free sequences. The question of Hausdorff dimension is a little more delicate.

**Theorem 2.1.** *(i) Any product–free sequence has upper density less than* 1, *but there exist product–free sequences with density greater than* $1-\epsilon$ *for any* $\epsilon > 0$.

*(ii) The set of product–free sequences has Hausdorff dimension* 1.

**Proof.** (i) Suppose that a product–free sequence $S$ with least element $a$ has upper density $1-c$. Choose $n$ such that at least $(1-c-\epsilon)an$ numbers in $[1,an]$ belong to $S$. In particular, at least $(1-ac-a\epsilon)n$ numbers in $[1,n]$ belong to $S$. But, if $x$ is one of these numbers, then $ax \notin S$. So

$$(c+\epsilon)an \geq (1-ac-a\epsilon)n$$

whence $c \geq 1/2a - \epsilon$. Since $\epsilon$ was arbitrary, $c \geq 1/2a$.

The second assertion follows from a result of the second author (10). Let $V_t(n)$ denote the number of prime factors of $n$ not exceeding $t$. Then, for any $\epsilon > 0$, there exists $t$ such that

$$S_t = \{n : (2/3)\log\log t \leq V_t(n) < (4/3)\log\log t\}$$

has density at least $1-\epsilon$. Of course, $S_t$ is product–free, and is the required set.

(ii) This follows immediately from (i).

## 2.2. Pairwise products distinct, and related constraints

There are a number of constraints satisfied by the sequence of primes. These include

(i) $a_i a_j \neq a_k a_\ell$ for $i, j, k, \ell$ distinct;
(ii) all partial products distinct;
(iii) $a_i \nmid a_j a_k$ for $i \notin \{j,k\}$.

The most stringent condition of this sort is that no term in the sequence divides the product of all the others. Trivially there are $2^{\pi(n)}$ such sequences, where $\pi(n)$ is the number of primes not exceeding $n$. This estimate can be improved:

**Theorem 2.2.** *The number of sequences*

$$a_1 < \dots < a_t \leq n$$

*for which no term divides the product of all the others, is at least* $2^{(c+o(1))\pi(n)}$, *where*

$$c = \Big( \sum_{r \geq 1} \log(r+1)/r(r+1) \Big) / \log 2 = 1.814\dots .$$

**Proof.** For each prime $p$ with $\sqrt{n} < p \leq n$, include at most one of the numbers $tp$, $1 \leq t \leq n/p$; the resulting sequence satisfies the constraint, since each term has a prime factor dividing none of the others. Thus the number of sequences is at least

$$\prod_{r \leq \sqrt{n}} (r+1)^{\pi(n/r) - \pi(n/(r+1))} = 2^{(c+o(1))\pi(n)},$$

where $c$ is as claimed (using the Prime Number Theorem).

For sequences with pairwise products distinct, this bound is exact, apart from the $o(1)$:

**Theorem 2.3.** *The number of sequences*

$$a_1 < \dots < a_t \leq n$$

*with* $a_i a_j \neq a_k a_\ell$ *for* $i, j, k, \ell$ *distinct, is* $2^{(c+o(1))\pi(n)}$, *where* $c$ *is as in Theorem* 2.2.

**Proof.** We partition any such sequences $S$ into three subsequences $S_1, S_2, S_3$, where:

(i) $S_1$ consists of all terms divisible by a prime $p > n^{2/3}$ which divides no other term in $S$;

(ii) $S_2$ consists of all terms $a_i$ divisible by a prime $p > n^{2/3}$ which divides another term $a_j$ (necessarily also in $S_2$);

(iii) $S_3$ consists of all terms whose prime factors do not exceed $n^{2/3}$.

We bound above the numbers of choices for $S_1, S_2$, and $S_3$; the product of these three bounds is an upper bound for the number of sequences $S$.

For $S_1$, the argument for Theorem 2.2 gives the upper bound $2^{(c+o(1))\pi(n)}$.

For $S_2$, we write each term (uniquely) as $a_i = p_i u_i$, where $p_i$ is prime and $p_i > n^{2/3}$. Two distinct $p$'s cannot be associated with the same two $u$'s, since

$$p_1 u_1 \cdot p_2 u_2 = p_1 u_2 \cdot p_2 u_1 .$$

Hence the number of terms in the sequence is at most the number of pairs $(u_1, u_2)$ with $u_1, u_2 < n^{1/3}$, hence is less than $n^{2/3}$; thus there are at most $n^{n^{2/3}}$ choices for $S_2$.

For $S_3$, we use an old result of the second author (6), according to which, if $m \leq n$ has no prime factor $p > n^{2/3}$, then $m = uv$, where $u$, $v \leq n^{2/3}$. Thus $S_3$ is determined by the graph $G$ with vertex set $(1, n^{2/3})$, in which $u$ and $v$ are joined whenever $uv \in S_3$. As above, this graph contains no 4–cycles. Also, any edge has one end in $(1, n^{\frac{1}{2}})$. So $G$ is the union of a $C_4$–free graph on $n^{\frac{1}{2}}$ vertices and a $C_4$–free bipartite graph on $n^{\frac{1}{2}}$ and $n^{2/3}$ vertices. Standard results in extremal graph theory (see (2),

Chapter 6) bound the number of edges of $G$ by $n^{3/4}+n^{5/6}$. Hence the number of choices for $G$, and hence for $S_3$, does not exceed

$$n^{2(n^{3/4}+n^{5/6})}.$$

For a related result, see Erdös (9).

## 3. Divisibility and common factors

### 3.1. Requiring divisibility ($a_i \mid a_j$ for $i<j$)

Let $f(n)$ be the number of sequences $a_1 < \ldots < a_t \leq n$ for which $a_i \mid a_j$ whenever $i<j$. For $n \geq 2$, the number of such sequences with $a_t = n$ is just twice the number of ordered factorizations of $n$ (the factor 2 because we can choose whether or not to include 1 in the sequence). Like the divisor function $d(n)$, this function is very irregular; but its sum function (our $f(n)$) is much better behaved:

**Theorem 3.1.** $f(n) \sim cn^{\alpha}$, *where* $\alpha$ *is the real number greater than* 1 *for which* $\zeta(\alpha) = 2$, *where* $\zeta$ *is the Riemann zeta-function* (*so* $\alpha = 1.7286\ldots$).

This is due to Kalmár (13); see also Erdös (7,8).

Note that the number of factorizations of $n$ is equal to the number of perfect partitions of $n-1$ (Riordan (16), p. 123); the sequence appears in Sloane (18) under this name.

### 3.2. Forbidding divisibility ($a_i \nmid a_j$ for $i \neq j$)

The interval $(\frac{1}{2}n, n]$ gives a sequence of size $\lceil \frac{1}{2}n \rceil$ with no member dividing another. So there are at least $2^{\lceil \frac{1}{2}n \rceil}$ such sequences. Moreover, there is no longer sequence; for, with each odd $d \leq n$, at most one of $d$, $2d$, $4d$, ... can be included.

We conjecture that the number $f(n)$ of sequences

$$a_1 < \dots < a_t \leq n$$

with $a_i \nmid a_j$ for $i \neq j$, is

$$f(n) = (c+o(1))^n$$

for some constant $c$; in other words, $f(n)^{1/n}$ tends to the limit $c$ as $n \rightarrow \infty$. Numerical evidence supports the conjecture, and suggests that the convergence is quite rapid. In addition, we have:

**Proposition 3.2.** $c_1^n \leq f(n) \leq c_2^n$ *for sufficiently large* $n$, *where* $c_1 = 1.55967...$, $c_2 = 1.59...$ .

**Proof.** We indicate how to establish exponential bounds better than the trivial $2^{\frac{1}{2}n} \leq f(n) \leq 2^n$; the improvement to the values given above is computational and not enlightening.

For the lower bound, we can choose neither or one of each pair $\{x,2x\}$ with $n/3 < x \leq n/2$, and also choose whether or not to include each of the remaining $n/3$ numbers above $\frac{1}{2}n$. Thus

$$f(n) \geq (2^{1/3}3^{1/6})^n$$

(and there are this many sequences whose smallest member exceeds $n/3$, if 6 divides $n$.) The bound in the proposition is obtained by replacing $n/3$ by $n/5$.

For the upper bound, we modify the argument used above to bound the number of terms in the sequence. For each odd number $x$ with $n/2 < x \leq n$, we can choose whether to include $x$; for each odd $x$ with $n/4 < x \leq n/2$, we can choose to include $x$, $2x$ or neither; and so on. This gives an upper bound of about $c^n$, where

$$c = \prod_{r \geq 2} r^{1/2^r} = 1.66169... .$$

The upper bound in the proposition is obtained by using the prime 3 as well as 2.

The exponential lower bound raises the question of Hausdorff dimension. But this is zero, since any divisor–free sequence has lower density zero (a result of the second author (5)).

### 3.3. Any two terms coprime

As in 2.2, any sequence of primes satisfies this constraint, so there are at least $2^{\pi(n)}$ such sequences. We show that this is an underestimate by a factor of $e^{c\sqrt{n}}$.

**Theorem 3.3.** *The number $f(n)$ of sequences*

$$a_1 < \ldots < a_t \leq n$$

*for which $(a_i, a_j) = 1$ for $i \neq j$, satisfies*

$$e^{(\frac{1}{2}+o(1))\sqrt{n}} \cdot 2^{\pi(n)} \leq f(n) \leq e^{(2+o(1))\sqrt{n}} \cdot 2^{\pi(n)}.$$

**Proof.** For the lower bound, choose first an even number $2k$ of primes less than $\sqrt{n}$, where $k = (1+o(1))\sqrt{n}/\log n$. Partition them into $k$ pairs (there are

$$(2k)!/2^k k! = e^{(\frac{1}{2}+o(1))\sqrt{n}}$$

ways of doing this), and take the $k$ products of the pairs. Then adjoin to the sequence any subset of the remaining primes (in

$$2^{\pi(n)-2k} = 2^{\pi(n)} e^{o(\sqrt{n})}$$

ways).

For the upper bound, note that the number of terms of such a sequence which are not prime is at most $t = \pi(\sqrt{n})$, since these terms have distinct prime factors not exceeding $\sqrt{n}$. These terms can be chosen in at most

$$\sum_{i=0}^{t} \binom{n}{i} = e^{(2+o(1))\sqrt{n}}$$

ways.

### 3.4. No two terms coprime

Any sequence of even numbers satisfies this constraint, so there are at least $2^{\lfloor \frac{1}{2}n \rfloor}$ sequences. The following argument, due to Carl Pomerance, gives an upper bound. We are grateful to him for his permission, nay, encouragement to include it here.

**Lemma 3.4.** *Let $m_1$, $m_2$, ... $m_{\lceil \frac{1}{2}n \rceil}$ be the odd numbers in $[1,n]$, ordered so that*

$$\varphi(m_1)/m_1 \leq \varphi(m_2)/m_2 \leq \dots ,$$

*where $\varphi$ is Euler's function. Let $N_i$ be the number of even integers in $[1,n]$ that are coprime to $m_i$. Then $N_i \geq i$ for $i \leq \lceil \frac{1}{2}n \rceil - 1$.*

**Proof.** For $i \leq \lceil \frac{1}{2}n \rceil - 2$, this follows directly from Theorem 2 in Pomerance and Selfridge (14). For $i = \lceil \frac{1}{2}n \rceil - 1$, we have $n \geq 3$ and $m_i = p$, the largest prime not exceeding $n$. Then by Bertrand's Postulate, $p$ does not not divide any even number in $[1,n]$, so $N_i = \lfloor \frac{1}{2}n \rfloor \geq i$.

**Theorem 3.5.** *The number $f(n)$ of sequences*

$$a_1 < \dots < a_t \leq n$$

*with $(a_i, a_j) > 1$ for every $i, j$ satisfies*

$$2^{\lfloor \frac{1}{2}n \rfloor} \leq f(n) \leq n \cdot 2^{\lfloor \frac{1}{2}n \rfloor}$$

*for all $n \geq 1$.*

**Proof.** Using the notation of the lemma, let $f_i(n)$ be the number of sequences counted by the theorem such that $i$ is the largest subscript of any $m_j$ in the sequence. Then for $i = \lceil \frac{1}{2}n \rceil$, we have $m_i = 1$, so that $f_i(n) = 0$. Suppose that $i \leq \lceil \frac{1}{2}n \rceil - 1$. If $S$ is a sequence counted by $f_i(n)$, then the set of odd members of $S$ is contained in $\{m_1, \dots, m_i\}$ and contains $m_i$ , while by the lemma, the set of even members of $S$ is contained in a set of at most $\lfloor\frac{1}{2}n\rfloor - i$ even numbers. Thus

$$f_i(n) \leq 2^{i-1} \cdot 2^{\lfloor\frac{1}{2}n\rfloor - i} = 2^{\lfloor\frac{1}{2}n\rfloor - 1}$$

for $1 \leq i \leq \lceil\frac{1}{2}n\rceil - 1$. Moreover, if $f_0(n)$ is the number of sequences in the theorem with only even members, then clearly

$$f_0(n) = 2^{\lfloor\frac{1}{2}n\rfloor}.$$

Thus

$$2^{\lfloor\frac{1}{2}n\rfloor} \leq f(n) \leq 2^{\lfloor\frac{1}{2}n\rfloor}(1 + (\lceil\frac{1}{2}n\rceil - 1)/2),$$

which proves the theorem.

Theorem 3.5 shows that the set of infinite sequences with no two terms coprime has dimension $\frac{1}{2}$. We conjecture that almost all sequences (in the sense of $\frac{1}{2}$-dimensional measure) consist of even numbers.

## 4. Miscellaneous problems

### 4.1. $a_i + a_j \nmid a_i a_j$

A sequence consisting of odd numbers and powers of 2 only satisfies this constraint; there are $\sim n \cdot 2^{\frac{1}{2}n}$ such sequences. It has been conjectured (see (11)) that the number of terms in any sequence satisfying the constraint is $(\frac{1}{2}+o(1))n$; we further conjecture that the number of sequences is $2^{(\frac{1}{2}+o(1))n}$, and that the set of infinite sequences has dimension $\frac{1}{2}$.

A related problem concerns the stronger constraint $a_i + a_j \nmid 2a_i a_j$. It is conjectured that the number of terms in such a sequence is $o(n)$.

Numerical evidence suggests that, if these conjectures are true, then the convergence is quite slow.

### 4.2. No k–term arithmetic progression

Szemeredi's celebrated theorem (19) asserts that the maximum number $r_k(n)$ of terms in such a sequence is $o(n)$. Because of the difficulty in obtaining a precise estimate for $r_k(n)$, our question here should be: How does the number of sequences compare with $2^{r_k(n)}$? In particular, is it $2^{(1+o(1))r_k(n)}$? Proposition 1.4 shows that

$$\limsup f(n)/2^{r_k(n)} = \infty .$$

### 4.3. $\sum(1/a_i) \leq s$

We conjecture that the number $f(n)$ of sequences

$$a_1 < \ldots < a_t \leq n$$

with $\sum(1/a_i) \leq s$ satisfies $f(n)^{1/n} \to c(s)$ as $n \to \infty$, for some $c(s)$. This should not be too difficult.

Any sequence of numbers between about $n/e^s$ to $n$ satisfies the constraint. So certainly

$$\liminf f(n)^{1/n} \geq 2^{1-e^{-s}}.$$

### References


[1] *P. Billingsley*, Hausdorff dimension in probability theory I, Illinois J. Math. **4** (1960), 187–209; II, ibid. **5** (1961), 291–298.

[2] *B. Bollobás*, Extremal Graph Theory, Academic Press, London, 1978.

[3] *N.J. Calkin*, Thesis, University of Waterloo, 1988.

[4] *P.J. Cameron*, Portrait of a typical sum–free set, Surveys in Combinatorics 1987 (ed. C.A. Whitehead), 13–42, London Math. Soc. Lecture Notes 123, Cambridge Univ. Press, Cambridge, 1988.

[5] *P. Erdös*, Note on sequences of integers no one of which is divisible by any other, J. London Math. Soc. **10** (1935), 126–128.

[6] *P. Erdös*, On sequences of integers no one of which divides the product of two others and on some related problems, Isvestija Nautsno–Issl. Inst. Mat. i Meh. Tomsk, **2** (1938), 74–82.

[7] *P. Erdös*, On some asymptotic formulas in the theory of the "factorisatio numerorum", Ann. Math. **42** (1941), 989–993.

[8] *P. Erdös*, Corrections to two of my papers, Ann. Math. **44** (1943), 647–651.

[9] *P. Erdös*, On some applications of graph theory to number–theoretic problems, Publ. Ramanujan Inst. **1** (1969), 131–136.

[10] *P. Erdös*, Some unconventional problems in number theory, Astérisque, **61** (1979), 73–82.

[11] *P. Erdös* and *R.L. Graham*, Old and New Problems and Results in Combinatorial Number Theory, Monogr. 28, L'Enseignement Math., Univ. Geneva, 1980.

[12] *H. Halberstam* and *K.F. Roth*, Sequences, Springer–Verlag, New York, 1983.

[13] *L. Kalmár*, Acta Litt. ac Scient. Szeged, **5** (1930), 95–107.

[14] *C. Pomerance* and *J.L. Selfridge*, Proof of D.J. Newman's coprime mapping conjecture, Mathematika, **27** (1980), 69–83.

[15] *K. Mahler*, On a special functional equation, J. London Math. Soc. **15** (1940), 115–123.

[16] *J. Riordan*, An Introduction to Combinatorial Analysis, Wiley, New York, 1958.

[17] *C.A. Rogers*, Hausdorff Measures, Cambridge Univ. Press, Cambridge, 1970.

[18] *N.J.A. Sloane*, A Handbook of Integer Sequences, Academic Press, New York, 1973.

[19] *E. Szermerédi*, On sets of integers containing no $k$ integers in arithmetic progression, Acta. Arith. **27** (1975), 199–245.

---

School of Mathematical Sciences, Queen Mary College,
Mile End Road, London E1 4NS, U.K.

Mathematical Institute of the Hungarian Academy of Sciences,
Realtanoda u. 13–15, H–1364, Budapest, Hungary

# On the Ramanujan–Nagell Equation and its Generalizations

*Edward L. Cohen*

**Dedicated to the memory of Trygve Nagell (1895-1988)**

## 1. Introduction

A history of the Ramanujan-Nagell diophantine equation $x^2 + 7 = 2^n$ is presented in this article. This diophantine equation was first stated by S. Ramanujan [55]. T. Nagell [50] was the first one to show that there were **exactly five** positive integral solutions. Since 1948, many mathematicians have obtained different proofs of what is now called the Ramanujan-Nagell equation. In addition, the several generalizations of this equation given by various authors are examined.

## 2. The Posing and Solving of the Problem

The original Ramanujan equation, which first was proposed in [55] as question 464, stated "$(2^n - 7)$ is a perfect square for the values 3, 4, 5, 7, 15 of $n$. Find other values." That is, the only positive integral solutions of $x^2 + 7 = 2^n$ known to Ramanujan were

$$(n, x) = (3,1), (4,3), (5,5), (7,11), (15,181).$$

Right after Ramanujan [55], [56] had stated the problem, Sanjana & Trevedi [57] showed there were no other solutions to the equation for $n \leq 75$. They used a Pellian equation; however their methods gave no treatment of estimates of an upper bound on $(n, x)$, thereby making extensive numerical calculations ineffective.

These five solutions happened to be the only ones, but that was not proved until 1948 by Nagell [50] in another context. In 1943, W. Ljunggren [38] had proposed this same problem in Norwegian without knowing the problem had already been stated. In 1948, T. Nagell [50] published a short two-page solution to the problem of Ljunggren [38], but unfortunately it was also in Norwegian. It was not until 11 years later that number theorists were fully aware of the existence of a solution—not by Nagell, but by others who were publishing abundantly on the problem. Nagell, in his remarkable elementary number theory book [51] published in 1951 states the Ramanujan problem as an elementary exercise after studying imaginary quadratic fields (Chapter VII, page 272, #165). Obviously it had not been noticed.

We will abbreviate Ramanujan-Nagell by RN and the Ramanujan-Nagell equation by RNE. We denote a **second-order linear recurrent sequence**, or, for short, a **binary linear recurrence** by *blr* since it is frequently used.

## 3. The Earlier Proofs (1956-1962)

In 1959, Skolem, Chowla & Lewis [62] solved the RNE. Here, 7-adic numbers are used according to the method of Skolem [61]. Also, in 1959, Shapiro & Slotnick [59] solved the equation $n^2 + n + 2 = 2^{k+1}$ to show that there does not exist a nontrivial perfect binary error-correcting code. This equation is equivalent to the RNE by a standard transformation. They used the algebraic number field $\mathbb{Q}(\sqrt{-7})$ and the *blr* $a_k = -a_{k-1} - 2a_{k-2}$.

In 1960, four more papers appeared involving the RNE. Browkin & Schinzel [15] in a generalization reminded us that their proof in 1956 [14] of the four solutions to $2^x - 1 = \frac{1}{2}y(y+1)$ was equivalent to the RNE, thus giving the **second** verification of the solutions to the RNE. Chowla, Dunton & Lewis [17], in a Norwegian journal, gave another proof using $\mathbb{Q}(\sqrt{-7})$ and a highest power argument. Apéry [4] showed that if $A \equiv 7 \pmod 8$, $A$ an integer $> 0$ then the equation $x^2 + A = 2^n$ has at most two solutions except for the case $A = 7$. In this case reliance was placed on

[62]. The proof uses $p$-adic fields and imaginary quadratic fields. Also, in 1960, Nagell [52], reading his paper on 13 January 1960, repeated his 1948 proof in English using $\mathbb{Q}(\sqrt{-7})$. It was printed on 5 August 1960, as mathematicians were becoming aware that he was the originator.

The last of these earlier proofs appeared in 1962 when Mordell [47] proved the RN problem by using $\mathbb{Q}(\sqrt[3]{7})$ and $\mathbb{Q}(\sqrt[3]{28})$. His proof is longer than those preceding. Mordell notes that "The theorem was proved by Chowla, Lewis, and Skolem...in [62]. Professor Nagell now informs me that he published (in Norwegian) a simple proof of the theorem in [50]." This is typical of the papers of the period 1956—1962. From the references in these articles, and the fact that Nagell prepared to republish his paper [52] in 1960, there was complete confusion on the history of the RN problem. No individual or group working on the problem was particularly aware that others were also solving the problem. Some of the articles mentioned previously as well as some of those in the next section have generalizations. These will be discussed in §5.

## 4. Proofs Post 1962

In 1966, a simpler version of Nagell's [50], [52] proof was given by Hasse [31]. The proof used $\mathbb{Q}(\sqrt{-7})$ and a highest power argument that was extended in his paper as well as in [19], [24], [25], [29]. In 1971, Mordell repeated the argument of Nagell-Hasse [50], [31] in his book, Diophantine Equations [48], pages 205—206. An elementary book by Stewart & Tall [63], pages 99—102, also used essentially the same proof just before it gave some algebraic number-theory exercises. There seemed to be a lull in the proofs of the RNE after 1966. Mead [42], in 1973, linked the RNE to differential algebra. In 1977, Hunt & van der Poorten [32] used an improved Gelfond–Baker [45] inequality, which yields effectively computable bounds, with a widely available computer software package to prove the RNE. In their book on error-correcting codes, MacWilliams & Sloane [41], page 182, used the same equation as in [59] to prove the RNE. As a specialized case, Beukers [8] in 1979 proved the RNE using hypergeometric polynomials and sharp bounds. His work is also published in [9],

[11], [12]. Also a new type of proof was presented in 1984 by Mignotte [43], who uses the fact that the ring $\mathbb{Z}[\sqrt{2}]$ has an infinite number of units. His proof is relatively short. Another proof was given in 1987 by Johnson [34], who uses the field $\mathbb{Q}(\sqrt{-7})$ and an idea of Beukers [10] on *blr* 's. In another paper, Johnson [35] discusses a generalized equation, which reduces to the RNE as a corollary.

## 5. Generalizations of the RNE

In the previous two sections we have discussed papers which have proofs of the RNE although this may not have been their main concern. We now discuss generalizations of the RNE that occur in these as well as some additional RNE-type papers. Most of the generalizations involve RN extensions of the form $Cx^2 + Dy^2 = z^n$. There is a huge amount of literature on this subject, dating back even to Dickson, Vol. II [26], Chapters XX—XXIII. Therefore, we will discuss only those parts that pertain fundamentally to extensions of the RNE. We will proceed chronologically unless the subjects are closely linked.

1959: Skolem, Chowla & Lewis [62] show that if $A$ is an odd rational integer $\not\equiv 1 \pmod 8$, then $2^n + A = x^2$ has at most one rational integer solution $(n, x)$. If there is a solution then $0 \leq n \leq 2$. Also discussed is the number of times a certain sequence related to a *blr* associated with the equation $x^2 + 7y^2 = 2^{n+2}$ can attain a value. Townes [65] further studies this *blr* to obtain specific results.

1960: Apéry [5] uses in the case of $x^2 + A = p^n$ the same methods of [4]. Browkin & Schinzel [15] discuss solutions of $2^n - D = y^2$ with $0 < D \leq 150$ under certain conditions for $D$.

1961: Lewis [37] discusses the number of solutions to $x^2 + 7M^2 = N^y$ $(M \geq 1, N \geq 2)$, $x^2 + 7 = N^y$ (N odd), and $x^2 + 7M^2 = y^n$ $(M \geq 1, n \geq 3)$. He uses $p$-adic arguments and $\mathbb{Q}(\sqrt{-7})$.

1964: Ljunggren [39] shows that the diophantine equation $x^2+7=y^z$, $z>1$, is impossible in rational integers $x, y, z$ if $y$ is an odd integer. Cohen [18] used the RNE to show that there are no perfect double error-correcting codes on 4 symbols. Other references to the theory of coding in this article appear in [1], [13], [41], [59], [70].

1966: Hasse [31] presents a thorough survey of the literature on the equation $x^2+D=y^k$, especially when $y=2$, $D$ odd; further results are obtained on this type of equation.

1967: Schinzel [58] proves theorems on the number of solutions of $x^2-d=2^m\,(p^m)$, where $d<0$, odd, $\neq 1-2^k\,(d<0$, odd, $p$ a prime factor of $1-4d)$.

1972: Cohen [19], [20], [22], proves that $x^2+D=p^k$ (p prime), where $p=\frac{D+1}{4}$, has no solutions, $D\geq 19$, $D\equiv 3 (\mathrm{mod}\ 8)$. This seems a natural extension of the RNE. Alter & Kubota [2] extend the results of [19] to obtain all $D\equiv 3(\mathrm{mod}\ 8)$ where there is no solution and show that for specific $D$ there are at most two solutions. Ljunggren [40], Cohen [19], [21], [24], [25] give several proofs that the equation $x^2+11=3^k$ has as the only solution $(k,x)=(3,4)$. The proof in [19] is incomplete; its completion occurs in [24], [25]. Here the proof is given by a modified Nagell-Hasse [50], [31] argument. The proof by Cohen [21] uses the result proved by Mordell [46] that the only solutions to the equation $x^2+11=y^3$ are given by $(x,y)=(4,3)$ and $(58,15)$. Alter & Kubota [3] also prove this result on the equation $x^2+11=3^k$. There is an error noticed by the reviewer [23] and corrected by Kubota [36]. In [53], Pethö showed the same result by putting an upper bound of 89 on $k$. Also solved is the equation $x^2+11=4\cdot 3^k$ with positive integral solutions $(k,x)=(1,1), (2,5), (5,31)$.

1972/1973: A curious problem appeared in the American Math. Monthly [27] in 1972 that was like the RNE for a Fibonacci-type sequence: "Let $F_n$ be the $n$th term of the sequence defined by

$$F_n=-F_{n-1}-2F_{n-2},\quad F_1=1, F_2=-1.$$

Prove that $2^{n+1} - 7F_{n-1}^{2}$ is a perfect square." It was solved in 1973 [28] by a number of different arguments. This is the same *blr* that Shapiro & Slotnick [59] used with the same initial conditions.

1975: Brown [16] proved by a different congruence argument than Nagell [49] that the only solutions to the related equation $x^2 + 3 = y^k$ $(k \geq 2;\ x, y \in \mathbb{Z})$ are $(x, y, k) = (\pm 1, \pm 2, 2)$. Bender & Herzberg [6] have considered bounds on the number of integral solutions to the equation $ax^2 + D = N^z$ and $ax^2 + D = z^n$, where $a, n, N$, and $D$ are positive integers. They also have a survey article [7] on diophantine equations associated with the positive definite quadratic form $ax^2 + by^2$ wherein the bounds on solutions of many of the types of generalized RN equations are discussed.

1977: Tanahashi [64] extends the results of Cohen [20] to the equation $x^2 + D^m = p^n$ and studies the equations $x^2 + 7^m = 2^n$ and $x^2 + 11^m = 3^n$, showing, e.g., that, for $(D, p) = (7,2)$, the only solution with $m > 1$ is $(x, m, n) = (13,3,9)$. The other five solutions come from the RNE. Toyoizumi [66] also showed this; and for $y^2 + D^m = 2^n$ under restricted conditions with $D \neq 7$, but with $D \equiv 7 \pmod 8$ and squarefree, reduced $m$ to 1 if the equation is to have solutions. Toyoizumi [67] further considers the equation $y^2 + D^m = p^n$, showing, that if $D \equiv 1, 2 \pmod 4$ and $p$ satisfy some strict conditions, then $m = 1$ unless $(D, p) = (2,3)$. In this case, the only solutions are given by $(y, m, n) = (1,1,1)$, $(5,1,3)$, $(1,3,2)$, $(7,5,4)$.

1978: Ennola [29] discusses the RNE, the generalized RNE $x^2 + 7 = y^q$ and an analogous equation $7x^2 + 1 = y^q$. If $y = 2$, the latter is solved using a modification of the Nagell–Hasse [48] method.

1979: Beukers [8], [9], [11], [12] proves that the equations $x^2 - D = 2^n$ and $x^2 - D = p^n$ $(D > 0, p$ an odd prime not dividing $D)$ have at most four solutions using hypergeometric polynomials. This is the first treatment concerning *-D* in a thorough manner. (A few remarks and congruences about *-D* were mentioned by Hasse [31] and Browkin & Schinzel [15].) Classes of examples are given where there are exactly 3 and 4 solutions. The method of Beukers [11] is used in Tzanakis & Wolfskill

[70] to show that $y^2 = 4q^n + 4q + 1$ is impossible if $q = 3^f$, $f > 9$, and $n \geq 3$. Some other results in [70] give solutions for the equations $y^2 = 4 \cdot a^n + b$, where $(a,b) = (3,13), (3,517), (21,85), (307,1229)$. Inkeri [33] simplifies a proof of Alter [1] on an analogous equation $2y^2 = 7^k + 1$, and gives an argument much like that of Ljunggren [40], [21] on the equation $x^2 + 11 = 3^n$.

1983: Bremner *et al.* [13] noted that the RNE had shown up in the theory of coding [41] and use similar diophantine equations to classify certain kinds of 2-weight ternary codes. Tzanakis [68] considers the equation $x^2 - D = 2^k$ for $0 < D \leq 100$. The cases of interest are $D = 17, 41, 73, 89, 97$. A complete solution to these $D$ is given using *blr* 's and algebraic number fields, whereas Beukers [11] applied hypergeometric methods.

1984: Mignotte [44] developed an algorithm for solving RN–type equations using continued fractions and *blr* 's.

1986: Pethö & de Weger [54] consider the generalized RNE

$$x^2 + k = p_1^{z_1} p_2^{z_2} \cdots p_t^{z_t}.$$

Evertse [30] has shown that this generalized RNE has no more than $3 \times 7^{4t+6}$ solutions. It is shown [54] that the only nonnegative integers $x$ such that $x^2 + 7$ has no prime divisors greater than 20 are the 16 in their table. Shorey and Tijdeman's book [60], Exponential Diophantine Equations, pages 137—138, gives references to the solutions of $x^2 + 7 = 2^n$ and similar equations [50], [52], [62], [14], [15], [17], [37], [47], [25], [33], [13], [68], [69], [44], [4], [5], [31]. These papers use elementary and algebraic number theory. The authors state that "Algebraic methods may not suffice to solve an equation (of the form $f(x, y) = kp_1^{z_1} p_2^{z_2} \cdots p_t^{z_t}$ $(x, y \in \mathbb{Z}, z_1, z_2, \ldots, z_t$ nonnegative integers)) with both $x$ and $y$ variable." On page 181, it is concluded by one of their theorems that equations like $x^2 + 7 = y^m$ and $7x^2 + 1 = y^m$ $(m > 2;\ x, y > 1)$ have only finitely many solutions.

Johnson [35] applies *blr* 's to find solutions for equations of the form $X^2 + (4q - 1)Y^2 = 4q^n$. The RNE is a consequence when $q = 2$, $Y = 1$.

In the above papers in §5, I have only included results about the generalizations of the RNE. In general, I have not discussed various other aspects of these papers.

## References

[1] *R. Alter*, On the nonexistence of close-packed double Hamming error-correcting codes on $q = 7$ symbols. J. Comput. Syst. Sci. **2** (1968), 169—176. MR 39 #1227

[2] *R. Alter*, and *K.K. Kubota*, The diophantine equation $x^2 + D = p^n$. Pacific J. Math. **46** (1973), 11—16. MR 48 #2063

[3] *R. Alter*, and *K.K. Kubota*, The Diophantine equation $x^2 + 11 = 3^n$ and a related sequence, J. Number Theory **7** (1975), 5—10. MR 51 #344

[4] *R. Apéry*, Sur une équation diophantienne. C.R. Acad. Sci. Paris Sér. **A251** (1960), 1263—1264. MR 22 #10951

[5] *R. Apéry*, Sur une équation diophantienne, C.R. Acad. Sci. Paris Sér. **A251** (1960), 1451—1452. MR 22 #10950

[6] *E.A. Bender* and *N.P. Herzberg*, Some diophantine equations related to the quadratic form $ax^2 + by^2$. Bull. Amer. Math. Soc. **81** (1975), 161—162. MR 53 #7936

[7] *E.A. Bender* and *N.P. Herzberg*, Some Diophantine equations related to the quadratic form $ax^2 + by^2$. Studies in Algebra and Number Theory, 219—272, Advances in Math., Suppl. Studies 6, Academic Press, New York, 1979. MR 80c:00011, MR 80g: 10020

[8] *F. Beukers*, The Generalized Ramanujan-Nagell Equation. Dissertation, Matematisch Instituut der Rijkuniversiteit te Leiden, Leiden, 1979, 57 pages. MR 80j: 10022, Zbl.412.10009

[9] *F. Beukers*, The generalized Ramanujan-Nagell equation. Proceedings of the Queen's Number Theory Conference, Queen's Papers in Pure and Applied Mathematics #**54**, 257—267, Queen's University, Kingston, Ontario, Canada, August 1980. MR 82j: 10003

[10] *F. Beukers*, The multiplicity of binary recurrences. Compositio Math. **40** (1980), 251—267. MR 81g: 10019

[11] *F. Beukers*, On the generalized Ramanujan-Nagell equation I. Acta Arith. **38** (1980/81), 389—410. MR 83a:10028a

[12] *F. Beukers*, On the generalized Ramanujan-Nagell equation II. Acta Arith. **39** (1981), 113—123. MR 83a:10028b

[13] *A. Bremner*, *R. Calderbank*, *P. Hanlon*, *P. Morton* and *J. Wolfskill*, Two-weight ternary codes and the equation $y^2 = 4 \times 3^\alpha + 13$. J. Number Theory **16** (1983), 212—234. MR 84m: 10011

[14] *G. Browkin* and *A. Schinzel*, Sur les nombres de Mersennes qui sont triangulaires. C.R. Acad. Sci. Paris **242** (1956), 1780—1781. MR 17, 1055d

[15] *J. Browkin* and *A. Schinzel*, On the equation $2^n - D = y^2$. Bull. Acad. Polon. Sci. Math. Astronom. Phys. **8** (1960), 311—318. MR 24 #A82

[16] *E. Brown*, Diophantine equations of the form $x^2 + D = y^n$. J. Reine Angew. Math. **274/275** (1975), 385—389. MR 51 #10221

[17] *S. Chowla*, M. *Dunton* and *D.J. Lewis*, All integral solutions of $2^n - 7 = x^2$ are given by $n = 3,4,5,7,15$. Norske Vid. Selsk. Forh. (Trondheim) **33**, nr 8 (1960), 37—38. MR 26 #6118

[18] *E.L. Cohen*, A note on perfect double error-correcting codes on $q$ symbols. Information and Control, **7** (1964), 381—384. MR 29 #5656

[19] *E.L. Cohen*, On Diophantine Equations of the Form $x^2 + D = p^k$, $D \equiv 3 (\mathrm{mod}\, 4)$. Dissertation, M.I.T., Cambridge, Mass., 1971, 45 pages.

[20] *E.L. Cohen*, Sur certaines équations diophantiennes quadratiques. C.R. Acad. Sci. Paris Sér. **A274** (1972), 139—140. MR 45 #169

[21] *E.L. Cohen*, Sur l'équation diophantienne $x^2 + 11 = 3^k$. C.R. Acad. Sci. Paris Sér. **A275** (1972), 5—7. MR 46 #3445

[22] *E.L. Cohen*, On diophantine equations of the form $x^2 + D = p^k$. Enseignement Math. (2), **20** (1974), 235—241. MR 50 #12913

[23] *E.L. Cohen*, Review of [3]. MR 51 #344 (1976), 48, Mathematical Reviews, Amer. Math. Soc., Providence, RI.

[24] *E.L. Cohen*, The diophantine equation $x^2 + 11 = 3^k$ and related questions. Math. Scand. **38** (1976), 240—246. MR 54 #12636

[25] *E.L. Cohen*, Sur une équation diophantienne de Ljunggren. Ann. Sc. Math. Québec **2** (1978), 109—112. Zbl.397.10010

[26] *L.E. Dickson*, History of the Theory of Numbers, Vol. II. Carnegie Institute of Washington, #256, Washington, D.C., 1920 (reprinted by Chelsea Publishing Co., New York, 1952, 1966). MR 39 #6807b

[27] Elementary problem E 2367 (proposed). Amer. Math. Monthly **79** (1972), 772.

[28] Elementary problem E 2367 (solved). Amer. Math. Monthly **80** (1973), 696.

[29] *V. Ennola*, $J$-fields generated by roots of cyclotomic integers. Mathematika **25** (1978), 242—250. MR 80f:10016

[30] *J.-H. Evertse*, On equations in $S$-units and the Thue–Mahler equation. Invent. Math. **75** (1984), 561—584. MR 85f:11048

[31] *H. Hasse*, Über eine diophantische Gleichung von Ramanujan-Nagell und ihre Verallgemeinerung. Nagoya Math. J. **27** (1966), 77—102. MR 34 #136

[32] *D.C. Hunt* and *A.J. van der Poorten*, Solving diophantine equations $x^2 + d = d^u$. (Preprint) (1977), School of Mathematics, The University of New South Wales, Kensington, N.S.W. 2033, Australia.

[33] *K. Inkeri*, On the diophantine equations $2y^2 = 7^k + 1$ and $x^2 + 11 = 3^n$. Elem. Math. **34** (1979), 119—121. MR 80j:10024

[34] *W. Johnson*, The diophantine equation $X^2 + 7 = 2^n$. Amer. Math. Monthly **94** (1987), 59—62. MR 88b:11015.

[35] *W. Johnson*, Linear recurrences and Ramanujan–type Diophantine equations. (Preprint) (June 1985), Department of Mathematics, Bowdoin College, Brunswick, ME (to appear).

[36] *K.K. Kubota*, Private communication, dated 13 May 1975.

[37] *D.J. Lewis*, Two classes of Diophantine equations. Pacific J. Math. **11** (1961), 1063—1076. MR 25 #3005

[38] *W. Ljunggren*, Oppgave nr 2. Norsk Matematisk Tidsskrift **25** (1943), 29.

[39] *W. Ljunggren*, On the diophantine equation $Cx^2 + D = y^n$. Pacific J. Math. **14** (1964), 585—596. MR 28 #5035

[40] *W. Ljunggren*, Private communication, dated 20 March 1972.

[41] *F.J. MacWilliams* and *N.J.A. Sloane*, The Theory of Error–Correcting Codes, I. North–Holland Mathematical Library, Vol. 16, 1—369, North–Holland Publishing Co., Amsterdam, 1977. MR 57 #5408a, Zbl.369.94008

[42] *D.G. Mead*, The equation of Ramanujan–Nagell and $[y^2]$. Proc. Amer. Math. Soc. **41** (1973), 333—341. MR 48 #6067

[43] *M. Mignotte*, Une nouvelle résolution de l'équation $x^2 + 7 = 2^n$. Rend. Semin. Fac. Sci. Università Cagliari **54**, no. 2 (1984), 41—43. MR 87e:11048

[44] *M. Mignotte*, On the automatic resolution of certain Diophantine equations. Proceedings EUROSAM 84 (Cambridge 1984), Lecture Notes in Computer Science #174, Springer-Verlag, Berlin, 1984, 378—385. MR 86e:11021, MR 85k:00009

[45] *M. Mignotte*, and *M. Waldschmidt*, Linear forms in two logarithms and Schneider's method. Publ. Math. de l'Université Pierre et Marie Curie (Paris VI) **5** (1977); Math. Ann. **231** (1977/78), 241—267. MR 57 #242.

[46] *L.J. Mordell*, The Diophantine equation $y^2 - k = x^3$. Proc. London Math. Soc. (2), **13** (1914), 60—80. (Dickson [26], p. 538, reference 56.)

[47] *L.J. Mordell*, The diophantine equation $2^n = x^2 + 7$. Ark. Mat. **4** (1962), 455—460. MR 26 #74, Zbl.106, 36

[48] *L.J. Mordell*, Diophantine Equations. Pure and Applied Mathematics 30, Academic Press, New York, London, 1969. MR 40 #2600, Zbl.188, 345

[49] *T. Nagell*, Sur l'impossibilité de quelques équations à deux indéterminées. Norsk Mat. Forenings Skr. Ser. 1, nr 13 (1923), 65—82.

[50] *T. Nagell*, Løsning til oppgave nr 2, {cf. [38]}. Norsk Matematisk Tidsskrift **30** (1948), 62—64.

[51] *T. Nagell*, Introduction to Number Theory. John Wiley & Sons, New York, Almqvist & Wiksell, Stockholm, 1951. MR 13, 207b, Zbl.42, 267

[52] *T. Nagell*, The diophantine equation $x^2 + 7 = 2^n$. Ark. Mat. **4** (1961), 185—187. MR 24 #A83, Zbl.103, 30

[53] *A. Pethö*, On the solutions of the diophantine equation $G^n = p^z$. Proceedings EUROCAL 85, Vol. 2 (Linz, Austria) Lecture Notes in Computer Science #204, Springer–Verlag, Berlin, 1985, 503—512. MR 87f:11021, MR 87a:68007

[54] *A. Pethö* and *B.M. de Weger*, Products of prime powers in binary recurrence sequences, I: The hyperbolic case, with an application to the generalized Ramanujan–Nagell equation. Math. Comp. **47** (1986), 713—727. MR 87m:11027a

[55] *S. Ramanujan*, Question 464. J. Indian Math. Soc. 5 (1913), 120.

[56] *S. Ramanujan*, Collected Papers, 327. Cambridge University Press, Cambridge, 1927 (reprinted by Chelsea Publishing Co., New York, 1962).

[57] *K.J. Sanjana* and *T.P. Trevedi*, Question 464 (solution). J. Ind. Math. Soc. **5** (1913), 227—228.

[58] *A. Schinzel*, On two theorems of Gelfond and some of their applications. Acta Arith. **13** (1967/68), 177—236. MR 36 #5086

[59] *H.S. Shapiro* and *D.L. Slotnick*, On the mathematical theory of error-correcting codes. IBM J. Res. Develop. **3** (1959), 25—34. MR 20 #5092

[60] *T.N. Shorey* and *R. Tijdeman*, Exponential Diophantine Equations. Cambridge Tracts in Mathematics, Vol. 87, Cambridge University Press, Cambridge, 1986. MR 88h:11002, Zbl.606.10011

[61] *Th. Skolem*, Ein Verfahren zur Behandlung gewisser exponentialer Gleichungen und diophantischer Gleichungen. 8de Skand. Mat. Kongr. Förh. (Stockholm 1934), 163—168, Ohlsons Boktryckeri, Lund, 1935. Zbl. 11, 392

[62] *Th. Skolem, S. Chowla* and *D.J. Lewis*, The diophantine equation $2^{n+2}-7=x^2$ and related problems. Proc. Amer. Math. Soc. **10** (1959), 663—669. MR 22 #25

[63] *I.N. Stewart* and *D.O. Tall*, Algebraic Number Theory. Chapman and Hall Mathematical Series, Chapman & Hall, London, 1979 (2nd ed., 1987). MR 81g:12001, MR 88d:11001, Zbl.413.12001

[64] *K. Tanahashi*, On the Diophantine equations $x^2+7^m=2^n$ and $x^2+11^m=3^n$. J. Predent. Fac. Gifu College Dent. **1977**, no. 3, 77—79. MR 82j:10033

[65] *S.B. Townes*, Notes on the Diophantine equation $x^2+7y^2=2^{n+2}$. Proc. Amer. Math. Soc. **13** (1962), 864—869. MR 26A #76

[66] *M. Toyoizumi*, On the Diophantine equation $y^2+D^m=2^n$. Comment. Math. Univ. St. Pauli **27** (1978/79), no. 2, 105—111. MR 82m:10031

[67] *M. Toyoizumi*, On the Diophantine equation $y^2+D^m=p^n$. Acta Arith. **42** (1983), 303—309. MR 85c:11033

[68] *N. Tzanakis*, On the Diophantine equation $y^2-D=2^k$. J. Number Theory **17** (1983), 144—164. MR 86c:11018

[69] *N. Tzanakis*, The complete solution in integers of $X^3+2Y^3=2^n$. J. Number Theory **19** (1984), 203—208. MR 85k:11014

[70] *N. Tzanakis* and *J. Wolfskill*, On the Diophantine equation $y^2=4q^n+4q+1$. J. Number Theory **23** (1986), 219—237. MR 88b:11016

---

Dept. of Mathematics, University of Ottawa, Ottawa, ON K1N 6N5, CANADA.

# The Uniform Group Under Change of Rings

*F.R. DeMeyer and G.R. Greenfield* [1]

Let $R$ be a commutative ring. The class $|A|$ in the Brauer group $\mathcal{B}(R)$ represented by the Azumaya $R$-algebra $A$ is in the *uniform subgroup* $\mathcal{U}(R)$ provided both of the following conditions hold.

(i) The group of $n$th roots of unity in $R$ is cyclic of order $n$, where $n$ is the exponent of $|A|$ in $\mathcal{B}(R)$.

(ii) If $\in_n$ is a primitive $n$th root of unity in $R$ and $\sigma$ is in $\mathrm{Aut}(R)$, then $|{}_{\sigma}A| = |A|^{\nu}$ if and only if $\sigma(\in_n) = (\in_n)^{\nu}$.

The action of $\mathrm{Aut}(R)$ on $\mathcal{B}(R)$ appearing in condition (ii) can be found in [2]. This definition of the uniform group, due to R. Mollin and the second author [5], gives a generalization to commutative rings of the concept of classes in the Brauer group of an algebraic number field with uniformly distributed Hasse invariants, which in turn, was originally formulated and intensively studied by R. Mollin [9], [10], [11] in connection with the Schur subgroup.

Let $S$ be a commutative $R$-algebra. The tensor product map induces a natural homomorphism from $\mathcal{B}(R)$ to $\mathcal{B}(S)$ whose restriction to $\mathcal{U}(R)$ gives a homomorphism to $\mathcal{B}(S)$. During his address at the First Conference of the Canadian Number Theory Association, the second author posed some open questions concerning the behaviour of $\mathcal{U}(-)$ under the change of ring map. Here we seek to answer some of

---

1 Research partially supported by UR Faculty Research Grant #88–083.

these questions. We would like to thank R. Mollin for extending to us the privilege of attending the CNTA's First Conference.

In [5], R. Mollin and the second author proved $\mathcal{U}(R) \to \mathcal{U}(R/M)$ is an isomorphism for $R$ a complete local ring of characteristic zero with maximal ideal $M$, and in [4], R. Mollin and the first author calculated $\mathcal{U}\left(R[t,t^{-1}]\right)$ for certain perfect fields $R$. In this note we consider the problem of relating $\mathcal{U}(R)$ to either $\mathcal{U}(K)$ or $\mathcal{U}(R[x])$ for an integral domain $R$ with quotient field $K$. In the former case, we give an example (the coordinate ring of a real elliptic curve) of a Noetherian integrally closed domain $R$ with quotient field $K$, for which the image of the natural map from $\mathcal{U}(R)$ to $\mathcal{B}(K)$ does not lie in $\mathcal{U}(K)$. Thus, $\mathcal{U}(-)$ is not a functor even when restricted to integral domains. On the positive side, we are able to show that if $R$ has quotient field $K$ and $P$ is the prime subfield, then $\mathcal{U}(R) \to \mathcal{U}(K)$ is well defined when $R$ is the integral closure in $K$ of $R \cap P$. Further when $R$ is regular, this map is a monomorphism. (For a discussion of regular domains see [1], or p. 67 of [12].) In the latter case, our main result is that $\mathcal{U}(R) \cong \mathcal{U}(R[x])$ when $R$ is a field. Note that $\mathcal{B}(R)$ is not isomorphic to $\mathcal{B}(R[x])$ when $R$ is not a perfect field ([1], Theorem 7.5). We also obtain isomorphisms between $\mathcal{U}(R)$ and $\mathcal{U}(R[x])$ for some special regular rings $R$. We conclude with some additional remarks about rings of finite characteristic.

**Example.** Let $\mathbb{R}$ be the field of real numbers, and let $R$ be the ring $\mathbb{R}[x,y]/(y^2 - x(x-1)(x-w))$ where $w \neq 1, -1, 1/2, 2$ or $0$ is in $\mathbb{R}$. $R$ is the coordinate ring of a nonsingular irreducible real elliptic curve and $\mathrm{Aut}_{\mathbb{R}}(R) = \mathbb{Z}_2$ with generating automorphism $\sigma$ given by $\sigma(x) = x$ and $\sigma(y) = -y$ ([7], Corollary 4.7 of IV). Following 6.2 on p. 46 of [7], for any $a \in R$ set $N(a) = a\sigma(a)$. Then $N(a)$ is in $\mathbb{R}[x]$, $N(1) = 1$, and $N(ab) = N(a)N(b)$ whence the units of $R$ are precisely the nonzero elements of $\mathbb{R}$. If $\tau$ is in $\mathrm{Aut}(R)$, $\tau$ preserves the units group, so $\tau|_{\mathbb{R}}$ is in $\mathrm{Aut}(\mathbb{R})$ and therefore $\tau$ is in $\mathrm{Aut}_{\mathbb{R}}(R)$. Thus $\mathrm{Aut}(R) = \mathbb{Z}_2$ also. From [3], $\mathcal{B}(R) = \mathbb{Z}_2 \oplus \mathbb{Z}_2$. Let $K$ be the quotient field of $R$ and let $Y_1$ and $Y_2$ be the two real components (in Euclidean topology) of the curve $y = x(x-1)(x-w)$. There exist rational functions $a_i$ $(i = 1,2)$ with $a_i$ negative on $Y_i$ and positive on $Y_{i+1 \bmod 2}$. Again from [3], for $i = 1,2$ the symbol algebras $(-1, a_i)$ over $K$ contain unramified maximal orders representing classes generating $\mathcal{B}(R)$ in $\mathcal{B}(K)$. Since no automorphism of $R$ exchanges the $Y_i$, $\mathcal{B}(R) = \mathcal{U}(R) = \mathbb{Z}_2 \oplus \mathbb{Z}_2$. But there is an automorphism of $K$ exchanging the projective completion of the $Y_i$, namely $\sigma$ defined by the action $\sigma(x) = w/x$ and $\sigma(y) = wy/x^2$, so the intersection of $\mathcal{U}(R)$ with $\mathcal{U}(K)$ has order 2. Thus, the restriction of the natural map $\mathcal{B}(R) \to \mathcal{B}(K)$ to $\mathcal{U}(R)$ has an image which does not lie in $\mathcal{U}(K)$.

We now reveal a situation where passing to the quotient field under $\mathcal{U}(-)$ does behave nicely. Its full measure would apply, for example, to the case where the quotient field $K$ is an algebraic number field whose class number is one.

**Proposition 1.** *Let $R$ be an integral domain with quotient field $K$, and let $P$ be the prime subfield of $K$. If $R$ is the integral closure in $K$ of $R \cap P$, then $\mathcal{U}(R) \to \mathcal{U}(K)$ is a well–defined homomorphism, If, in addition, $R$ is regular, then $\mathcal{U}(R) \to \mathcal{U}(K)$ is a monomorphism.*

**Proof.** Let $|A|$ represent an element in $\mathcal{B}(R)$ of order $n$, let $\epsilon_n$ be a primitive $n$th root of unity in $R$, and set $B = A \otimes K$. Let $\sigma$ be in $\mathrm{Aut}(K)$ and suppose $\sigma(\epsilon_n) = (\epsilon_n)^\nu$. Since $R$ is the integral closure of $R \cap P$ and $\sigma_{R \cap P} = \mathrm{id}|_{R \cap P}$, it follows that $\sigma|_R$ is in $\mathrm{Aut}(R)$. By assumption, $|{}_\sigma A| = |A|^\nu$, and the map from $\mathcal{B}(R)$ to $\mathcal{B}(K)$ is always a homomorphism, so we obtain $|{}_\sigma B| = |B|^\nu$, which verifies the map from $\mathcal{U}(R)$ to $\mathcal{U}(K)$ is well–defined. The second assertion follows immediately from the fact (Theorem 7.2 of [1]) that $\mathcal{B}(R) \to \mathcal{B}(K)$ is a monomorphism whenever $R$ is regular.

We next turn our attention to the homomorphism from $\mathcal{U}(R)$ to $\mathcal{B}(R[x])$. Note that when $R$ is an integral domain, the units group of $R[x]$ equals the units group of $R$, and hence the group of roots of unity of these two rings must coincide. With this in mind, we deal first with the case where $R$ is a field.

**Theorem 1.** *If $R$ is a field, then $\mathcal{U}(R) \cong \mathcal{U}(R[x])$.*

**Proof.** When $R$ is a field, the units group of $R$ is the multiplicative group of nonzero elements. Thus if $\sigma$ is in $\mathrm{Aut}(R[x])$, $\sigma|_R$ is in $\mathrm{Aut}(R)$. Evidently, the restriction of the natural map $\mathcal{B}(R) \to \mathcal{B}(R[x])$ to $\mathcal{U}(R)$ induces a monomorphism $\mathcal{U}(R) \to \mathcal{U}(R[x])$, and the restriction of the natural map $\mathcal{B}(R[x]) \to \mathcal{B}(R)$ induced by sending $x$ to zero yields an epimorphism $\mathcal{U}(R[x]) \to \mathcal{U}(R)$. If $\mathcal{B}'(R[x])$ (respectively $\mathcal{U}'(R[x])$) is the kernel of $\mathcal{B}(R[x]) \to \mathcal{B}(R)$ (respectively $\mathcal{U}(R[x]) \to \mathcal{U}(R)$) then $\mathcal{B}'(R[x]) = (0)$ when $R$ is perfect (Theorem 7.5 of [1]), so $\mathcal{U}'(R[x]) = (0)$ when $R$ is perfect. If $R$ has characteristic $p > 0$ then by Proposition 7.6 of [1], every element of $\mathcal{B}'(R[x])$ has order a power of $p$. But if $R$ is a field of characteristic $p$, and $n$ is a power of $p$, then $R$ cannot contain a primitive $n$th root of unity, so in this case $\mathcal{U}'(R[x]) = (0)$ also. This completes the proof.

**Remark.** It is also true that $\mathcal{U}(R) \cong \mathcal{U}(R[[x]])$ when $R$ is a field of characteristic zero, because $\mathcal{B}(R) \cong \mathcal{B}(R[[x]])$ (see [6]) and so as in the proof of the Theorem we would obtain $\mathcal{B}'(R[[x]]) = (0)$. This observation is implicit in [5], Theorem 3.1.

To obtain further results we will restrict our attention to regular rings, and attempt to refine a key ingredient of the proof of our Theorem—controlling the automorphisms of $R[x]$ through the units group of $R$. The following proposition is easy, but it will serve as a catalyst for the exposition to follow.

**Proposition 2.** *If $R$ is a regular domain of characteristic zero such that for every $\sigma$ in $\mathrm{Aut}(R[x])$, $\sigma|_R$ is in $\mathrm{Aut}(R)$, then $\mathcal{U}(R) \cong \mathcal{U}(R[x])$.*

**Proof.** Since $R$ is regular, Proposition 7.7 of [1] establishes that the natural map $\mathcal{B}(R[x]) \to \mathcal{B}(R)$ is an isomorphism. It is clear then that the map $\mathcal{U}(R[x]) \to \mathcal{U}(R)$ is a well-defined monomorphism. The hypothesis about automorphisms guarantees that this map is surjective. In other words, $|A|$ is in $\mathcal{U}(R)$ if and only if $|A \otimes R[x]|$ is in $\mathcal{U}(R[x])$.

We now furnish two applications of the previous Proposition, beginning with the context we first encountered in Proposition 1.

**Corollary 1.** *Let $R$ be a regular domain of characteristic zero with quotient field $K$. Let $\mathbb{Q}$ be the prime subfield of $K$. If $R$ is the integral closure in $K$ of $R \cap \mathbb{Q}$, then $\mathcal{U}(R) \cong \mathcal{U}(R[x])$.*

**Proof.** Let $\sigma$ be in Aut($R[x]$). Then $\sigma$ extends uniquely to an automorphism $\tilde{\sigma}$ of $K(x)$. By hypothesis, every element of $K$ is algebraic over $\mathbb{Q}$. Since $\sigma|_{\mathbb{Q}} = \mathrm{id}|_{\mathbb{Q}}$ and the elements in $K(x)$ which are algebraic over $\mathbb{Q}$ are necessarily in $K$, $\tilde{\sigma}|_K$ is in Aut($K$). In particular, $\tilde{\sigma}|_R = \sigma|_R : R \to R$. But $R$ is the integral closure of a subring $\mathbb{Q}$, therefore $\sigma|_R$ is in Aut($R$) so we are done by Proposition 2.

**Corollary 2.** *Let $R = K[t, t^{-1}]$ where $K$ is a field of characteristic zero. Then $\mathcal{U}(R) \cong \mathcal{U}(R[x])$.*

**Proof.** Since $R$ is just the ring $K[t]$ localized at the multiplicative set consisting of powers of $t$, it is routine to check that $R$ is regular. Since Units($R$) $= K^* \times \langle t \rangle$, where $K^*$ is the group of nonzero elements of $K$, Units($R$) generates $R$, and it follows therefore that any automorphism of $R[x]$ must restrict to an automorphism of $R$. Once more, an application of Proposition 2 completes the proof.

In [4] it was shown that $\mathcal{U}(K[t, t^{-1}])$ is nontrivial when $K$ is a finite field of odd characteristic. Concerning rings of finite characteristic, more is known. If $K$ is a field of characteristic $p$, $m > 1$, and $R = K[t_1, \ldots, t_m]$ then by Theorem 3.7 of [8], all classes in $\mathcal{B}(R)$ have $p$-power order, whence $\mathcal{U}(R)$ must be trivial as $R$ has no $p$th root of unity. If we assume $K$ is finite, then by Proposition 5.6 of [8], $\mathcal{B}(K[t])$ is $p$-torsion, so here too the roots of unity condition forces $\mathcal{U}(R)$ to be trivial, $R = K[t]$. Alternatively, when $K$ is finite, $\mathcal{B}(K)$ itself is trivial ([12], p. 56) so, modulo notation, Theorem 1 above could force the same result. Finally, if $R = \mathbb{Z}_{p^2}[t, t^{-1}]$ where $p$ is an odd prime then the principal ideal $M = (p)$ satisfies $M^2 = (0)$ so $\mathcal{U}(R) \cong \mathcal{U}(R/M) = \mathcal{U}(\mathbb{Z}_p[t, t^{-1}]) = \mathbb{Z}_2$ giving an example of a ring of finite characteristic which is not a domain but for which $\mathcal{U}(R)$ is nontrivial.

## References


[1] *M. Auslander* and *O. Goldman*, The Brauer group of a commutative ring. Trans. Amer. Math. Soc. **97** (1960), 367—409.

[2] *F.R. DeMeyer*, An action of the automorphism group of a commutative ring on its Brauer group. Pacific J. Math. **97** (1981), 327—338.

[3] *F.R. DeMeyer* and *M.A. Knus*, The Brauer group of a real curve. Proc. Amer. Math. Soc. **57** (1976), 227—232.

[4] *F.R. DeMeyer* and *R.A. Mollin*, Invariants of group rings. J. Algebra (to appear).

[5] *G.R. Greenfield* and *R.A. Mollin*, The uniform distribution group of a commutative ring. J. Algebra **108** (1987), 179—187.

[6] *P.A. Griffith*, The Brauer group of $A[t]$. Math. Zeitschrift **147** (1976), 79—86.

[7] *R. Hartshorne*, Algebraic Geometry. Graduate Texts in Math. v. 549, Springer Verlag, New York 1977.

[8] *M.A. Knus*, *M. Ojanguren* and *D.J. Saltman*, On Brauer groups of characteristic *p*. Brauer Groups, Evanston 1975, Lecture Notes in Math. v. 549, Springer Verlag, New York 1976, 25—49.

[9] *R.A. Mollin*, $U(K)$ for quadratic fields $K$. Comm. in Algebra **4** (1976), 747—759.

[10] *R.A. Mollin*, Algebras with uniformly distributed invariants. J. Algebra **44** (1977), 271—282.

[11] *R.A. Mollin*, Uniform distribution and the Schur subgroup. J. Algebra **42** (1976), 261—277.

[12] *M. Orzech* and *C. Small*, The Brauer Group of Commutative Rings, Lecture Notes in Pure and Applied Math. v. 11, Marcel Dekker, New York 1975.


---


Dept. of Mathematics, Colorado State University, Fort Collins, Colorado 80523, USA.

Department of Mathematics and Computer Science, University of Richmond, Richmond, Virginia 23173, USA.

# Sieve Auxiliary Functions

*Harold G. Diamond* [1], *H. Halberstam and H–E. Richert* [1]

## 1. Introduction

This is a survey of two families of auxiliary functions that occur in sieve theory. It is based in part on work in our series of papers on sieve theory [2], [3a], [3b]; some results from the doctoral dissertation of F. Wheeler [11]; and ideas suggested in H. Iwaniec's seminal paper Rosser's Sieve [7], where, to our knowledge, these functions appear for the first time.

For $\kappa$ a real parameter, $\kappa > 1$, we define functions $p_\kappa$ and $q_\kappa$ on $(0,\infty)$. We omit the $\kappa$ when it is clearly understood. These functions satisfy difference differential equations (D.D.E.'s) of advanced type

$$(up(u))' = \kappa p(u) - \kappa p(u+1) \tag{1}$$

$$(uq(u))' = \kappa q(u) + \kappa q(u+1) \tag{2}$$

or, equivalently,

$$(u^{1-\kappa} p(u))' = -\kappa u^{-\kappa} p(u+1) \tag{3}$$

$$(u^{1-\kappa} q(u))' = \kappa u^{-\kappa} q(u+1) \tag{4}$$

The functions also satisfy some mild side conditions that will be discussed in section 2.

The $p$ and $q$ functions occur in Iwaniec's article as adjoint functions for his upper and lower bound sieve functions, $F$ and $f$, which satisfy D.D.E.'s with retarded arguments. He obtains the expressions

1 Research supported in part by a grant from the National Science Foundation.

$$up(u)(F+f)(u)+\kappa\int_{u-1}^{u} p(x+1)(F+f)(x)dx=2$$

$$uq(u)(F-f)(u)-\kappa\int_{u-1}^{u} q(x+1)(F-f)(x)dx=0,$$

which provides the basis for further analysis. The number $\kappa$ is, roughly, the average number of residue classes per prime in the sequence being sifted.

The $q$ function has a largest positive root, which we denote by $\rho_\kappa$, or briefly, $\rho$. This number plays a key role in the Rosser–Iwaniec sieve, where $\rho+1$ is the "lower sifting limit," that is, the largest number $u$ at which $f(u)=0$. To the left of this point Rosser's sieve gives only the trivial lower bound of zero. The lower sifting limit of the DHR sieve is smaller than $\rho+1$ for each $\kappa>1$, which is advantageous for arithmetic applications. The existence of $\rho_\kappa$ and its estimates are nevertheless vital in this theory as well.

The following direct application of $p$ functions, found by Wheeler [11], is reminiscent of the occurrence of the Dickman function in largest prime factor problems. Let $P_1(n)$ be the largest prime factor of $n$ and $P_2(n)$ the second largest prime factor. An unweighted form of Wheeler's estimate is

$$\sum_{\substack{1\le n\le x \\ P_2(n)\le P_1(n)^{1/u}}} 1 = e^{\gamma}p_1(u)x + O(x/\log x).$$

Here and elsewhere $\gamma$ denotes Euler's constant.

Tabular data of values of $p$ and $q$ for various $\kappa$ and $u$ have been given by te Riele [10] and Wheeler [11].

## 2. The $p$ and $q$ Functions

The solution on a half line of a D.D.E. with retarded argument depends on the values of the function in an initial interval; it is continued forward by use of the D.D.E. (*cf* [6]). In the present situation of advanced arguments, we seek solutions that have normalized polynomial–like behaviour at infinity, i.e., $p(u) \sim u^a$ and $q(u) \sim u^b$ for suitable $a$ and $b$ as $u \to \infty$.

The precise asymptotic behaviour of the functions is determined by the D.D.E. and the polynomial–like condition. For $p$ we have from (1) that

$$up(u) + \kappa\int_{u-1}^{u} p(x)dx = \text{constant}.$$

The condition that $p(u) \sim u^a$ at $\infty$ yields

$$u^{a+1} + \kappa u^a \sim \text{constant}$$

as $u \to \infty$. It follows that $p(u) \sim u^{-1}$ and the integral equation above should read

$$up(u) + \kappa\int_{u}^{u+1} p(x)dx = 1. \tag{5}$$

For $q$, we have from (2)

$$(u^{1-2\kappa}q(u))' = \kappa u^{-2\kappa}\{q(u+1) - q(u)\}$$

and $q(u+1) \sim q(u)$ at $\infty$ by the polynomial–like condition. It follows that

$$\frac{(u^{1-2\kappa}q(u))'}{u^{1-2\kappa}q(u)} = o\left(\frac{1}{u}\right) \qquad (u \to \infty)$$

and upon integrating, $q(u) = u^{2\kappa-1+o(1)}$.

Thus $q(u) \sim u^{2\kappa-1}$ as $u \to \infty$.

We now give explicit functions $p$ and $q$ which satisfy (1) and (2) respectively on $(0,\infty)$ and which have normalized polynomial–like behaviour at infinity. The representations use the entire function

$$\text{Ein}\, z := \int_0^z (1 - e^{-t})t^{-1}dt\,, \quad z \in \mathbb{C}.$$

On $\mathbb{C}\backslash(-\infty,0]$ we have the formula [1]

$$\operatorname{Ein} z = \log z + \gamma + \int_z^\infty e^{-t} t^{-1} dt .$$

We have for $u > 0$

$$p_\kappa(u) = \int_0^\infty e^{-\kappa \operatorname{Ein} x - xu} dx \tag{6}$$

and

$$q_\kappa(u) = \frac{\Gamma(2\kappa)}{2\pi i} \int_C z^{-2\kappa} e^{uz + \kappa \operatorname{Ein}(-z)} dz . \tag{7}$$

Here $\Gamma$ is the Euler gamma function and $C$ the path from $-\infty$ back to $-\infty$ which surrounds the negative real axis in the positive sense.

Differentiation under the integral signs is justified by the exponential rate of convergence of the integrands. The estimation of the integrals for $u \to \infty$ is achieved by use of Laplace's method [8]; we find that

$$p(u) \sim u^{-1}, \quad q(u) \sim u^{2\kappa - 1}$$

as required.

When $2\kappa$ is an integer, $q_\kappa$ turns out to be a polynomial (of degree $2\kappa - 1$ of course!). We list the first few polynomial $q$'s here.

$$q_1(u) = u - 1$$

$$q_{3/2}(u) = u^2 - 3u + 3/2$$

$$q_2(u) = u^3 - 6u^2 + 9u - 8/3$$

$$q_{5/2}(u) = u^4 - 10u^3 + 30u^2 - 85u/3 + 55/12$$

$$q_3(u) = u^5 - 15u^4 + 75u^3 - 145u^2 + 90u - 18/5 .$$

Because of the complexity of the integrals, particularly that of $q$, we shall often use the D.D.E.'s rather than the explicit formulas to establish properties of $p$ and $q$.

We conclude this section with some remarks on smoothness and uniqueness. One can show by Laplace transform techniques that the solutions of (1) and (2) are unique in the class of normalized polynomial–like functions. A solution of (1) or (2) different from (6) or (7) would doubtless be very wild.

A solution on a half line of a D.D.E. with retarded argument may well fail to have a first derivative at the right end of the initial interval. It would then have no second derivative at the right end of the second interval, etc. [6]. In contrast, $p$ and $q$ as given in (6) and (7) are clearly analytic on the half plane $\{\mathrm{Re}\, u > 0\}$. This is reasonable, since the value of one of these functions is determined by an infinite number of integrations of (3) or (4), and each integration is a smoothing operation.

## 3. Properties of $p$

We show here that the function $p$ is quite simply behaved and easy to estimate. In particular, it is everywhere positive, decreasing and convex.

**Proposition 1.** $(-1)^{\nu} p^{(\nu)}(u) > 0,\ u > 0,\ \nu = 0,1,2,\ldots$ .

**Proof.**

$$(-1)^{\nu} p^{(\nu)}(u) = \int_0^{\infty} e^{-\kappa \operatorname{Ein} x} x^{\nu} e^{-xu}\, dx > 0.$$

■

Since $p(u) > p(u+1)$, it follows from (1) that $up(u) \uparrow$. We use these facts and equation (5) to obtain bounds for $p$.

**Proposition 2.** *For all* $u > 0$

$$(u+\kappa)^{-1} < p(u) < \{u + \kappa u \log(1 + \frac{1}{u})\}^{-1} < \{u + \kappa - \frac{\kappa}{2u}\}^{-1}.$$

**Proof.** $1 = up(u) + \kappa \int_u^{u+1} p(t)\, dt < (u+\kappa)\, p(u).$

Also

$$1 = up(u) + \kappa \int_u^{u+1} tp(t)\, t^{-1}\, dt > up(u)\, \{1 + \kappa \log \frac{u+1}{u}\}.$$

The last inequality in the proposition comes from the estimate $\log(1 + a^{-1}) > a^{-1} - (2a^2)^{-1}$.

■

We can measure the quality of these estimates by forming the relative difference

$$\Delta := \frac{p^{\text{upper}} - p^{\text{lower}}}{p^{\text{lower}}} = \frac{\left(u + \kappa - \frac{\kappa}{2u}\right)^{-1} - (u+\kappa)^{-1}}{(u+k)^{-1}} = \frac{\kappa}{2u\left(u+\kappa-\frac{\kappa}{2u}\right)}.$$

For example, for $\kappa = 2$ and $u = 5$, $\Delta = 1/34 < .03$.

To give sharper estimates we obtain further monotonicity relations for $p$.

**Proposition 3.** *Log* $p$ *is convex on* $(0,\infty)$.

**Proof.** If we differentiate (6) and apply the Cauchy Schwarz inequality, we obtain

$$p'(u)^2 < p(u)\,p''(u) \qquad (u > 0).$$

Then

$$(\log p)'' = \{pp'' - (p')^2\}/p^2 > 0.$$

■

**Proposition 4.** $-p'(u)/p(u) > \dfrac{1}{\kappa + u}$, *for all* $u > 0$.

**Proof.**

$$p(u) + up'(u) = \kappa p(u) - \kappa p(u+1)$$

$$1 + \frac{up'(u)}{p(u)} = \kappa - \kappa \exp \int_u^{u+1} p'(t)/p(t)\,dt < \kappa - \kappa \exp(p'(u)/p(u))$$

since $p'/p$ is increasing. Let $\delta = -p'(u)/p(u)$. Then

$$1 - u\delta < \kappa - \kappa \exp(-\delta) < \kappa\delta,$$

and so $1 < \delta(\kappa + u)$.

■

**Corollary 4.** $(\kappa + u)\,p(u) \downarrow$ *on* $0 < u < \infty$.

**Proof.** $((u+\kappa)p(u))' < 0$.

■

We use the last result to obtain a sharper lower bound for $p$.

**Proposition 5.** *For* $u > 0$

$$p(u) > \{u + \kappa(u + \kappa)\log(1 + \frac{1}{u + \kappa})\}^{-1}.$$

**Proof.** For $t > u$ we have

$$(t + \kappa)\, p(t) < (u + \kappa)\, p(u).$$

If we combine this inequality with (5) we obtain

$$1 = up(u) + \kappa\int_u^{u+1} p(t)\, dt < p(u)\,\{u + \kappa\int_u^{u+1} \frac{u + \kappa}{t + \kappa}\, dt\}.$$

■

For an improved upper bound for $p$ we establish a dual to Corollary 4.

**Proposition 6.** *Let* $\delta > 0$. *Then* $(u + \kappa - \delta)p(u)$ *is increasing on* $\kappa/\delta - 1 \le u < \infty$.

**Proof.**

$$\{(u + \kappa - \delta)\, p(u)\}' = \kappa\, p(u) - \kappa\, p(u + 1) + \frac{\kappa - \delta}{u} up'(u)$$

$$= \frac{1}{u}\{p(u)(\kappa u + (\kappa - \delta)(\kappa - 1)) - p(u + 1)(\kappa u + \kappa(\kappa - \delta))\}$$

$$> \frac{p(u)}{u}\{\kappa u + \kappa^2 - \kappa - \delta\kappa + \delta - \frac{(\kappa u + \kappa^2 - \kappa\delta)(\kappa + u)}{\kappa + u + 1}\}$$

$$= \frac{p(u)}{u(\kappa + u + 1)}\{u\delta + \delta - \kappa\} \ge 0 \quad \text{for} \quad u \ge \kappa/\delta - 1.$$

■

In particular, $(u + \kappa - 1)p(u) \uparrow$ for $u \ge \kappa - 1$.

By a now familiar argument we obtain

**Proposition 7.** *Let* $\delta > 0$. *If* $u \ge \kappa/\delta - 1$ *then*

$$p(u) < \{u + \kappa(u + \kappa - \delta)\log(1 + \frac{1}{u + \kappa - \delta})\}^{-1}.$$

The relative error of the simpler estimates for $p_2(5)$ was about .03; the last two estimates yield

$$.145572 < p_2(5) < .145699$$

and a relative error of about .001.

## 4. Properties of $q$

Since $q_\kappa(u) \sim u^{2\kappa-1}$ as $u \to \infty$, $q_\kappa$ is ultimately positive for each $\kappa \ge 1$. We begin by showing that each $q_\kappa$ has one or more real positive roots. Define $\rho = \rho_\kappa$ by

$$\rho_\kappa = \inf\{u \ge 0 : q_\kappa(x) > 0 \text{ for all } x > u\}.$$

**Proposition 8.** *For each* $\kappa \ge 1$, $\rho_\kappa > 0$.

**Proof.** Since $q_1(u) = u - 1$ has a positive root, we can assume $\kappa > 1$. On $\rho \le u < \infty$ we have

$$uq'(u) = (\kappa - 1)q(u) + \kappa q(u+1) > 0.$$

Thus $q$ is increasing on $[\rho,\infty)$. Integrating the D.D.E. (2) yields

$$(u+1)q(u+1) = uq(u) + \kappa\int_u^{u+2} q(t)\,dt\,. \tag{8}$$

If $\rho = 0$, then for each $u > 0$ we would have

$$(u+1)q(u+1) > \kappa\int_{u+1}^{u+2} q(t)\,dt > \kappa q(u+1),$$

and this relation is false for $u \le \kappa - 1$. ■

As we noted earlier, estimates of $\rho_\kappa$ as a function of $\kappa$ are needed for sieve applications. Reasonably sharp estimates (and in some cases the method of estimation itself) are valid only on limited ranges of $\kappa$. Estimates of $\rho_\kappa$ for many intervals of length 1/2 are given in [4]. As $\kappa \to \infty$, $\rho_\kappa \sim c\kappa$, where $c = 3.591121...$ satisfies $c\log(c/e) = 1$ [7]. It is shown in [3b] that $\rho_\kappa < c - 2c/(1+c)$. Here is a small table of approximate values of $\rho_\kappa$:

| $\kappa$ | 1 | 1.5 | 2 | 3 | 5 | 10 |
|---|---|---|---|---|---|---|
| $\rho_k$ | 1 | 2.366 | 3.834 | 6.919 | 13.541 | 31.066 |

We have seen that $q$ and $q'$ are positive on $(\rho,\infty)$. To obtain further information about $q$, particularly estimates of $\rho$, we want estimates on further derivatives.

**Proposition 9.** *$q_\kappa^{(\nu)}(u) > 0$ on $\rho \le u < \infty$ for $1 \le \nu < 2\kappa$.*

**Proof.** The argument is made inductively on $\nu$. We give the case $\nu = 2$ explicitly (for $\kappa > 1$); the other cases are done analogously.

For $\rho \le u < \infty$ we have

$$uq'(u) = (\kappa - 1)q(u) + \kappa q(u+1) > 0,$$

and, in case $\kappa \ge 2$,

$$uq''(u) = (\kappa - 2)q'(u) + \kappa q'(u+1) > 0.$$

Now suppose $1 < \kappa < 2$.

The function $q''(u)$ is positive for all sufficiently large $u$, since $q(u) \sim u^{2\kappa-1}$ and the preceding two D.D.E.'s imply that $q'(u) \sim (2\kappa - 1)u^{2\kappa-2}$ and $q''(u) \sim (2\kappa - 1)(2\kappa - 2)u^{2\kappa-3}$. Suppose $q''(u) \le 0$ for some $u \ge \rho$. Then there exists a largest such value, say $u^*$. We have

$$0 \ge u^* q''(u^*) = (\kappa - 2)q'(u^*) + \kappa q'(u^* + 1),$$

so

$$q'(u^* + 1) \le \frac{2-\kappa}{\kappa} q'(u^*) < q'(u^*).$$

It follows that $q''(u_1) < 0$ for some $u_1 > u^*$. $\rightarrow\leftarrow$.

■

**Corollary 9.** *$\rho_\kappa \ge 2\kappa - 1$ for all $\kappa \ge 1$.*

**Proof.** The estimate is exact for $\kappa = 1$, since $q_1(u) = u - 1$. For $\kappa > 1$, take $u = \rho$ in (8) and recall that $q$ is convex. We get

$$(\rho + 1)q(\rho + 1) = \kappa \int_\rho^{\rho+2} q(t)\,dt > 2\kappa q(\rho + 1).$$

Since $q(\rho + 1) > 0$, the result follows.

■

The following further derivative estimates for $q$ will be used to give an upper bound for $\rho_\kappa$.

**Proposition 10.** *Let* $\nu \in \mathbb{Z}, \nu > 2\kappa - 1$ *and* $2\kappa \notin \mathbb{Z}$. *Then* $\operatorname{sgn} q^{(\nu)}(u) = (-1)^{[2\kappa]-\nu}$ *for all* $u > 0$.

**Proof.** If we differentiate (7) $\nu$ times, we obtain for $u > 0$

$$q^{(\nu)}(u) = \frac{\Gamma(2\kappa)}{2\pi i} \int_{\mathcal{C}} z^{\nu - 2\kappa} e^{uz + \kappa \operatorname{Ein}(-z)} dz .$$

As before, $\mathcal{C}$ is the path from $-\infty$ back to $-\infty$ which surrounds the negative real axis in the positive sense. The integral along a path $\{z = \varepsilon e^{i\theta}, -\pi < \theta \le \pi\}$ goes to zero with $\varepsilon$ if $\nu > 2\kappa - 1$. If we collapse the contour onto the negative real axis (and change the integration variable) we obtain

$$q^{(\nu)}(u) = \frac{\Gamma(2\kappa)\{e^{-\pi i(\nu - 2\kappa)} - e^{\pi i(\nu - 2\kappa)}\}}{2\pi i} \int_{t=0}^{\infty} t^{\nu - 2\kappa} e^{-ut + \kappa \operatorname{Ein} t} dt$$

$$= \sin(\pi(2\kappa - \nu)) \frac{\Gamma(2\kappa)}{\pi} \int_0^{\infty} t^{\nu - 2\kappa} e^{-ut + \kappa \operatorname{Ein} t} dt .$$

The sign of this expression is that of $\sin(\pi(2\kappa - \nu))$, i.e., $(-1)^{[2\kappa - \nu]}$. ■

**Proposition 11.** *For* $1 < \kappa \le 3/2$, $\rho_\kappa < 3\kappa - 2$.

**Proof.** For $u > \rho$ we have

$$q(u) > 0, \ q'(u) > 0, \ q''(u) > 0, \ q'''(u) \le 0.$$

Apply Taylor's formula twice and subtract:

$$q(\rho + 2) = q(\rho + 1) + q'(\rho + 1) + \frac{1}{2} q''(\rho + 1) + \frac{1}{6} q'''(\xi)$$

$$0 = q(\rho) = q(\rho + 1) - q'(\rho + 1) + \frac{1}{2} q''(\rho + 1) - \frac{1}{6} q'''(\eta),$$

$$q(\rho + 2) = 2q'(\rho + 1) + \text{neg} < 2q'(\rho + 1). \tag{9}$$

Substitute this inequality into the D.D.E. for $q$:

$$(\rho+1)q'(\rho+1) = (\kappa-1)q(\rho+1) + \kappa q(\rho+2)$$
$$< (\kappa-1)q(\rho+1) + 2\kappa q'(\rho+1),$$

or (recalling that $\rho + 1 - 2\kappa > 0!$),

$$\cdot q'(\rho+1) < \frac{(\kappa-1)}{\rho+1-2\kappa} q(\rho+1). \tag{10}$$

Also, by convexity,

$$2q(\rho+1) < q(\rho+2) + q(\rho) = q(\rho+2).$$

If we combine the last inequality with (9) and (10) we get

$$2q(\rho+1) < q(\rho+2) < 2q'(\rho+1) < \frac{2(\kappa-1)}{\rho+1-2\kappa} q(\rho+1),$$

and hence $\rho < 3\kappa - 2$.

Note that the inequality is sharp at $\kappa = 1$.

■

We are going to obtain another lower estimate for $\rho$ that is sharper than that of Corollary 9 for larger values of $\kappa$. This bound will be established by use of equation (4). We begin by showing that the right hand side of (4) is convex for $\kappa \geq 2$.

**Proposition 12.** *Let* $\kappa \geq 2$. *Then* $u^{-\kappa} q(u+1)$ *is convex on* $\rho < u < \infty$.

**Proof.** For $\kappa = 2$,

$$\{u^{-2} q_2(u+1)\}'' = 8u^{-4} > 0.$$

For $\kappa > 2$, two applications of the D.D.E. (2) yield

$$u^{\kappa+2}\{u^{-\kappa}q(u+1)\}'' = \kappa^2 q(u+3) - 3\kappa q(u+2) - 3\kappa q'(u+2)$$
$$+ 2q(u+1) + 4q'(u+1) + q''(u+1). \tag{11}$$

By Taylor's Theorem and the positivity of $q$, $q'$, $q''$, $q'''$, and $q^{(4)}$ on $(\rho,\infty)$ we have

$$\kappa^2 q(u+3) > \{q(u+2) + q'(u+2) + \frac{1}{2}q''(u+2) + \frac{1}{6}q'''(u+2)\}\kappa^2,$$

$$2\{q(u+1) + q'(u+1) + \frac{1}{2}q''(u+1) + \frac{1}{6}q'''(u+2)\} > 2q(u+2),$$

and

$$2\{q'(u+1)+q''(u+2)\}>2q'(u+2).$$

If we insert the first of the inequalities in (11) we obtain

$$u^{\kappa+2}\{u^{-\kappa}q(u+1)\}''>A+B,$$

where

$$A=(\kappa^2-3\kappa)q(u+2)+(\kappa^2-3\kappa)q'(u+2),$$

$$\begin{aligned}B&=\frac{\kappa^2}{2}q''(u+2)+\frac{\kappa^2}{6}q'''(u+2)+2q(u+1)+4q'(u+1)+q''(u+1)\\&>\{2q(u+1)+2q'(u+1)+q''(u+1)+\frac{1}{3}q'''(u+2)\}\\&\qquad+\{2q'(u+1)+2q''(u+2)\}\\&>2q(u+2)+2q'(u+2).\end{aligned}$$

(The last inequality follows from the last two Taylor estimates.) Thus

$$u^{\kappa+2}\{u^{-\kappa}q(u+1)\}''>(\kappa^2-3\kappa+2)\{q(u+2)+q'(u+2)\}>0.$$

■

**Proposition 13.** *Let* $\kappa\geq 2$. *Let* $B=2.84305987...$ *be the solution of* $B^{-1}e^{1/B}=1/2$. *Then* $\rho_\kappa>B\kappa-1.87$.

**Proof.** By integrating (4) and using the convexity result above we obtain

$$(\rho+2)^{1-\kappa}q(\rho+2)=\kappa\int_\rho^{\rho+1}u^{-\kappa}q(u+1)\,du>2\kappa(\rho+1)^{-\kappa}q(\rho+2)$$

or

$$\rho+2>2\kappa\left(\frac{\rho+2}{\rho+1}\right)^{\kappa}=2\kappa\exp\{\kappa\log\frac{\rho+2}{\rho+3/2}-\kappa\log\frac{\rho+1}{\rho+3/2}\}.$$

Applying the estimate

$$\log(1+a)-\log(1-a)>2a\qquad(0<a<1)$$

we get

$$\rho+2>2\kappa\exp\left\{\frac{\kappa}{\rho+3/2}\right\}$$

or

$$\left(\frac{\kappa}{\rho+3/2}\right)\exp\left(\frac{\kappa}{\rho+3/2}\right)<\frac{1}{2}+\frac{1}{4\kappa}\cdot\frac{\kappa}{\rho+3/2}.$$

Let $f(x) = xe^x$. Note that $f, f'$ are increasing on $(0,\infty)$. If we let $\kappa/(\rho + 3/2) = \delta$ and recall that $f(1/B) = 1/2$, then by the mean value theorem we obtain

$$(\delta - 1/B) f'(1/B) < f(\delta) - f(1/B) < \delta/(4\kappa).$$

Now

$$f'(1/B) = e^{1/B}(1 + 1/B) = (B + 1)/2,$$

so

$$\delta\left\{\frac{B+1}{2} - \frac{1}{4\kappa}\right\} < \frac{B+1}{2B},$$

and finally

$$\rho > B\kappa - \frac{B}{2B+2} - \frac{3}{2} > 2.84305\ldots\kappa - 1.86989\ldots .$$

■

This estimate is quite sharp near $\kappa = 2$. Indeed, $\rho_2 \doteq 3.834$ and the estimate is $\rho_2 > 3.816$.

## 5. Further Results and Problems

We conclude with a brief mention of some solved and unsolved problems involving the $p$ and $q$ functions. Not surprisingly, in view of the relative complexity, the outstanding problems concern only $q$.

Most applications of $q$ require estimates of ratios of $q$ at two points, rather than just evaluation at a single point. The reason for this is because $q$ usually occurs in homogeneous expressions, e.g., the adjoint equation for $F - f$ mentioned in section 1. For such applications we require monotonicity and numerical estimates for $q'/q$.

There are, however, instances in which $q$ by itself has to be estimated. This done following a method of Iwaniec. We write $\exp\{\kappa\,\mathrm{Ein}(-z)\}$ as a Taylor polynomial of degree $> 2\kappa - 1$ plus a remainder term and substitute this into the explicit formula (7) for $q$. The contour integral involving the Taylor polynomial can be evaluated explicitly by Hankel's formula, yielding a polynomial–like "main term" for $q$ of the form $u^{2\kappa-1} + a_1 u^{2\kappa-2} + \ldots + a_\nu u^{2\kappa-\nu-1}$. The remaining integral can have its contour collapsed onto the negative real axis by the argument used in the proof of Proposition 10, and is estimated as an error term.

Sieve estimates are sought that are valid for all values of $\kappa$ lying in some range. In some arguments it is necessary to evaluate functions explicitly by computer. For these estimates to hold for a range of $\kappa$'s, we need to know the behaviour of various functions, including $p_\kappa$ and $q_\kappa$ as functions of $\kappa$. In particular, we require estimates of $\partial p/\partial\kappa$ and $\partial q/\partial\kappa$. The first of these expressions can be given in a reasonably convenient way [5]. The partial derivative of $q$ is computed by expressing $q$ as a polynomial–like part plus an integral remainder and differentiating with respect to $\kappa$. The result is very complicated, as is the actual behaviour of $\partial q/\partial\kappa$, and this problem is still under investigation.

## References


[1] *M. Abramowitz* and *I. Stegun,* Handbook of Mathematical Functions. Dover, New York, 1965.

[2] *H. Diamond, H. Halberstam* and *H.–E. Richert,* Combinatorial sieves of dimension exceeding one. J. Number Theory **28** (1988), 306—346.

[3a,b] *H. Diamond, H. Halberstam* and *H.E. Richert,* A boundary value problem for a pair of differential delay equations related to sieve theory, I, II. (In preparation).

[4] *F. Grupp,* On zeros of functions satisfying certain differential difference equations. Acta Arith. **51** (1988), 247—268.

[5] *F. Grupp* and *H.E. Richert,* Notes on functions connected with the sieve. Analysis **8** (1988), 1—24.

[6] *J. Hale,* Theory of Functional Differential Equations, 2nd ed. Springer, New York, 1977.

[7] *H. Iwaniec,* Rosser's sieve. Acta Arith. **36** (1980), 171—202.

[8] *F.W.J. Olver,* Asymptotics and Special Functions. Academic Press, New York, 1974.

[9] *D.A. Rawsthorne,* Improvements in the small sieve estimate of Selberg by iteration. Ph.D. thesis, University of Illinois, Urbana, 1980.

[10] *H.J.J. te Riele,* Numerical solution of two coupled non–linear equations related to the limits of Buchstab's iteration sieve. Math Centrum Amsterdam Afd. NW 86/80, 15pp.

[11] *F. Wheeler,* On two differential difference equations arising in analytic number theory. Ph.D. thesis, University of Illinois, Urbana, 1988.

---

Mathematics Department, University of Illinois, 1409 West Green Street, Urbana, IL 61801, USA.

Mathematics Department, University of Illinois, 1409 West Green Street, Urbana, IL 61801, USA.

Mathematics Department, University of Ulm, Oberer Eselsberg, 7900 Ulm, West Germany.

# Discriminants and Class Numbers of Indefinite Integral Quadratic Forms

*A.G. Earnest*

## 1. Introduction

This paper will focus on the relationship between two invariants of an indefinite integral quadratic form—the discriminant and the class number. Specifically, the question to be addressed here is how large the discriminant of such a form with rank exceeding two must be in order for the form to have a prescribed class number.

Due to the lack of uniformity of terminology in the literature on this subject, it will be helpful to specify at the outset the conventions of notation and terminology to be used here. An integral quadratic form of rank $n$ shall mean a quadratic form $f(x) = f(x_1, \dots, x_n)$ such that $f(x) \in \mathbb{Z}$ for all $x \in \mathbb{Z}^n$, and $\det F \neq 0$, where $F$ is the $n \times n$ matrix

$$\left( \frac{\partial^2 f}{\partial x_i \, \partial x_j} \right).$$

Such a form is said to represent the integer $\alpha$ (primitively) if $f(x) = \alpha$ for some $x \in \mathbb{Z}^n$ (with $\gcd(x_1, \dots, x_n) = 1$), and is indefinite if it represents both positive and negative integers. The notion of discriminant which will be used here is that from Watson's book [15], p. 2; in particular, when $n = 3$, the discriminant $d(f)$ of the form is given by $d(f) = -\frac{1}{2} \det F$. The genus of an integral quadratic form $f$ consists of all forms which are equivalent to $f$ over $\mathbb{R}$ and over the $p$-adic completions $\mathbb{Z}_p$ of $\mathbb{Z}$ for all primes $p$. Finally, the class number $h(f)$ of an integral quadratic form $f$ is the number of integral equivalence classes of forms in the genus of $f$. When $f$ is indefinite, $h(f)$ is known to be a power of 2.

Throughout this paper, $\mathcal{F}$ will denote the set of all indefinite ternary (i.e., rank three) integral quadratic forms. In order to quantify the relationship between the discriminant and class number for such forms, define a function $\Delta$ on the set of all positive integers by

$$\Delta(t) = \min\left\{ |d(f)| : f \in \mathcal{F},\ h(f) = 2^t \right\}.$$

The results in this paper are of two types. In section 3, two constructions are described which yield upper bounds for $\Delta(t)$ in terms of sets of primes satisfying certain conditions; moreover, the exact values of $\Delta(1)$ and $\Delta(2)$ are produced. Detailed proofs of the results in this section will appear elsewhere [4]. Then the derivation of a weak lower bound for $\Delta(t)$ will be sketched in section 4. The results in that section are stated and proved in the setting of quadratic lattices over arbitrary algebraic number fields. This generality has been adopted primarily to emphasize the role of the ideal class group of the underlying field, a feature which is not manifest when working only over the field of rational numbers. When specialized to the classical case, the argument used to prove Theorem 1 yields the lower bound $\Delta(t) \geq 4P(t)^3$, where $P(t)$ is the product of the first $t$ rational primes.

This paper constitutes an expanded version of a talk presented by the author at the First Conference of the Canadian Number Theory Association in Banff, Alberta in April, 1988.

## 2. Historical Background and Examples

From his tabulation of indefinite ternary quadratic forms, Eisenstein observed in 1851 [6] that in this setting the class number always seemed equal to one. Although Meyer in 1896 published criteria under which such a form would have class number $2^t$ for any positive integer $t$ [12], pp. 317–318, over fifty years later it remained unknown whether any indefinite integral quadratic form of rank exceeding two could indeed have class number exceeding one [7], p. 189.

The first such example was apparently contained in a letter from Siegel to Jones in the early 1950's [14], p. 392. As described by Jones [8], p. 408, the proof that the class number is at least two for that example consisted of showing the existence of integers primitively represented by some form in the genus of the given form, but not by the form itself. To illustrate this method, consider the integers of the type $u^2$, with 17 not dividing $u$, which are primitively represented by the form $f(x,y,z) = x^2 + 17y^2 - 17^2z^2$. While it can be shown by standard local arguments that all such integers are represented by the genus of $f$ (hence, by some form in the genus), $f$ primitively represents no $u^2$ for which the Legendre symbol $(17:u) = -1$. For suppose, on the contrary, that there exist relatively prime integers $x$, $y$ and $z$ for which $f(x,y,z) = u^2$, where $(u:17) = -1$. Then

$$17(y^2 - 17z^2) = (u - x)(u + x), \tag{1}$$

and it may be assumed, by changing the sign of $x$ if necessary, that $17 \nmid (u - x)$; say $u = x + 17s$, $s \in \mathbb{Z}$. Since $(u + x:17) = (2x + 17s:17) = (2x:17) = (x:17) = (u:17) = -1$, it follows that $u + x$ is exactly divisible by an odd power of some prime $p$ for which $(p:17) = -1$. But then $(u:17) = -1$, which together with $y^2 \equiv 17z^2 \pmod{p}$, implies that $p \mid y$, $p \mid z$, and the left hand side of (1) is exactly divisible by an even power of $p$. Hence, $p \mid (u - x)$ and it follows that $p$ also divides $x$, contradicting the primitivity of the representation.

In a paper by Jones and Hadlock which appeared in 1953 [9], the basic idea illustrated by this example was generalized to give sufficient conditions under which a form in $\mathcal{F}$ has class number exceeding one. A concrete example used in that paper to show the applicability of these conditions is the form $f(x,y,z) = 289x^2 - 882y^2 - 146z^2 + 828yz + 224xz$ of discriminant $5{,}018{,}112 = 2^9 \cdot 3^4 \cdot 11^2$.

This method of using representation properties to prove that certain forms have class number exceeding one does not give any additional information regarding the actual value of the class number. Moreover, it is not always possible to demonstrate the inequivalence of two forms in a genus in $\mathcal{F}$ solely on the basis of representation properties. For example, when $p$ is a prime congruent to 1 modulo 8, the class number of the form $f(x,y,z) = -x^2 + p^2y^2 + p^4z^2$ is two, but $f$ primitively represents every integer which is primitively represented by its genus [3], Example 3.2.

A means for the explicit calculation of the class number of an indefinite integral quadratic form of rank exceeding two was ultimately provided by the theory of the spinor genus introduced by Eichler in 1952 [5]. Using this machinery, such a class number can be computed as the index in the idèle group of $\mathbb{Q}$ of a suitable subgroup determined by the images of the local spinor norm mappings (see e.g., [13], 102:7 and 104:5). That this class number is always a 2–power follows from the fact that the corresponding quotient group is an elementary abelian 2–group.

Full accounts of the spinor genus theory from somewhat differing points of view can be found in the books of Watson [15], O'Meara [13], and Cassels [1]. Using this theory, it is routine to calculate the class numbers of the following two examples, the significance of which will be seen in the next section. First, the form $x^2 + xy - y^2 + 25z^2$ of discriminant 125 has class number two. A representative form in the other equivalence class in the genus is $x^2 + 5y^2 + 5yz - 5z^2$ [15], p. 116. Finally, the form $x^2 - 34y^2 + 18{,}496z^2$ of discriminant $2{,}515{,}456 = 2^9 17^3$ has class number four [4].

## 3. Constructions and Minimality

A key feature common to all of the examples mentioned in the preceding section is the divisibility of their discriminants by the cube of at least one prime. While the class number of a primitive positive definite integral quadratic form grows rapidly with the size of either the discriminant or rank of the form [11], the class number of an indefinite form is intimately connected with the prime factorization of the discriminant. Indeed, a necessary condition for an indefinite integral quadratic form of rank $n$ to have class number exceeding one is that its discriminant be divisible by the $n(n-1)/2$–th power of at least one odd prime, or by a sufficiently large power of 2 [10], Satz 5. In particular, there exist primitive indefinite integral quadratic forms of arbitrarily large rank and discriminant which have class number one.

Generally, for an indefinite integral quadratic form to have class number $2^t$, $t > 1$, it is necessary not only that the discriminant be divisible by sufficiently large powers of at least $t$ distinct primes, but also that these primes interact with each other in rather specific ways. In this section, two sets of conditions are described which, when imposed on a set of $t$ primes, are sufficient to guarantee the existence of a form in $\mathcal{F}$ of class number $2^t$ and discriminant explicitly given in terms of the primes in the set. Proofs of the assertions made here can be found in [4].

For a set $S$ of primes, let $P(S) = \prod_{p \in S} p$. Let $( : )$ denote the Legendre symbol.

**Proposition 1.** *Let $S$ be a set consisting of $t$ primes such that $p \equiv 1 \pmod 4$ for all $p \in S$ and $(p:q) = 1$ for all $p \neq q \in S$. Then there exists $f \in \mathcal{F}$ such that $d(f) = P(S)^3$ and $h(f) = 2^t$.*

**Proposition 2.** *Let $S$ be a set consisting of $t-1$ primes such that $p \equiv 1 \pmod 8$ for all $p \in S$ and $(p:q) = 1$ for all $p \neq q \in S$. Then there exists $f \in \mathcal{F}$ such that $d(f) = 2^9 P(S)^3$ and $h(f) = 2^t$.*

**Remark 1.** The proofs of these propositions consist of several steps. First, a subgroup $H$ of the idèle group $J_{\mathbb{Q}}$ of $\mathbb{Q}$ is defined utilizing the $p-$components for the primes $p$ in the set $S$. It is then shown that, under the conditions imposed on the primes in $S$, the index of $H$ in $J_{\mathbb{Q}}$ is exactly $2^t$. Next, local forms are constructed $p-$adically, such that the images of all the spinor norm mappings on the local rotation groups give precisely the subgroup $H$. Finally, consistency of the local invariants is shown in order to guarantee the existence of a global form having the given $p-$adic forms as localizations, and thus having class number $2^t$.

**Remark 2.** Taking $S = \{5\}$ in Proposition 1 yields the existence of a form $f \in \mathcal{F}$ with $d(f) = 125$ and $h(f) = 2$. Similarly, taking $S = \{17\}$ in Proposition 2 yields the existence of a form $f \in \mathcal{F}$ with $d(f) = 2^9 17^3$ and $h(f) = 4$. Explicit examples of such forms were given at the end of the preceding section.

The constructions described in the above propositions are of interest due to the fact that they appear to be quite efficient in terms of the size of the discriminant required to achieve a prescribed class number. In order to make this statement more precise, define $\pi(t)$ and $\bar{\pi}(t)$, for a natural number $t$, to be the minima of the values $P(S)$ and $8P(S)$ taken over all sets $S$ of primes satisfying the conditions of Proposition 1 and Proposition 2, respectively.

**Proposition 3.** $\Delta(t) \leq \min\{\pi(t)^3, \bar{\pi}(t)^3\}$, *with equality for $t = 1$ and $t = 2$.*

The values these functions take on for small values of $t$ are summarized in the table below.

| $t$ | $\pi(t)$ | $\boldsymbol{\pi}(t)$ | $\Delta(t)$ |
|---|---|---|---|
| 1 | 5 | 8 | 125 |
| 2 | 145 | 136 | 2,515,456 |
| 3 | 11,713 | 12,104 | ? |

**Remark 3.** The inequality in Proposition 3 follows immediately from Propositions 1 and 2. Equality for $t = 1$ and $t = 2$ was established in [4] by *ad hoc* arguments which become unwieldy already for the determination of $\Delta(3)$. The author does not know whether strict inequality in Proposition 3 holds for any value of $t$.

**Remark 4.** It is evident from the above values that no inequality between $\pi(t)$ and $\boldsymbol{\pi}(t)$ holds for all values of $t$. It would be interesting to know whether one of these functions is asymptotically dominant over the other.

**Remark 5.** The computations of $\Delta(1)$ and $\Delta(2)$ show that the examples at the end of Section 2 have the smallest possible discriminants (in absolute value) among all forms in $\mathcal{F}$ of class number two and four, respectively.

## 4. Lower Bounds

In this final section, quadratic lattices over the rings of integers of arbitrary algebraic number fields will be considered. In this context, the notation and terminology of O'Meara's book [13] will be adopted, except where otherwise indicated.

Throughout this discussion, let $F$ be a fixed algebraic number field with ring $R$ of algebraic integers, and let $S$ be the set of all nonarchimedean prime spots on $F$. For brevity, the term "indefinite quadratic lattice" will be used for an $R$-lattice on a regular $F$-quadratic space $V$ which is indefinite in the sense that the localization $V_p$ is isotropic for at least one archimedean prime spot $p$ on $F$. Such a lattice $L$ will be referred to as integral if its scale ideal $sL$ is contained in $R$. In this context, the role analogous to that of the discriminant of a quadratic form is played be the volume ideal $vL$ of a lattice $L$. The size of this volume ideal will be measured by its absolute norm $N(vL)$.

**Theorem 1.** *Let L be an indefinite integral quadratic lattice of rank $n \geq 3$ over the ring R of integers of the algebraic number field F. Then there exists a constant $\lambda = \lambda(F)$, depending only on F, such that $N(\nu L) \geq (\lambda h^+(L))^{n(n-1)/2}$.*

**Proof.** By [13], 102:7, the proper class number $h^+(L)$ is given by the group index $(J_F : P_D J_F^L)$, where $D$ consists of the set of elements of $\dot{F}$ which are positive at all real spots $p$ at which the underlying quadratic space is anisotropic [13], 101:8. It is convenient to introduce the following subgroups of the idèle group $J_F$:

$$J_F^e = \{j \in J_F : j_p \in U_p \dot{F}_p^2 \text{ for all } p \in S\},$$

$$\hat{J}_F^L = \{j \in J_F : j_p \in \theta(O^+(L_p)) \cap U_p \dot{F}_p^2 \text{ for all } p \in S\}.$$

Now

$$\begin{aligned}(J_F : P_D J_F^L) &= (J_F : P_F J_F^L)(P_F J_F^L : P_D J_F^L)\\ &\leq (\dot{F} : D)(J_F : P_F J_F^L)\\ &\leq (\dot{F} : D)(J_F : P_F \hat{J}_F^L)\\ &= (\dot{F} : D)(J_F : P_F J_F^e)(P_F J_F^e : P_F \hat{J}_F^L)\\ &\leq (\dot{F} : D)(J_F : P_F J_F^e)(J_F^e : \hat{J}_F^L).\end{aligned}$$

Here $(J_F : P_F J_F^e) = (C_F : C_F^2)$, where $C_F$ denotes the ideal class group of $F$ ([2], Lemma 1). To estimate the factor $(J_F^e : \hat{J}_F^L)$, let $\Gamma = \Gamma(L)$ be the set of nondyadic prime spots $p$ in $S$ for which $U_p \dot{F}_p^2 \not\subset \theta(O^+(L_p))$, and let $\gamma = \gamma(L) = |\Gamma(L)|$. Then $(U_p \dot{F}_p^2 : \dot{F}_p^2) = 2$ for all $p \in \Gamma$ and

$$(J_F^e : \hat{J}_F^L) \leq \prod_{p|2} (U_p \dot{F}_p^2 : \dot{F}_p^2) \prod_{p \in \Gamma} 2 = 2^\gamma \mu,$$

where $\mu = \mu(F)$ is a constant depending only on $F$. Combining the above yields the following inequality

$$\lambda h^+(L) \leq 2^\gamma, \tag{2}$$

where $\lambda = \lambda(F)$ is a constant depending only on $F$.

Now consider the volume ideal $vL$. Write $vL = \prod_{p \in S} p^{e(p)}$, where the nonnegative integers $e(p)$ equal zero for almost all $p \in S$. Then

$$\begin{aligned} N(vL) &= \prod_{p \in S} N(p)^{e(p)} \geq \prod_{p \in S} 2^{e(p)} \\ &\geq \prod_{p \in \Gamma} 2^{e(p)} \geq \prod_{p \in \Gamma} 2^{n(n-1)/2} \\ &= (2^{\gamma})^{n(n-1)/2}, \end{aligned}$$

the last inequality following from [10], Satz 5. Combining the resulting inequality with (2) yields the desired result.

**Remark 6.** The factors $(F^{\dot{}} : D)$ and $\prod_{p|2} (U_p \dot{F}_p^2 : \dot{F}_p^2)$ appearing in the above argument can be bounded from above by constants depending only on the degree $m = [F : \mathbb{Q}]$. Thus, when the field $F$ is allowed to vary, one can obtain, for lattices $L$ as in Theorem 1, an inequality of the type

$$N(vL) \geq \left( \frac{\eta h^{+}(L)}{(C_F : C_F^2)} \right)^{n(n-1)/2},$$

where $\eta = \eta(m)$ is a constant depending only on the degree $m$.

**Remark 2.** Minor modifications of the proof of Theorem 1 produce the lower bound for $\Delta(t)$ given in Section 1. Let $f \in \mathcal{F}$, let $L$ be a quadratic lattice corresponding to $f$, and let $\Omega(L)$ denote the set consisting of the rational primes $p$ such that $\theta(O^{+}(L_p)) \not\supseteq U_p \dot{\mathbb{Q}}_p^2$. Since $(F^{\dot{}} : D) = (C_F : C_F^2) = 1$ for $F = \mathbb{Q}$,

$$h(f) = h^{+}(L) = (P_{\mathbb{Q}} J_{\mathbb{Q}}^{e} : P_{\mathbb{Q}} J_{\mathbb{Q}}^{\Lambda L}) \leq 2^{|\Omega(L)|}.$$

Thus, $|\Omega(L)| \geq t$ whenever $f \in \mathcal{F}$ and $h(f) = 2^t$. So, for such $f$,

$$|d(f)| = 4N(vL) \geq 4 \prod_{p \in \Omega(L)} p^3 \geq 4P(t)^3.$$

It follows that $\Delta(t) \geq 4P(t)^3$, as asserted.

## References

[1] *J.W.S. Cassels,* Rational Quadratic Forms. Academic Press, London, (1978).

[2] *A.G. Earnest,* Spinor genera of unimodular $\mathbb{Z}$–lattices in quadratic fields. Proc. Amer. Soc. **64** (1977), 189—195.

[3] *A.G. Earnest,* Congruence conditions on integers represented by ternary quadratic forms. Pacific J. Math. **90** (1980), 325—333.

[4] *A.G. Earnest,* Minimal discriminants of indefinite ternary quadratic forms having specified class number. Mathematika **35** (1988), 95—100.

[5] *M. Eichler,* Quadratische Formen und orthogonale Gruppen. Springer–Verlag, Berlin–Göttingen–Heidelberg (1952).

[6] *G. Eisenstein,* Tabelle der reducirten positiven quadratischen Formen. nebst den Resultaten neuerer Forschungen. J. reine angew. Math. **41** (1851), 140—190.

[7] *B.W. Jones,* The Arithmetic Theory of Quadratic Forms. Carus Math. Monographs, Buffalo (1950).

[8] *B.W. Jones,* Quasi–genera of quadratic forms. J. Number Theory **9** (1977), 393—412.

[9] *B.W. Jones* and *E.H. Hadlock,* Properly primitive ternary indefinite quadratic genera of more than one class. Proc. Amer. Math. Soc. **4** (1953), 539—543.

[10] *M. Kneser,* Klassenzahlen indefiniter quadratischer Formen in drei oder mehr Veranderlichen. Arch. Math. **7** (1956), 323—332.

[11] *W. Magnus,* Über die Anzahl der in einem Geschlecht enthaltenen Klassen von positiv–definiten quadratischen Formen. Math. Ann. **114** (1937), 465—475; **115** (1938), 643—644.

[12] *A. Meyer,* Über indefinite ternäre quadratische Formen. J. reine angew. Math. **116** (1896), 307—325.

[13] *O.T. O'Meara,* Introduction to Quadratic Forms. Springer–Verlag, Berlin–Göttingen–Heidelberg (1963).

[14] *O.T. O'Meara,* Hilbert's eleventh problem: the arithmetic theory of quadratic forms. Mathematical developments arising from Hilbert problems (ed. F.E. Browder). Proc. Symposia Pure Math. **28** (1976), 379—400.

[15] *G.L. Watson,* Integral Quadratic Forms. Cambridge Univ. Press, Cambridge (1960).

---

Department of Mathematics, Southern Illinois University, Carbondale, Illinois 62901–4408, USA.

# On a Conjecture of Roth and Some Related Problems, II

*P. Erdös and A. Sárközy* [1]

## 1. Introduction

Throughout this paper we use the following notations:

$\mathbb{N}$ denotes the set of natural numbers, and we write $\mathbb{N}_M = \{1,2,\dots,M\}$. The cardinality of the finite set $S$ is denoted by $|S|$. If $S$ is a given set and $\mathcal{A}_1,\mathcal{A}_2,\dots,\mathcal{A}_k$ are subsets of $S$ with

$$S = \bigcup_{i=1}^{k} \mathcal{A}_i\,, \quad \mathcal{A}_i \cap \mathcal{A}_j = \phi \text{ for } i \neq j\,,$$

then $\{\mathcal{A}_1,\mathcal{A}_2,\dots,\mathcal{A}_k\}$ will be called a partition of $S$, and the subsets $\mathcal{A}_1,\mathcal{A}_2,\dots,\mathcal{A}_k$ will be referred to as classes. If $S$ is a given set, $\{\mathcal{A}_1,\mathcal{A}_2,\dots,\mathcal{A}_k\}$ is a fixed partition of it (in other words, a $k$–colouring of $S$ is given), $f(x_1,x_2,\dots,x_t\}$ is a given function, $n \in \mathbb{N}$ and

$$f(a_1,a_2,\dots,a_t) = n \tag{1}$$

with $a_1,a_2,\dots,a_t$ belonging to the same class, then this will be called a monochromatic representation of $n$ in the form (1).

If $\mathcal{A}$ is an infinite subset of $\mathbb{N}$, and we write $A(M) = \left|\mathcal{A} \cap \mathbb{N}_M\right|$, then

$$\bar{d}(\mathcal{A}) = \limsup_{M \to +\infty} \frac{A(M)}{M}$$

and

---

1 Research partially supported by Hungarian National Foundation for Scientific Research grant no. 1811.

$$\underline{d}(\mathcal{A}) = \lim_{M \to +\infty} \inf \frac{A(M)}{M}$$

are called the upper asymptotic density and lower asymptotic density of $\mathcal{A}$, respectively. If $\bar{d}(\mathcal{A}) = \underline{d}(\mathcal{A})$, then this is called the asymptotic density of $\mathcal{A}$ and it is denoted by $d(\mathcal{A})$.

## 2. Preliminaries

K.F. Roth conjectured (see [3] and [5], p. 112) that for $k \in \mathbb{N}$, $M > M_o(k)$ and for every partition $\{\mathcal{A}_1, \mathcal{A}_2, \ldots, \mathcal{A}_k\}$ of $\mathbb{N}_M$, there are more than $cM$ integers not exceeding $M$ (where $c$ is a positive absolute constant independent of $k$) which have a monochromatic representation in the form $a + a'$ with $a \neq a'$. (Note that if $a = a'$ is allowed, then this is trivial since every even number not exceeding $2M$ has a monochromatic representation in the form $a + a$.) In Part I [4], V.T. Sós and the authors of this paper proved the conjecture of Roth in the following sharper form:

If $k \in \mathbb{N}$, $k \geq 2$ and $M > M_o(k)$, then for every partition $\{\mathcal{A}_1, \mathcal{A}_2, \ldots, \mathcal{A}_k\}$ of $\mathbb{N}_M$, all but $3M^{1-2^{-k-1}}$ even integers not exceeding $M$ have a monochromatic representation in the form

$$a + a' \text{ with } a \neq a'. \tag{2}$$

Moreover, we proved several further theorems on monochromatic representations in the form (2). Furthermore, we extended the original problem by studying monochromatic representations in the form (1) in the special case when $f(x_1, x_2, \ldots, x_t)$ is a linear polynomial.

In this paper, our goal is to study the analogous questions on products $aa'$ instead of sums $a + a'$. In other words, we will estimate the number of those integers which have a monochromatic representation in the form $aa'$ for a given partition of $\mathbb{N}_M$ or $\mathbb{N}$. (Clearly, here we may drop the condition of $a \neq a'$.)

## 3. Distinct Monochromatic Products

First we will show that here the situation is completely different in the sense that the number of the distinct monochromatic products not exceeding $M$ can be $o(M)$:

**Theorem 1.** (i) *For* $\varepsilon > 0$, $M \in \mathbb{N}$, $M > M_o(\varepsilon)$ *there is a partition* $\{\mathcal{A}_1, \mathcal{A}_2\}$ *of* $\mathbb{N}_M$ *such that the number of the distinct monochromatic products* $aa'$ *not exceeding* $M$ *is less than* $M(\log M)^{-\alpha+\varepsilon}$ *where* $\alpha = 1 - \frac{\log(e \log 2)}{\log 2}$ $\approx 0.0860$.

(ii) *If* $k \in \mathbb{N}$, $\varepsilon > 0$, $M \in \mathbb{N}$, *and* $M > M_o(\varepsilon)$ *then for every partition* $\{\mathcal{A}_1, \mathcal{A}_2, \ldots, \mathcal{A}_k\}$ *of* $\mathbb{N}_M$, *the number of the distinct monochromatic products* $aa'$ *not exceeding* $M$ *is greater than* $M(\log M)^{-\beta-\varepsilon}$ *where* $\beta = 2\log 2 - 1 \approx 0.03863$.

There is a gap between the lower and upper bounds. Surely, the upper bound gives the right order of magnitude and it could be shown by the method used in [2] that the number of the distinct monochromatic products not exceeding $M$ is greater than $M(\log M)^{-\alpha-\varepsilon}$. However, since we will prove a much sharper result on the logarithmic density of the monochromatic products, we will not work out the details here.

**Proof.** (i) Let $\mathcal{A}_1 = \{n : n \in \mathbb{N}, n \leq M^{1/2}\}$ and $\mathcal{A}_2 = \{n : n \in \mathbb{N}, M^{1/2} < n \leq M\}$. Then the only monochromatic products $aa'$ not exceeding $M$ are the products with $a \in \mathcal{A}_1$, $a' \in \mathcal{A}_1$, i.e., with $a \leq M^{1/2}$, $a' \leq M^{1/2}$. By a theorem of Erdös [2], the number of the distinct products $ij$ with $i \leq x, j \leq x$ is less than $x^2(\log x)^{-\delta+\varepsilon}$. Using this theorem with $x = M^{1/2}$, the result follows.

(ii) One of the classes, say $\mathcal{A}_1$, contains at least $[M^{1/2}/k]$ of the integers not exceeding $M^{1/2}$. Pomerance and Sárközy [6] proved that if $\varepsilon > 0$, $x \in \mathbb{N}$, $x > x_o(\varepsilon)$, $\mathcal{A}, \mathcal{B} \subset \mathbb{N}_x$, $|\mathcal{A}| > \varepsilon x$ and $|\mathcal{B}| > \varepsilon x$, then the number of the distinct products $ab$ with $a \in \mathcal{A}$, $b \in \mathcal{B}$ is greater than $x^2(\log x)^{-\beta-\varepsilon}$. Using this theorem with $[M^{1/2}]$, $\mathcal{A}_1$ and $\mathcal{A}_1$ in place of $x$, $\mathcal{A}$ and $\mathcal{B}$ respectively, the result follows.

## 4. Logarithmic Density

While the number of the distinct monochromatic products can be relatively small, their logarithmic density must be large:

**Theorem 2.** *There is a positive absolute constant* $c$ *such that if* $k \in \mathbb{N}$, $M \in \mathbb{N}$ $M > M_0(k)$, $\{\mathcal{A}_1, \mathcal{A}_2, \dots, \mathcal{A}_k\}$ *is a partition of* $\mathbb{N}_M$ *and* $\mathcal{B}$ *denotes the set of integers that have a monochromatic representation in the form* $aa'$, *then*

$$\sum_{b \in \mathcal{B}} \frac{1}{b} > \frac{c}{k} \log M .$$

**Proof.** For a finite set $\mathcal{D} \in \mathbb{N}$, we write

$$S(\mathcal{D}) \sum_{d \in \mathcal{D}} \frac{1}{d} .$$

For $n = p_1^{\alpha_1} p_2^{\alpha_2} \dots p_r^{\alpha_r} \in \mathbb{N}$, let $\nu(n)$ denote the number of prime factors of $n$ counted with multiplicity:

$$\nu(n) = \alpha_1 + \alpha_2 + \dots + \alpha_r .$$

Finally, for $t \in \mathbb{N}$, $\mathcal{D} \in \mathbb{N}$ we write

$$\mathcal{D}(t) = \{d : d \in \mathcal{D},\ \nu(d) = t\}.$$

First we are going to show that for every $t \in \mathbb{N}$ we have

$$\sum_{i=1}^{k} (S(\mathcal{A}_i(t)))^2 \le \binom{2t}{t} S(\mathcal{B}(2t)). \tag{3}$$

On the left hand side we have

$$\sum_{i=1}^{k} \sum_{a \in \mathcal{A}_i(t)} \frac{1}{a} \sum_{a' \in \mathcal{A}_i(t)} \frac{1}{a'} = \sum_{i=1}^{k} \sum_{a \in \mathcal{A}_i(t)} \sum_{a' \in \mathcal{A}_i(t)} \frac{1}{aa'} .$$

Writing

$$b = aa' = p_1^{\alpha_1} p_2^{\alpha_2} \dots p_r^{\alpha_r} , \tag{4}$$

for every term of this sum we have

$$\nu(b) = \nu(a) + \nu(a') = t + t = 2t$$

so that $b \in \mathfrak{B}(2t)$. For a fixed $b \in \mathfrak{B}(2t)$, denote the number of the monochromatic representation of $b$ in the form $aa'$ with $\nu(a) = \nu(a') = t$ by $f(b)$ so that

$$\sum_{i=1}^{k} (S(\mathcal{A}_i(t)))^2 = \sum_{b \in \mathfrak{B}(2t)} \frac{f(b)}{b}. \tag{5}$$

If $b \in \mathfrak{B}(2t)$ is fixed, $a$, $a'$ satisfy (4) and $\nu(a) = \nu(a') = t$, then $a$ is of the form $a = p_1^{\beta_1} p_2^{\beta_2} \dots p_r^{\beta_r}$ where

$$\begin{cases} 0 \le \beta_i \le \alpha_i & \text{for } i = 1,2,\dots,r, \\ \sum_{i=1}^{r} \beta_i = t; \end{cases} \tag{6}$$

$b$ and $a$ determine $a'$ uniquely. Thus $f(b)$ is less than or equal to the number of solutions of the system (6). Denoting the number of solutions of the system

$$0 \le \beta_i \le \alpha_i \quad \text{for } i = 1,2,\dots,r,$$

$$\sum_{i=1}^{r} \beta_i = u$$

by $g(u)$, in view of $\alpha_1 + \alpha_2 + \dots + \alpha_r = 2t$ clearly we have

$$\sum_{u=0}^{2t} g(u) x^u = \prod_{i=1}^{r} (1 + x + \dots + x^{\alpha_i}). \tag{7}$$

For $i = 1,2,\dots,r$, every coefficient of $1 + x + \dots + x^{\alpha_i}$ is less than or equal to the corresponding coefficient of $(1 + \alpha)^{\alpha_i}$ (and all the coefficients are non–negative). Thus every coefficient of the polynomial in (7) is less than or equal to the corresponding coefficient in the polynomial

$$\prod_{i=1}^{r} (1 + x)^{\alpha_i} = (1 + x)^{\alpha_1 + \alpha_2 + \dots + \alpha_r} = (1 + x)^{2t} = \sum_{u=0}^{2t} \binom{2t}{u} x^u.$$

Hence

$$g(u) \le \binom{2t}{u} \quad \text{for every } u,$$

in particular,

$$g(t) = \binom{2t}{u}. \tag{8}$$

In view of $f(b) \le g(t)$ (for every $b \in \mathfrak{B}(2t)$), (18) follows from (5) and (8).

Since $\{\mathcal{A}_1, \mathcal{A}_2, \dots, \mathcal{A}_k\}$ is a partition of $\mathbb{N}_M$, thus

$$\sum_{i=1}^{k} S(\mathcal{A}_i(t)) = S(\mathbb{N}_M(t))$$

so that by the Cauchy–Schwarz inequality we have

$$\sum_{i=1}^{k} (S(\mathcal{A}_i(t)))^2 \ge \frac{1}{k}(S(\mathbb{N}_M(t)))^2. \tag{9}$$

It follows from (3) and (9) that

$$S(\mathfrak{B}(2t)) \ge \left(k\binom{2t}{t}\right)^{-1} (S(\mathbb{N}_M(t)))^2. \tag{10}$$

Let us write $x = \frac{1}{2}\log\log M$ and $Y = (\log\log M)^{1/2}$. For $t \ne t'$, $\mathfrak{B}(2t)$ and $\mathfrak{B}(2t')$ are disjoint, hence (10) implies

$$S(\mathfrak{B}) \ge \sum_{X<t<X+Y} S(\mathfrak{B}(2t)) \ge \frac{1}{k} \sum_{X<t<X+Y} \binom{2t}{t}^{-1} (S(\mathbb{N}_M(t)))^2. \tag{11}$$

It remains to estimate $\binom{2t}{t}^{-1} (S(\mathbb{N}_M(t)))^2$ for $X < t < X+Y$.

By a well-known theorem [7], [8] for $\varepsilon > 0$, $M > M_o(\varepsilon)$ and $1 \le t \le (2-\varepsilon)\log\log M$ we have

$$|\{n : n \in \mathbb{N}_M, \nu(n) = t\}| > c(\varepsilon)\frac{M}{\log M}\frac{(\log\log M)^{t-1}}{(t-1)!}$$

where $c(\varepsilon)$ is a positive constant depending on $\varepsilon$. It follows by partial summation that there is a positive absolute constant $c_1$ such that for large $M$ and $1 \le t \le \log\log M$ we have

$$S(\mathbb{N}_M(t)) > c_1 \frac{(\log\log M)^t}{t!}.$$

Thus by Stirling's formula, for large $M$ and $X < t < X + Y$ we have

$$\binom{2t}{t}^{-1} (S(\mathbb{N}_M(t)))^2 > c_2 \frac{(t!)^2}{(2t)!} \frac{(\log\log M)^{2t}}{(t!)^2} \tag{12}$$

$$> c_3 t^{-1/2} \left(\frac{e}{2t}\right)^{2t} (\log\log M)^{2t}$$

$$= c_3 t^{-1/2} \exp\left(2t\left(1 + \log\frac{\log\log M}{2t}\right)\right)$$

$$= c_3 t^{-1/2} \exp\left(2t\left(1 + \log\left(1 - \frac{2t - \log\log M}{2t}\right)\right)\right)$$

$$> c_4 t^{-1/2} \exp\left(2t\left(1 - \frac{2t - \log\log M}{2t} - c_5\left(\frac{2t - \log\log M}{2t}\right)^2\right)\right)$$

$$= c_4 t^{-1/2} \exp\left(\log\log M - c_5\left(\frac{2t - \log\log M}{2t}\right)^2\right)$$

$$> c_6 (\log\log M)^{-1/2} \exp(\log\log M - c_7)$$

$$= c_8 (\log M)(\log\log M)^{-1/2}$$

($c_2, c_3, \ldots$ denote positive absolute constants). It follows from (11) and (12) that for large M

$$S(\mathcal{B}) > \frac{1}{k} \sum_{X < t < X+Y} c_8 (\log M)(\log\log M)^{-1/2}$$

$$> c_9 \frac{1}{k} Y (\log M)(\log\log M)^{-1/2} > c_{10} \frac{1}{k} \log M$$

and this completes the proof of the theorem.

## 5. Upper Asymptotic Density

Of course, Theorem 2 implies

**Corollary 1.** *There is a positive absolute constant* $c$ *such that if* $k \in \mathbb{N}$ *and* $\{\mathcal{A}_1, \mathcal{A}_2, \ldots, \mathcal{A}_k\}$ *is a partition of* $\mathbb{N}$, *then the upper asymptotic density of the set of the distinct monochromatic products* $aa'$ *is greater than* $\frac{c}{k}$.

This result is best possible apart from the value of the constant $c$ as the following theorem shows:

**Theorem 3.** *There is an absolute constant* $c$ *such that for every* $k \in \mathbb{N}$ *there is a partition* $\{\mathcal{A}_1, \mathcal{A}_2, \ldots, \mathcal{A}_k\}$ *of* $\mathbb{N}$ *with the property that the upper asymptotic density of the sequence of the distinct monochromatic products* $aa'$ *is less that* $\frac{c}{k}$.

**Proof.** Let $k$ be a fixed positive integer. Let $\delta$ be a positive number which is small enough in terms of $k$ and it will be fixed later. Let us write

$$\alpha_i = \frac{2}{k-i} \quad \text{for } i = 1,2,\ldots,k-1.$$

Let $\mathcal{P}_1, \mathcal{P}_2, \ldots, \mathcal{P}_{k-1}$ be pairwise disjoint sets of distinct prime numbers greater than $p_0$ (where $p_0 = p_0(k)$ is large enough in terms of $k$) and satisfying

$$\left| \sum_{p \in \mathcal{P}_i} \frac{1}{p} - \alpha_i \right| < \delta, \tag{13}$$

and write $\prod_{p \in \mathcal{P}_i} p = P_i$ (for $i = 1,2,\ldots,k-1$). Let

$$\mathcal{A}_1 = \{n : n \in \mathbb{N},\ (n, P_1) > 1\},$$

$$\mathcal{A}_i = \{n : n \in \mathbb{N},\ (n, P_1 P_2 \ldots P_{i-1}) = 1,\ (n, P_i) > 1\}$$
$$\text{for } i = 2,3,\ldots k-1$$

and

$$\mathcal{A}_k = \{n : n \in \mathbb{N},\ (n, P_1 P_2 \ldots P_{i-1}) = 1\}.$$

Obviously, $\{\mathcal{A}_1, \mathcal{A}_2, \ldots, \mathcal{A}_k\}$ is a partition of $\mathbb{N}$.

Clearly, the asymptotic density of the monochromatic products $aa'$ with $a \in \mathcal{A}_i$, $a' \in \mathcal{A}_i$ exists for each of $i = 1,2,\ldots,k$; let us denote this density by $\beta_i$.

It is easy to see that the asymptotic densities of the sets

$$\{n : n \in \mathbb{N},\ (n, P_1 P_2 \ldots P_{i-1}) = 1\} \quad (\text{for } i = 2,3,\ldots k)$$

and

$$\{n : n \in \mathbb{N}, \text{ there are } n_1, n_2 \in \mathbb{N} \text{ with } n = n_1 n_2, \quad (n_1, P_i) > 1,\ (n_2, P_i) > 1\} \quad (\text{for } i = 1,2,\ldots k-1) \tag{14}$$

exist; let us denote these densities by $\varphi_i$ and $\psi_i$, respectively. Clearly,

$$\beta_1 = \psi_1,\ \beta_i = \varphi_i \psi_i \ \text{ for } i = 2,3,\ldots,k-1,\ \beta_k = \psi_k. \tag{15}$$

Obviously,

$$\varphi_i = \prod_{j=1}^{i=1} \prod_{p \in \mathcal{P}_j} \left(1 - \frac{1}{p}\right) \quad (\text{for } i = 2,3,\ldots k). \tag{16}$$

Furthermore, if $n$ belongs to the set (14), then there exist $p, q$ with $p \in \mathcal{P}_i$, $q \in \mathcal{P}_i$, $pq|n$. Thus

$$\psi_i \le \sum_{p \in \mathcal{P}_i} \frac{1}{pq} = \left(\sum_{p \in \mathcal{P}_i} \frac{1}{p}\right)^2 \quad (\text{for } i = 1,2,\ldots,k-1). \tag{17}$$

If $\delta \to 0$ and $p_0 \to +\infty$, then from (16)

$$\lim \varphi_i = \lim \exp\left(\sum_{j=1}^{i-1} \sum_{p \in \mathcal{P}_j} \log\left(1 - \frac{1}{p}\right)\right) = \tag{18}$$

$$= \lim \exp\left(-\sum_{j=1}^{i-1} \sum_{p \in \mathcal{P}_j} \frac{1}{p}\right)$$

$$= \exp\left(-\sum_{j=1}^{i-1} \alpha_j\right) \quad (\text{for } i = 2,3,\ldots,k).$$

For sufficiently small $\delta$, it follows from (13) and (17) that

$$\psi_i < 2\alpha_i^2 \text{ (for } i = 1,2,\dots,k-1). \tag{19}$$

It follows from (15), (18) and (19) that if $\delta$ is sufficiently small and $p_0$ is sufficiently large, then the asymptotic density of the monochromatic products $aa'$ is

$$\sum_{i=1}^{k} \beta_i < 3\left(\sum_{i=1}^{k-1} \alpha_i^2 \exp\left(-\sum_{j=1}^{i-1} a_j\right) + \exp\left(-\sum_{j=1}^{k-1} a_j\right)\right). \tag{20}$$

Here we have

$$\exp\left(-\sum_{j=1}^{i-1} a_j\right) = \exp\left(-\sum_{j=1}^{i-1} \frac{2}{k-j}\right) \tag{21}$$

$$= \exp\left(-2 \sum_{u=k-i+1}^{k-1} \frac{2}{u}\right) < \exp(-2(\log k - \log(k-i+1) + c_1)$$

$$= c_2 \frac{(k-i+1)^2}{k^2} \quad \text{(for } i = 1,2,\dots,k).$$

By (20) and (21), the asymptotic density of the monochromatic products is

$$\sum_{i=1}^{k} \beta_i < c_3\left(\sum_{i=1}^{k-1} \frac{2}{(k-1)^2} \frac{(k-i+1)^2}{k^2} + \frac{1}{k^2}\right)$$

$$< c_4\left(\sum_{i=1}^{k-1} \frac{1}{k^2} + \frac{1}{k^2}\right) = \frac{c_4}{k}$$

and this completes the proof of the theorem.

## 6. Lower Asymptotic Density

By Corollary 1, for every partition of $\mathbb{N}$ the upper asymptotic density of the distinct monochromatic products is $> \frac{c}{k}$. In this section, we are going to study their lower asymptotic density.

Let $\{\mathcal{A}_1, \mathcal{A}_2, \dots, \mathcal{A}_k\}$ be a partition of $\mathbb{N}$, and denote the smallest element of $\mathcal{A}_i$ by $a_i$. Then for every $a \in \mathcal{A}_i$, $a \leq M/a_i$, we have $aa_i \leq M$ and $aa_i$ is a monochromatic product. It follows that the number of distinct monochromatic products not exceeding $M$ is

$$\geq \frac{1}{k}\left(\frac{M}{\max_{i=1,2,\ldots,k} a_i}\right),$$

so that their lower asymptotic density is $\geq (k \max_{i=1,2,\ldots,k} a_i)^{-1}$; thus the lower asymptotic density cannot be 0. On the other hand, we will show that for $k \geq 3$ the lower asymptotic density (contrary to the upper asymptotic density) can be arbitrarily small:

**Theorem 4.** *For every* $\varepsilon > 0$ *there is a partition* $\{\mathcal{A}_1, \mathcal{A}_2, \mathcal{A}_3\}$ *of* $\mathbb{N}$ *such that the lower asymptotic density of the sequence of the distinct monochromatic products* $aa'$ *is less than* $\varepsilon$.

Note that in the finite case discussed in Section 3, we constructed a partition consisting of two classes only with the property that the number of the distinct monochromatic products is small; here in the infinite case we need three classes for the same purpose. Unfortunately, we have not been able to decide whether it is necessary to take three classes here. In other words, we have not been able to answer the following question: for every $\varepsilon > 0$, does there exist a partition $\{\mathcal{A}_1, \mathcal{A}_2\}$ if $\mathbb{N}$ such that the lower asymptotic density of the sequence of the distinct monochromatic products $aa'$ is less than $\varepsilon$?

**Proof.** The proof will be based on the following result of Erdös [1]:

**Lemma 1.** *For every* $\delta > 0$, *there exist positive numbers* $\eta = \eta(\delta)$ *and* $N_o = N_o(\delta) \in \mathbb{N}$ *such that if* $N \geq N_o$ *and* $M \geq M_1(\delta, N) \in \mathbb{N}$, *then the integers lying in the interval* $[N^{1-\eta}, N]$ *have less than* $\delta M$ *distinct multiples not exceeding* $M$.

(In other words, the density of the distinct multiples of the integers lying in $[N^{1-\eta}, N]$ is $\leq \delta$.)

We are going to define the classes $\mathcal{A}_1, \mathcal{A}_2, \mathcal{A}_3$ recursively.

Let $L_1 = N_o(\varepsilon/4)$ (where $N_o(\delta)$ is the function defined in Lemma 1), and let

$$\mathcal{A}_1 \cap \mathbb{N}_{L_1} = \mathbb{N}_{L_1}, \mathcal{A}_2 \cap \mathbb{N}_{L_1} = \mathcal{A}_3 \cap \mathbb{N}_{L_1} = \phi.$$

If $L_1, L_2, \dots, L_{i-1}, \mathcal{A}_1 \cap \mathbb{N}_{L_{i-1}}, \mathcal{A}_2 \cap \mathbb{N}_{L_{i-1}}$ and $\mathcal{A}_3 \cap |N|_{L_{i-1}}$ (where $i \geq 2$) have been defined, then we define $L_i$ and the intersections of $\mathcal{A}_1, \mathcal{A}_2$ and $\mathcal{A}_3$ in the following way:

Let

$$\gamma_i = \min(\eta(\varepsilon/2^i), 1/2)$$

and

$$L_i = \max\left(\left(\frac{6}{\varepsilon}L_{i-1}\right)^{1/\gamma_i}, N_o(\varepsilon/2^i), \max_{2 \leq j \leq i-1} M_1(\varepsilon/2^j, L_j)\right) \tag{22}$$

(where $\eta(\delta)$, $N_o(\delta)$ and $M_1(\delta, N)$ are defined in Lemma 1). Let

$$\mathcal{A}_1 \cap \{L_{i-1}+1, L_{i-1}+2, \dots, L_i\} = \{n : n \in \mathbb{N}, L_{i-1} < n \leq L_i^{1/2}\},$$

$$\mathcal{A}_2 \cap \{L_{i-1}+1, L_{i-1}+2, \dots, L_i\} = \{n : n \in \mathbb{N}, L_i^{1/2} < n \leq L_i^{1-\gamma_i}\}$$

and

$$\mathcal{A}_3 \cap \{L_{i-1}+1, L_{i-1}+2, \dots, L_i\} = \{n : n \in \mathbb{N}, L_i^{1-\gamma_i} < n \leq L_i\}.$$

Clearly, $\mathcal{A}_1, \mathcal{A}_2, \mathcal{A}_3$ are pairwise disjoint and their union is $\mathbb{N}$, so that $\{\mathcal{A}_1, \mathcal{A}_2, \mathcal{A}_3\}$ is a partition of $\mathbb{N}$. Let $E_1(n)$, $E_2(n)$ and $E_3(n)$ denote the number of the distinct products $aa'$ not exceeding $n$ with $a, a' \in \mathcal{A}_1$, $a, a' \in \mathcal{A}_2$ and $a, a' \in \mathcal{A}_3$, respectively, and let us denote the total number of the distinct monochromatic products not exceeding $n$ by $E(n)$, so that

$$E(n) \leq E_1(n) + E_2(n) + E_3(n) \text{ (for every } n \in \mathbb{N}). \tag{23}$$

Now we are going to estimate $E(L_i)$ (for $i = 2,3,\dots$).

If $a, a' \in \mathcal{A}_1 \cap \mathbb{N}_{L_i}$, then $a, a' \leq L_i^{1/2}$, hence, by the result of Erdös [2] used also in the proof of the first half of Theorem 1, we have

$$E_1(L_i) \leq o((L_i^{1/2})^2) = o(L_i) < \frac{\varepsilon}{6}L_i \quad \text{(for large } i\text{)}. \tag{24}$$

If $a, a' \in \mathcal{A}_2 \cap \mathbb{N}_{L_i}$, and $aa' \le L_i$, then clearly at least one of $a$ and $a'$ is $\le L_i^{1-\gamma_i}$. Thus in view of (22),

$$E_2(L_i) \le L_{i-1} L_i^{1-\gamma_i} \le \frac{\varepsilon}{6} L_i^{\gamma_i} L_i^{1-\gamma_i} = \frac{\varepsilon}{6} L_i .$$

Assume finally that $a, a' \in \mathcal{A}_3$ and $aa' \le L_i$. Then clearly, $aa'$ is the multiple of an integer $n$ satisfying $L_i^{1-\gamma_i} < n \le L_j$ for some $2 \le j \le i-1$. By Lemma 2 and in view of (22), for fixed $j$ these integers have less than $\frac{\varepsilon}{2^j} L_i$ distinct multiples not exceeding $L_i$, thus

$$E_3(L_i) \le \sum_{j=2}^{i-1} \frac{\varepsilon}{2^j} L_i < \frac{\varepsilon}{2} L_i . \tag{25}$$

It follows from (23, (24), (25) and (26) that

$$\frac{E(L_i)}{L_i} < \frac{\varepsilon}{6} + \frac{\varepsilon}{6} + \frac{\varepsilon}{2} < \varepsilon \quad \text{for } i = 2,3,\ldots$$

and this completes the proof of the theorem.

## 7. An Open Problem

The number of the distinct monochromatic products not exceeding $M$ may depend strongly on the distribution of the small numbers (small in terms of $M$) among the classes of the given partition. To eliminate this dependence, one might like to exclude the monochromatic products $aa'$ with $\min(a, a') < f(aa')$ where $f(x)$ tends to infinity "slowly". We note that our results can be extended easily to this case. In fact, we can show that both (ii) in Theorem 1 and Theorem 2 are still valid if we exclude the monochromatic products $aa'$ with $\min(a, a') < o((\log\log aa')^{1/2})$.

Furthermore, the construction given in the proof of Theorem 4 can be modified easily so that it should give the following result: For every $f(x) \to +\infty$ there is a partition $\{\mathcal{A}_1, \mathcal{A}_2, \mathcal{A}_3\}$ of $\mathbb{N}$ such that the asymptotic density of the distinct monochromatic products $aa'$ with $\min(a, a') > f(aa')$ is 0. The problem for two classes, of course, remains open here, too.

## 8. Conclusion

So far we have studied monochromatic representations in the form (1) in the special cases when $f(x_1, x_2, \dots, x_t)$ is a linear polynomial and $t = 2$, $f(x_1, x_2) = x_1 x_2$. One might like to extend the problem by studying general polynomials $f(x_1, x_2, \dots, x_t)$. In particular, what can be said on quadratic polynomials?

Let $\mathfrak{R}(f)$ denote the set of the integers $n$ that can be represented in the form $f(a_1, a_2, \dots, a_t) = n$ with $a_1, a_2, \dots, a_t \in \mathbb{N}$, denote the degree of $f$ by $d_f$, and for a partition $\mathcal{P} = \{\mathcal{A}_1, \mathcal{A}_2, \dots, \mathcal{A}_k\}$ let $S(f, \mathcal{P})$ denote the set of those integers which have a monochromatic representation in the form (1). Is it true that for every polynomial $f(x_1, x_2, \dots, x_t)$ and every partition $\mathcal{P} = \{\mathcal{A}_1, \mathcal{A}_2, \dots, \mathcal{A}_k\}$ of $\mathbb{N}$ we have

$$\lim_{m \to +\infty} \sup \frac{|S(f, \mathcal{P}) \cap \mathbb{N}_M|}{|\mathfrak{R}(f) \cap \mathbb{N}_M|} > c = c(d_f, k) \quad (> 0)?$$

## References

[1] *P. Erdös,* Generalizations of a theorem of Besicovitch. J. London Math. Soc. **11** (1936), 92—98.

[2] *P. Erdös,* An asymptotic inequality in the theory of numbers. Vestnik Leningrad. Univ. **15** (1960), 41—49 (in Russian).

[3] *P. Erdös,* Some unsolved problems. MTA MKI Közl. **6** (1961), 221—254.

[4] *P. Erdös, A. Sárközy* and *V.T. Sós,* On a conjecture of Roth and some related problems, I. Coll. Math. Soc. J. Bolyai, to appear.

[5] *R.K. Guy,* Unsolved Problems in Number Theory. Springer–Verlag, 1981.

[6] *C. Pomerance* and *A. Sárközy,* On products of sequences of integers. Number Theory, Coll. Math. Soc. J. Bolyai, to appear.

[7] *L.G. Sathe,* On a problem of Hardy on the distribution of integers having a given number of prime factors. J. Indian Math. Soc. **17** (1953), 63—141 and **18** (1954), 27—81.

[8] *A. Selberg,* Note on a paper by L.G. Sathe. J. Indian Math. Soc. **18** (1954), 83—87.

---

Mathematical Institute of the Hungarian Academy of Sciences, Budapest,
Reáltonada v. 13-15, H-1053, HUNGARY.

# Some New Identities Involving the Partition Function $p(n)$

*J. Fabrykowski and M.V. Subbarao*

Here we give some partition identities of a recursive nature. They resemble the well known Euler recursions for the partition function $p(n)$.

## 1. Introduction

Let $p(n)$ denote as usual, the number of unrestricted partitions of $n$. Throughout this paper, $\varphi(x)$ denotes the Euler product defined by

$$\varphi(x) = \prod_{n=1}^{\infty}(1 - x^n), \quad |x| < 1$$

It is well known, as first proved by Euler, that

$$\begin{aligned}\varphi(x) &= \sum_{-\infty}^{\infty}(-1)^k x^{3k^2+k/2} \\ &= 1 + \sum_{1}^{\infty}(-1)^k (x^{3k^2+k/2} + x^{3k^2-k/2}) \qquad (1)\end{aligned}$$

and

$$1/\varphi(x) = \sum_{n=0}^{\infty} p(n)x^n, \quad p(0) = 1. \qquad (2)$$

We use the convention that $p(n) = 0$ whenever $n$ is a negative integer.

From the above two relations we get the well known recursion relation for $p(n)$:

$$p(n) = \sum_{k\geq 1} (-1)^{k}\left\{p\left(n - \frac{3k^{2} - k}{2}\right) + p\left(n - \frac{3k^{2} + k}{2}\right)\right\} \tag{3}$$

The above formula has been proven to be useful in investigation of the problem of parity of $p(n)$. For example in 1959 O. Kolberg using (3) proved that $p(n)$ takes both even and odd values, each of them infinitely often. This result is a special case of an old conjecture of M. Newman (1960); which states that for all $m \geq 2$ $p(n) \equiv r \pmod m$, $0 \leq r \leq m-1$, has infinitely many solutions in $n$. It has been proven for $m = 2, 5, 7, 13, 17, 19, 29, 31, 65$ and $121$. Also it is a special case of another conjecture of M.V. Subbarao (1966) which says that for all integers $a \geq 1$, each of the congruences: $p(an + b) \equiv 0 \pmod 2$, $p(an + b) \equiv 1 \pmod 2$ has, for each $b$ $(0 \leq b \leq a - 1)$, infinitely many solutions. So far it is known to be true for $a = 1$, 2, 4, 8 and 16. See [1].

In this paper we shall obtain some recursion identities for $p(n)$ which are believed to be new. For this purpose we need to utilize, in addition to the Euler expansion of $\varphi(x)$, the following identities due to Jacobi.

$$\varphi^{3}(x) = \sum_{k=0}^{\infty}(-1)^{k}(2k+1)x^{k(k+1)/2}, \qquad |x| < 1 \tag{4}$$

and the Triple Product Identity:

$$\varphi(z^{2})\prod_{n=1}^{\infty}(1 + yz^{2n-1})(1 + y^{-1}z^{2n-1}) = \sum_{k=-\infty}^{\infty} y^{k} z^{k^{2}} \tag{5}$$

where $|z| < 1$, $y \neq 0$.

## 2. New Recursion Identities for $p(n)$

Analogous to the Euler recursion formula (3), we shall prove:

**Theorem 1.**

$$p(2n+1) + \sum_{k>0}\{p(2n+1-(8k^{2}-2k)) + p(2n+1-(8k^{2}+2k))\}$$

$$= p(2n) + \sum_{k>0}\{p(2n-(8k^{2}-6k)) + p(2n-(8k^{2}+6k))\} \tag{6}$$

$$p(2n)+\sum_{k>0}(-1)^k\{p(2n-(3k^2-k))+p(2n-(3k^2+k))\}$$
$$=p(n)+\sum_{k>0}\{p(n-(4k^2-k))+p(n-(4k^2+k))\} \tag{7}$$

$$p(2n+1)+\sum_{k>0}(-1)^k\{p(2n+1-(3k^2-k))+p(2n+1-(3k^2+k))\}$$
$$=p(n)+\sum_{k>0}\{p(n-(4k^2-3k))+p(n-(4k^2+3k))\} \tag{8}$$

**Proof.** By applying the Jacobi triple product identity (5) we have:

$$\sum_{-\infty}^{\infty}x^{2n^2-n}=\prod_{n=1}^{\infty}(1-x^{4n})(1+x^{4n-3})(1+x^{4n-1})$$
$$=\prod_{n=1}^{\infty}(1+x^{2n-1})(1-x^{4n})$$
$$=\prod_{n=1}^{\infty}\frac{(1+x^{4n-2})(1-x^{4n})}{1-x^{2n-1}}$$
$$=\prod_{n=1}^{\infty}\frac{1-x^{2n}}{1-x^{2n-1}}$$
$$=\frac{\prod_{n=1}^{\infty}(1-x^{2n})^2}{\prod_{n=1}^{\infty}1-x^n}$$
$$=\varphi^2(x^2)\sum_{n=0}^{\infty}p(n)x^n$$
$$=\varphi^2(x^2)\sum_{n=0}^{\infty}p(2n)x^{2n}+\varphi^2(x^2)\sum_{n=0}^{\infty}p(2n+1)x^{2n+1} \tag{9}$$

Now

$$\sum_{n=-\infty}^{\infty}x^{2n^2-n}=\sum_{n=-\infty}^{\infty}x^{8n^2-2n}+\sum_{n=-\infty}^{\infty}x^{8n^2-6n+1} \tag{10}$$

From (9) and (10) we obtain

$$\varphi^2(x^2)\sum_{n=0}^{\infty} p(2n)x^{2n} = \sum_{n=-\infty}^{\infty} x^{8n^2-2n} \tag{11}$$

and

$$\varphi^2(x^2)\sum_{n=0}^{\infty} p(2n+1)x^{2n+1} = \sum_{n=-\infty}^{\infty} x^{8n^2-6n+1} \tag{12}$$

that is

$$\varphi^2(x^2)\sum_{n=0}^{\infty} p(2n+1)x^{2n} = \sum_{n=-\infty}^{\infty} x^{8n^2-6n} \tag{13}$$

To prove (6) we eliminate the $\varphi^2(x^2)$ term from (11) and (13) and equate coefficients of like powers of $x$.

To prove (7) we rewrite (11) in the form:

$$\varphi(x^2)\sum_{n=0}^{\infty} p(2n)x^{2n} = \frac{1}{\varphi(x^2)}\sum_{k=-\infty}^{\infty} x^{8k^2-2k}$$

and using (1) and (2) we derive:

$$\sum_{k=-\infty}^{\infty}(-1)^k x^{3k^2+k}\sum_{n=0}^{\infty} p(2n)x^{2n} = \sum_{n=0}^{\infty} p(n)x^{2n}\sum_{k=-\infty}^{\infty} x^{8k^2-2k} \tag{14}$$

now (7) follows from (14) on equating coefficients of $x$. Similarly (8) follows from (13).

## 3. Further Identities For $p(n)$

Let $N$ be a non-negative integer and let $k, l, r, s$ be integers. Define:

$$a_N = \sum_{48N+5=4(6k+1)^2+(24l+1)^2} (-1)^k ,$$

$$b_N = \sum_{48N+5=4(6r+1)^2+(24s+7)^2} (-1)^r ,$$

$$c_N = \sum_{48N+29=4(6k+1)^2+(24l+19)^2} (-1)^k ,$$

$$d_N = \sum_{48N+29=4(6r+1)^2+(24s+13)^2} (-1)^r ,$$

Then the following holds:

**Theorem 2.**

$$\sum_{\substack{v\geq 0,\ m\geq 0 \\ \frac{1}{2}v(v+1)+2m=N}} (-1)^v (2v+1)p(4m) + \sum_{\substack{v\geq 0,\ m\geq 0 \\ \frac{1}{2}v(v+1)+2m=N-1}} (-1)^v (2v+1)p(4m+2) = a_N - b_N . \tag{15}$$

$$\sum_{\substack{v\geq 0,\ m\geq 0 \\ \frac{1}{2}v(v+1)+2m=N}} (-1)^v (2v+1)p(4m+1) + \sum_{\substack{v\geq 0,\ m\geq 0 \\ \frac{1}{2}v(v+1)+2m=N-1}} (-1)^v (2v+1)p(4m+3) = c_N - d_N . \tag{16}$$

**Proof.** We use the formulas (1.5), (1.7) and (3.7) as given in [2]. Thus

$$\sum_0^\infty p(4m)x^{2m} + \sum_0^\infty p(4m+2)x^{2m+1} = \frac{\varphi(x^2)}{\varphi^3(x)}\varphi(x^{24})A_1(x), \tag{17}$$

where

$$\varphi(x^{24})A_1(x) = \sum_{-\infty}^\infty x^{l(12l+1)} - \sum_{-\infty}^\infty x^{(3s+1)(4s+1)} \tag{18}$$

Combining the above formulas we obtain

$$\varphi^3(x)\left\{\sum_0^\infty p(4m)x^{2m} + \sum_0^\infty p(4m+2)x^{2m+1}\right\}$$

$$= \sum_{k,l=-\infty}^\infty (-1)^k x^{3k^2+k+12l^2+l} - \sum_{r,s=-\infty}^\infty (-1)^r x^{3r^2+r+12s^2+7s+1}$$

$$= \sum_{N=0}^\infty (a_N - b_N)x^N . \tag{19}$$

Now (15) follows on equating the coefficients of like powers of $x$ in (19).

The proof of (16) actually follows the lines of the previous case, but we do not have a formula analogous to (18), but which we now develop. Using the formulas (1.6) and (1.8) in [2] we have:

$$\sum_{0}^{\infty} p(4m+1)x^{2m} + \sum_{0}^{\infty} p(4m+3)x^{2m+1} = \frac{\varphi(x^2)}{\varphi^3(x)}\varphi(x^{24})A_3(x) \tag{20}$$

where

$$A_3(x) = \prod_{m=1}^{\infty}(1+x^{24m-17})(1+x^{24m-7}) - x^2\prod_{m=1}^{\infty}(1+x^{24m-23})(1+x^{24m-1}) \tag{21}$$

(See [2], page 348.) We need to have $\varphi(x^{24})A_3(x)$ expressed in the form of an infinite series. Let us put $y = x^{-5}$, $z = x^{12}$ in (5). Then

$$\varphi(x^{24})\prod_{m=1}^{\infty}(1+x^{24m-17})(1+x^{24m-7}) = \sum_{k=-\infty}^{\infty} x^{12k^2-5k} = \sum_{l=-\infty}^{\infty} x^{(l+1)(12l+7)} \tag{22}$$

Similarly letting $y = x^{-11}$ and $z = x^{12}$ we get

$$x^2\varphi(x^{24})\sum_{m=1}^{\infty}(1+x^{24m-23})(1+x^{24m-1}) = \sum_{k=-\infty}^{\infty} x^{12k^2-11k+2} = \sum_{s=-\infty}^{\infty} x^{(4s+3)(3s+1)}. \tag{23}$$

Combining (20), (21), (22) and (23) we derive

$$\varphi^3(x)\left\{\sum_{m=0}^{\infty} p(4m+1)x^{2m} + \sum_{m=0}^{\infty} p(4m+3)x^{2m+1}\right\} = \varphi(x^2)\left\{\sum_{l=-\infty}^{\infty} x^{(l+1)(12l+7)} - \sum_{m=0}^{\infty} x^{(4s+3)(3s+1)}\right\} \tag{24}$$

Now (16) follows on the lines of the proof of Theorem 2.

**Corollary 1.** *If* $48N+5$ *is not a sum of two squares then:*

$$\sum_{\substack{v\ge 0,\ m\ge 0\\ \frac{1}{2}v(v+1)+2m=N}} p(4m) + \sum_{\substack{v\ge 0,\ m\ge 0\\ \frac{1}{2}v(v+1)+2m=N-1}} p(4m+3) \equiv 0 \pmod 4 \tag{25}$$

*If* $48N+29$ *is not a sum of two squares then:*

$$\sum_{\substack{v\ge 0,\ m\ge 0\\ \frac{1}{2}v(v+1)+2m=N}} p(4m+1) + \sum_{\substack{v\ge 0,\ m\ge 0\\ \frac{1}{2}v(v+1)+2m=N-1}} p(4m+3) \equiv 0 \pmod 4 \tag{26}$$

**Proof.** Since $\varphi^3(x) \equiv \sum_{n=0}^{\infty} x^{n(n+1)/2} \pmod 4$, then (25) and (26) follow immediately from (15) and (16) respectively.

**Remark 1.** We apply (6) to provide a new proof of Kolberg's result, that is $p(n)$ takes both even and odd values, each of them infinitely often.

Assume first that $p(n)\equiv 0 \pmod 2$ for all $n\ge t$, and let $2n=8t^2+6t$. Then $p(8t^2+6t+1)+\ldots+p(8t+1)+p(4t+1)\equiv 0\pmod 2$ and $p(8t^2+6t)+\ldots$ $\ldots+p(12t)+p(0)\equiv 1\pmod 2$ contradiction.

If $p(n)\equiv 1\pmod 2$ for all $n\ge t$, then we set $2n=8t^2+2t$ and:

$$p(8t^2+2t+1)+\sum_{0<k\le t-1}\{p(8t^2+2t+1-8k^2+2k)$$
$$+p(8t^2+2t+1-8k^2-2k)\}+\{p(4t+1)+p(1)\}\equiv 1\pmod 2$$

since we have an odd number of odd terms. On the other hand:

$$p(8t^2+2t)+\sum_{0<k\le t-1}\{p(8t^2+2t-8k^2+6k)$$
$$+p(8t^2+2t-8k^2-6k)\}+p(8t)\equiv 0\pmod 2$$

since now, there are an even number of odd terms.

Added in Proof: F. Garavan and D. Stanton (unpublished) have recently verified the Subbarao conjecture for $a = 3,5$ and 10.

## References

[1] *M.V. Subbarao,* On the parity of the partition function. Congressus Numeratium **56** (1987), 265—275.

[2] *M.V. Subbarao* and *V.V. Subrahmanyasastri,* A transformation formula for products arising in partition theory. The Rocky Mountain J. Math. Vol. 6 (2), 1976, 345—356.

---

Department of Mathematics, University of Manitoba, Winnipeg, MN R3T 2N2, CANADA
Department of Mathematics, University of Alberta, Edmonton, AB T6G 2G1, CANADA

# What is Special about 195? Groups, $n$th Power Maps and a Problem of Graham

*R.W. Forcade and A.D. Pollington*

## 1. Introduction

Galovich and Stein [4] define a logarithm to be a one-to-one function $f : \{1,\ldots,k\} \rightarrow C(k)$, where $C(k)$ denotes the cyclic group of order $k$, such that $f(xy) = f(x) + f(y)$ when $1 \leq xy \leq k$. More generally we will say that a subset of the integers $S$, with $|S| = k$ is **groupable** if there is a group $G$, $|G| = k$, and a map $\varphi: S \rightarrow G$ which is one-to-one and, if $x, y, z \in S$ with $z = xy$, then $\varphi(z) = \varphi(x) + \varphi(y)$. We call such a map a 'group map' and conjecture [6] that if $S = \{1,2,\ldots,k\}$ then $G$ must be Abelian. Hickerson [7], and independently Chandler [2], have shown that if such a non-Abelian group exists then its centre has index 4, 6 or 8. Hickerson also shows that if $G$ is non–Abelian then $|G| > 9696$. This order is far too large for the range of our computer search technique. Thus we leave the question of non–Abelian group maps open.

Group maps for Abelian $G$ have applications to questions concerning tilings, the splitting of groups, the existence of group characters modulo a prime $p$, $p \equiv 1 \pmod{k}$, which separate $1,2,\ldots,k$ and to a number–theoretic conjecture of Graham [5].

We shall describe the connection with Graham's problem and the results of a computer search which showed that there is no logarithm for $k = 195$, but that such a logarithm exists for every $k$ less than 195. We describe some properties of functions which count the number of such group maps.

## 2. Power Maps and Primes

If, for $n = 1$ or $2$, $nk + 1$ is a prime, $p$ say, with primitive root $g$, then the map $\varphi(i) = j$ where

$$i \equiv g^{j} \pmod{p} \text{ for } 1 \le i \le k\text{, and } 1 \le j \le k-1$$

is a logarithm. If $n = 1$ the group $G$ is simply the integers modulo $p$. If $n = 2$, *then* G is the group of least quadratic residues modulo $p$. This was first noticed by Stein [12].

In fact if $p = kn + 1$ is a prime and $1^n, 2^n, \ldots, k^n$ are distinct modulo $p$ then, since the $n$ th power subgroup of $\mathbb{Z}_p^*$, the multiplicative group of the field of $p$ elements, is the cyclic group of order $k$, a logarithm for $k$ is induced by the map

$$i \to i^{n} \pmod{p} \quad 1 \le i \le k$$

or, as above,

$$i \equiv g^{j} \pmod{p} \to j \pmod{k} \quad \text{for} \quad 1 \le i \le k$$

Rewriting the cyclic group of order $k$ as a group of $k$ th roots of unity induces a Dirichlet character modulo $p$ which separates 1,2,...,k.

Galovich and Stein [4] call a logarithm induced by an $n$ th power map a KM–logarithm. If $k$ is odd then every logarithm is a KM–logarithm but if $k$ is even the situation is more complicated as we see from the following theorem of Kummer and Mills (see Stein [12]):

**Theorem 1.** *(Kummer and Mills). Let $p_1, \ldots, p_r$ be distinct primes and let $b_1, \ldots, b_r \in C(k)$. There is a prime $p$ and homomorphism $\varphi: \mathbb{Z}_p^* \to C(k)$ such that $\varphi(p_i) = b_i$ for $1 \le i \le r$ if and only if*

$$k \text{ is odd,} \tag{1}$$

*or*

$$k = 2m \text{ where } m \text{ is odd and} \tag{2}$$

(a) *for each $p_i$ such that $p_i \equiv 1 \pmod{4}$ and $p_i | m$, $b_i$ is even and*

(b) *for all $p_i \equiv 3 \pmod{4}$ the corresponding $b_i$ all have the same parity,*

*or*

$$k = 4m \text{ and for each } p_i | m,\ b_i \text{ is even.} \tag{3}$$

*Moreover, if there is one such prime* $p$ *for which a k–character exists with prescribed values at* $p_1, p_2, \ldots, p_r$ *then there is an infinite number.*

Elliott obtains the Kummer and Mills theorem and calculates the density of this set of primes $p$.

## 3. Connection With a Problem of Graham

Graham conjectured and Szegedy [13], and independently Alexandru [1], proved, at least for large $n$, that if $a_1, a_2, \ldots, a_n$ is a set of $n$ natural numbers then

$$\max_{i,j} \frac{a_i}{(a_i, a_j)} \geq n.$$

In 1969, Marica and Schonheim [8] proved Graham's conjecture for sets of square–free integers by using properties of relative compliments of sets. In order to study the general problem we were led to recast the problem in terms of lattice points. A completely factored positive integer can be identified with a lattice point whose $n$th coordinate is the multiplicity of the $n$th prime. E.g.,

$$80 = 2^4 5 \rightarrow (4, 0, 1, 0, \ldots, 0).$$

Graham's conjecture can be expressed in terms of the set of lattice points

$$F_n = \left\{ (e_1, \ldots, e_t) \middle| p_1^{e_1} p_2^{e_2} \cdots p_t^{e_t} \leq n \right\}, \quad \text{where } t = \pi(n).$$

Graham's conjecture, for $n + 1$, will be true if and only if any sequence comprised of numbers which are products of primes $\leq n$ satisfies the following statement about $F_n$:

There is no set $F$, of $n+1$ lattice points in $\mathbb{Z}^t$, such that

$$F - F \subset F_n - F_n \quad \text{where } F - F \text{ denotes } \{a - b : a \in F, b \in F\}.$$

The existence of a group map for $n$ would give a map from $F_n$ to a commutative group $G$, which is one-to-one on $F_n$ and which preserves the internal additions. Because $F_n$ contains the generators of $\mathbb{Z}^t$, this map can easily be extended to a homomorphism $\psi : \mathbb{Z}^t \rightarrow G$ which is one–to–one and onto when restricted to $F_n$. Note that if $f_1 \neq f_2 \in F_n$ then $\psi(f_1) \neq \psi(f_2)$ and $\psi(f_1 - f_2) \neq 0$.

Let $F$ contain $n+1$ lattice points. Then it has two elements $f_1, f_2$ with $\psi(f_1) = \psi(f_2)$ so $\psi(f_1 - f_2) = 0$ but $f_1 - f_2 \neq 0$. Hence $f_1 - f_2 \notin F_n - F_n$. Thus Graham's conjecture for $n+1$ follows from the existence of a group map for $n$.

## 4. Counting Group Maps

It is clear that if a group map exists then there are in fact many realizations of the map. For any map

$$\psi : \{1,2,\ldots,k\} \to C(k)$$

to be a group map we need only that $\psi(ij) = \psi(i) + \psi(j)$ if $1 \leq ij \leq k$. Thus if $\frac{k}{n+1} < p, q \leq \frac{k}{n}$ are distinct primes then their images under $\psi$ may be exchanged to obtain a new group map $\psi'$. This observation yields at least

$$\sum_{n=1}^{\infty} \left( \pi\left(\frac{k}{n}\right) - \pi\left(\frac{k}{n+1}\right) \right)!$$

distinct group maps, which by the prime number theorem and Stirling's formula is of order $A^k$ for some $A > 1$. The actual number of group maps for $n = 2,\ldots,11$ can be found by setting $n = \zeta$ in the polynomials given at the end of this section. In this section we will show, using an inclusion–exclusion argument, that these maps can be counted by finding certain polynomials.

Let $F$ be a finite subset of $\mathbb{Z}^t$, and let $G$ be a cyclic group. Define $\rho(F, G)$ to be the number of homomorphisms $\gamma : \mathbb{Z}^t \to G$ which are one–to–one on $F$. (If $\mathrm{card}(F) > \mathrm{card}(G)$ then $\rho(F, G) = 0$.)

Let

$$D_F = (F - F) \setminus \{(0)\} = \{x - y \mid x, y \in F, x \neq y\}.$$

Then $\rho(F, G)$ is just the number of homomorphisms which are non–zero on $D_F$. This observation allows us to count $\rho(F, G)$. Using inclusion–exclusion we obtain:

**Lemma 1.** $$\rho(F, G) = \sum_{S \subset D_F} (-1)^{\mathrm{card}(S)} \lambda_G(S)$$

*where the sum is taken over all subsets of* $D_F$ *, including the empty set, and* $\lambda_G(S)$ *denotes the number of homomorphisms* $\gamma : \mathbb{Z}^t \to G$ *which are zero on S.*

Now, let $H(G)$ denote the group of homomorphisms $\gamma : \mathbb{Z}^t \to G$. For each non-empty $S \subset D_F$, let $G^{\mathrm{card}(S)}$ denote a direct sum of $\mathrm{card}(S)$ copies of $G$, and define a homomorphism $t_{SG} : H(G) \to G^{\mathrm{card}(S)}$ by

$$t_{SG}(\gamma) = (\gamma(s_1), \ldots, \gamma(s_k))$$

where $s_1, \ldots, s_k$ are the elements of $S$ and $k = \mathrm{card}(S)$. Then

$$\lambda_G(S) = \mathrm{card}(H(G))/\mathrm{card}(\mathrm{Im}(t_{SG}))$$

and

$$\lambda_G(S) = (\mathrm{card}(G))^t/\mathrm{card}(\mathrm{Im}(t_{SG})).$$

We would then like to count the image group $\mathrm{Im}(t_{SG}(\gamma))$.

If $G$ is the cyclic group of order then we write $\lambda_G(S) = \lambda(S)$. Any $\gamma \in H(G)$ can be represented by an element $(\gamma_1, \ldots, \gamma_t) \in \mathbb{Z}^t$, and the transformation $t_{SG}$ can be represented by a matrix $M_S$, whose columns are the transposed elements of $S$. Then $t_{SG} = (\gamma_1, \ldots, \gamma_t)M_S \pmod{\ }$.

Now let $N = AM_S B$ where $A$ and $B$ are integer unimodular square matrices ($t$ by $t$ and $k$ by $k$ respectively), and $N$ is the diagonalized (Smith normal) form of $M_S$ (see [10]). Then $(\gamma_1, \ldots, \gamma_t)M_S B = (\gamma_1, \ldots, \gamma_t)A^{-1}N$, from which we obtain

$$\mathrm{Im}(t_{SG}) \cong (q_1 G) \oplus \ldots (q_t G)$$

where the $q_i$ are the diagonal entries of the matrix $N$, thus

$$\mathrm{card}(\mathrm{Im}(t_{SG})) = \prod_i \frac{\ }{(q_i, \ )}.$$

Note that $q_i = d_i/d_{i-1}$ where $d_0 = 1$ and $d_i$ is the gcd of all $i$ by $i$ sub-determinant's of $M_S$ for $1 \le i \le \mathrm{rank}(M_S)$ or simply $q_i = q_i(S)$, since the $q_i$ are invariant factors of the matrix formed by the elements of $S$.

Combining the previous formulas for $\lambda_G(S)$ and $\mathrm{card}(\mathrm{Im}(t_{SG}))$, we have

$$\lambda_G(S) = {}^{t-r(S)} \prod_{1 \le i \le r(S)} (q_i(S), \ ).$$

Where $r(S)$ denotes the rank of $S$ (i.e., of $M_S$), and

$$\rho(F, G) = \sum_{S \subset D_F} (-1)^{\mathrm{card}(S)} \ {}^{t-r(S)} \prod_{1 \le i \le r(S)} \gcd(q_i(S), \ ).$$

Note that the standard convention, that an empty product is 1, gives the correct value, ${}^{t}$, for $\lambda(\emptyset)$.

Using the fact that $k = \sum_{d | k} \varphi(d)$, where $\varphi$ denotes the Euler totient function, we have

$$\rho(F, G) = \sum_{S \subset D_F} (-1)^{\mathrm{card}(S)} \ {}^{t-r(S)} \prod_{1 \le i \le r(S)} \ \sum_{d | (l, q_i(S))} \varphi(d).$$

Since $q_i(S) | q_{i+1}(S)$ for $i < r(S)$ (a property of the Smith normal form), we can arrange the product to give the following.

**Theorem 2.** *Given a finite set $F$, and a cyclic group $G$, of order ,*

$$\rho(F, G) = \sum_{d |} P_{F,d}(\ ),$$

*where*

$$P_{F,d}(l) = \sum_{S \subset D_F} (-1)^{\mathrm{card}(S)} \ {}^{t-r(S)} B_d(S)$$

*and*

$$B_d(S) = \sum \prod_{1 \le i \le r(S)} \varphi(d_i)$$

*where the latter sum is taken over all* $r(S)$–tuples $(d_1, \ldots, d_{r(S)})$ *for which*

$$\mathrm{lcm}[d_1, \ldots, d_{r(S)}] = d,$$

*and* $d_i | q_i(S)$ *for* $i = 1, \ldots, r(S)$.

Note that $B_1(\emptyset) = 1$ and $B_d(\emptyset) = 0$ when $d > 1$.

Clearly, each $P_{F,d}$ ( ) is a polynomial in . In view of Theorem 2, the *Cyclic group conjecture* is equivalent to asserting that

$$\sum_{d|} P_{F,d}(\ ) > 0$$

where $F = F_n$, $t = \pi(n)$ and $= n$, for each positive integer $n$.

The first few polynomials are:

$$Q_2(\zeta) = \zeta - 1$$
$$Q_3(\zeta) = (\zeta - 2)(\zeta - 1)$$
$$Q_4(\zeta) = (\zeta - 3)(\zeta - 2)$$
$$Q_5(\zeta) = (\zeta - 4)(\zeta - 3)(\zeta - 3)$$
$$Q_6(\zeta) = (\zeta - 5)(\zeta^2 - 6\zeta + 10)$$
$$Q_7(\zeta) = (\zeta - 6)(\zeta - 5)(\zeta - 4)(\zeta - 1)$$
$$Q_8(\zeta) = (\zeta - 7)(\zeta - 6)(\zeta - 4)(\zeta - 3)$$
$$Q_9(\zeta) = (\zeta - 8)(\zeta - 7)^2(\zeta - 3)$$
$$Q_{10}(\zeta) = (\zeta - 9)(\zeta^3 - 21\zeta^2 - 6\zeta + 10)$$
$$Q_{11}(\zeta) = (\zeta - 10)(\zeta - 9)(\zeta^2 - 17\zeta + 73)$$

Obviously, $Q_n(\zeta)$ is of degree $\pi(n)$. That $(\zeta - n + 1)$ is always a factor is easily shown. Are there any more patterns in these polynomials $Q_n(\zeta)$?

## 5. Computational Results

We have checked for the existence of a logarithm for every number up to 205 with the only failure being at 195. The method used was a back-tracking algorithm.

The mapping $\psi$ is fully determined by the values it assigns to the primes $p \leq k$, since $\psi(xy) = \psi(x) + \psi(y)$, with the addition in $C(k)$. The algorithm assigns elements of $C(k)$ to each prime in turn, back-tracking whenever the current set of prime assignments causes two composite numbers to have the same image, The algorithm continues until all possibilities are exhausted or a logarithm has been found.

Note that in assigning 2 an image in $C(k)$ we need only try images which divide $k$. For if $i \rightarrow \psi(i)$ is a consistent map then so is

$$i \to t\,\psi(i) \pmod k$$

for $(t\,,k)=1$. As $\psi(2)$ runs through the divisors of $k$, the numbers $t\,\psi(2)$, $(t\,,k)=1$ run through $1,2,...,k-1$. This observation considerably reduces the running time of the algorithm.

In the cases where the group map exists we were always able to find one for which $\psi(2)=1$. In other words, 2 could always be assigned to a generator of the image group.

This is an exponential time algorithm in $k$. If assignments exist they are found relatively quickly but in the case of 195, the first failure of the group conjecture, the program ran for seventy hours of CPU time on a VAX 8600 at Bellcore before confirming that 195 is a counter–example.

We now believe that if neither $k+1$ or $2k+1$ is a prime and $k$ has a large number of prime factors then it is likely that $k$ will fail to have a logarithm. With this in mind we checked the next number of this type with three prime factors, which is 255. It too fails to have a logarithm. If our speculations are correct, then the answer to the question of the title (at least in this context) is "nothing," and we can attribute the existence of a logarithm to what Richard Guy, MAA address 1987, has called the "Strong law of small numbers."

Although we know that there exist integers for which there is no logarithm we could ask whether there is a KM–logarithm whenever a logarithm exists. The answer is again no. Calculations on a computer show that there is a logarithm for 184 but no KM-logarithm.

Here is a mapping for 201:

| Prime | 2 | 3 | 5 | 7 | 11 | 13 | 17 | 19 | 23 | 29 | 31 | 37 | 47 |
|---|---|---|---|---|---|---|---|---|---|---|---|---|---|
| Image | 1 | 32 | 150 | 159 | 135 | 124 | 112 | 39 | 27 | 141 | 146 | 75 | 17 |
| Prime | 43 | 47 | 53 | 59 | 61 | 67 | 71 | 73 | 79 | 83 | 89 | 97 | |
| Image | 79 | 164 | 11 | 20 | 55 | 88 | 8 | 25 | 50 | 91 | 121 | 133 | |

Note that the primes between 100 and 201 may be freely assigned to any remaining image after the products of primes less than 100 have been assigned.

**Acknowledgment.** We would like to thank David Bailey for the use of the Cray II at Ames–NASA, where the values up to 194 were first verified and who was the first to detect a problem with 195, and B. Gopinath for the use of the VAX 8600 at Bellcore.

## References

[1] *Z. Alexandru* , On a conjecture of Graham. J. Number Theory **27** (1987), 33—40.

[2] *K. Chandler* , Is a group formed by redefining multiplication on the integers 1,...,*n* necessarily Abelian? Preprint (Dalhousie University).

[3] *P.D.T.A. Elliot*, The distribution of power residues and certain related results. Acta Arith. **17** (1970), 141—159.

[4] *S. Galovich* and S. *Stein*, Splitting of Abelian groups by integers. Aequationes Math. **22** (1981), 249—267.

[5] *R.L. Graham* , Unsolved problem 5749. Amer. Math. Monthly **77** (1970), 775.

[6] *R. Guy*, ed., A group of two problems in groups. Amer. Math. Monthly **92** (1986), 119—121.

[7] *D. Hickerson* , Private communication.

[8] *J. Marica* and *J. Schonheim*, Differences of sets and a problem of Graham. Canad. Math. Bul. **12** (1969), 635—637.

[9] *W.H. Mills*, Characters with preassigned values. Canad. J. Math. **15** (1963), 169—171.

[10] *M. Newman* , Integral Matrices. Academic Press, 224 pages.

[11] *S. Stein*, Algebraic tiling. Amer. Math. Monthly **81** (1974), 445—462.

[12] *S. Stein*, Factoring by subsets. Pacific J. Math. **22** (1967), 523—541.

[13] *M. Szegedy* , The solution of Graham's common divisor problem. Combinatorica **6** (1986), 67—71.

---

Department of Mathematics, Brigham Young University, Provo, Utah 84602, USA.

# Modular Elliptic Curves and Diophantine Problems

*Dorian Goldfeld* [1]

## 1. Introduction

Let $E$ be an elliptic curve, defined over $\mathbb{Q}$, given in Weierstrass normal form

$$\begin{aligned} E: y^2 &= x^3 - ax - b \\ &= (x - e_1)(x - e_2)(x - e_3). \end{aligned}$$

The discriminant of $E$ is defined to be $D = (e_1 - e_2)^2 (e_1 - e_3)^2 (e_2 - e_3)^2$. Two elliptic curves given in Weierstrass normal form will be isomorphic if and only if they are equivalent under a rational transformation of type $x \to u^2 x$, $y \to u^3 y$ with $u \in \mathbb{Q}$, and $u$ unequal to 0. Under this transformation $a$ is transformed to $u^{-4}a$ and $b$ is transformed to $u^{-6}b$. Similarly, $D$ is transformed to $u^{-12}D$.

We say $E$ is in minimal Weierstrass normal form or is a minimal Weierstrass model over $\mathbb{Q}$ if among all isomorphic Weierstrass models for $E$ (with $a, b \in \mathbb{Z}$) we have that $D$ is minimized.

If the cubic $x^3 - ax - b = (x - e_1)(x - e_2)(x - e_3)$ has three distinct real roots, then the real points of $E$ (denoted $E(\mathbb{R})$) has two nonsingular connected components which are symmetric with respect to the $x$-axis. Although $E(\mathbb{R})$ is nonsingular, it may very well happen that $E(\mathbb{F}_p)$ (where $\mathbb{F}_p$ is the finite field of $p$ elements) is singular. It is not hard to see that this can only happen for primes $p | D$, and such primes are called primes of bad reductions. A measure for the amount of bad reduction

---

[1] This work was done while the author was partially supported by a grant from the Vaughn Foundation.

is given by the conductor of the elliptic curve. The conductor is denoted by the symbol $N$ and is defined as follows:

$$N = \prod_{p \mid D} p^{e(p)}$$

where for $p$ unequal to 2 or 3, $e(p) = 1$ if the singularity is a node, curve with two distinct lines at the singular point;

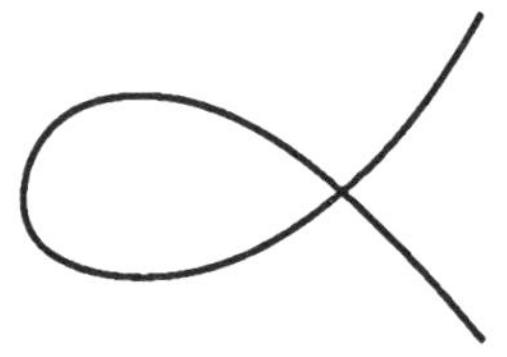

while $e(p) = 2$ if the singularity is a cusp, curve with one tangent at the singular point;

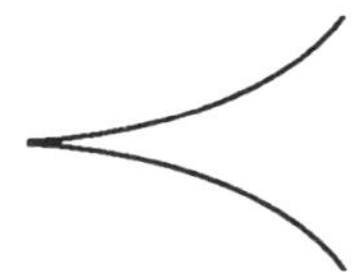

and in the remaining cases of $p = 2,3$, $e(p)$ is absolutely bounded. An elliptic curve is said to be semi–stable if it never has bad reduction of cuspidal type, and in this case $N$ is always the square–free part of $D$.

In a remarkable series of papers [2], [3], G. Frey constructed minimal semi–stable elliptic curves over $\mathbb{Q}$. Let me briefly describe Frey's construction. Let $A, B, C \in \mathbb{Z}$ with $A \equiv 0 \pmod{32}$, $B \equiv 1 \pmod 4$, $(A, B) = 1$, and $A + B + C = 0$. Consider the elliptic curve

$$E_{A,B} : y^2 = x(x - A)(x + B).$$

A normal Weierstrass form for $E$ is given by

$$E_{A,B} : y^2 = x^3 - \alpha x + \beta \tag{1}$$

where we have

$$\alpha = \frac{1}{3}(A^2 + B^2 + AB), \qquad \beta = \frac{1}{27}(A + B)(2A^2 + 2B^2 + 5AB)$$

and $\alpha, \beta \in \mathbb{Z}$ if and only if $A \equiv B \pmod 3$. Frey shows that this curve is semi–stable. Moreover, in the case $A \equiv B \pmod 3$, since $(\alpha, \beta) = 1$, $E_{A,B}$ is in minimal Weierstrass form with discriminant $A^2B^2C^2$. On the other hand, if $A \not\equiv B \pmod 3$, then the simple transformation $x \to \frac{1}{9}x$, $y \to \frac{1}{27}y$, gives a minimal Weierstrass normal form with discriminant $3^{12}A^2B^2C^2$. Note that our definition of minimal Weierstrass normal form is different from the usual notion of minimal model over $\mathbb{Z}$. Frey shows that a minimal model for $E_{A,B}$ over $\mathbb{Z}$ is given by the curve

$$y^2 + xy = x^3 + \frac{A - B - 1}{4}x^2 - \frac{AB}{16}x$$

with minimal discriminant $\frac{A^2B^2C^2}{256}$.

A surprisingly novel idea of Frey is to suggest that if the Fermat equation

$$u^p + v^p + w^p = 0$$

has nontrivial solution in rational integers $u, v, w$ for $p > 2$ then the elliptic curve (1) with $A = u^p$, $B = v^p$, $C = w^p$ cannot exist as a minimal Weierstrass model. Using this approach and earlier work of Mazur [10], and Serre [14], [15], Ribet [13] has recently shown that Fermat's last theorem would follow from the conjecture of Taniyama and Weil which is described in the next section. I shall not discuss Ribet's theorem in this article, but focus instead on another approach of Frey [3] based on a conjecture of Szpiro [17], [18], (1983).

Let

$$E : y^2 = x^3 - ax - b$$

be an elliptic curve with $a, b \in \mathbb{Z}$, $D$ nonzero, in minimal Weierstrass form. Let $N$ be the conductor of $E$.

**Conjecture 1.** *(Szpiro) There exists an absolute constant* $\kappa$ *(independent of N, D) such that*

$$D \leq N^\kappa.$$

A stronger form of this conjecture states that if $E$ is also semi–stable then:

**Conjecture 2.** *(Szpiro) For every* $\epsilon > 0$ *there exists a constant* $c(\epsilon)$ *depending only on* $\epsilon$ *such that*

$$D \leq c(\epsilon) N^{6+\epsilon}.$$

Applying this to the Frey curve (1), for example, yields the inequality

$$p^{6+\epsilon},$$

and this proves Fermat's last theorem for all sufficiently large exponents $p$. On the basis of the above example, Masser and Osterlé [12] (1985) conjectured the following.

**Conjecture 3.** *For rational integers* $A, B, C$ *with* $A + B + C = 0$

$$\sup(|A|,|B|,|C|) << \prod_{p \mid ABC} p^{1+\epsilon},$$

*where the* $\ll$ *– constant depends at most on* $\epsilon > 0$.

In fact, Conjecture 3 with $\sup(|A|,|B|,|C|)$ replaced by $|ABC|^{\frac{1}{3}}$ follows from Conjecture 2. We also remark that Conjecture 1 should hold over any number field with a constant $\kappa$ depending at most on the field. Recently, Hindry and Silverman [6] showed that Lang's conjecture on the lower bound for the height of non–torsion points on an elliptic curve defined over a number field follows from Conjecture 1, and more recently, Frey [4], under the assumption of Conjecture 1 gave a bound for the order of a torsion point on an elliptic curve defined over a number field. If Szpiro's conjecture is proven, this would generalize an unconditional result of Mazur [9] which says that a torsion point on an elliptic curve defined over $\mathbb{Q}$ can be of order at most twelve.

## 2. The Conjecture of Taniyama and Weil

We now consider the elliptic curve

$$E : y^2 = 4x^3 - ax - b \tag{2}$$

where for simplicity we assume that $4x^3 - ax - b = 4(x - e_1)(x - e_2)(x - e_3)$ and the three roots $e_1 < e_2 < e_3$ are real. The periods of $E$ (denoted $\Omega_1, \Omega_2$ ) are defined by the integrals

$$\Omega_1 = 2\int_{e_3}^{+\infty} \frac{dx}{\sqrt{4x^3 - ax - b}}$$

$$\Omega_2 = 2\int_{e_2}^{e_3} \frac{dx}{\sqrt{4x^3 - ax - b}}$$

where $\Omega_1$ is real and $\Omega_2$ is pure imaginary. Let $D = a^3 - 27b^2$ be the discriminant of $E$. It is well known that $E$ can be parametrized by doubly periodic functions

$$x = \wp(z)$$

$$y = \wp'(z)$$

where

$$\wp'(z) = -2 \sum_{m,n \in \mathbb{Z}} \frac{1}{(z + m\Omega_1 + n\Omega_2)^3},$$

and this is just the generalization of the well known parametrization of the circle $x^2 + y^2 = 1$ by the trigonometric functions $x = \cos z$ , $y = \sin z$ .

The Taniyama–Weil conjecture in its simplest form states that every elliptic curve $E$ defined over $\mathbb{Q}$, in minimal form and with conductor $N$, can be parametrized by modular functions for the group (see [11])

$$\Gamma_o(N) = \{\begin{pmatrix} a & b \\ c & d \end{pmatrix} \mid a, b, c, d \in \mathbb{Z}, \quad ad - bc = 1, \quad c \equiv 0 (\mathrm{mod}\, N)\}.$$

That is to say there exist meromorphic functions $\alpha(z)$, $\beta(z)$ with $z$ in the upper half plane satisfying

$$\alpha\left(\frac{az+b}{cz+d}\right) = \alpha(z)$$

$$\beta\left(\frac{az+b}{cz+d}\right) = \beta(z)$$

for all $\begin{pmatrix} a & b \\ c & d \end{pmatrix} \in \Gamma_o(N)$. Moreover, the curve

$$y^2 = 4x^3 - ax - b$$

can be parametrized by

$$x = \alpha(z)$$

$$y = \beta(z).$$

We shall now explicitly construct $\alpha(z)$, $\beta(z)$, assuming they exist.

Let

$$f(z) = \sum_1^\infty a(n) e^{2\pi i n z}$$

be a cusp of weight 2 for $\Gamma_o(N)$ so that

$$f\left(\frac{az+b}{cz+d}\right) = (cz+d)^2 f(z).$$

We assume that $f$ is normalized so that $a(1) = 1$, $a(n) \in \mathbb{Z}$ for $n \leq 1$, and that

$$a(mn) = a(m)a(n)$$

for $(m,n) = 1$.

Let $X_o(N)$ be the modular curve of the compactified Riemann surface obtained from factoring the upper half plane by $\Gamma_o(N)$. By a theorem of Shimura [16], there exists an elliptic curve $E$ which we may take to be (2) and a covering map $\phi$, normalized so that $\phi(i\infty) = 0$,

$$\begin{array}{c} X_o(N) \\ \big\downarrow \phi \\ E \end{array}$$

so that $f(z)dz$ is the pullback under $\phi$ of a differential one–form on $E$.

Let

$$F(\tau) = -2\pi i \int_{\tau}^{i\infty} f(z)dz$$
$$= \sum_{1}^{\infty} \frac{a(n)}{n} e^{2\pi i n \tau}$$

be the antiderivative of $f$. For $\begin{pmatrix} a & b \\ c & d \end{pmatrix} \in \Gamma_o(N)$ let us consider the Shimura map

$$\begin{pmatrix} a & b \\ c & d \end{pmatrix} \mapsto F\left(\frac{a\tau + b}{c\tau + d}\right) - F(\tau). \tag{3}$$

By the fundamental theorem of calculus

$$\frac{\partial}{\partial \tau}\{F\left(\frac{a\tau + b}{c\tau + d}\right) - F(\tau)\} = 0,$$

so the right side of (3) is independent of $\tau$. We now define

$$H\left(\begin{pmatrix} a & b \\ c & d \end{pmatrix}\right) = F\left(\frac{a\tau + b}{c\tau + d}\right) - F(\tau)$$

to be the Shimura map.

Since for $\alpha_1, \alpha_2 \in \Gamma_o(N)$ we have

$$H(\alpha_1 \alpha_2) = F(\alpha_1(\alpha_2 \tau)) - F(\alpha_2 \tau) + F(\alpha_2 \tau) - F(\tau)$$
$$= H(\alpha_1) + H(\alpha_2)$$

we see that $H$ is a homomorphism of $\Gamma_o(N)$. In fact if the pullback $\phi^*(f(z)dz)$ is the standard differential on $E$ then

$$H(\alpha) = 2\pi i \int_{\tau}^{\alpha\tau} f(z)dz$$

must lie in the homology of $X_o(N)$ and hence in the homology of $E$. It follows that $H$ is a homomorphism from $\Gamma_o(N)$ onto the lattice

$$\Lambda = \{m\Omega_1 + n\Omega_2 \mid m, n \in \mathbb{Z}\}$$

of periods of $E$ which is just the abelian group of rank 2 isomorphic to $\mathbb{Z}\times\mathbb{Z}$.

We can now give the desired parametrization of $E: y^2 = 4x^3 - ax - b$. Let us define

$$\alpha(z) = \wp(F(z)) = \wp\left(\sum_{n=1}^{\infty}\frac{a(n)}{n}e^{2\pi inz}\right)$$

$$\beta(z) = \wp'(F(z)) = \wp'\left(\sum_{n=1}^{\infty}\frac{a(n)}{n}e^{2\pi inz}\right),$$

where $\wp$ is the Weierstrass $\wp$-function. We have

$$\begin{aligned}\alpha\left(\frac{az+b}{cz+d}\right) &= \wp\left(F\left(\frac{az+b}{cz+d}\right)\right)\\ &= \wp\left(F(z) + H\left(\begin{pmatrix} a & b\\ c & d\end{pmatrix}\right)\right)\\ &= \wp(F(z))\\ &= \alpha(z)\end{aligned}$$

since $H\left(\begin{pmatrix} a & b\\ c & d\end{pmatrix}\right) \in \Lambda$. Similarly for $\beta(z)$.

## 3. Properties of Shimura Maps

The Shimura map $H: \Gamma_o(N) \to \Lambda$ as defined in the previous section satisfies the following properties:

**Property 1.** *$H$ is a homomorphism from $\Gamma_o(N)$ on to the period lattice $\Lambda$ of the elliptic curve $E$.*

**Property 2.** *For $\begin{pmatrix} a & b\\ c & d\end{pmatrix} \in \Gamma_o(N)$, we have $H\left(\begin{pmatrix} a & -b\\ -c & d\end{pmatrix}\right) = \overline{H\left(\begin{pmatrix} a & b\\ c & d\end{pmatrix}\right)}$.*

**Proof.** Let $\sigma = \begin{pmatrix} i & 0\\ 0 & -i\end{pmatrix}$ with $i = \sqrt{-1}$. Then we have

$$\sigma\begin{pmatrix} a & b\\ c & d\end{pmatrix}\sigma^{-1} = \begin{pmatrix} ai & bi\\ -ci & -di\end{pmatrix}\begin{pmatrix} -i & 0\\ 0 & i\end{pmatrix} = \begin{pmatrix} a & -b\\ -c & d\end{pmatrix}.$$

Since the Fourier coefficients of $f$ are real it follows that

$$F(\sigma\bar{z}) = F(\sigma^{-1}\bar{z}) = F(-\bar{z}) = \overline{F(z)}.$$

Hence, replacing $\tau$ by $\sigma\bar{\tau}$, we have

$$H\left(\begin{pmatrix} a & -b \\ -c & d \end{pmatrix}\right) = F\left(\sigma\begin{pmatrix} a & b \\ c & d \end{pmatrix}\sigma^{-1}\tau\right) - F(\tau)$$
$$= F\left(\sigma\begin{pmatrix} a & b \\ c & d \end{pmatrix}\bar{\tau}\right) - F(\sigma\bar{\tau})$$
$$= \overline{H\left(\begin{pmatrix} a & b \\ c & d \end{pmatrix}\right)}.$$

**Property 3.** *For each positive square-free integer* $N$, *there exists* $\epsilon_N = \pm 1$ *such that for all* $\begin{pmatrix} a & b \\ c & d \end{pmatrix} \in \Gamma_o(N)$, *we have*

$$H\left(\begin{pmatrix} d & -\frac{c}{N} \\ -bN & a \end{pmatrix}\right) = \epsilon_N\, H\left(\begin{pmatrix} a & b \\ c & d \end{pmatrix}\right).$$

**Proof.** Let $\omega = \begin{pmatrix} 0 & \frac{1}{\sqrt{N}} \\ -\sqrt{N} & 0 \end{pmatrix}$ so that

$$\omega\begin{pmatrix} a & b \\ c & d \end{pmatrix}\omega^{-1} = \begin{pmatrix} d & -\frac{c}{N} \\ -bN & a \end{pmatrix}.$$

It follows that

$$H\left(\begin{pmatrix} d & -\frac{c}{N} \\ bN & a \end{pmatrix}\right) = H\left(\omega\begin{pmatrix} a & b \\ c & d \end{pmatrix}\omega^{-1}\right)$$
$$= F\left(\omega\begin{pmatrix} a & b \\ c & d \end{pmatrix}\omega^{-1}\tau\right) - F(\tau)$$
$$= L + M + N$$

where

$$L = F\left(\omega\begin{pmatrix} a & b \\ c & d \end{pmatrix}\omega^{-1}\tau\right) - \epsilon_N\, F\left(\begin{pmatrix} a & b \\ c & d \end{pmatrix}\omega^{-1}\tau\right)$$

$$M = \epsilon_N\, F\left(\begin{pmatrix} a & b \\ c & d \end{pmatrix}\omega^{-1}\tau\right) - \epsilon_N\, F(\omega^{-1}\tau)$$

$$N = \epsilon_N\, F(\omega^{-1}\tau) - F(\tau).$$

By the functional equation $F(\tau) = \epsilon_N\, F(\omega\tau)$, we have $L = 0$, and $N = 0$. The result follows.

**Property 4.** *Let* $\sigma_p = \begin{pmatrix} p & 0 \\ 0 & 1 \end{pmatrix}$ *and* $\sigma_j = \begin{pmatrix} 1 & j \\ 0 & p \end{pmatrix}$ *for* $j = 0, 1, \ldots, (p-1)$. *Assume that* $\begin{pmatrix} a & b \\ c & d \end{pmatrix}$, $\sigma_k \begin{pmatrix} a & b \\ c & d \end{pmatrix} \sigma_k^{-1} \in \Gamma_o(N)$ *for* $k = 0, 1, \ldots, p$. *(this will be the case if* $p|b$, $p|c$, *and* $p|(d-a)$.*) Then for* $p$ *a rational prime not dividing* $N$ *we have*

$$\sum_{k=1}^{p} H\left(\sigma_k \begin{pmatrix} a & b \\ c & d \end{pmatrix} \sigma_k^{-1}\right) = a(p) H\left(\begin{pmatrix} a & b \\ c & d \end{pmatrix}\right)$$

*where* $a(p)$ *is the* $p$th *Fourier coefficient of* $f(z)$.

**Proof.** We make use of the properties of the Hecke operator

$$T_p = \sum_{k=0}^{p} \sigma_k$$

and the fact that the differential one form $f(z)dz$ is an eigenfunction of $T_p$ with eigenvalue $a(p)$

$$T_p(f(z)dz) = a(p) f(z)dz.$$

From the definition of $H$ we see that

$$\sum_{k=0}^{p} H\left(\sigma_k \begin{pmatrix} a & b \\ c & d \end{pmatrix} \sigma_k^{-1}\right) = \sum_{k=0}^{p} \left[ \int_{\sigma_k \alpha\tau_o}^{i\infty} f(z)dz - \int_{\sigma_k \tau_o}^{i\infty} f(z)dz \right]$$

after putting $\alpha = \begin{pmatrix} a & b \\ c & d \end{pmatrix}$, and $\tau_o = \sigma_k^{-1}\tau$. It follows that

$$\begin{aligned} \sum_{k=0}^{p} H\left(\sigma_k \begin{pmatrix} a & b \\ c & d \end{pmatrix} \sigma_k^{-1}\right) &= \left( \int_{\alpha\tau_o}^{i\infty} - \int_{\tau_o}^{i\infty} \right) \sum_{k=0}^{p} f(\sigma_k z) d(\sigma_k z) \\ &= a(p) \left( \int_{\alpha\tau_o}^{i\infty} - \int_{\tau_o}^{i\infty} \right) f(z)dz \\ &= a(p) H(\alpha) \end{aligned}$$

by the properties of the $p$th Hecke operator.

**Property 5.** $H\left(\begin{pmatrix} 1 & 1 \\ 0 & 1 \end{pmatrix}\right) = 0.$

**Proof.** By definition $H\left(\begin{pmatrix} 1 & 1 \\ 0 & 1 \end{pmatrix}\right) = F(\tau + 1) - F(\tau)$. Since

$$F(\tau) = \sum_{n=1}^{\infty} \frac{a(n)}{n} e^{2\pi i n \tau}$$

which is periodic in $\tau$ we easily see that $F(\tau + 1) = F(\tau)$.

## 4. Equivalent Forms of Szpiro's Conjecture

We now give some equivalent forms of Szpiro's Conjecture 2. Let

$$\Delta(z) = e^{2\pi i z} \prod_{n=1}^{\infty} \left(1 - e^{2\pi i n z}\right)^{24}$$

be the Ramanujan cusp of weight twelve for the full modular group. Then since the discriminant $D$ of the elliptic curve

$$E : y^2 = x^3 - ax - b$$

can be expressed

$$D = \frac{\Delta\left(-\dfrac{\Omega_1}{\Omega_2}\right)}{2\pi^{12} \Omega_2^{12}}$$

and, without loss of generality we may assume that $\left|\frac{\Omega_1}{\Omega_2}\right| > 1$, we see that $\left|\Delta\left(-\frac{\Omega_1}{\Omega_2}\right)\right|$ is absolutely bounded from above by some fixed constant $c > 0$. It follows that we have $D < \frac{c}{\Omega_2^{12}}$. Hence, a lower bound of type

$$\Omega_2 \gg \frac{1}{N^{\kappa}} \tag{4}$$

for some fixed constant $\kappa > 0$ would give Szpiro's conjecture.

Now, since $e_1, e_2, e_3$, are roots of $4x^3 - ax - b = 0$ with $a, b$ integers and $y^2 = 4x^3 - ax - b$ is a Frey curve, we see that $|e_i - e_j| \gg 1$ for $1 \le i < j \le 3$. Consequently the discriminant $D$ satisfies $D \gg |e_i - e_j|^2$ for $i \ne j$. Hence

$$\Omega_2 = 2\int_{e_2}^{e_3} \frac{dx}{\sqrt{4(x-e_1)(x-e_2)(x-e_3)}}$$

$$\geq \frac{1}{\sqrt{e_3 - e_1\,(e_3 - e_2)}} \int_{e_2}^{e_3} dx$$

$$\geq \frac{1}{\sqrt{e_3 - e_1}}$$

$$\gg D^{-\frac{1}{4}} .$$

So if Szpiro's conjecture is true, this yields a lower bound of type (4). Similarly for $\Omega_1$. It follows that Szpiro's conjecture is equivalent to lower bounds of type (4) for the periods of $E$.

If we assume the conjecture of Taniyama and Weil, then certain properties of the Shimura map $H : \Gamma_o(N) \to \Lambda$ as defined in section 3 can be shown to be equivalent to Szpiro's conjecture. We have the following conjecture.

**Conjecture 4.** *Let $N \to \infty$. There exists a fixed constant $\kappa > 0$ such that if $\begin{pmatrix} a & b \\ c & d \end{pmatrix} \in \Gamma_o(N)$ with $|a|, |b|, |c|, |d| \leq N^2$ then*

$$H\left(\begin{pmatrix} a & b \\ c & d \end{pmatrix}\right) = m\Omega_1 + n\Omega_2$$

*with*

$$|m|, |n| \ll N^{\kappa}.$$

Assuming the Taniyama–Weil conjecture, it can be shown that Conjecture 4 is equivalent to Szpiro's Conjecture 1. Moreover, the assumption that $|a|, |b|, |c|, |d| \leq N^2$ can be replaced by the simpler assumption that $|c| \leq N^2$. This is because $\begin{pmatrix} 1 & 1 \\ 0 & 1 \end{pmatrix}$ is in the kernel of $H$ which implies that we can always arrange $|a|, |b|, |d| \leq |c|$ after a suitable left or right multiplication by upper triangular matrices in $\Gamma_o(N)$. On the basis of numerical evidence, however, it seems we may take $\kappa$ in Conjecture 4 arbitrarily small as $(N \to \infty)$ if we restrict ourselves to matrices $\begin{pmatrix} a & b \\ c & d \end{pmatrix}$ (satisfying $|c| \leq N^2$) which form a minimal set of generators for $\Gamma_o(N)$, but this seems hopelessly difficult to prove at the present time.

To prove Szpiro's conjecture, it suffices to assume the existence of a homomorphism $H : \Gamma_o(N) \to \Lambda$, satisfying Properties 1 to 5, and in addition satisfying Conjecture 4. In this context, Conjecture 4 is a conjecture concerning a group homomorphism between a non–abelian group of rank $\approx \frac{N}{6}$, (namely, $\Gamma_o(N)$), and a free abelian group of rank 2. A matrix $\begin{pmatrix} a & b \\ c & d \end{pmatrix} \in \Gamma_o(N)$ will be termed close to the identity if $|a|, |b|, |c|, |d|$ are small. Conjecture 4 says that if $\begin{pmatrix} a & b \\ c & d \end{pmatrix}$ is close to the identity, then its image under $H$ is close to the origin in the lattice $\Lambda$, (implying that $H$ has properties analogous to a continuous function). A proof of Conjecture 4 should make strong use of Property 5 (Hecke operators).

We now give a sketch of the proof of the equivalence of Conjectures 1 and 4. Let

$$f(z) = \sum_{1}^{\infty} a(n) e^{2\pi i n z}$$

be the normalized Hecke newform of weight 2 associated to $E$. Then we have for $\alpha \in \Gamma_o(N)$

$$H(\alpha) = \sum_{1}^{\infty} \frac{a(n)}{n} \left[ e^{2\pi i n \alpha(\tau)} - e^{2\pi i n \tau} \right],$$

which is independent of $\tau$ in the upper half of the plane. If we define

$$L_f(s, \theta) = \sum_{1}^{\infty} \frac{a(n)}{n^s} e^{2\pi i n \theta}$$

and

$$H_s(\alpha) = \sum_{1}^{\infty} \frac{a(n)}{n^{1+s}} \left[ e^{2\pi i n \alpha(\tau)} - e^{2\pi i n \tau} \right],$$

then letting $\tau \to i\infty$ and $s \to 0$ it follows that

$$H(\alpha) = H_o(\alpha, i\infty) = L_f\left(1, \frac{a}{c}\right). \tag{5}$$

To obtain further information about $H(\alpha)$, we need the functional equation of $L_f\left(s, \frac{a}{c}\right)$. This is obtained as follows. Let us put $z = -\frac{d}{c} + iy$, and $\alpha = \begin{pmatrix} a & b \\ c & d \end{pmatrix}$. Then we have $\alpha(z) = \frac{a}{c} + \frac{i}{c^2 y}$. It follows that

$$
\begin{aligned}
c^s\Gamma(s)L_f\left(s,-\tfrac{d}{c}\right) &= \int_0^\infty f\left(-\tfrac{d}{c}+iy\right)(cy)^s\frac{dy}{y} \\
&= \int_0^{\frac{1}{c}} f(z)(cy)^s\frac{dy}{y} + \int_{\frac{1}{c}}^\infty f(z)(cy)^s\frac{dy}{y} \\
&= \int_1^\infty f\left(\frac{a+iy}{c}\right)y^{2-s}\frac{dy}{y} + \int_1^\infty f\left(\frac{-d+iy}{c}\right)y^s\frac{dy}{y}
\end{aligned}
$$

which gives the functional equation

$$
c^s\Gamma(s)L_f\left(s,-\tfrac{d}{c}\right) = c^{2-s}\Gamma(2-s)L_f\left(2-s,\tfrac{a}{c}\right)
$$

where $ad \equiv 1(\mathrm{mod}\ c)$. The usual convexity argument then gives

$$
\left|L_f\left(1,\tfrac{a}{c}\right)\right| \ll c^{\frac{1}{2}+\epsilon}. \tag{6}
$$

For $\begin{pmatrix} a & b \\ c & d \end{pmatrix} \in \Gamma_o(N)$ with $|c| < N^2$, let us choose

$$
\alpha = \begin{pmatrix} a & b \\ c & d \end{pmatrix}\begin{pmatrix} a & -b \\ -c & d \end{pmatrix}.
$$

If $H\left(\begin{pmatrix} a & b \\ c & d \end{pmatrix}\right) = m\Omega_1 + n\Omega_2$, then by Property 2, we have that

$$
H(\alpha) = 2n\,\Omega_2.
$$

It then follows from this and equations (5), (6) that

$$
|n\Omega_2| \ll N^{2+\epsilon}. \tag{7}
$$

But if Szpiro's conjecture is true, then

$$
|n\Omega_2| \gg \frac{1}{N^\kappa}
$$

for some $\kappa > 0$. The inequality (6) yields

$$
|n| \ll N^{\kappa+2+\epsilon}.
$$

A similar argument also works for the $m-$component of $H$. So we have shown that Szpiro's conjecture implies Conjecture 4.

To show that Conjecture 4 implies Conjecture 1 is more difficult. Let us define $\chi : \mathbb{Z}/q\mathbb{Z} \to \{\pm 1\}$ to be a real primitive Dirichlet character (mod $q$). Consider the twisted $L$-series

$$L_f(s,\chi) = \sum_{1}^{\infty} \frac{a(n)\chi(n)}{n^s}.$$

If

$$G(\chi) = \sum_{a=1}^{q} \chi(a) e^{2\pi i \frac{a}{q}}$$

denotes the Gauss sum, then by the standard argument

$$G(\chi) L_f(s,\chi) = \sum_{b=1}^{q} \chi(b) L_f(s, \frac{b}{q}).$$

For any two integers $b, q$ satisfying $(q, N) = 1$, $0 < b < q$, and $(b, q) = 1$ we can always choose suitable integers $a, c$ so that $\gamma = \begin{pmatrix} a & b \\ c & q \end{pmatrix}$ lies in $\Gamma_o(N)$. We then have

$$\sum_{b=1}^{q} \chi(b) H_s(\gamma, \tau) = \sum_{b=1}^{q} \chi(b) \sum_{1}^{\infty} \frac{a(n)}{n^{1+s}} \left[ e^{2\pi i n \gamma(\tau)} - e^{2\pi i n \tau} \right].$$

Letting $\tau \to 0$ and $s \to 0$, yields

$$\sum_{b=1}^{q} \chi(b) H(\gamma) = \sum_{b=1}^{q} \chi(b) H_o(\gamma, 0) = \sum_{b=1}^{q} \chi(b) L_f(1, \frac{b}{q})$$

since

$$\sum_{b=1}^{q} \chi(b) = 0.$$

It follows that

$$\sum_{b=1}^{q} \chi(b) H\left(\begin{pmatrix} a & b \\ c & q \end{pmatrix}\right) = G(\chi) L_f(1, \chi). \tag{8}$$

If $\chi(-1) = -1$, so that $\chi$ is an odd character, then the substitution $b \mapsto -b$, $c \mapsto -c$ does not change the value of the left side of equation (8) since we can sum over any set of residues (mod $q$). But by Property 2 of the homomorphism $H$ this implies that $G(\chi)L_f(1,\chi)$ must be pure imaginary, and hence must be an integral multiple of the imaginary period $\Omega_2$.

Now, by a theorem of Waldspurger [19], (see Kohnen [7]) it follows that $L_f(1,\chi)$ is the square of a Fourier coefficient of a cusp form of weight $\frac{3}{2}$. Applying the Rankin–Selberg method, as in Kohnen and Zagier's proof [8] of the Goldfeld–Viola conjecture [5] on mean values of $L_f(1,\chi)$ one obtains

$$\sum_{q \ll N^2} L_f(1,\chi) \sim N^2.$$

Since $|G(\chi)| = \sqrt{q}$, it follows that for some twist $\chi$ with conductor $q \ll N^2$

$$G(\chi)L_f(1,\chi) \gg N.$$

Consequently, if we assume Conjecture 4, there is an integer $m$ satisfying $m \ll N^{2+\kappa}$ for some fixed $\kappa > 0$ such that

$$m\Omega_2 \gg N.$$

We then obtain that

$$\Omega_2 \gg N^{-1-\kappa},$$

and as shown earlier, this implies Conjecture 1.

In conclusion, I should like to focus on yet another equivalence to Szpiro's Conjecture 2. Kohnen [7] has shown that associated to a normalized newform $f(z)$ of weight 2 for $\Gamma_o(N)$ there is a cusp form $g(z)$ of weight $\frac{3}{2}$ for $\Gamma_o(4N)$ whose $q$th Fourier coefficient $c(q)$ is given by

$$\frac{c(q)^2}{\langle g, g\rangle} = \frac{2^{\upsilon(N)}\sqrt{q}\,L_f(1,\chi)}{\pi\langle f, f\rangle} \tag{9}$$

where $\upsilon(N)$ denotes the number of prime factors of $N$ and for cusp forms $f_1, f_2$ of weight $k \in \frac{1}{2}\mathbb{Z}$ for $\Gamma = \Gamma_o(M)$

$$\langle f_1, f_2\rangle = \frac{1}{[\Gamma_o(1) : \Gamma]} \int_{\Gamma\backslash H} f_1(z)\overline{f_2(z)}\, y^k \frac{dx\, dy}{y^2}$$

denotes the Peterson inner product (here $H$ is the upper half plane).

Clearly, the left hand side of (9) is independent of the normalization of $g$. Let us normalize $g$ so that $c(q) \in \mathbb{Z}$ for all $q$ and

$$G(\chi)L_f(1,\chi) = c(q)^2 \Omega_2.$$

Szpiro's conjecture is then equivalent to the bound

$$\langle g, g \rangle \ll N^c$$

for some fixed constant $c > 0$. This follows easily from (9) by the estimate

$$\frac{1}{[\Gamma_o(1) : \Gamma_o(N)]} \ll \langle f, f \rangle \ll 1. \tag{10}$$

To prove (10) note that

$$\begin{aligned}
\int_{\Gamma_o(N)\backslash H} |f(z)|^2 dx\, dy &\geq \int_1^\infty \int_0^1 |f(z)|^2 dx\, dy \\
&= \int_1^\infty \sum_1^\infty |a(n)|^2 e^{-4\pi n y}\, dy \\
&= \sum_1^\infty \frac{a(n)^2 e^{-4\pi n}}{4\pi n} \\
&\gg 1
\end{aligned}$$

since $a(1) = 1$.

On the other hand, if we let $d(n)$ denote the number of divisors of an integer $n$, then the Fourier coefficients of $f$ at an arbitrary cusp (see [1]) are bounded by $\sqrt{n}\, d(n)$. It follows that

$$\int_{\Gamma_0(N)\backslash H} |f(z)|^2 dx\, dy = \sum_{\gamma \in \Gamma_0(N)\backslash\Gamma_0(1)} \int_{\Gamma_0(1)\backslash H} |f(\gamma z)|^2 \operatorname{Im}(\gamma z)^2 \frac{dx\, dy}{y^2}$$

$$\ll \sum_{\gamma} \int_{\frac{\sqrt{3}}{2}}^{\infty} \int_0^1 |f(\gamma z)|^2 \operatorname{Im}(\gamma z)^2 \frac{dx\, dy}{y^2}$$

$$\ll [\Gamma_0(1) : \Gamma_0(N)] \sum_{n=1}^{\infty} n d(n)^2 e^{-2\pi\sqrt{3}\, n}$$

$$\ll [\Gamma_0(1) : \Gamma_0(N)]$$

We have seen that for an integral weight modular form $f$ with relatively prime rational integer Fourier coefficients, it is possible to give an absolute bound for $\langle f, f\rangle$ which is independent of the level. This is due to the fact that the $n$th Fourier coefficient $a(n)$ is bounded by $\sqrt{(n)}\, d(n)$. If we knew that $|a(n)| \le Cn^{\theta}$ for constants $C$, $\theta$ independent of $n$ (but possibly $C$ depending on $N$) then by the properties of the Hecke operators we would have

$$a(p) \approx 2a(p^M)^{\frac{1}{M}}$$

for rational primes $p$. Letting $M \to \infty$, it follows by a simple argument that $|a(n)| \le d(n) n^{\theta}$; and in effect, the constant $C$ drops out of the picture. In the half integral weight case, however, this does not happen because there are not enough Hecke operators.

## References

[1] *P. Deligne* , La conjecture de Weil, I. Publ. Math. IHES, **43** (1974).273—308.

[2] *G. Frey* , Rationale Punkte auf Fermatkurven und getwistete Modulkurven. J. Reine. Angew. Math. **331** (1982), 185—191.

[3] *G. Frey* , Links between stable elliptic curves and certain diophantine equations. Annales Universiatis Saraviensis, Vol 1, No. 1 (1986), 1—39.

[4] *G. Frey* , Links between elliptic curves and the solutions of the equation A - B = C. Saarebrüken RFA, (preprint).

[5] *D. Goldfeld* and *C. Viola* , Mean values of L–functions associated to elliptic, Fermat and other curves at the centre of the critical strip. J. Number Theory **11** (1979), 305—320.

[6] *M. Hindry* and *J. Silverman*, The canonical height and elliptic curves. (Preprint) 1987.

[7] *W. Kohnen*, Fourier coefficients of modular forms of half–integral weight. Math. Ann. **271** (1985), 237—268.

[8] *W. Kohnen* and *D. Zagier*, Values of $L-$series of modular forms at the centre of the critical strip. Invent. Math. **64** (1981), 175—198.

[9] *B. Mazur*, Modular curves and the Eisenstein ideal. Inst. Hautes Études Sci. Publ. Math. **47** (1977), 33—186.

[10] *B. Mazur*, Letter to J–F. Mestre.

[11] *B. Mazur* and *P. Swinnerton–Dyer*, Arithmetic of Weil curves. Invent. Math. **25** (1974), 1—61.

[12] *J. Osterlé*, Nouvelles approaches du Théorème Fermat. Sem. Bourbaki, no. 694 (1987–88), 694–01 — 694–21.

[13] *K. Ribet*, Lectures on Serre's conjectures. MSRI preprint (1987).

[14] *J–P. Serre*, Lettre á J–F. Mestre (13 Août 1985). (To appear in Current trends in arithmetic.)

[15] *J–P. Serre*, Sur les representations modulaires de degré 2 de $\mathrm{Gal}(\bar{Q}/Q)$. Duke Math. J. **54** (1987), 179—230.

[16] *G. Shimura*, On the factors of the jacobian variety for a modular function field *J.* Math. Soc. Japan **25** (1973), 523—544.

[17] *L. Szpiro*, Seminaire sur les pinceaux de courbes de genre au moins deux. Astérisque, exposé no. 3, **86** (1981), 44—78.

[18] *L. Szpiro*, Présentation de la théorie d'Arakélov. Contemp. Math. **67** (1987), 279—293.

[19] *J–L. Waldspurger*, Sur les coefficients de Fourier des formes modulaires de poides demi–entier. J. Math. Pures Appl. **60** (1981), 375—484.

---

Dept. of Mathematics, Columbia University, New York City, NY 10027, USA.

# Some Conjectures Related to Fermat's Last Theorem

*Andrew Granville*

**Dedicated to Harry Bennett on the occasion of his 90th birthday.**

## 1. Introduction

In about 1637, Fermat claimed to have proved that for all integers $n \geq 3$, there do not exist integers $x$, $y$ and $z$ that satisfy

$$x^n + y^n = z^n \quad \text{and} \quad xyz \neq 0. \tag{1}_n$$

This still unproved assertion, known as "Fermat's Last Theorem", has eluded the efforts of many great mathematicians (see Ribenboim's book [36] for an excellent introduction), although the many attacks to solve it have inspired much important mathematics.

The first partially successful approach was begun, in 1823, by Sophie Germain [24] who used local methods and ingenious combinatorial ideas to state results on the equation

$$x^n + y^n + z^n = 0 \quad \text{and} \quad \gcd(n, xyz) = 1 \tag{2}_n$$

for odd integers $n$. She proved that if $n$ and $2n + 1$ are both prime then $(2)_n$ has no solutions; this can be generalized as follows:

**Lemma 1.** ([17], [21]) *Suppose that $m$ is a given positive even integer and $S_m$ is the set of primes that divide some non–zero norm of the sum of three $m$-th roots of unity. Suppose further that $p$ and $q = mp + 1$ are both odd primes with $p$ not a divisor of $m$ and $q \notin S_m$. Then, for $n = p$ ($m$ not a multiple of 3), $p^2$ ($3|m$), we have that $(2)_n$ has no solutions in integers $x$, $y$ and $z$.*

In 1954, Ankeny and Erdös [2] used Sophie Germain's approach to show that $(2)_n$ has solutions for $o(N)$ exponents $n \leq N$. More recently, Adleman and Heath–Brown [1], used Lemma 1 together with Fouvry's work on the Brun–Titchmarsh theorem [8] to show that $(2)_p$ has no solutions for infinitely many primes $p$. In their paper they made a number of conjectures in analytic number theory that, if proved, would in conjunction with Lemma 1, prove successively stronger theorems on the First Case of Fermat's Last Theorem (i.e., the assertion that $(2)_n$ has no solutions with $n \geq 2$).

The second successful approach was started by Kummer [22], in 1847, who introduced the concept of divisors (or ideals) and showed that $(1)_p$ has no solutions for all "regular" primes $p$. In 1857 Kummer [23] examined $(2)_p$ in more detail and established that if $(2)_p$ does have solutions then a complicated set of $p-2$ congruence conditions involving Bernoulli numbers and Euler polynomials must be satisfied. It was not until 1909 that Wieferich [42] ingeniously derived the following result from Kummer's congruences.

**Lemma 2.** *If* $(2)_p$ *has solutions in integers* $x, y, z$ *then*

$$2^p \equiv 2 \pmod{p^2}.$$

The "Fermat quotient" is defined as $q_p(2) = \dfrac{2^{p-1} - 1}{p}$, so that the conclusion of Lemma 2 can be rewritten as "$p$ divides $q_p(2)$". The question of whether or not $p$ divides $q_p(2)$ (or, indeed $q_p(a)$, for any integer $a \geq 2$) seems to be particularly difficult, and in Section 2 we shall examine various related conjectures. For $a = 2$, our total knowledge is that, of the primes $p < 6 \cdot 10^9$, $p$ divides $q_p(2)$ only for $p = 1093$ and $p = 3511$ ([25]). If we assume that the "probability" that $p$ divides $q_p(2)$ is $1/p$ (note that $p$ divides exactly one of $q_p(2)$, $q_p(2+p), \ldots$, $q_p(2+p(p-1))$ as $q_p(2+kp) \equiv q_p(2) - k/2 \pmod{p}$) then, of the primes $p \leq x$, one expects that for about $\sum_{p \leq x} 1/p \sim c + \log\log x$ such primes we have that $p$ divides $q_p(2)$: As has been pointed out by a number of people at this conference, $\Sigma\, 1/p$ will always be less than 4 if we sum over all the primes that we know, ever will know, (or even wish to know), and so finding two primes for which $p$ divides $q_p(2)$ is not so bad!

In 1910, Mirimanoff [31] tidied up Wieferich's long and difficult proof and extended the result to:

**Lemma 3.** *If* $(2)_p$ *has solutions in integers x, y, z then*

$$3^p \equiv 3 \pmod{p^2}.$$

In rapid succession, a number of authors proved the next few criteria (i.e., $5^p \equiv 5 \pmod{p^2}$, $7^p \equiv 7 \pmod{p^2}$, etc.), but it was not until 1914 that Frobenius [10] made an attempt to give an "algorithm" to determine each successive criteria: However, his algorithm is difficult to implement and the paper contains numerous errors. In 1917 Pollaczek [34] gave different, mostly correct proofs to determine an algorithm that allowed him to prove:

**Lemma 4.** *If* $p$ *is sufficiently large and* $(2)_p$ *has solutions in integers x, y, z then* $q^p \equiv q \pmod{p^2}$, *for each prime* $q \leq 31$.

In 1931 Morishima [33], adding a few ideas to Frobenius's paper, claimed to have extended the criteria up to $q \leq 43$. However, the only proof he gave of this was to state that the computations can be done "*In analoger Weise*" to the way in which they were done up to 31. This is far from a proof, and far from a trivial assertion.

Gunderson, in his Ph.D. thesis under the supervision of Rosser [18], pointed out a number of technical errors in Morishima's paper. (Similar errors appear in the papers of both Frobenius [10] and Pollaczek [34]). Gunderson corrected these errors using some ingenious ideas. In 1988, using a more combinatorial and less algebraic approach, the author and Monagan [16] reproved the many technical theorems of Frobenius *et al.* We also proved a succession of stronger technical results and extended the above Lemmas to:

**Lemma 5.** *If* $(2)_p$ *has solutions in integers x, y, z then* $q^p \equiv q \pmod{p^2}$ *for each prime* $q \leq 89$.

We also made two conjectures that, if proved, would imply that the first case of Fermat's Last Theorem is true. We shall examine these and related conjectures in Section 3.

The third important approach is due to Furtwangler [11] who used local class field theory to derive criteria, again in terms of Fermat quotients. It seems, however, that there is little to add to this approach following the recent paper of Azuhata [3]. Using this approach Hellegouarch [20] showed that if $(2)_n$ has solutions, where $n = p^t$ ($p$ prime, $t \geq 1$) then $p^{2t}$ divides both $2^p - 2$ and $3^p - 3$. Using this I proved in my Ph.D. thesis [15]:

**Theorem 1.** *For any odd prime* $p$, $(2)_n$ *has no solutions when* $n = p^t$ *and* $t \geq p^{1/2}/\log p$.

This improves on the many previous bounds given for $t$ (e.g., $t \geq \phi(p-1) \log 3 / \log p$ —see [17]).

A number of recent approaches have come from the perspective of algebraic geometry; to wit, those of Faltings [6], Frey, Ribet and Serre (see [9]). It is not my intention to discuss these here except to state the important theorem of Faltings:

**Lemma 6.** *For any* $n \geq 3$, *there are only finitely many triples of integers* $(x, y, z)$ *that are coprime and satisfy* $(1)_n$.

Heath–Brown [19] and I [13] deduced from Lemma 6 that $(1)_n$ has solutions for only $o(x)$ exponents $n \leq x$.

Stewart and Tijdeman [40] observed that Lemma 6, together with the so–called "*abc* conjecture", implies that there are only finitely many solutions $(x, y, z, n)$ to $(1)_n$ with $\gcd(x, y, z) = 1$ and $n \geq 3$. We shall discuss this further in Section 2.

## 2. Fermat Quotient and Powerful Numbers

We start this section with the "trendy" conjecture of number theory, due to Oesterlé and Masser [29]:

**Conjecture 1.** (The "abc conjecture") *Suppose that* $a$, $b$ *and* $c$ *are positive integers satisfying*

$$a + b = c \tag{3}$$

with $$\gcd(a, b, c) = 1.$$

*Let* $G = G(a,b,c)$ *be the product of the primes dividing* $abc$, *each to the first power. For all* $\varepsilon > 0$, *there exists a constant* $k = k(\varepsilon)$ *such that* $c < kG^{1+\varepsilon}$.

(See de Weger [4] for some interesting computational information.)

Actually Oesterlé originally conjectured the existence of a constant $T$ for which $c < G^T$ and this was sharpened by Masser. Recently Stewart and Tijdeman [40] proved a result in this direction, which they tell me can now be sharpened to $c < \exp(kG^{1+\varepsilon})$. Note that if $x$ and $y$ are integers and $x + y\sqrt{d}$ is a unit of $\mathbb{Q}(\sqrt{d})$, for any squarefree $d \geq 2$, and $e + f\sqrt{d} = (x + y\sqrt{d})^{2d}$ then $e^2 - df^2 = 1$ where $d$ divides $f$; therefore $a = 1$, $b = df^2$, $c = e^2$ is a solution of (3) with $G(a,b,c) \leq ef \leq c/\sqrt{d}$, so that the exponent in Conjecture 1 certainly can't be improved.

Assume only Oesterlé's conjecture (i.e., $c < G^T$). Suppose that we have a solution x, y, z of $(1)_n$ with $n \geq 3T$. Let $a = x^n$, $b = y^n$, $c = z^n$ in (3) so that $G(a,b,c)^T \leq (xyz)^T < z^{3T} < z^n = c$, giving a contradiction. Thus $n < 3T$, and so, by Lemma 6, we have only finitely many quadruples $(x,y,z,n)$ satisfying $(1)_n$ with $\gcd(x,y,z) = 1$ and $n \geq 3$.

We shall later relate Conjecture 1 directly to Fermat quotients.

As we stated in the introduction, there is very little known about the "$p$-divisibility" of Fermat quotients. To illustrate this we state a number of conjectures (some of these are well-known though perhaps they have never all appeared together before). We shall suppose that $a$ is some fixed integer, with $a \geq 2$.

**Conjecture 2a.** *There is an odd prime* $p$ *which divides* $q_p(a)$.

**Conjecture 2b.** *There is an odd prime* $p$ *which doesn't divide* $q_p(a)$.

**Conjecture 3.** *There are infinitely many primes* $p$ *for which* $p$ *divides* $q_p(a)$.

For each integer $m \geq 2$,

**Conjecture 4a)$_m$.** *There are only finitely many primes* $p$ *for which* $p^m$ *divides* $q_p(a)$.

For each integer $m \geq 1$,

**Conjecture 4b)$_m$.** *There are infinitely many primes $p$ for which $p^m$ does not divide $q_p(a)$.*

**Conjecture 5a.** *Conjecture 4a)$_m$ holds for some integer $m = m(a)$.*

**Conjecture 5b.** *Conjecture 4b)$_m$ holds for some integer $m = m'(a)$.*

If we assume that $p^m$ divides $q_p(a)$ with "probability" $1/p^m$ then it is easy to give a heuristic justification to each of the above conjectures.

We shall determine a number of interrelations between these conjectures, and with some others below. We first note some trivial relations between the conjectures above: If Conjecture 4a)$_m$ holds then Conjectures 4b)$_m$ and 5a hold, as well as 4a)$_n$ for each $n \geq m$. If 4b)$_m$ holds then 5b) holds as well as 4b)$_n$ for each $n \geq m$. Also Conjecture 3 implies 2a, Conjecture 4b)$_1$ implies 2b and Conjecture 5a implies 5b with $m'(a) \leq m(a)$.

Taking $a = 2$ in Conjecture 4b)$_1$ implies the theorem of Adleman and Heath–Brown (that $(2)_p$ has no solutions for infinitely many primes $p$), by Lemma 2. Moreover, by Hellegouarch's theorem (mentioned in the introduction), we see that Conjecture 4a)$_2$ implies that $(2)_{p^2}$ has no solutions for all but finitely many primes $p$; and, by Faltings' theorem (Lemma 6), this implies that there are only finitely many $(n, x, y, z)$ satisfying $(2)_n$ with $n$ divisible by a square and $\gcd(x, y, z) = 1$.

By generalizing an argument of Puccioni [35], I was able to show in [12] that, for any $m \geq 1$, Conjecture 4a)$_{m+1}$ implies Conjecture 4b)$_m$. (In other words if $p^m \mid q_p(a)$ for all but finitely many primes $p$, then $p^{m+1} \mid q_p(a)$ for infinitely many primes $p$.

We now make a sequence of seemingly unrelated conjectures:

**Conjecture 6.** (Erdös [5], Mollin and Walsh [32]) *There are only finitely many triples of consecutive powerful numbers.* (Note that $n$ is called powerful if $p^2$ divides $n$ whenever $p$ divides $n$).

As noted by Mollin and Walsh, if $n-1$, $n$ and $n+1$ are all powerful numbers then 4 divides $n$ (as an integer $\equiv 2 \pmod 4$) can't be powerful) and so $n^2-1$ is powerful if and only if $n-1$ and $n+1$ are both powerful (as $\gcd(n-1, n+1)=1$). Therefore the following is equivalent to Conjecture 6.

**Conjecture 6a.** *There are only finitely many even powerful numbers $n$ such that $n^2-1$ is also powerful.*

In [38] Ribenboim stated the even weaker (take $n=m^2$ above):

**Conjecture 6b.** *There are only finitely many even integers $m$ such that $m^4-1$ is also powerful.*

If $A$ is a fixed even integer then we can take $n=A^r$ in Conjecture 6a and deduce the even weaker conjecture:

**Conjecture 7a.** *For every even integer $A$ there are infinitely many values of $n$ for which $A^n-1$ is not powerful.*

Actually, as $(A^{2n}-1)=(A^n-1)(A^n+1)$, we see that Conjecture 7a also follows from

**Conjecture 7b.** *For every even integer $A$ there are infinitely many values of $n$ for which $A^n+1$ is not powerful.*

The "link" between Conjectures 2–5 and Conjectures 6–7 comes from the following argument, which is similar to that given in [14] (we shall show that Conjecture 7a implies Conjecture $4b_1$): If Conjecture $4b_1$ is false then $p$ divides $q_p(a)$ for all $p>p_0$. Set

$$t=\prod_{p\le p_0}\phi(p^2)$$

and $A=a^t$. It is easy to show that $A^n-1$ is a powerful number for all positive integers $n$ (consider the prime divisors $p>p_0$ and $p\le p_0$ separately), and this contradicts Conjecture 7a.

In 1953, Mahler [28] proved that as $x, y\to\infty$, the largest prime factor of $x^2+y^3\to\infty$. We make an analogous conjecture:

**Conjecture 8.** *The largest prime factor of* $1 + x^2y^3$ *tends to infinity as* $x + |y|$ *tends to infinity.*

It is easy to show that Conjecture 8 implies Conjecture 7b: If 7b is false then $A^n + 1$ is powerful for all sufficiently large $n$, i.e., $A^n + 1 = x^2(-y)^3$ for some $x$ and $y$; and, as $n \to \infty$, this contradicts Conjecture 8.

We now show how the "*abc*" Conjecture implies both Conjecture 6a and 8:

If $n$ and $n^2 - 1$ are both powerful then, by taking $a = 1$, $b = n^2 - 1$ and $c = n^2$ in (3), we get $G \leq \sqrt{(bn)} < n^{3/2}$. Therefore, $n^2 < kn^{3/2+\varepsilon}$ by Conjecture 1, which bounds $n$, and so Conjecture 6a holds.

If $x$ and $y$ are integers for which the largest prime factor of $1 + x^2y^3$ is $\leq t$, then take $a = 1$, $b = x^2y^3$ in (3), so that $G \leq xyT$, where $T$ is the product of primes $\leq t$. Therefore $x^2y^3 \leq c\,(xy)^{1+\varepsilon}$ where $c = kT^{1+\varepsilon}$, by Conjecture 1, which bounds $xy$ and thus $x + |y|$. Conjecture 8 follows.

In a very recent paper Silverman [39] deduced a quantitative result on the $p$–divisibility of Fermat quotients from the "*abc*" Conjecture: If Conjecture 1 holds then, for any $a \geq 2$, there are $\gg \log x$ primes $p \leq x$ for which $p$ does not divide $q_p(a)$. Actually a weaker quantitative result can be deduced from Conjecture 6a.

There are many fascinating connections between these conjectures and questions on Fermat and Mersenne numbers (see Gary Walsh's forthcoming master's thesis and also [37] and [38]); and between Fermat quotients and Bernoulli numbers (see Emma Lehmer's paper [26]).

It is also of interest to determine, for each odd prime $p$, an upper bound on the least integer $a = a(p)$ for which $p$ does not divide $q_p(a)$. By D.H. Lehmer's computations ([25]) we know that $a(p) \leq 3$ whenever $p < 6 \cdot 10^9$; and H.W. Lenstra ([27]) has asserted that this is probably always the case. We are less ambitious:

**Conjecture 9.** *There exists an integer* $N$ *such that for all odd primes* $p$*, there is a positive integer* $a \leq N$ *for which* $p$ *does not divide* $q_p(a)$. (I.e., $a(p) \leq N$ for all odd primes $p$).

**Conjecture 9a.** (H.W. Lenstra [27]) *We may take $N = 3$ in Conjecture 9.*

In fact Lenstra [27] has shown that $a(p) \le 4\log^2 p$ (subsequently reproved independently by Fouché [7]). The proof is elegant:

Define $S(x, y)$ to be the set of positive integers $\le x$, free of prime factors greater than $y$. If $p$ divides $q_p(a)$ for every $a \le y$ then, as $q_p(\bullet)$ is an additive function (mod $p$), we see that $p$ divides $q_p(a)$ for any $a \in S(p^2, y)$. However, as $p$ divides *exactly* one of $q_p(a)$, $q_p(a+p), \ldots, q_p(a+(p-1)p)$, we see that $|S(p^2, y)| \le p$; and so, by choosing $y$ sufficiently large (i.e., $y = 4\log^2 p$) we get a contradiction.

By considering the set $S^*(p, y)$ of quotients $m/n$ of coprime integers $m, n$ from $S(p, y)$ it is possible to improve the above to $a(p) \le \log^2 p$; and, by a similar method, Tanner and Wagstaff [41] have shown, as a corollary to Lemma 5, that $(2)_n$ has no solutions for $n \le 1.564 \times 10^{17}$. More recently, Coppersmith [43] has come up with a new method that gives $n \le 7.568 \times 10^{17}$; however, in general, Coppersmith's method also gives $a(p) \le \log^2 p$.

As we shall see in the next section, we would like to improve these results to

**Conjecture 10.** *For any constant $c > 0$, if $p$ is sufficiently large then $a(p) \le c(\log p)^{1/4}$.*

This would seem to require a genuinely new idea. I have been unable to prove even the existence of infinitely many primes $p$ for which $a(p) \le (\log p)^{2-\varepsilon}$, for some $\varepsilon > 0$.

We now give

**Proof of Theorem 1.** Hellegouarch [20] showed that if $(2)_n$ has solutions then $p^{2t}$ divides both $2^{p-1} - 1$ and $3^{p-1} - 1$. If $x = 2^a 3^b$ or $1/2^a 3^b$ or $2^a/3^b$ or $3^b/2^a$ where $a$ and $b$ are non-negative integers then it is easy to show that $x^{p-1} \equiv 1 (\mathrm{mod}\ p^{2t})$; and, if both numerator and denominator are $< p^t$ then these integers are distinct (mod $p^{2t}$). Gunderson [18] showed that the number of such integers is

$$\geq 1 + \frac{3\log^2 n - (\log 12)\log n}{\log 2 \log 3}. \tag{4}$$

However, as there are exactly $p-1$ distinct solutions $(\text{mod } p^{2t})$ of $x^{p-1} \equiv 1 (\text{mod } p^{2t})$, we have a contradiction if the quantity in (4) is $\geq p$. This clearly occurs if $t \geq p^{1/2}/\log p$.

## 3. Some Matrices

The purpose of this section is to expand upon the conjectures given in [16]. First I will give a vague outline of the previously mentioned method of Frobenius *et al.*: For integers $a, b, c$ with $c > 0$ and $\gcd(b, c) = 1$ define $\alpha(a, b, c)$ ($\beta(a, b, c)$) to be the least positive (non–negative) residue of $a/b$ (mod $c$).

Let $A_n(t)$ be the $2n$ by $n$ matrix with $(i, j)$th entry

$$t^{\alpha(j,n,i)} \text{ (if } \gcd(ij, n) = 1\text{)}, \quad 0 \text{ (otherwise).}$$

If $(x, y, z)$ is a solution of $(2)_p$ then define $H = H(x, y, z)$ to be the set of congruence classes (mod $p$) of $-x/y$, $-y/x$, $-y/z$, $-z/y$, $-x/z$, $-z/x$. Note that if $t \in H$ then

$$H = \{t, 1-t, t^{-1}, 1-t^{-1}, t/(t-1), 1/(1-t)\}. \tag{5}$$

The main theorem of Frobenius *et al.* states

**Lemma 7.** *Suppose that $t \in H$ and $n$ is a positive integer for which*

(i) *The matrix $A_m(t)$ has rank $\phi(m)$ in the ring $\mathbb{Z}/p\mathbb{Z}$, for each $m$ in the range $1 \leq m \leq n$.*

(ii) *$t$ has order $\geq 2n+1$ (mod $p$).*

*Then $p^2$ divides $q^p - q$ for all primes $q \leq 2n+1$.*

Gunderson [18] gave the first correct proof that $t$ cannot have order 3, 4 or 6 (mod $p$). From this and a result of Pollaczek one can derive:

**Lemma 8.** *There exists a constant $c_1 > 0$ such that if $(2)_p$ has solutions then there exists $t \in H$ which has order $> c_1(\log p)^{1/2}$ (mod $p$).*

Lemma 8 makes it easy to satisfy Lemma 7 (ii) and so the real difficulty in implementing Lemma 7 is in proving that the criteria in Lemma 7 (i) hold for each successive integer $m$. In practice, we do not know much about $t \in H$ except that it is not $\equiv 0 \pmod{p}$ and can have reasonably high order (by Lemma 8). Thus we have to prove Lemma 7 (i) by taking determinants of $\phi(m)$ by $\phi(m)$ submatrices of $A_m(X)$ and examining these polynomials.

Let us suppose that $A_m(t)$ does indeed have rank $\phi(m)$ in $\mathbb{C}$ for any complex number $t$, except when $t \in U$ $(= \{0\} \cup \{\text{the roots of unity}\})$. We shall prove that there exists a constant $c_2 > 0$ such that if $\log p > c_2 m^4$ then $A_m(t)$ has rank $\phi(m)$ in $\mathbb{Z}/p\mathbb{Z}$:

First note that as each entry of $A_m(t)$ has degree $\leq 2m$ thus

$$\text{Any subdeterminant of } A_m(t) \text{ has degree } \leq 2m^2. \tag{6}$$

Now suppose that $A_m(t)$ has rank $< \phi(m)$ in $\mathbb{Z}/p\mathbb{Z}$. Then each non–zero $\phi(m)$ by $\phi(m)$ subdeterminant $D$ of $A_m(x)$ is divisible by an irreducible polynomial $f_D(x)$ such that $f_D(t) \equiv 0 \pmod{p}$. By hypothesis, either $f_D(x)$ is a cyclotomic polynomial, or we get two distinct polynomials $g_1$ and $g_2$ with $g_1(t) \equiv g_2(t) \equiv 0 \pmod{p}$.

If $f$ is a cyclotomic polynomial then, by (6), $t$ has order $\leq 2m^2 \ll (\log p)^{1/2}$, which contradicts Lemma 8 if $c_2$ is chosen sufficiently small.

As the matrix $A_m(t)$ has got monomial entries with coefficients 1, we see that, for any subdeterminant $D$, the sum of the absolute values of the coefficients is bounded by $m!$. By a method of Mignotte [30] this means that for any $g$ dividing $D$ we have

$$\|g\|_2 \leq 2^{2m^2} m!$$

(by (6)), where $\|g\|_2 = \left( \sum_{i=0}^{d} g_i^2 \right)^{1/2}$ and $g(x) = \sum_{i=0}^{d} g_i x^i$.

We have two such (distinct) polynomials $g_1$ and $g_2$, and as they have no common root, we know that $p$ divides their resultant ( as $p$ divides $g_1(t)$ and $g_2(t)$). Therefore, by using the standard bounds for the determinant of a matrix we have

$$p \le \|g_1\|_2^{\deg g_2} \|g_2\|_2^{\deg g_1} \le (2^{2m^2} m!)^{4m^2} < \exp(c_3 m^4)$$

giving a contradiction.

Observing that we have already chosen $t$ with order $\ge 2n+1 \pmod p$ we see that we have proved the following:

**Theorem 2.** *Suppose that $A_m(t)$ has rank $\phi(m)$ in $\mathbb{C}$ for any complex number $t$, not in $U$, and for any $m$, $1 \le m \le n$. There exists a constant $c_2 > 0$ such that if $p > \exp(c_2 n^4)$ and $(2)_p$ has solutions then $p^2$ divides $q^p - q$ for each $q \le 2n+1$.*

(In [16] the constant $c_2$ is given explicitly.) So what we really wish to prove is

**Conjecture 11.** *For any complex number $t$, $t \notin U$, and for any positive integer $n$, the matrix $A_n(t)$ has rank $\phi(n)$.*

As a corollary to Theorem 2 we have

**Corollary 1.** *If Conjectures 10 and 11 are true then the first case of Fermat's Last Theorem is true for all sufficiently large exponents.*

Suppose that $\gcd(m,n)=1$ and consider using the Euclidean algorithm in $\mathbb{Z}[t,x]$ to eliminate the variable $x$ from $1-x^m$ and $1-tx^n$. It is easy to see that there exist polynomials $U_m(x)$ and $V_m(x)$ of degree $\le n-1$, $\le m-1$, respectively, such that

$$t^m - 1 = (1-x^m)U_m(x) - (1-tx^n)V_m(x) \tag{7}$$

Explicitly we can show that

$$U_m(x) = \sum_{i=0}^{n-1} t^{\alpha(i,n,m)} x^i$$

and

$$V_m(x) = \sum_{j=0}^{m-1} t^{\beta(j,n,m)} x^j.$$

We thus see that the entries of $A_n(t)$ appear in a natural way as the coefficients of $U_m(x)$!!

It may well turn out to be easier to approach Conjecture 11 by considering the following equivalent Conjecture:

**Conjecture 12.** *Let $B_n(t)$ be the $2n$ by $n$ matrix with $(i, j)$th entry*

$$\frac{1}{(t^i - \alpha^{ij})}$$

*where $\alpha$ is a primitive n-th root of unity. For every positive integer $n$ and complex number $t \notin U$, the matrix $B_n(t)$ has full rank.*

**Theorem 3.** *Conjecture 11 holds if and only if Conjecture 12 holds.*

A proof of Theorem 3 can be found in [15]; the main idea comes from substituting $x = \tau$ in (7) for $\tau^n = t^{-1}$, so that $U_m(\tau) = -(1 - t^m)/(1 - \tau^m)$. Letting $\rho = \tau_0^{-1}$, for a fixed root $\tau_0$ of $X^n = t^{-1}$, we see that if $\tau = \tau_0 \alpha^j$ then

$$U_m(\tau)/\rho^m (t^m - 1) = 1/(\rho^m - \alpha^{jm})$$

and so we can compare the matrices $B_n(\rho)$ and $A_n(t)$.

## References


[1] *L.M. Adleman* and *D.R. Heath-Brown,* The first case of Fermat's last theorem. Invent. Math., **79** (1985), 409—416.

[2] *N.C. Ankeny* and *P. Erdös,* The insolubility of classes of diophantine equations. Amer. J. Math., **76** (1954), 488—496.

[3] *T. Azuhata,* On Fermat's Last Theorem. Acta Arith., **45** (1985), 19—27.

[4] *B. de Weger,* Solving exponential diophantine equations using lattice basis reduction algorithms. J. Number Theory, **26** (1987), 325—367.

[5] *P. Erdös,* Problems and results on consecutive integers. Eureka, **38** (1975/6), 3—8.

[6] *G. Faltings,* Endlichkeitssätze für Abelsche Varietaten über Zahlkonpern. Invent. Math., **73** (1983), 349—366.

[7] *W.L. Fouché,* On the Kummer–Mirimanoff congruences. Quart. J. Math., **37** (1986), 257—261.

[8] *E. Fouvry,* Théorème de Brun–Titchmarsh; application au théorème de Fermat. Invent. Math., **79** (1985), 383—407.

[9] *G. Frey,* Links between stable elliptic curves and certain diophantine equations. Ann. Univ. Saraviensis, **1** (1986), 40pp.

[10] *G. Frobenius,* Über den Fermatschen Satz III. Sitzungsber. Akad. d. Wiss. zu Berlin, (1914), 653—681.

[11] *P. Furtwängler,* Letster Fermatschen Satz und Eisentstein'sches Reziprositätsgesetz. Sitzungsber. Akad. d. Wiss. Wien. Abt. IIa, **121** (1912) 589—592.

[12] *A. Granville,* Refining the conditions on the Fermat quotient. Math. Proc. Camb. Phil. Soc., **98** (1985), 5—8.

[13] *A. Granville,* The set of exponents for which Fermat's Last Theorem is true, has density one. C.R. Math. Acad. Sci. Canada, **7** (1985), 55—60.

[14] *A. Granville,* Powerful numbers and Fermat's Last Theorem. C. R. Math. Acad. Sci. Canada, **8** (1986), 215—218.

[15] *A. Granville,* Diophantine equations with varying exponents (with special reference to Fermat's Last Theorem). Thesis, Queen's University, (1987).

[16] *A. Granville* and *M.B. Monagan,* The First Case of Fermat's Last Theorem is true for all prime exponents up to 714,591,416,091,389. Trans. Amer. Math. Soc., **306** (1988), 329—359.

[17] *A. Granville* and *B. Powell,* Sophie Germain Type criteria for Fermat's Last Theorem. Acta Arith., **50** (1988), 265—277.

[18] *N.G. Gunderson,* Derivation of Criteria for the First Case of Fermat's Last Theorem and the Combination of these Criteria to Produce a New Lower Bound for the Exponent. Thesis, Cornell University, (1948).

[19] *D.R. Heath–Brown,* Fermat's Last Theorem for "almost all" primes. Bull. London Math. Soc., **17** (1985), 15—16.

[20] *Y. Hellegouarch,* Courbes Elliptique et Equation de Fermat. Thesis, Besançon, (1972).

[21] *M. Krasner,* A propos du critère de Sophie Germain–Fertwängler pour le premier cas du théorème de Fermat, Mathematica Cluj, **16** (1940), 109—114.

[22] *E.E. Kummer,* Beweis des Fermat'schen Satzes der Unmöglichkeit von $x^\lambda + y^\lambda = z^\lambda$ für eine unendliche Anzahl Primzahlen $\lambda$. Monatsber. Akad. d. Wiss., Berlin, (1847) 132—139, 140—141, 305—319.

[23] *E.E. Kummer,* Einige Sätze über die aus den Wurzeln der Gleichung $\alpha^\lambda = 1$ gebildeten complexen Zahlen, für den Fall dass die Klassenzahl durch $\lambda$ theilbar ist, nebst Anwendungen derselben auf einen weiteren Beweis des letztes Fermat'schen Lehrsatzes. Math. Abhandl. Akad. d. Wiss., Berlin, (1857), 41—74.

[24] *A.M. Legendre,* Sur quelques objets d'analyse indéterminée et particuliérement sur le théorème de Fermat. Mém. de l'Acad. des Sciences, Institut de France, **6** (1823) 1—60.

[25] *D.H. Lehmer,* On Fermat's quotient, base two. Math. Comp., **36** (1981), 289—290.

[26] *E. Lehmer,* On congruences involving Bernoulli numbers and the quotients of Fermat and Wilson. Ann. of Math., **39** (1938), 350—360.

[27] *H.W. Lenstra Jr.,* Miller's primality test. Inform. Proc. letters, **8** (1979), 86—88.

[28] *K. Mahler,* On the greatest prime factor of $ax^m + by^m$. Neiuw. Arch. Wiskunde, **1** (1953), 113—122.

[29] *D.W. Masser,* Open problems. Proc. Symp. Analytic Number Thy., *W.W.L. Chen* (ed.), London: Imperial Coll., (1985).

[30] *M. Mignotte,* Some useful bounds. Algebra, Symbolic and Algebraic Computation, New York, Springer–Verlag, (1983), 259—263.

[31] *D. Mirimanoff,* Sur le dernier théorème de Fermat. C. R. Acad. Sci. Paris, **150** (1910), 204—206.

[32] *R.A. Mollin* and *P.G. Walsh,* A note on powerful numbers, quadratic fields and the Pellian. C. R. Math. Acad. Sci. Canada, **8** (1986), 109—111.

[33] *T. Morishima,* Über die Fermatsche Quotienten. Jpn. J. Math., **8** (1931), 159—173.

[34] *F. Pollaczek,* Über den grossen Fermat'schen Satz. Sitzungsber. Akad. d. Wiss. Wien, Abt. IIa, **126** (1917), 45—59.

[35] *S. Puccioni,* Un teorema per una resoluzioni parziali del famoso problema di Fermat. Archimede, **20** (1968), 219—220.

[36] *P. Ribenboim,* 13 Lectures on Fermat's Last Theorem. New York, Springer–Verlag, (1979).

[37] *P. Ribenboim,* The Book of Prime Number Records. New York, Springer–Verlag, (1988).

[38] *P. Ribenboim,* Impuissants devant les puissances. Expo. Math., **6** (1988), 3—28.

[39] *J. Silverman,* Wieferich's Criterion and the *abc*–Conjecture. J. Number Theory, **30** (1988), 226—237.

[40] *C.L. Stewart* and *R. Tijdeman,* On the Oesterlé–Masser Conjecture. Monatsh. Math., **102** (1986), 251—257.

[41] *J.W. Tanner* and *S.S. Wagstaff Jr.,* New bound for the first case of Fermat's Last Theorem. Preprint.

[42] *A. Wieferich,* Zum letzten Fermat'schen Theorem. J. reine u. angew. Math., **136** (1909), 293—302.

[43] *D. Coppersmith,* Fermat's Last Theorem (Case 1) and the Weiferich Criterion. Preprint.

---

Department of Mathematics, University of Toronto, Toronto, Ontario, CANADA M5S 1A1.

# Canadian Number Theory Association
# Unsolved Problems 1988

*Richard K. Guy*

By way of introduction to the problems presented at Banff, I list some recent results which partially answer questions raised in [11], and which will appear in a second edition.

Jeff Young and Duncan Buell [32] have shown that $F_{20} = 2^{2^{20}} + 1$ is composite, so the smallest Fermat number whose status is currently unknown is $F_{22}$.

Although it is quite well known that $2^{86243} - 1$, $2^{132049} - 1$ and $2^{216091} - 1$ are primes, it may not be so well known that *not* all smaller Mersenne numbers have been tested. Indeed, it was recently announced in the *Los Angeles Times* that Walter Colquitt has discovered the prime $2^{110503} - 1$.

There is a faint hope that the largest known prime might be other than a Mersenne prime. The three largest known primes which are not Mersenne primes are $7 \cdot 2^{54486} + 1$, $8423 \cdot 2^{55157} + 1$ and $8423 \cdot 2^{59877} + 1$. These were found by Buell and Young [3] in their investigation of the Sierpinski problem: find the least $k$ for which $k \cdot 2^n + 1$ is composite for all exponents $n$ (see B21 in [11]). The answer is probably $k = 78557$, but there remain the 35 possibilities

| | | | | | | | | | |
|---|---|---|---|---|---|---|---|---|---|
| 4847 | 5297 | 5359 | 7013 | 10223 | 13787 | 19249 | 21181 | 22699 | 24737 |
| 25819 | 27653 | 27923 | 28433 | 33661 | 34999 | 39781 | 44131 | 46157 | 46187 |
| 46471 | 47897 | 48833 | 50693 | 54767 | 55459 | 59569 | 60443 | 60541 | 63017 |
| 65567 | 67607 | 69109 | 74191 | 74269 | | | | | |

for none of which is there a prime with $n \le 50000$. See also [15].

In answer to a question in A3 in [11], Rotkiewicz [24] has proved that a **repunit**, $(10^n - 1)/9$, is never a perfect cube for $n > 1$.

Jeff Young and James Fry, in a letter dated 87–09–01, announced the discovery of three arithmetic progressions of 20 primes. The smallest of these is

$$214861583621 + 1943 \times 9699690k \qquad (0 \le k \le 19)$$

The record length for an A.P. of **consecutive** primes remains at 6. Even as I write this, a letter arrives from Stephane Vandermergel, who has discovered 16 consecutive primes of shape $4k + 1$; these are **not** in A.P., of course! Their first six digits are 207622, and the final three are 273, 297, 301, 313, 321, 381, 409, 417, 421, 489, 501, 517, 537, 549, 553, 561.

Harry Nelson [21] has collected the $100.00 prize that Martin Gardner offered to the first discoverer of a 3x3 magic square whose nine entries are consecutive primes. These are also not in A.P.: the central prime is 1480028171 and the others are this ±12, ±18, ±30 and ±42. He found more than 20 other such squares.

Randall Rathbun, in an 86–08–20 letter, notes a **Cunningham chain** of length 8, i.e., a set of primes, $p_i$, $p_{i+1} = 2p_i + 1$ $(1 \le i \le 7)$, with $p_1 = 19099919$. More recently still, Harry Nelson, in answer to a problem [13], has found a chain of nine, starting at 85864769.

Maier and Pomerance [18] have improved the constant in Rankin's theorem, that the gap, $d_n = p_{n+1} - p_n$, between consecutive primes is infinitely often greater than

$$(c + o(1)) \ln n \, \ln\ln n \, \ln\ln\ln\ln n \,/\, (\ln\ln\ln n)^2$$

from $c = e^\gamma$ to $c = c_0 e^\gamma$ where $c_0 \approx 1.31256$ is the solution of the equation

$$4/c_0 - e^{-4/c_0} = 3.$$

Aaron Potler and Jeff Young, in an 87–09–25 letter, announce a gap of 778 in the primes following 42842283925351; this is the earliest gap of this length.

Andrew Odlyzko (letter dated 87–08–03) has checked Gilbreath's conjecture, that successive absolute differences of the sequence of primes always start with 1, for all primes up to $10^{10}$. He agrees with Hallard Croft and others that this is not so much a property of the primes, as of any sequence consisting of 2 and odd numbers which do not increase too rapidly nor have large gaps. To reach $10^{10}$, he needed only to examine the first 350 differences.

Graeme Cohen [7] has shown that an odd perfect number contains a component (prime power divisor) greater than $10^{20}$.

Herman te Riele, in an 87–05–15 letter, announced an amicable pair whose members were prime to 6,

$$5 \cdot 7^2 \cdot 13 \cdot 19^3 \cdot 181 \cdot 11^2 \cdot 17 \cdot 23 \cdot 37 \begin{cases} 101 \cdot 8643 \cdot 1947938229 \\ 365147 \cdot 47303071129 \end{cases}$$

and whose 33 decimal digits were much less than the 48 to 55 of the four previously known specimens.

An example with 36 decimal digits:

$$5^4 7^3 11^3 13^2 17^2 19 \cdot 61^2 97 \cdot 307 \begin{cases} 140453 \cdot 85857199 \\ 56099 \cdot 214955207 \end{cases}$$

has been produced by Battiato and Borho [1], who also found another 14 such pairs. In an earlier paper [23], te Riele had found all 1427 amicable pairs whose lesser members are less than $10^{10}$. If $A(x)$ is the number of such pairs less than $x$, he remarks that $A(x)(\ln x)^3 / x^{1/2}$ "remains very close to 174.6." He also asks: Is there an odd amicable pair with one member, but not both, divisible by 3?

Ernst Selmer [28] has produced two volumes on the "local postage stamp" problem, but there remains a variety of open questions.

Henry Ibstedt (preprint, 88–02–23) has done extensive calculations on **Göbel sequences**, $x_n = (1 + x_0^m + x_1^m + \ldots + x_{n-1}^m)/n$, for $m = 2, 3, \ldots, 11$ and not only for $x_0 = 1$, but also for $2 \le x_0 \le 11$. For $x_0 = 1$ he finds that the first non-integer value of $x_n$ is that of rank $n$, where

$m =$ 2 3 4 5 6 7 8 9 10 11

$n =$ 43 89 97 214 19 239 37 79 83 239

David Boyd and Alf van der Poorten have a forthcoming paper on this topic.

News spread rapidly about the disproof by Noam Elkies [10] of Euler's conjecture, which even hit the national newspapers. The infinite family of solutions, of which the smallest member is $2682440^4 + 15365639^4 + 18796760^4 = 20615673^4$, comes from an elliptic curve given by $u = -5/8$ in the parametric solution of $x^4 - y^4 = z^4 + t^2$ given by Dem'janenko [9]. A much smaller solution,

$$95800^4 + 217519^4 + 414560^4 = 422481^4,$$

corresponding to $u = -9/20$, has since been found by Roger Frye [see 22].

By way of introduction to the problems, CNTA 88:01 to CNTA 88:22, presented at the closing session of the Conference, 88–04–27, here is one asked by Mike Filaseta and Emil Grosswald at the 1988 Illinois Number Theory Conference:

Is $P(n(n+1))$ always greater than $\min(n - p_k, p_{k+1} - n)$, where $P(x)$ is the largest prime factor of $x$, and $p_k$, $p_{k+1}$ are consecutive primes such that $p_k < n \le p_{k+1}$?

**CNTA 88:01** (Allen Freedman) Given $n$ points $a_1, a_2, \ldots, a_n$, in a rectangle of unit area, one of them being the lower left corner, can you always find $n$ non-overlapping rectangles, $R_1, R_2, \ldots, R_n$, with sides parallel to the original rectangle, and $a_i$ as the lower left corner of $R_i$, $1 \le i \le n$, whose total area is at least $1/2$? In fact $\ge (n+1)/2n$? The best that is known is $1/n$.

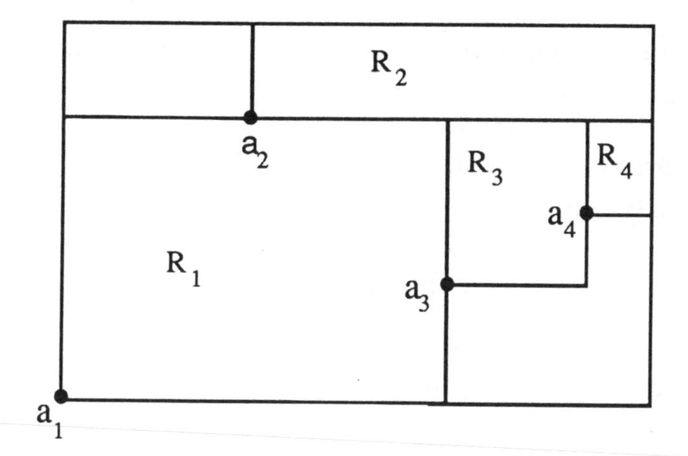


**CNTA 88:02** (Erdös & Nicolas) $\tau(n)$ is the number of divisors of $n$ and $1 = d_1 < d_2 < \ldots < d_{\tau(n)}$ is the set of divisors of $n$. Prove that for $n > 5040$,

$$\tau(n) > \sum_{1 \le i < j \le \tau(n)} \frac{1}{d_j - d_i} \tag{1}$$

The inequality (1) fails for $n = 5040$, which appears to be the greatest counterexample. It seems that

$$\frac{1}{\tau(n)} \sum_{1 \le i < j \le \tau(n)} \frac{1}{d_j - d_i} \to 0 \qquad \text{if } \tau(n) \to \infty$$

**CNTA 88:03** (P. Erdös) Let $p_1 < p_2 < \ldots < p_t$ be any set of $t$ primes. Is it true that

$$\sum_{1 \le i < j \le t} \frac{1}{p_j - p_i} < ct \tag{2}$$

for some absolute constant $c$? The inequality (2) easily follows from Bruns' method if the $p_i$ are the first $t$ primes.

I once conjectured that if $a_1 < a_2 < \ldots < a_t$ is any sequence of integers for which any interval of length $x$ contains, for every $x$, fewer than $cx/\ln x$ of the $a_i$, then

$$\sum \frac{1}{a_j - a_i} < ct \tag{3}$$

But Ruzsa found an ingenious counterexample. In fact he showed that nothing better than the trivial upper bound, $ct \ln \ln t$ holds.

The prime number theorem implies, by a simple averaging process, that if $2, 3, \ldots, p_t$ are the first $t$ primes, then

$$\sum_{1 \le i < j \le t} \frac{1}{p_j - p_i} > t + o(t).$$

From Hoheisel's theorem it easily follows that

$$\liminf_{k \to \infty} \sum \frac{1}{p_k - p_i} > 0 \tag{4}$$

and in fact it is very likely that the lim inf in (4) is 1, but this is probably unattackable. I could not disprove that

$$\limsup \sum \frac{1}{p_k - p_i} = 1, \tag{5}$$

but it seems certain that the lim sup in (5) is $\infty$. Perhaps

$$\limsup \sum_{p_i < p_k} \frac{1}{p_k - p_i} (\ln \ln \ln k)^{-1} = 1. \tag{6}$$

Let $p_1, p_2, \ldots, p_t$ be any set of $t$ primes. Is it then true that there is an absolute constant $C$ such that for at least one of the primes $p_i$ we have

$$\sum_{\substack{j=1 \\ j \ne i}}^{t} \frac{1}{p_i - p_j} < C \ ? \tag{7}$$

It is easy to see that (7) is equivalent to (2).

**CNTA 88:04** (P. Erdös) Let $x_1, x_2, \ldots, x_n$ be $n$ points in the plane. Denote by $d(x_i, x_j)$ the distance from $x_i$ to $x_j$. Assume that if $d(x_i, x_j) \ne d(x_u, x_v)$, then

$$|d(x_i, x_j) - d(x_u, x_v)| \ge 1 \tag{8}$$

and that $d(x_i, x_j) \ge 1$. Denote by $D(x_1, \ldots, x_n)$ the diameter of $x_1, x_2, \ldots, x_n$. Is it true that (8) implies

$$D(x_1,\dots,x_n) > cn\,? \tag{9}$$

In a problem of mine [Elem. Math., 1981] I asked if (8) implies $D(x_1,\dots,x_n) > cn^{2/3}$. Kanold [Elem. Math., 1982] proved that $D(x_1,\dots,x_n) > cn^{3/4}$. An old conjecture of mine states that the number of distinct distances determined by $n$ points is $> \frac{cn}{\sqrt{\ln n}}$, which would imply $D(x_1,\dots,x_n) > \frac{cn}{\sqrt{\ln n}}$.

Perhaps if (8) holds, then

$$D(x_1,\dots,x_n) \geq n - 1 \tag{10}$$

with equality in (10) just if the points are all on one line. For small values of $n$, (10) certainly doesn't hold, and in fact Makai determined the exact value of $D(x_1,\dots,x_n)$ for $n \leq 6$.

Pomerance observed that to every $\varepsilon > 0$ there are $n$ points, no three on a line, which satisfy (8) and for which $D(x_1,\dots,x_n) < n - 1 + \varepsilon$. On the other hand, if we assume that the points are in general position, i.e., no three on a line and no four on a circle, then (8) almost certainly implies

$$D(x_1,\dots,x_n) > n^{1+c}. \tag{11}$$

Perhaps in this case the number of distinct distances determined by $x_1, x_2, \dots, x_n$ is $> n^{1+c}$, which, of course, would imply (11). Perhaps, in (11), $c \geq 1$, and I would like to find a nontrivial upper bound for $D(x_1,\dots,x_n)$ if the $x_i$ are in general position.

If we assume that all the distances are integers, then (8) is automatically satisfied. I suspect then in this case that $D(x_1,\dots,x_n) > n^k$ for every $k$, if $n > n_0$.

Perhaps the following two variants could be investigated. Assume that (8) only holds if the four points are distinct, or alternatively, assume that (8) only holds if $i = u$. What inequality can one obtain for $D(x_1,\dots,x_n)$? I hope that (9) remains true, In fact the following conjectures are perhaps of interest.

Let $x_1, x_2, \dots, x_n$ be $n$ points in the plane, and denote by $d(x_i)$ the number of distinct distances among the $d(x_i, x_j)$. An old, and for the moment hopeless, conjecture of mine states that, for some $i$,

$$d(x_i) > cn / \sqrt{\ln n} \tag{12}$$

and perhaps even

$$\sum_{i=1}^{n} d(x_i) > cn^2 / \sqrt{\ln n} \tag{13}$$

Similarly one could conjecture that the number of pairwise disjoint pairs $(x_{u_i}, x_{v_i})$, $1 \le i \le t_n$, for which all the distances $d(x_{u_i}, x_{v_i})$ are distinct, is $> cn / \sqrt{\ln n}$, and, in fact, that there are $c'n$ such ( $cn / \sqrt{\ln n}$ )–tuples.

Another very old conjecture of mine states that if $x_1, x_2, \ldots, x_n$ are $n$ points in the plane, and if $h(n)$ denotes the maximum number of times the same distance can occur among the $d(x_i, x_j)$, then

$$h(n) < n^{1+c/\ln\ln n} \tag{14}$$

and the lattice points show that (14), if true, is best possible. Is it true that the pairs $(x_i, x_j)$ can be decomposed into $h(n)$ classes so that, in the same class, all distances should be distinct? This last conjecture is perhaps easy to prove, or disprove.

**CNTA 88:05** (A. Schinzel) Do there exist absolute constants $A$ and $B$ with the following property? For all vectors $\bar{n} \in \mathbb{Z}^k$ there exist vectors $\bar{p}_1, \ldots, \bar{p}_l \in \mathbb{Z}^k$ such that $\bar{n} \in \sum_{i=1}^{l} u_i \bar{p}_i$ and

$$1) \quad \prod_{i=1}^{l} h(\bar{p}_i) \le A h(\bar{n})^{\frac{k-l}{k-1}} \qquad , u_i \in \mathbb{Q},$$

$$2) \quad \prod_{i=1}^{l} h(\bar{p}_i) \le B h(\bar{n})^{\frac{k-l}{k-1}} \qquad , u_i \in \mathbb{Z}.$$

where $h(\bar{n}) = \max |n_i|$. See [4], [5], [25], [26].

**CNTA 88:06** (James P Jones) Does there exist an algorithm, in time of order $\ln n$, to find all pairs $(x, y)$ of non–negative integers such that $n = x^2 + y^2$? See [12].

**CNTA 88:07** (James P. Jones) Does there exist a polynomial $P(n, x_1, \ldots, x_8)$ in eight unknowns which defines primes in the sense that for every natural number $n$,

$$n \text{ is prime just if } \exists\, x_1, \ldots, x_8 \ [P(n, x_1, \ldots, x_8) = 0]?$$

It was proved by Matijasevich [19] that there is a polynomial $P(n, x_1, \ldots, x_9)$ of the above kind. It follows (using a device of Hilary Putnam) that the set of primes is identical with the set of non-negative values of a polynomial $Q(x_0, x_1, \ldots, x_9)$ in ten variables:

$$n \text{ is prime just if } (\exists\, x_0, x_1, \ldots, x_9)\,[n = Q(x_0, x_1, \ldots, x_9)].$$

*cf.* [6] for an early statement of a problem of this type.

**CNTA 88:08** (Andrew Granville) Conjectures concerning Fermat quotients:

1) For all $a \geq 2$, there is a prime $p$, $p \nmid a$, such that $p^2 \mid a^p - a$.
2) For all $a \geq 2$, there is a prime $p$, such that $p^2 \nmid a^p - a$.

1') For all $a \geq 2$, there are infinitely many primes $p$ such that $p^2 \mid a^p - a$.

2') For all $a \geq 2$, there are infinitely many primes $p$ such that $p^2 \nmid a^p - a$.

3) There is an $N$ such that for all odd primes $p$, there is an $a \leq N$ with $p^2 \nmid a^p - a$. Lenstra [17] suggested that $N$ might even be 3.
4) For all primes $p > p_0$, there is an $a < (\ln p)^{1/4}$ with $p^2 \nmid a^p - a$. Lenstra and others have shown that one can take $a \ll (\ln p)^2$. Can the exponent 2 be improved?

For further comments on the above problems, see Andrew Granville's paper elsewhere in these Proceedings.

**CNTA 88:09** (Andrew Granville) Erdös has conjectured that there do not exist three consecutive powerful numbers. Can you prove that there does not exist $t$ such that $2^{tn} - 1$ is powerful for all $n \geq 1$? Note that if $n = 1, 2$, then, since $(2^t - 1, 2^t + 1) = 1$, we would have $2^t - 1$, $2^t$, $2^t + 1$ each powerful.

**CNTA 88:10** (Denis Richard) Write $a \sim b$ if $a$ and $b$ have the same set of prime divisors. Prove that if $x^{2^n} - 1 \sim y^{2^n} - 1$ for all $n$, then $x = y$.

**CNTA 88:11** (Denis Richard) Let $a$ be a fixed integer. Define $S = \{x \in \mathbb{N} : x \sim a \text{ and } x + 1 \sim a + 1\}$. Størmer [29], and see also [16], showed that the cardinality of $S$ is finite. is it possible that $x \in S$, $y \in S$, $x \neq y$ and both $x$ and $y$ squares? Compare B19 in [11] $x = 2^n - 2$, $y = 2^n(2^n - 2)$, $x + 1 = 2^n - 1$, $y + 1 = (2^n - 1)^2$.

**CNTA 88:12** (Denis Richard) Let $S : \mathbb{N} \to \mathbb{N}$ be the successor function $S(x) = x + 1$ and $x \perp y$ just if $x, y$ are relatively prime. Can you define any of the following predicates, using just the symbols $\forall$, $\exists$, $\neg$ (negation), $\Rightarrow$ (implies), $\Leftrightarrow$ (equivalent), $S$, and $\perp$?

a) $M(p, q, x) \Leftrightarrow (p \text{ prime}, q \text{ prime}, x = pq)$

b) $RES(p, x) \Leftrightarrow (p \text{ prime}, \exists y \ (y^2 \equiv x \pmod p))$

c) $POW(y) \Leftrightarrow \exists x \ \exists \alpha \ (y = x^{\alpha})$

d) $POW(x, y) \Leftrightarrow \exists x \ (y = x^{\alpha})$

e) $\square(y) \Leftrightarrow \exists x \ (y = x^2)$

f) $\square(x, y) \Leftrightarrow (y = x^2)$

Solution of any of a), b), c) or d) will solve the Woods–Erdös conjecture (B29 in [11]):

$$\exists k \ \forall x \ \forall y \ (x = y \Leftrightarrow x \sim y \ \& \ x+1 \sim y+1 \ \& \ldots \& \ x+k \sim y+k).$$

**CNTA 88:13** (Attila Pethö) Let $\{G_n\}$ be a $k$th order linear recursive sequence,

$$I(G) = \left\{ \{n_i\} \subseteq \mathbb{Z}^+, n_1 < n_2 < \ldots \ \& \ G_{n_i} \middle| G_{n_{i+1}} \right\}$$

$$\omega(G) = \inf_{\{n_i\} \in I(G)} \liminf_{i \to \infty} \frac{n_{i+1}}{n_i}$$

Conjecture: If $k \geq 2$ and $\{G_n\}$ is non-degenerate, then $\omega(G) \geq 2$. Can prove it in the special case $T_0 = T_1 = 0$, $T_2 = 1$, $T_{n+1} = T_n + T_{n-1} + T_{n-2}$: if $n \geq 0$, then $\omega(T) \geq \frac{5}{2}$.

**CNTA 88:14** (Attila Pethö) Buchmann and Pethö [2] found that in the field $\mathbb{Q}(\vartheta)$ with $\vartheta^7 = 3$, the integer

$$10 + 9\vartheta + 8\vartheta^2 + 7\vartheta^3 + 6\vartheta^4 + 5\vartheta^5 + 4\vartheta^6$$

is a unit. Does there exist, for any $n \geq 2$, a unit $\varepsilon$ of degree $n$ such that there exists an integer basis $\omega_1, \ldots, \omega_n$ of $\mathbb{Q}(\varepsilon)$ with the property that if $\varepsilon = \sum_{i=1}^{n} x_i \omega_i$, then the sequence $x_i, \ldots, x_n$ is an arithmetic progression with positive difference?

**CNTA 88:15** (Philip Parker, via Gerry Myerson) Call an $n$ by $n$ matrix $A$ **integralizable** if there is a matrix $P$ such that $P^{-1}AP$ has integer entries. Conditions for a matrix to be integralizable were known to Morgan Ward. Note that $A$

does not necessarily have rational entries. When are two matrices $A$ and $B$ simultaneously integralizable?

**CNTA 88:16** (Gerry Myerson) Prove that it's just as hard to find the $n$th digit of $\pi$ as it is to find the first $n$ digits of $\pi$. Generalize: e.g., to finding the $n$th partial quotient of a continued fraction.

**CNTA 88:17** (James P Jones) Wolstenholme [31] proved that if $n \geq 5$ is prime, then $\binom{2n}{n} \equiv 2 \pmod{n^3}$, equivalently, $\binom{2n-1}{n} \equiv 1 \pmod{n^3}$. The converse is known to hold up to $n = 283686649$ (with $n^2$ in place of $n^3$). See [14], [27].

**CNTA 88:18** (from Baton Rouge via Andrew Granville) If $x+1, \ldots, x+n$ are $n$ consecutive integers containing no primes or prime powers, i.e., every integer contains at least two distinct prime factors, can you colour the primes red and blue so that every integer in the interval has at least one red and one blue prime factor?

During the conference, Langevin proved that if $n$ is a positive integer and $x > \exp \exp n$, then any such interval can be so coloured; he used methods of transcendental number theory.

**CNTA 88:19** (H. Shapiro, via Gerry Myerson) Does $p > q > 1$, $q \nmid p$ imply that $\lfloor (p/q)^3 \rfloor$ is composite for infinitely many $n$?

**CNTA 88:20** (Ron Evans) Let $n, k, a$ be non–negative integers with $nk$ even, and $a$ odd. Define

$$f(q) = (1+q)^{\lfloor (k+1)/2 \rfloor} \prod_{\nu=1}^{k} \frac{1-(-q)^{a(n+\nu)}}{1-(-q)^{\nu}}$$

in $\mathbb{Z}[q]$. Are all coefficients of $f(q)$ non-negative?

The answer is "yes" if $a = 1$ (Stanton–Zeiberger).

[In an 88–06–20 letter, Ron Evans announced that he had used a computer to show that the conjecture is false when $k = 3$; however, the problem is open for $k > 3$.]

**CNTA 88:21** (M.J. DeLeon) Is it true that if $a$, $b$, $c$, $k$ are positive integers such that $a \neq b$, $c \neq ab - 1$, $ab + c^2 - 1 = bck$ and $a$ divides $c^2 - 1$, then $k$ divides $a$? This conjecture is related to Carlitz four–tuples [8].

Added in proof: This conjecture has since been proved by Waterhouse [30].

**CNTA 88:22** (R.A. Mollin, H.C. Williams) Let $d \equiv i \pmod 4$ for $i = 2,3$ and put $b = b(d) = \sqrt{\frac{d+2-i}{2}}$. Suppose that the Legendre symbol $(d/p) = -1$ for all odd primes $p < b(d)$. Under this hypothesis we conjecture that $d \in \{3, 6, 7, 10, 11, 14, 15, 23, 26, 35, 38, 47, 62, 83, 122, 143, 167, 227, 362, 398\}$. (From the results of [20] the conjecture follows under the assumption of the generalized Riemann hypothesis.)

## References


[1] *S. Battiato* and *W. Borho*, Are there odd amicable pairs not divisible by three? Math. Comp. **50** (1988), 633—637.

[2] *J. Buchmann* and *A. Pethö*, Computation of independent units by Dirichlet's method. Math. Comp. (to appear).

[3] *D.A. Buell* and *J. Young*, Some large primes and the Sierpinski problem. Math. Comp. (submitted).

[4] *S. Chatedus*, A note on a decomposition of integer vectors. Bull. Polon. Acad. Sci. Math. (1988), (to appear).

[5] *S. Chatedus* and *A. Schinzel*, A decomposition of integer vectors II, a volume in memory of Jaroslav Tagamlitzky. Serdica, Sofia, Bulgaria, 1988.

[6] *S. Chowla*, The Riemann hypothesis and Hilbert's tenth problem. Proc. 1963 Number Theory Conf., Boulder, 13—15.

[7] *G. Cohen*, On the largest component of an odd perfect number. J. Austral. Math. Soc. A **42** (1987), 280—286.

[8] *M. J. DeLeon*, Carlitz four-tuples. Fibonacci Quart. **26** (1988), 224—232.

[9] *V. A. Dem'janenko*, L. Euler's conjecture. (Russian), Acta Arith. **25** (1973/74) 127—135; MR **50** #12912.

[10] *N. Elkies*, On $A^4 + B^4 + C^4 = D^4$. Math. Comp. **51** (1988), 825—835..

[11] *R.K. Guy*, Unsolved Problems in Number Theory. Springer Problem Books in Mathematics, (1981).

[12] *K. Hardy* and *K.S. Williams* , A deterministic algorithm for solving the equation $n = u^2 + mv^2$ in coprime integers $u$, $v$. Math. Comp., (to appear).

[13] *R.I. Hess* , Problem 1591, Chained primes. J. Recreational Math. **19** (4) (1987), 325.

[14] *W. Johnson* , Irregular primes and cyclotomic invariants. Math. Comp. **29** (1975), 113—120; MR **51** #12781.

[15] *W. Keller,* The least prime of the form $k \cdot 2^n + 1$ for certain values of $k$. Abstract 88k–11–184, Amer. Math. Soc. Abstracts, **9** (1988), 417—418.

[16] *D. H. Lehmer* , On a problem of Størmer. Illinois J. Math. **8** (1964), 57—79; MR **28** #2072.

[17] *H.W. Lenstra* , Miller's primality test. Inform. Process. Lett. **8** (1979), 86—88; MR 80c : 10008.

[18] *H. Maier* and *C. Pomerance* , Unusually large gaps between consecutive primes. (submitted).

[19] *Yu.V. Matijasevich* , Primes are enumerated by a polynomial in 10 variables. Zap. Naukn. Sem. Leningrad. Otdel. Mat. Inst. Steklov (LOMI) **68** (1977), 62—82, 144—145; MR **58** #21534. English translation: J. Soviet Math. **15** (1981), 33—44.

[20] *R.A. Mollin* and *H.C. Williams* , Quadratic non–residues and prime–producing polynomials. (to appear: Canad. Math. Bull.).

[21] *H.L. Nelson* , A consecutive prime 3x3 magic square. J. Recreational Math., (to appear).

[22] *I. Peterson* , Science News, **133**, #5 (88–01–30), 70; & v. Barry Cipra in Science, & James Gleick, New York Times, Sunday 88–04-17.

[23] *H. J. J. te Riele* ,Computation of all the amicable pairs below $10^{10}$. Math. Comp. **47** (1986), 361—368 & S9—S40.

[24] *A. Rotkiewicz* , Note on the diophantine equation $1 + x + x^2 + \ldots + x^n = y^m$. Elem. Math. **42** (1987), 76.

[25] *A. Schinzel* , A decomposition of integer vectors I. Bull. Polon. Acad. Sci. Math., (1987).

[26] *A. Schinzel* , A decomposition of integer vectors III. Bull. Polon. Acad. Sci. Math., (1988).

[27] *J. L. Selfridge* and B. W. *Pollack* , Fermat's last theorem is true for any exponent up to 25000. Abstract #608—138, Notices Amer. Math Soc. **11** (1964), 97.

[28] *E.S. Selmer*, The Local Postage Stamp Problem, Part1: General Theory; Part 2: the Bases $A_3$ and $A_4$. Univ. of Bergen, Norway, (1986), ISSN 0332—5047.

[29] *C. Størmer*, Quelques théorèmes sur l'équation de Pell $x^2 - Dy^2 = \pm 1$ et leurs applications. Skrifler Videns–selsk. (Christiana) I. Mat. Natur.–Kl., (1897), no 2, 48 pp.

[30] *W.C. Waterhouse,* The determination of all Carlitz sets. (To appear).

[31] *J. Wolstenholme*, On certain properties of prime numbers. Quart. J. Math **5** (1862), 35—39.

[32] *J. Young* and *D. Buell*, The twentieth Fermat number is composite. Math. Comp. **50** (1988), 261—263.

---

Dept. of Mathematics, The University of Calgary, Calgary, AB T2N 1N4, CANADA

# The Partial Zeta Functions of a Real Quadratic Number Field Evaluated at s = 0

*David R. Hayes*

## Introduction

For base field $\mathbb{Q}$, let $f > 0$ be a "conductor" and let $\mathfrak{b} = b\mathbb{Z}$ be an integral ideal which is prime to $f$. We take $b > 0$. The narrow ideal class zeta-function of $\mathfrak{b}$ modulo $f$ is defined for $Re(s) > 1$ by the Dirichlet series

$$\zeta_f(s,\mathfrak{b}) = \sum_r \frac{1}{r^s} \quad (r > 0, \quad r \equiv b \pmod{f}). \tag{1.1}$$

As is well-known, $\zeta_f(s,\mathfrak{b})$ exists as a holomorphic function on the whole complex $s$-plane except for a simple pole at $s = 1$. For base field $k$ a real quadratic number field of discriminant $D$, the narrow ideal class zeta-functions are defined in a similar way. Let $\mathcal{O}_k$ be the ring of integers of $k$, and let $\mathfrak{e} = \mathcal{O}_k$ be the unit ideal of $\mathcal{O}_k$. For a given *integral* ideal $\mathfrak{f}$ of $k$, let $I(\mathfrak{f})$ be the group of fractional ideals which are prime to $\mathfrak{f}$. Then $I(\mathfrak{f})$ contains the subgroup

$$P_{\mathfrak{f}}^{+} = \{x\,\mathcal{O}_k \colon x \in k, \quad x >> 0, \quad x \equiv 1 \pmod{\mathfrak{f}}\},$$

where $x >> 0$ means that $x$ is totally positive (i.e., $x$ is positive in the completions of $k$ at both of its archimedean places). For $\mathfrak{b} \in I(\mathfrak{f})$, let $C\ell_{\mathfrak{f}}^{+}(\mathfrak{b})$ be its coset in the narrow ideal class group $G_{\mathfrak{f}}^{+} = I(\mathfrak{f})/P_{\mathfrak{f}}^{+}$. The narrow ideal class zeta-function of $\mathfrak{b}$ modulo $\mathfrak{f}$ is defined for $Re(s) > 1$ by the Dirichlet series

$$\zeta_{\mathfrak{f}}(s,\mathfrak{b}) = \sum_{\mathfrak{a}} \frac{1}{\mathbb{N}\mathfrak{a}^s} \quad (\mathfrak{a} \in C\ell_{\mathfrak{f}}^{+}(\mathfrak{b}), \quad \mathfrak{a} \text{ integral}) \tag{1.2}$$

where $\mathbb{N}\mathfrak{a} = \#\mathcal{O}_k/\mathfrak{a}$ is the norm of the integral ideal $\mathfrak{a}$. The zeta-function $\zeta_{\mathfrak{f}}(s,\mathfrak{b})$ exists as a holomorphic function on the whole complex $s$-plane except for a simple pole at $s = 1$.

Let $E^+$ be the group of totally positive units in $\mathcal{O}_k$, and let $E^+_{\mathfrak{f}} = \{\eta \in E^+ : \eta \equiv 1 \pmod{\mathfrak{f}}\}$. By Dirichlet's unit theorem, $E^+$ is a free abelian group of rank 1. Since $\mathcal{O}_k/\mathfrak{f}$ is finite, the index $m(\mathfrak{f}) = \#(E^+/E^+_{\mathfrak{f}})$ is finite. The main result of this paper, Theorems 3.2 and 6.4 below, provide a geometric interpretation for the value of the function

$$Z_{\mathfrak{f}}(s,\mathfrak{b}) = \zeta_{\mathfrak{f}}(s,\mathfrak{b}) - \frac{m(\mathfrak{f})}{\mathbb{N}\mathfrak{f}^s} \cdot \zeta_{\mathfrak{e}}(s,\mathfrak{b}\mathfrak{f}^{-1}) \tag{1.3}$$

at $s = 0$ for *integral* ideals $\mathfrak{b}$. If $k$ has narrow class number $h^+$, then $\zeta_{\mathfrak{e}}(0,\mathfrak{b}\mathfrak{f}^{-1})$ takes at most $h^+$ values for all the choices of $\mathfrak{f}$ and $\mathfrak{b}$. Therefore, Theorems 3.2 and 6.4 provide a geometric interpretation for $\zeta_{\mathfrak{f}}(0,\mathfrak{b})$ itself up to a "correction" which is computed from a fixed finite set of values.

In 1968, Siegel [6] derived a formula for $\zeta_{\mathfrak{f}}(0,\mathfrak{b})$ in terms of the first two Bernoulli polynomials

$$B_1(x) = x - \frac{1}{2} \quad \text{and} \quad B_2(x) = x^2 - x + \frac{1}{6}\,. \tag{1.4}$$

We state a version of Siegel's formula in §2. Because this marvelous formula computes $\zeta_{\mathfrak{f}}(0,\mathfrak{b})$ in coordinates relative to a choice of basis elements for the ideal $\mathfrak{b}^{-1}\mathfrak{f}$, it gives one no immediate insight into the "meaning" of the partial zeta-value at $s = 0$. In fact, Siegel's formula illustrates a common dilemma in mathematics which has a certain analogy with the Heisingberg uncertainty principle of physics: In order to make computations, one must introduce coordinates. However, the choice of coordinates injects additional, and often arbitrary, information into the calculations. If the attributes of the coordinate system remain visible in the final results of the computation, they can obscure the meaning of the results.

The idea that $Z_{\mathfrak{f}}(0,\mathfrak{b})$ has a geometrical interpretation evolved from some calculations of function field partial zeta–functions. The values of these zeta–functions at $s = 0$ (see [2], §4) have the form $N - E$, where $N$ is a count of lattice points in a bounded domain and where $E$ is a rational number which gives the "expected value" of the count $N$. This expected value is the measure of the bounded domain divided by the volume of the lattice, both being computed in Haar measure.

The same sort of interpretation can be given for the value at $s = 0$ of the partial zeta–function (1.1). For $x \in \mathbb{R}$ we write $x = [x] + \langle x \rangle$, where $[x]$ is the greatest integer in $x$ and $\langle x \rangle$ is the fractional part of $x$. As one knows (see, e.g. [4])

$$\zeta_{\mathrm{f}}(0,b\mathbb{Z}) = -B_1(\langle\tfrac{b}{f}\rangle) = \tfrac{1}{2} - \langle\tfrac{b}{f}\rangle \,. \tag{1.5}$$

Since the Riemann zeta–function $\zeta(s)$ equals $-\frac{1}{2}$ at $s = 0$, we have

$$\zeta_{\mathrm{f}}(0,b\mathbb{Z}) - \zeta(0) = 1 - \langle\tfrac{b}{f}\rangle = (1 + [\tfrac{b}{f}]) - \tfrac{b}{f} \,. \tag{1.6}$$

As the reader can easily check, $N = 1 + [b/f]$ is the number of points in the translated lattice $M = 1 + \mathbb{Z}\cdot(f/b)$ which lie in the half–open interval $J = (0,1]$. Since the volume of $\mathbb{Z}\cdot(f/b)$ is $f/b$, the quantity $E = b/f$ is just the length of $J$ divided by that volume.

## 2. Siegel's formula for $\zeta_{\mathfrak{f}}(0,\mathfrak{b})$

One can specify the free parameters in Siegel's formula in various ways. We will state a version of the formula which can be efficiently derived from the formulas of Shintani (see [3], §4). Let $\{1,\omega\}$ be a basis for $\mathcal{O}_{\mathrm{k}}$. We choose the order of the real factors in the standard embedding $e: x \longrightarrow (x^{(1)},x^{(2)}) \in \mathbb{R}^2$ of $x \in k$ into Euclidean 2–space so that

$$\det\begin{bmatrix} 1 & 1 \\ \omega^{(1)} & \omega^{(2)} \end{bmatrix} = \sqrt{D}, \tag{2.1}$$

is positive. Let $\epsilon_+$ be the generator of $E^+$ which has slope greater than 1 (i.e., $\epsilon_+^{(2)} > \epsilon_+^{(1)}$) in the embedding $e$, and let $\epsilon_{\mathfrak{f}} = \epsilon_+^{m(\mathfrak{f})}$. Then $\epsilon_{\mathfrak{f}}$ is the generator of $E_{\mathfrak{f}}^+$ with slope greater than 1. Given an integral ideal $\mathfrak{b} \in I(\mathfrak{f})$, let $\{\rho,\tau\}$ be a $\mathbb{Z}$-basis of the lattice $L = e(\mathfrak{b}^{-1}\mathfrak{f})$ in $\mathbb{R}^2$ with $\rho$ totally positive. We assume that $\{\rho,\tau\}$ is oriented in the sense that

$$\mathrm{vol}(L) = \det\begin{bmatrix} \rho^{(1)} & \rho^{(2)} \\ \tau^{(1)} & \tau^{(2)} \end{bmatrix} \tag{2.2}$$

is positive. Now $\epsilon_{\mathfrak{f}}$ acting by multiplication on $\mathbb{R}^2$ through the embedding $e$ is an $\mathbb{R}$-linear map which acts as an automorphism on the lattice $L$. Therefore, if $\mathrm{Mat}_{\mathfrak{f}}$ is the matrix of $\epsilon_{\mathfrak{f}}$ in the oriented basis $\{\rho,\tau\}$, then

$$\mathrm{Mat}_{\mathfrak{f}} = \begin{bmatrix} a & b \\ c & d \end{bmatrix} \in SL_2(\mathbb{Z}). \tag{2.3}$$

We have

$$\begin{aligned} \epsilon_{\mathfrak{f}}\,\rho &= a\rho + c\tau \\ \epsilon_{\mathfrak{f}}\,\tau &= b\rho + d\tau \end{aligned} \tag{2.4}$$

with $ad - bc = 1$. Further, $c > 0$ because

$$c = \frac{\det\begin{bmatrix} \rho^{(1)} & \epsilon_{\mathfrak{f}}^{(1)}\rho^{(1)} \\ \rho^{(2)} & \epsilon_{\mathfrak{f}}^{(2)}\rho^{(2)} \end{bmatrix}}{\mathrm{vol}(L)} = \frac{\rho^{(1)}\rho^{(2)}(\epsilon_{\mathfrak{f}}^{(2)} - \epsilon_{\mathfrak{f}}^{(1)})}{\mathrm{vol}(L)}. \tag{2.5}$$

Let $w_0, z_0 \in \mathbb{Q}$ be the coordinates of $e(1)$ in the basis $\{\rho,\tau\}$:

$$e(1) = w_0\rho + z_0\tau. \tag{2.6}$$

Now Siegel's formula may be stated as follows:

$$\begin{aligned} \zeta_{\mathfrak{f}}(0,\mathfrak{b}) = {} & \frac{a+d}{4\,c} \cdot [B_2(\langle -cw_0\rangle) + B_2(\langle cz_0\rangle)] \\ & - \sum_r B_1\left(\left\langle \frac{ar}{c} - w_0\right\rangle\right) \cdot B_1\left(\left\langle \frac{r}{c} + z_0\right\rangle\right) \end{aligned} \tag{2.7}$$

where $r$ runs through any complete residue system modulo $c$.

In general, the value of $c$ in Siegel's formula is quite large. For example ([3], §4), when $k = \mathbb{Q}(\sqrt{13})$ and $\mathfrak{f} = 17\,\mathcal{O}_k$, we have $c \geq 55{,}602{,}393$.

## 3. The special value $Z_\mathfrak{f}(0,\mathfrak{b})$

Let $\mathscr{C}^+$ be the open cone of points in $\mathbb{R}^2$ with strictly positive coordinates (i.e., the open first quadrant in $\mathbb{R}^2$); and let $\overset{*}{\Delta}_\mathfrak{f}$ be the closure in $\mathscr{C}^+$ of the open triangular region with vertices at the origin, at $e(1)$ and at $e(\epsilon_\mathfrak{f})$. Thus $\overset{*}{\Delta}_\mathfrak{f} \cup \{(0,0)\}$ is compact, but $\overset{*}{\Delta}_\mathfrak{f}$ is not itself compact. Let $J_1$ be the open line segment from the origin to $e(1)$ in $\mathbb{R}^2$, and let $J_\mathfrak{f}$ be the half-open line segment from $e(1)$ (inclusive) to $e(\epsilon_\mathfrak{f})$ (exclusive). We define $\Delta_\mathfrak{f}$ to be the half-open triangular region

$$\Delta_\mathfrak{f} = \text{Interior}(\overset{*}{\Delta}_\mathfrak{f}) \cup J_1 \cup J_\mathfrak{f} \tag{3.1}$$

(see Figure 1). The *weight function* $wt_\mathfrak{f} : \mathscr{C}^+ \longrightarrow \mathbb{R}$ is defined for $\alpha \in \mathscr{C}^+$ by

$$wt_\mathfrak{f}(\alpha) = \begin{cases} 1 & \text{if } \alpha \in \Delta_\mathfrak{f} - J_\mathfrak{f} \\ \frac{1}{2} & \text{if } \alpha \in J_\mathfrak{f} \\ 0 & \text{if } \alpha \notin \Delta_\mathfrak{f} \end{cases} .$$

We may now state our main result.

**Theorem 3.2.** *Let $L = e(\mathfrak{b}^{-1}\mathfrak{f})$, a lattice in $\mathbb{R}^2$. Then*

$$Z_\mathfrak{f}(0,\mathfrak{b}) = \sum_{\lambda \in \mathbf{1}+L} wt_\mathfrak{f}(\lambda) - \frac{\text{area}(\Delta_\mathfrak{f})}{\text{vol}(L)} \tag{3.3}$$

*where $Z_\mathfrak{f}(s,\mathfrak{b})$ is the function* (1.3) *and where* $\mathbf{1} = (1,1)$.

The proof of this theorem is based on two of the Shintani formulas. The contribution of the first formula is worked out in §4 and that of the second in §5. It would be interesting to derive (3.3) directly from Siegel's formula (2.7). However, the technical details involved in dealing with two distinct coordinate systems, one for evaluating $\zeta_{\mathfrak{f}}(0,\mathfrak{b})$ and the other for evaluating $\zeta_{\mathfrak{e}}(0,\mathfrak{b}\mathfrak{f}^{-1})$, present unpleasant obstacles. The Shintani formulas also force one to choose a coordinate system, but one can arrange to work with a single system throughout. We proceed now to describe a choice of coordinates which is especially convenient for our purpose.

Given a fractional ideal $\mathfrak{a}$ of $k$, we define $\mu(\mathfrak{a}) \in \mathbb{Q}^*$ to be the positive generator of $\mathfrak{a} \cap \mathbb{Q}$, and we define $\nu(\mathfrak{a}) \in \mathbb{Q}^*$ by

$$\mathbb{N}\mathfrak{a} = \mu(\mathfrak{a}) \cdot \nu(\mathfrak{a}). \tag{3.4}$$

These functions are homogeneous under multiplication by positive $g \in \mathbb{Q}^*$:

$$\mu(g\mathfrak{a}) = g \cdot \mu(\mathfrak{a}) \quad \text{and} \quad \nu(g\mathfrak{a}) = g \cdot \nu(\mathfrak{a}). \tag{3.5}$$

For integral $\mathfrak{a}$, $\mu(\mathfrak{a}) \in \mathbb{Z}$ is the exponent of the additive group $\mathcal{O}_k/\mathfrak{a}$ and so divides $\mathbb{N}\mathfrak{a} = \#(\mathcal{O}_k/\mathfrak{a})$. Thus, $\nu(\mathfrak{a}) \in \mathbb{Z}$ also when $\mathfrak{a}$ is integral.

**Lemma 3.6.** *There is a rational number $y$ such that $\sigma = y + \nu(\mathfrak{a}) \cdot \omega \in \mathfrak{a}$. For any such $\sigma$, $\{\mu(\mathfrak{a}),\sigma\}$ is an oriented $\mathbb{Z}$–basis for $\mathfrak{a}$ relative to the embedding $\mathfrak{e}$.*

**Proof.** By (3.5), there is no loss of generality in assuming that $\mathfrak{a}$ is integral. Put $\mu = \mu(\mathfrak{a})$ and $\nu = \nu(\mathfrak{a})$, and let $\phi : \mathcal{O}_k \rightarrow \phi: \mathcal{O}_k/\mathfrak{a}$ be the canonical map. The element $\phi(1)$ generates a subgroup $A$ of order $\mu$. Since $A$ has index $\nu$ in $\mathcal{O}_k/\mathfrak{a}$, $\phi(\nu\omega) = \nu \cdot \phi(\omega) \in A$, which proves that $-\nu\omega \equiv y \pmod{\mathfrak{a}}$ for some $y \in \mathbb{Z}$. Let $M$ be the $\mathbb{Z}$–submodule of $\mathcal{O}_k$ generated by the ordered basis $\{\mu,\sigma\}$. Then $M \subseteq \mathfrak{a}$. On the other hand

$$\det\begin{bmatrix} \mu & \mu \\ \sigma^{(1)} & \sigma^{(2)} \end{bmatrix} = \mu\nu \cdot \det\begin{bmatrix} 1 & 1 \\ \omega^{(1)} & \omega^{(2)} \end{bmatrix} = \mathbb{N}\mathfrak{a} \cdot \sqrt{D},$$

from which we conclude that $\{\mu,\sigma\}$ is oriented and that $M = \mathfrak{a}$. □

Suppose that $\mathfrak{a} = \mathfrak{b}^{-1}\mathfrak{f}$, where $\mathfrak{f}$ and $\mathfrak{b}$ are the relatively prime integral ideals introduced in §2 above. Let $f = \mu(\mathfrak{f})$, a positive integer, and let $p = \mu(\mathfrak{b}^{-1}\mathfrak{f})$. Since $1 \in \mathfrak{b}^{-1}$, $f \in \mathfrak{b}^{-1}\mathfrak{f}$ and so $f = gp$ for some integer $g$. Thus,

$$p = \mu(\mathfrak{b}^{-1}\mathfrak{f}) = \frac{\mu(\mathfrak{f})}{g} = \frac{f}{g}. \tag{3.7}$$

**Lemma 3.8.** *The positive integers $f$ and $g$ are relatively prime.*

**Proof.** Let $d = (f,g)$. We will show that $\mathfrak{f}$ contains $f/d$, a contradiction unless $d = 1$. From $(f/g) \in \mathfrak{b}^{-1}\mathfrak{f}$, it follows that

$$(f/d)\mathfrak{b} \subseteq (g/d)\mathfrak{f} \subseteq \mathfrak{f}$$

which implies that $\mathfrak{f}$ divides $(f/d)\mathfrak{b}$. Since $\mathfrak{f}$ and $\mathfrak{b}$ are relatively prime, $\mathfrak{f}$ must divide $(f/d)\,\mathcal{O}_k$. □

By Lemma 3.6, $\mathfrak{b}^{-1}\mathfrak{f}$ has an oriented $\mathbb{Z}$–basis of the form $\{p,\sigma\}$. The matrix $\mathrm{Mat}_{\mathfrak{f}}$ (see (2.3)) of the unit $\epsilon_{\mathfrak{f}}$ in this basis has entries which are defined by

$$\begin{aligned} \epsilon_{\mathfrak{f}}\, p &= ap + c\sigma \\ \epsilon_{\mathfrak{f}}\, \sigma &= bp + d\sigma. \end{aligned} \tag{3.9}$$

For our calculations in §5, we require some information about the reduction of $\mathrm{Mat}_{\mathfrak{f}}$ modulo $\mathfrak{f}$. As the next lemma shows, $\mathrm{Mat}_{\mathfrak{f}}$ belongs to the subgroup $\Gamma_1(f)$ of the modular group $\Gamma$:

**Lemma 3.10.** *We have $c \equiv 0 \pmod{f}$ and $a \equiv d \equiv 1 \pmod{f}$.*

**Proof.** From the first equation in (3.9)

$$\epsilon_{\mathfrak{f}} - 1 = \left[\frac{a-1}{f}\right] f + \left[\frac{c}{f}\right] g\sigma. \tag{3.11}$$

Now $\{f,g\sigma\} = g \cdot \{p,\sigma\}$ is a $\mathbb{Z}$–basis for $g\mathfrak{b}^{-1}\mathfrak{f}$. It follows, therefore, from (3.11) and $ad - bc = 1$ that the assertions of this lemma are equivalent to $\epsilon_{\mathfrak{f}} - 1 \in g\mathfrak{b}^{-1}\mathfrak{f}$. We will show that $1 \in g\mathfrak{b}^{-1}$, and this will complete the proof since then $\epsilon_{\mathfrak{f}} - 1 \in \mathfrak{f} \subseteq g\mathfrak{b}^{-1}\mathfrak{f}$. Now because $1 \in \mathfrak{b}^{-1}$, $\mu(\mathfrak{b}^{-1}) = 1/t$ for some integer $t$. Therefore

$$\mathfrak{b}^{-1}\mathfrak{f} \subseteq \mathfrak{b}^{-1} \Longrightarrow g \,|\, t,$$

since $(f,g) = 1$ by Lemma 3.8. Thus $(1/g) \in \mathfrak{b}^{-1}$ or $1 \in g\mathfrak{b}^{-1}$ as required.

□

## 4. The sector $S$

For $\alpha = (\alpha^{(1)}, \alpha^{(2)}) \in \mathbb{R}^2$, we write

$$Tr(\alpha) = \alpha^{(1)} + \alpha^{(2)} \quad \text{and} \quad N(\alpha) = N\alpha = \alpha^{(1)}\alpha^{(2)}. \tag{4.1}$$

For $x \in k$, we write $x\alpha = (x^{(1)}\alpha^{(1)}, x^{(2)}\alpha^{(2)})$ for the action of $x$ on $\mathbb{R}^2$ through the embedding $e$. For $q \in \mathbb{Q}$, we put $q = e(q) = (q,q)$; and we let $M = 1 + L$ be the translation by 1 of the lattice $L = e(\mathfrak{b}^{-1}\mathfrak{f})$ of Theorem 3.2.

For any fractional ideal $\mathfrak{C}$, the integral ideals in the narrow ideal class $Cl_{\mathfrak{e}}^{+}(\mathfrak{C})$ of $\mathfrak{C}$ are the ideals $\mathfrak{a} = x\mathfrak{C}$ with

$$x >> 0 \quad \text{and} \quad x \in \mathfrak{C}^{-1}. \tag{4.2}$$

Therefore for $Re(s) > 1$,

$$\zeta_{\mathfrak{e}}(s,\mathfrak{C}) = \mathbb{N}\mathfrak{C}^{-s} \cdot \sum_{x}{}' \frac{1}{\mathbb{N}x^{s}},$$

where the prime on the summation sign indicates that $x$ is restricted by (4.2) and also by the condition that only one element $x$ appears from each orbit of $E^{+}$ acting on $\mathfrak{C}^{-1}$. Thus if $\mathscr{D}$ is any fundamental domain for $E^{+}$ acting through $e$ on $\mathscr{C}^{+}$, then

$$\zeta_{\mathfrak{e}}(s,\mathfrak{C}) = \mathbb{N}\mathfrak{C}^{-s} \cdot \sum_{\lambda} \frac{1}{N\lambda^{s}} \quad (\lambda \in e(\mathfrak{C}^{-1}) \cap \mathscr{D}).$$

By a slightly more complicated argument, one can show that for *integral* ideals $\mathfrak{b}$,

$$\zeta_{\mathfrak{f}}(s,\mathfrak{b}) = \mathbb{N}\mathfrak{b}^{-s} \cdot \sum_{\gamma} \frac{1}{N\gamma^{s}} \qquad (\gamma \in M \cap \mathscr{D}(\mathfrak{f})) \tag{4.3}$$

where $\mathscr{D}(\mathfrak{f})$ is any fundamental domain for the action of $E_{\mathfrak{f}}^{+}$ on $\mathscr{C}^{+}$. Since $\mathscr{D}(\mathfrak{f})$ is a union of $m(\mathfrak{f})$ fundamental domains for the action of $E^{+}$, we have also

$$m(\mathfrak{f}) \cdot \zeta_{\mathfrak{e}}(s,\mathfrak{b}\mathfrak{f}^{-1}) = \mathbb{N}(\mathfrak{b}\mathfrak{f}^{-1})^{-s} \cdot \sum_{\lambda} \frac{1}{N\lambda^{s}} \quad (\lambda \in L \cap \mathscr{D}(\mathfrak{f})). \tag{4.4}$$

One very convenient fundamental domain for the action of $E_{\mathfrak{f}}^{+}$ is $\mathscr{D}(\mathfrak{f}) = S$, where $S$ is the sector (see Figure 2)

$$S = \left\{ \alpha \in \mathscr{C}^{+} \colon 1 \leq \frac{\alpha^{(2)}}{\alpha^{(1)}} < \frac{\epsilon_{\mathfrak{f}}^{(2)}}{\epsilon_{\mathfrak{f}}^{(1)}} \right\}. \tag{4.5}$$

Let $\chi_S$ be the characteristic function of $S$ in $\mathscr{C}^{+}$. Then from (4.3) and (4.4) and the definition (1.3), we find the following representation for the function $Z_{\mathfrak{f}}(s,\mathfrak{b})$:

$$Z_{\mathfrak{f}}(s,\mathfrak{b}) = \mathbb{N}\mathfrak{b}^{-s} \left[ \sum_{\gamma \in M} \frac{\chi_S(\gamma)}{N\gamma^{s}} - \sum_{\lambda \in L} \frac{\chi_S(\lambda)}{N\lambda^{s}} \right]. \tag{4.6}$$

The first summation on the right in (4.6) decomposes into $\sum_1 + \sum_2$ where

$$\sum\nolimits_1 = \sum_{\gamma} \frac{1}{N\gamma^{s}} \qquad (\gamma \in M; \quad \gamma \in S - (1 + S))$$

and

$$\sum\nolimits_2 = \sum_{\gamma} \frac{1}{N\gamma^{s}} \qquad (\gamma \in M; \quad \gamma \in 1 + S).$$

Thus

$$Z_{\mathfrak{f}}(s,\mathfrak{b}) = \mathbb{N}\mathfrak{b}^{-s} \cdot (\textstyle\sum_1 + \sum_2 - \sum_3). \tag{4.7}$$

We note that

$$\Sigma_2 - \Sigma_3 = \sum_\lambda \left( \frac{1}{N(1+\lambda)^s} - \frac{1}{N\lambda^s} \right) \quad (\lambda \in L \cap S). \tag{4.8}$$

In the remainder of this section, we will evaluate the analytic continuation of $\sum_2 - \sum_3$ at $s = 0$ using a formula of Shintani. In §5, we will use another Shintani formula to evaluate $\sum_1$ at $s = 0$.

Let $\rho, \tau$ be totally positive basis vectors for $\mathbb{R}^2$. For $\alpha \in \mathscr{C}^+$, the double series

$$\sum_{m,n=0}^{\infty} \frac{1}{N(\alpha + m\rho + n\tau)^s} \tag{4.9}$$

converges when $Re(s) > 1$ and defines an analytic function in that domain. Shintani shows in [7], §1 (the case $r = 1$, $n = 2$ of Proposition 1) that (4.9) exists as a meromorphic function on the whole complex $s$-plane and takes at $s = 0$ the value

$$B_1(w)B_1(z) + \tfrac{1}{4}\left[B_2(w) \cdot \boldsymbol{Tr}(\rho/\tau) + B_2(z) \cdot \boldsymbol{Tr}(\tau/\rho)\right] \tag{4.10}$$

where $\alpha = w\rho + z\tau$ and where $B_1$ and $B_2$ are the Bernoulli polynomials (1.4).

We will apply the evaluation (4.10) with $\rho = \boldsymbol{p}$ and $\tau = \epsilon_{\mathfrak{f}}\boldsymbol{p}$. Then $\tau/\rho = \epsilon_{\mathfrak{f}}\mathbf{1}$, so that

$$\boldsymbol{Tr}(\rho/\tau) = \boldsymbol{Tr}(\tau/\rho) = \text{trace}(\text{Mat}_{\mathfrak{f}}) = a + d. \tag{4.11}$$

Let $L_0$ be the sublattice of $L$ which is generated by $\boldsymbol{p}$ and $\epsilon_{\mathfrak{f}}\boldsymbol{p}$. Since $L$ is generated by $\boldsymbol{p}$ and $\boldsymbol{e}(\sigma)$ it follows from (3.9) that

$$r\boldsymbol{e}(\sigma) = \frac{r}{c}\left(\epsilon_{\mathfrak{f}}\boldsymbol{p} - a\boldsymbol{p}\right) \tag{4.12}$$

runs through a set of coset representatives for $L/L_0$ as $r$ runs through any complete residue system modulo $c$. Therefore, $L/L_0$ is a cyclic group of order $c$. Since $ad - bc = 1$, $d$ is invertible modulo $c$. Replacing $r$ by $-dr$ in (4.12), we see that

$$\frac{adr}{c}\cdot p - \frac{rd}{c}\cdot \epsilon_f p = \frac{(1+bc)r}{c}\cdot p - \frac{rd}{c}\cdot \epsilon_f p$$

$$\equiv \frac{r}{c}p + \left\langle\frac{-rd}{c}\right\rangle \epsilon_f p \pmod{L_0}$$

runs through coset representatives for $L/L_0$ as $r$ runs modulo $c$. Put

$$\lambda = \frac{r}{c}p + \left\langle\frac{-rd}{c}\right\rangle \epsilon_f p + mp + n\epsilon_f p\ .$$

Then $\lambda$ takes every value in $L$ for $m,n \in \mathbb{Z}$ and $1 \leq r \leq c$. For these values of $r$, $\lambda \in S \Longleftrightarrow m > -r/c$ and $n \geq -\langle -rd/c\rangle \Longleftrightarrow m \geq 0$ and $n \geq 0$. Thus for any $\beta \in \mathscr{C}^+$,

$$\sum_{\lambda\in S\cap L} \frac{1}{N(\beta+\lambda)^s} = \sum_{r=1}^{c} \sum_{m,\,n=0}^{\infty} \frac{1}{N(\alpha_r+mp+n\epsilon_f p)^s}$$

where $\alpha_r = \beta + (r/c)p + \langle -rd/c\rangle\epsilon_f p$. Applying Shintani's evaluation (4.10) to the inner summation on the right above with first $\beta = 1$ and then $\beta = 0$, we find that at $s = 0$

$$\Sigma_2 = \sum_{r=1}^{c}\left\{B_1\left[\frac{1}{p}+\frac{r}{c}\right]B_1\left[\left\langle\frac{-rd}{c}\right\rangle\right] + \frac{a+d}{4}\left[B_2\left[\frac{1}{p}+\frac{r}{c}\right] + B_2\left[\left\langle\frac{-rd}{c}\right\rangle\right]\right]\right\}$$

and

$$\Sigma_3 = \sum_{r=1}^{c}\left\{B_1\left[\frac{r}{c}\right]B_1\left[\left\langle\frac{-rd}{c}\right\rangle\right] + \frac{a+d}{4}\left[B_2\left[\frac{r}{c}\right] + B_2\left[\left\langle\frac{-rd}{c}\right\rangle\right]\right]\right\}.$$

Thus at $s = 0$,

$$\Sigma_2 - \Sigma_3 = \frac{1}{p}\sum_{r=1}^{c} B_1\left[\left\langle\frac{-rd}{c}\right\rangle\right] + \frac{a+d}{4}\sum_{r=1}^{c}\left[B_2\left[\frac{1}{p}+\frac{r}{c}\right] - B_2\left[\frac{r}{c}\right]\right].$$

By the distribution property for $B_1$ (see [5], Chap. 2) the first summation on the right above equals $-1/2$. The value of the second sum is easily computed and is found to be $(c+p)/p^2$. Therefore,

$$\Sigma_2 - \Sigma_3 = -\frac{1}{2p} + \frac{(a+d)(c+p)}{4p^2} = \frac{ac+dc}{4p^2} + \frac{a+d-2}{4p} \tag{4.13}$$

at $s = 0$.

## 5. The strip $T$

Let

$$\Delta_* = \Delta_{\mathfrak{f}} - J_{\mathfrak{f}}$$

be $\Delta_{\mathfrak{f}}$ deprived of one of its sides. In order to evaluate $\sum_1$ at $s = 0$, we decompose the strip $S - (1 + S)$ as a disjoint union

$$S - (1+S) = \Delta_* \cup T, \tag{5.1}$$

where

$$T = \{u1 + ve(\epsilon_{\mathfrak{f}}): \ 0 < u \leq 1, \ 0 \leq v, \ 1 \leq u + v\}$$

is a half-open strip in $\mathscr{C}^+$ with vertices at $1$ and $e(\epsilon_{\mathfrak{f}})$ (see Figure 2). Let $\sum_T$ denote the value at $s = 0$ of the series

$$\sum_{\gamma \in M \cap T} \frac{1}{N\gamma^s}.$$

Then by (5.1), the value of $\sum_1$ at $s = 0$ is

$$\#(\Delta_* \cap M) + \textstyle\sum_T . \tag{5.2}$$

Our evaluation of $\sum_T$ requires a second Shintani formula ([7], §1, Proposition 1 with $r = n = 1$): For $\beta \in \mathscr{C}^+$,

$$\sum_{\ell=0}^{\infty} \frac{1}{N(\beta+\ell)^s}$$

exists as a meromorphic function on the whole complex $s$-plane and takes the value

$$\tfrac{1}{2}\,(1 - Tr(\beta)) \tag{5.3}$$

at $s = 0$. To apply this formula, we need a convenient description of the points $\gamma \in M$ in terms of the basis $\{p, \epsilon_{\mathfrak{f}} p\}$. Because $\mathrm{Mat}_{\mathfrak{f}} \in SL_2(\mathbb{Z})$, $(m,n) = (a\ell + bt, c\ell + dt)$ runs through $\mathbb{Z}^2$ when $(\ell,t)$ runs through $\mathbb{Z}^2$. Using (3.9), we may therefore write a typical element $\gamma \in M$ in the form

$$\gamma = 1 + mp + n\sigma = 1 + mp + \tfrac{n}{c}(\epsilon_f p - ap) = 1 - \tfrac{t}{c}p + \tfrac{n}{c}\,\epsilon_f p; \quad (5.4)$$

keeping always in mind that $n = c\ell + dt$.

**Lemma 5.5.** *The value $c/p = cg/f$ is an integer. Further*

$$\#(\Delta_* \cap M) = \sum_{t=1}^{c/p} \left[\frac{dt}{c}\right] - \left[\frac{(d-1)t}{c}\right]. \quad (5.6)$$

**Proof.** That $c/p$ is an integer follows from Lemma 3.10. Now the pairs $(t,n)$ such that $\gamma$ in (5.4) belongs to $\Delta_*$ are those satisfying $0 < t < c/p$ and $0 \leq n < t$. For fixed $t$, the number of $n = c\ell + dt$ satisfying these inequalities is the number of integers $\ell$ such that

$$0 \leq c\ell + dt < t \iff \frac{t(d-1)}{c} < -\ell \leq \frac{dt}{c}.$$

This number is $[dt/c] - [(d-1)t/c]$. □

**Lemma 5.7.** *Let $\delta$ be the integer such that $\delta f = \mathrm{GCD}(c,d-1)$. Then*

$$\#(J_f \cap M) = g\delta. \quad (5.8)$$

**Proof.** That $\delta$ is an integer follows from Lemma 3.10. The pairs $(t,n)$ such that $\gamma$ in (5.4) belongs to $J_f$ are those satisfying $0 \leq t < c/p$ and $n = t$. For fixed $t$, $t = dt + c\ell$ for some integer $\ell \iff c$ divides $(d-1)t \iff c/f\delta$ divides $t$. Since $p = f/g$, this happens for exactly $g\delta = (c/p)/(c/f\delta)$ values of $t$ in the range $0 \leq t < c/p$. □

The pairs $(t,n)$ such that $\gamma$ in (5.4) lies in the strip $T$ are those satisfying $0 \leq t < c/p$ and $n \geq t$; and $n \geq t \iff \ell \geq (1-d)c/t$. Therefore, $\sum_T$ is the value at $s = 0$ of the series

$$\sum_{t=0}^{(c/p)-1} \sum_{n \geq t} \frac{1}{N(1 - \frac{t}{c}p + \frac{n}{c}\,\epsilon_f p)^s}.$$

Putting $n = c\ell + dt$, we can decompose the inner summation above into the finite sum

$$\Sigma_t = \sum_{\gamma} \frac{1}{N(\gamma)^s} \quad \left(\frac{(1-d)t}{c} \leq \ell < 0\right) \tag{5.9}$$

plus $N(\epsilon_f p)^{-s}$ times the summation

$$\sum_{\ell=0}^{\infty} \frac{1}{N(\beta_t + \ell)^s},$$

where $\beta_t = \epsilon_f^{-1} p^{-1} + (td - t\epsilon_f^{-1})c^{-1}$. By the Shintani formula (5.3), this last last summation takes the value

$$\tfrac{1}{2}(1 - Tr(\beta_t)) = \tfrac{1}{2}\left[1 - \frac{a+d}{p} - \frac{(d-a)t}{c}\right] \tag{5.10}$$

at $s = 0$. Of course, the finite sum $\sum_t$ evaluates to $[t(d-1)/c]$ at $s = 0$. Therefore

$$\Sigma_T = \sum_{t=0}^{(c/p)-1} \left[\frac{t(d-1)}{c}\right] + \frac{1}{2} \sum_{t=0}^{(c/p)-1} \left(1 - \frac{a+d}{p} + \frac{(a-d)t}{c}\right). \tag{5.11}$$

The second summation in (5.11) (times 1/2) is easily evaluated and is found to equal

$$\frac{d-a+2c}{4p} - \frac{ac+3dc}{4p^2}. \tag{5.12}$$

For the first summation, we have

$$\sum_{t=0}^{(c/p)-1} \left[\frac{t(d-1)}{c}\right] = \sum_{t=1}^{(c/p)-1} \frac{t(d-1)}{c} - \left\langle\frac{t(d-1)}{c}\right\rangle$$

$$= \frac{cd-c}{2p^2} - \frac{d-1}{2p} - \sum_{t=0}^{(c/p)-1} \left\langle\frac{td_*}{c_*}\right\rangle$$

where $c_* = c/f\delta$ and $d_* = (d-1)/f\delta$. Arranging the last summation into classes modulo $c_*$ via $t = r + jc_*$, we compute

$$\sum_{t=0}^{(c/p)-1} \left\langle\frac{td_*}{c_*}\right\rangle = \sum_{r=0}^{c_*-1} \sum_{j=0}^{g\delta-1} \left\langle\frac{(r+jc_*)d_*}{c_*}\right\rangle$$

$$= g\delta \sum_{r=0}^{c_*-1} \left\langle\frac{rd_*}{c_*}\right\rangle = \frac{c}{2p} - \frac{g\delta}{2}$$

because $rd_*$ runs over a complete residue system modulo $c_*$. Putting these evaluations together with (5.12), we obtain

$$\sum\nolimits_T = \frac{2-a-d}{4p} - \frac{ac+dc}{4p^2} + \frac{g\delta}{2} - \frac{c}{2p^2}\,. \tag{5.13}$$

We may now complete the proof of Theorem 3.2. At $s = 0$, we have

$$Z_{\mathfrak{f}}(0,\mathfrak{b}) = \#(\Delta_* \cap M) + \sum\nolimits_T + \sum\nolimits_2 - \sum\nolimits_3$$

by (4.7) and (5.2). By (5.13) and (4.13), at $s = 0$, $\sum_T + \sum_2 - \sum_3 = (g\delta/2) - (c/2p^2)$; and we recognize $g\delta$ as $\#(J_{\mathfrak{f}} \cap M)$ by Lemma 5.7. Therefore, the theorem will follow if we prove that

$$\frac{\mathrm{area}(\Delta_{\mathfrak{f}})}{\mathrm{vol}(L)} = \frac{c}{2p^2}\,.$$

But by (3.9)

$$\mathrm{area}(\Delta_{\mathfrak{f}}) = \frac{1}{2}\cdot\det\begin{bmatrix} 1 & 1 \\ \epsilon_{\mathfrak{f}}^{(1)} & \epsilon_{\mathfrak{f}}^{(2)} \end{bmatrix}$$

$$= \frac{1}{2p^2}\cdot\det\begin{bmatrix} p & p \\ \epsilon_{\mathfrak{f}}^{(1)}p & \epsilon_{\mathfrak{f}}^{(2)}p \end{bmatrix}$$

$$= \frac{1}{2p^2}\cdot\det\begin{bmatrix} p & p \\ c\sigma^{(1)} & c\sigma^{(2)} \end{bmatrix}$$

$$= \frac{c}{2p^2}\cdot\mathrm{vol}(L)$$

as required.

This completes the proof of Theorem 3.2. Although a coordinate system figured in the proof in an essential way, the attributes of the coordinate system are not visible in the final result. If we wish to compute $Z_{\mathfrak{f}}(0,\mathfrak{b})$ using the coordinate system, then by (5.6) we can employ the result in the form

$$Z_{\mathfrak{f}}(0,\mathfrak{b}) = \sum_{t=1}^{c/p} \left( \left[\frac{dt}{c}\right] - \left[\frac{(d-1)t}{c}\right] \right) + \frac{g\delta}{2} - \frac{c}{2p^2} . \tag{5.14}$$

However from the point of view of computational efficiency, (5.14) is little better than Siegel's formula (2.7).

## 6. An application of Pick's Theorem

For positive $h$, let $\mathscr{P}_h$ be the half-open interval $(0,h]$ on the real line. Let $N_h$ be the number of points in the translated lattice $1 + \mathbb{Z} \cdot (f/b)$ which lie in $\mathscr{P}_h$. The reader can easily verify that (1.6) implies

$$\zeta_f(0,b\mathbb{Z}) - \zeta(0) = N_h - \frac{h}{f/b} \tag{6.1}$$

whenever $h \in 1 + \mathbb{Z}\cdot(f/b)$. In §1, we noted the case $h = 1$ of (6.1). In this section we state an analogue of (6.1) for $Z_{\mathfrak{f}}(0,\mathfrak{b})$.

Let $\bar{S}$ be the closure of $S$ in $\mathscr{C}^+$, and let $\mathscr{L}_1$ and $\mathscr{L}_2$ be the two boundary rays of $\bar{S}$. An *M-polygon* for the sector $S$ is a simple closed polygon $\mathscr{P}$ in $\bar{S} \cup \{(0,0)\}$ such that:

(1) The origin is a vertex of $\mathscr{P}$. (2) The other vertices of $\mathscr{P}$ all belong to $\boldsymbol{M}$. (3) One side of $\mathscr{P}$ is a line segment from $(0,0)$ to a point of $\boldsymbol{M}$ on $\mathscr{L}_1$. (4) Another side of $\mathscr{P}$ is a line segment from $(0,0)$ to a point of $\boldsymbol{M}$ on $\mathscr{L}_2$. The *M-weight* of the *M*-polygon $\mathscr{P}$ is defined by

$$wt_M(\mathscr{P}) = \frac{1}{2} B + C$$

where $B$ is the number of points of $\boldsymbol{M}$ on the boundary of $\mathscr{P}$ in $\mathscr{C}^+$ (so the origin is never included in the count $B$) and where $C$ is the number of points of $\boldsymbol{M}$ in the interior of $\mathscr{P}$.

**Lemma 6.2 (Pick's Theorem).** *Let $\mathcal{T}$ be any triangle in $\overline{S}$ whose vertices are points of $M$. Then*

$$area(\mathcal{T}) = (\tfrac{1}{2}B + C - 1)\cdot vol(L) \qquad (6.3)$$

*where $B$ is the number of points of $M$ on the boundary of $\mathcal{T}$ and where $C$ is the number of points of $M$ in the interior of $\mathcal{T}$.*

**Proof.** Any proof of Pick's Theorem for the standard lattice $\mathbb{Z}^2$ (see, e.g., [1], §13.5) is easily generalized to the translated lattice $M$.

□

We may now restate Theorem 3.2 with an arbitrary $M$–polygon for $S$ replacing $\Delta_f$:

**Theorem 6.4.** *Let $\mathcal{P}$ be an $M$–polygon for $S$. Then*

$$\frac{1}{2} + Z_f(0,\mathfrak{b}) = wt_M(\mathcal{P}) - \frac{\text{area}(\mathcal{P})}{\text{vol}(L)}\,. \qquad (6.5)$$

**Proof.** First take $\mathcal{P} = \Delta_f^* \cup \{0,0\}$, where $\Delta_f^*$ is the closure of $\Delta_f$ in $\mathscr{C}^+$. Then (6.5) is just another way of writing (3.3). In fact since each point $P \in M \cap J_1$ has a twin $\epsilon_f P \in \Delta_f^*$, these points contribute the same total weight to $wt_M(\mathcal{P})$ as they do to the summation on the right hand side of (3.3). The points in $M \cap J_f$ each contribute weight 1/2 to $wt_M(\mathcal{P})$ and the summation. This accounts for every point of $M$ in the boundary of $\mathcal{P}$ except for the point $e(\epsilon_f)$ which contributes an extra 1/2. Of course, each point of $M$ in the interior of $\mathcal{P}$ contributes weight 1 to both $wt_M(\mathcal{P})$ and the summation.

Now let $P$ be an arbitrary $M$–polygon for $S$. Let $\mathcal{T}$ be any triangle in $\overline{S}$ with vertices in $M$ which has at least one side which is a boundary segment of the polygon $\mathcal{P}$. Then by Lemma 6.2, Equation (6.5) is valid for $\mathcal{P}$ if and only if it is valid for either $\overline{\mathcal{P} - \mathcal{T}}$ or $\mathcal{P} \cup \mathcal{T}$. Since $\overline{S}$ is simply connected, we may construct a sequence of $M$–polygons $\mathcal{P}_0, \mathcal{P}_1, \ldots, \mathcal{P}_{n+1}$ such that $\mathcal{P}_0 = \Delta_f^* \cup \{(0,0)\}$ and $\mathcal{P}_{n+1} = \mathcal{P}$ and such that for $0 \leq i \leq n$, $\mathcal{P}_{i+1}$ is obtained from $\mathcal{P}_i$ by adding or removing a triangle $\mathcal{T}_i$ in $\bar{S}$ with vertices from $M$. □

The author thanks Glenn Stevens for bringing Pick's Theorem to his attention. Stevens was introduced to Pick's Theorem when, as a high-school student, he attended one of the summer institutes for prospective young mathematicians organized by Professor Arnold Ross at Ohio State University.

## 7. A note on computing $\zeta_{\mathfrak{f}}(0,\mathfrak{b})$

The computational cost in using (5.14) to compute $Z_{\mathfrak{f}}(0,\mathfrak{b})$ is proportional to $c/p = cg/f$. However, there are algorithms for computing $\zeta_{\mathfrak{f}}(0,\mathfrak{b})$ at a cost proportional to $\log(c)$. (See [3] or [8]). Clearly, any algorithm which computes $\zeta_{\mathfrak{f}}(0,\mathfrak{b})$ for all $\mathfrak{f}$ and $\mathfrak{b}$ will compute $Z_{\mathfrak{f}}(0,\mathfrak{b})$. The reverse is also true. To prove it, take $\mathfrak{f} = \mathfrak{p} = z\,\mathcal{O}_k$ where $\mathfrak{p}$ is a principal prime ideal of $k$ which is generated by a totally positive element $z \in \mathcal{O}_k$. Let $P^+(\mathfrak{p})$ be the group of principal ideals which may be generated by a totally positive element which is prime to $\mathfrak{p}$. Choose a set $R_1$ of integral ideals which represent the classes in $P^+(\mathfrak{p})/P^+_{\mathfrak{p}}$, and let $R$ be a set of totally positive generators for the ideals in $R_1$. Then from the theory of $L$-functions

$$\sum_{x \in R} \zeta_{\mathfrak{p}}(0,x\mathfrak{b}) = 0,$$

so that

$$\sum_{x \in R} Z_{\mathfrak{p}}(0,x\mathfrak{b}) = -m(\mathfrak{p}) \cdot \sum_{x \in R} \zeta_{\mathfrak{e}}(0,xz^{-1}\mathfrak{b})$$

$$= (1 - \mathbb{N}\mathfrak{p}) \cdot \zeta_{\mathfrak{e}}(0,\mathfrak{b}).$$

Since there exist many such primes $\mathfrak{p}$ (in fact, we can take $z = p$, where $p$ is any rational prime which is inert in $k/\mathbb{Q}$), any algorithm which computes $Z_{\mathfrak{f}}(0,\mathfrak{b})$ will compute $\zeta_{\mathfrak{f}}(0,\mathfrak{b})$.

From the geometry of the sector $\bar{S}$, there is an $M$-polygon $\mathcal{P}_*$ in $\bar{S}$ which is mininal in the sense that its interior contains no points of $M$. We leave the reader with the following question: Is there a simple algorithm for computing $\mathscr{P}_*$, and does it in conjunction with $\mathscr{P} = \mathscr{P}_*$ in Theorem 6.4 provide an algorithm for computing $Z_f(0,\mathfrak{b})$ at a cost proportional to $\log(c)$?

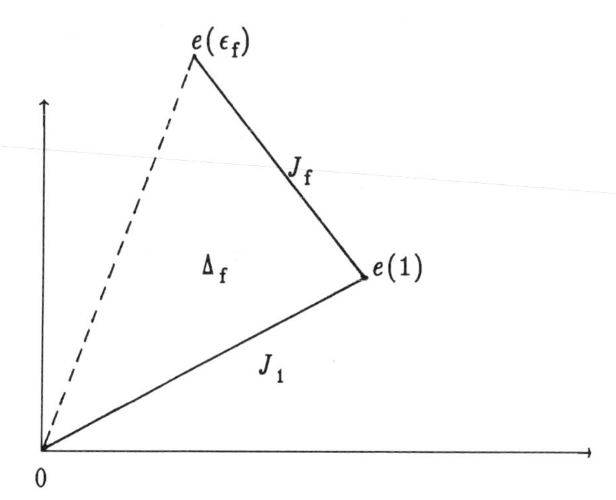


**Figure 1**

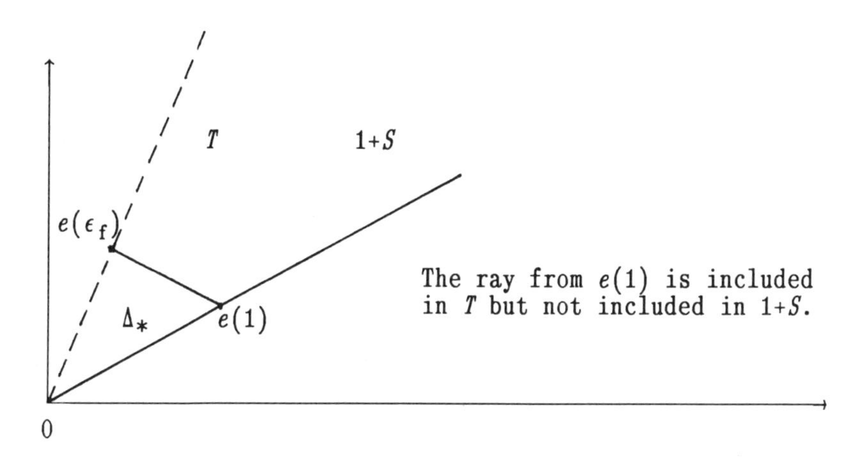


**Figure 2**

## References


[1] *H.S.M. Coxeter*, Introduction to Geometry (2nd ed.). John Wiley and Sons, New York (1980).

[2] *D. Hayes*, The refined $\mathfrak{p}$-adic abelian Stark conjecture in function fields. Inventiones Mathematicae, **94** (1988), 505–527.

[3] *D. Hayes*, Brumer elements over a real quadratic base field. Expositiones Mathematicae (to appear).

[4] *K. Ireland* and *M. Rosen*, A Classical Introduction to Modern Number Theory. Springer–Verlag, New York (1982).

[5] *S. Lang*, Cyclotomic Fields. Springer–Verlag, New York (1978).

[6] *C. L. Siegel*, Bernoullische Polynome und quadratischer zahlkörper. Nachr. Akad. Wiss. Göttingen, **2** (1968), 7–38.

[7] *T. Shintani*, On evaluation of zeta functions of totally real algebraic fields at non–positive integers. J. Fac. Sci. Univ. of Tokyo, **23** (1976), 393–417.

[8] *G. Stevens*, The Eisenstein measure and real quadratic fields. Proceedings of the International Conference on Number Theory, Université Laval, Quebec, 1987 (to appear).


---

Department of Mathematics and Statistics,
University of Massachusetts, Amherst, Massachusetts 01003, USA

# Characterizations of the Logarithm as an Additive Arithmetic Function

*Adolf Hildebrand*

## 1. Introduction

An additive arithmetic function is a real–valued function $f$ defined on the set $\mathbb{N}$ of positive integers and satisfying

$$f(nm) = f(n) + f(m), \tag{1}$$

whenever $n$ and $m$ are coprime positive integers. If (1) holds for all pairs $(n, m)$ of positive integers, then $f$ is called completely additive. Additive functions arise naturally in many number theoretic problems, and are studied systematically in probabilistic number theory. A typical example of an additive function is the function "number of prime factors of $n$."

The equation (1) is the familiar functional equation of the logarithm, and in this connection it is usually regarded as an equation for functions $f$ defined on the set of all positive real numbers and with the variables $n$ and $m$ denoting arbitrary positive real numbers. It is well–known that any continuous (or in some other sense "smooth") function $f : \mathbb{R}^+ \to \mathbb{R}$ satisfying (1) for all positive real numbers $n$ and $m$ must be of the form $f = \lambda \log$ with some constant $\lambda$. In view of this, one might ask whether similar results hold for functions $f : \mathbb{N} \to \mathbb{R}$ that satisfy (1). In other words, can one characterize the functions $f(n) = \lambda \log n \quad (n \in \mathbb{N})$ as additive arithmetic functions by appropriate regularity conditions?

This problem was first considered by Erdös [5] in a paper that appeared in 1946. Among other results on additive functions Erdös proved in this paper that an additive function that is either monotonic or satisfies $f(n+1) - f(n) = o(1)$ as $n \to \infty$ must be of the form $f = \lambda \log$ for some constant $\lambda$. He also made several conjectures concerning possible improvements and generalizations of his results. Stimulated by Erdös' paper, a number of authors have subsequently proved results of similar nature, and the subject has grown into an important subfield of probabilistic number theory.

The aim of this paper is to give a representative survey of results obtained in this connection, and to call attention to some unsolved problems. The paper is organized as follows: Section 2 describes more precisely the results and conjectures stated in Erdös' 1946 paper, and the progress made so far on these conjectures. Section 3 discusses various generalizations of these results. In Section 4 an analogous characterization problem for multiplicative arithmetic functions is considered. Finally, in Section 5, a more fundamental general problem concerning the distribution of values of additive functions is formulated and put into context.

## 2. Erdös' Paper

In his paper [5] Erdös showed that an additive function $f : \mathbb{N} \to \mathbb{R}$ satisfying

$$f(n+1) - f(n) \geq 0 \quad \text{for all} \quad n \in \mathbb{N} \tag{2}$$

or

$$f(n+1) - f(n) = o(1) \quad (n \to \infty) \tag{3}$$

must be of the form $f = \lambda \log$. Since trivially any function of this form satisfies (2) and (3), these results can be viewed as characterizations of the logarithm (and its multiples) among all additive functions. Erdös deduced his results from another, rather deep theorem proved in the same paper, but subsequently other authors have found relatively simple direct proofs; see, for example, Besicovitch [1].

In connection with his two results, Erdös stated several conjectures . One of them was that the condition (3) could be replaced by the weaker condition

$$\frac{1}{x} \sum_{n < x} |f(n+1) - f(n)| \to 0 \quad (x \to \infty). \tag{4}$$

This was later proved by Kátai [13]. An alternative proof of a slightly stronger form of this conjecture was given by Wirsing [20].

Erdös also conjectured that if an additive function $f$ satisfies the one–sided bound

$$f(n+1) - f(n) \geq K \quad \text{for all } n$$

for some constant $K$, then there exists a constant $\lambda$ such that the difference $|f(n) - \lambda \log n|$ is bounded uniformly in $n$. This conjecture was proved by Wirsing [19].

Finally, Erdös conjectured that the conditions (2) and (3) in his characterizations could be replaced, respectively, by

$$f(n+1) \geq f(n) \text{ on a set of density } 1 \tag{5}$$

and

$$f(n+1) - f(n) = o(1) \quad (n \to \infty \text{ through a set of density } 1) \tag{6}$$

Here the density of a set $A \subset \mathbb{N}$ is defined as

$$\lim_{x \to \infty} \frac{1}{x} \#\{n \in A : n \leq x\}.$$

The second of these two conjectures implies the above–quoted result of Kátai and Wirsing, since the condition (6) is weaker than (4). This conjecture has been established only very recently [10] as a corollary to a more general result on the limit distribution for $f(n+1) - f(n)$, when $f$ is an additive function. The first conjecture is still open, but it seems likely that the method of [10] can be adapted to prove this conjecture too.

## 3. Generalizations

Since the appearance of Erdös' paper, new characterizations of the logarithm have been found that generalize or sharpen Erdös' original results in a variety of ways. For example, Kátai [11] showed that Erdös' conditions (2) and (3) can be generalized to

$$\liminf_{n \to \infty} \Delta^k f(n) = 0 \quad \text{for some} \quad k \in \mathbb{N},$$

where $\{\Delta^k f(n)\}$ denotes the $k$ th difference sequence of the sequence $\{f(n)\}$. Another interesting generalization discovered by Kátai [14] states that if $f_1$ and $f_2$ are additive functions satisfying

$$f_1(n+1) - f_2(n) = o(1) \quad (n \to \infty),$$

then $f_1$ and $f_2$ must be equal to a common multiple of the logarithm. For further results of this type and related open problems see Kátai's survey paper [16].

The characterizations mentioned so far all involve the behaviour of an additive function at pairs of consecutive integers $(n, n+1)$ or, more generally, at $k$-tuples $(n, n+1, \dots n+k-1)$. Kátai [12] proposed the problem to obtain similar characterizations when $n$ and $n+1$ are replaced by two linear forms $an+b$ and $cn+d$. Specifically, Kátai asked for a characterization of additive functions $f$ satisfying, with fixed integers $a, b, c, d$, with $a, c > 0$,

$$f(an+b) - f(cn+d) = o(1) \quad (n \to \infty). \tag{7}$$

Elliott (see Chapter 13 in [3]) showed that if

$$D := ac(ad - bc) \neq 0,$$

then (7) implies that, with a suitable constant $\lambda$, $f(n) = \lambda \log n$ holds for all $n$ satisfying $(n, D) = 1$. This essentially solved Kátai's problem.

Concerning Erdös' condition (3), one might ask whether the bound "$o(1)$" can be replaced by a larger bound, but this is not the case, as Erdös had observed, However, in 1979, Wirsing [21] showed that for *completely* additive functions $f$ this bound can be relaxed quite substantially to

$$f(n+1) - f(n) = o(\log n) \quad (n \to \infty).$$

This is a best-possible result in the sense that the bound "$o(\log n)$" cannot be further relaxed to "$O(\log n)$." Its proof is much more intricate than that of earlier results of this kind, and based on a probabilistic argument using Markov chains.

A few years ago, Elliott [3] established the inequality

$$\min_{\lambda \in \mathbb{R}} \max_{n \leq x} |f(n) - \lambda \log n| \leq c_1 \max_{n \leq x^{c_2}} |f(n+1) - f(n)| \tag{8}$$

uniformly for all additive functions $f$ and real $x \geq 2$ with absolute constants $c_1$ and $c_2$. This inequality, which had been earlier conjectured by Ruzsa [18], is extremely powerful, since most of the previously obtained characterizations of the logarithm, including the above–mentioned deep result of Wirsing, can be derived from it. Elliott's result is the first quantitative estimate in connection with the characterization problem, and as such it represents a significant breakthrough.

Elliott derived (8) from another inequality involving second moments, namely

$$\min_{\lambda\in\mathbb{R}} \sum_{p\le x} \frac{|f(p)-\lambda\log p|^2}{p} \le c_1 \max_{x<y\le x^{c_2}} \frac{1}{y} \sum_{x<n\le y} |f(n+1)-f(n)|^2. \qquad (9)$$

In fact, he established a more general version of (9) with the differences $f(n+1)-f(n)$ replaced by $f(an+b)-f(cn+d)$, which he used to solve Kátai's problem on additive functions satisfying (7). The proof of (9), along with some background material, is given in [3]; it utilizes a variety of techniques from analytic and probabilistic number theory, as well as from functional analysis. A well–motivated outline of the proof can be found in Elliott's survey paper [4].

To conclude this section, we mention a characterization that stands somewhat apart from the other known characterizations in that it does not involve conditions on $f(n+1)-f(n)$ or similar linear combinations of additive functions. In [7] it was shown that an additive function $f$ is of the form $\lambda \log n$ if and only if it satisfies

$$\frac{1}{x} \sum_{\substack{n\le x \\ p^m \| n}} f(n) = \frac{1}{p^m}\left(1-\frac{1}{p}\right)\frac{1}{x} \sum_{n\le x} f(n) + o(1) \qquad (x\to\infty) \qquad (10)$$

for every (fixed) prime power of $p^m$. The condition (10) is of a quite different nature than, say, Erdős' conditions (2) and (3); it characterizes the functions $\lambda \log n$ as the only additive functions that are, in a certain sense, well–distributed over the sets $\{n\in\mathbb{N} : p^m \| n\}$.

## 4. A Multiplicative Analog

A real– or complex–valued function $f$ defined on the set of positive integers is called multiplicative if it is not identically zero and satisfies the functional equation

$$f(nm) = f(n)f(m) \qquad (11)$$

for coprime positive integers $n$ and $m$. If (11) holds for all pairs $(n, m)$ of positive integers, then $f$ is called completely multiplicative. An important subclass of multiplicative functions is given by the unimodular multiplicative functions, i.e., multiplicative functions $f$ that satisfy $|f| \equiv 1$ identically.

The equation (11) has as "trivial" solutions the functions

$$f(n) = n^s \quad (n \in \mathbb{N}), \tag{12}$$

where $s$ is a fixed complex number, and one might expect that these functions play a similar exceptional role among multiplicative functions as the functions $\lambda \log$ among additive functions. This raises the question: Can one characterize the functions (12) as multiplicative functions by imposing suitable regularity conditions on $f$? It turns out that this leads to problems that are much more difficult than those arising in the case of additive functions and to a large extent still unsolved.

For simplicity, we will henceforth consider only unimodular multiplicative functions $f$. In this case one might conjecture that Erdös' regularity condition (3), as well as the weaker conditions (4) and (6), force $f$ to be of the form (12). The converse implication that a unimodular function of the form (12) satisfies (3), (4) and (6), is obviously true.

That the condition (3) is sufficient, has been recently proved by Wirsing. The proof, which has not yet been published, proceeds by showing that $f$ can be written as $f = e^{2\pi i h}$, where $h$ is an additive function satisfying $h(n+1) - h(n) = o(1)$ as $n \to \infty$, and then applying Erdös' result to the function $h$.

The weaker conditions (4) and (6) are in the case of unimodular functions $f$ actually equivalent, as one can easily see. In contrast to the situation for additive $f$, it is not known whether for multiplicative $f$ these conditions are sufficient to imply (12). It seems likely that this is the case, but so far only partial results have been obtained. The problem is unsolved even in the case when $f$ assumes only the values $\pm 1$. In this case (12) obviously forces $f \equiv 1$, and the problem amounts to showing that

$$\limsup_{x \to \infty} \frac{1}{x} \# \{ n \leq x : f(n+1) = -f(n) \} > 0 \tag{13}$$

holds, unless $f \equiv 1$. An interesting example is furnished by the Liouville function $\lambda(n)$, defined as 1 if $n$ has an even number of prime factors, and -1 otherwise. This function is completely multiplicative and thus would fall into the above class of functions $f$, but it is not known whether (13) holds for $f = \lambda$.

The partial results known in connection with the above–mentioned conjecture involve either stronger conditions than (4) or additional hypotheses on the function $f$. Kátai [15] showed that a unimodular multiplicative function $f$ satisfying the condition

$$\sum_{n=1}^{\infty} \frac{|f(n+1) - f(n)|}{n} < \infty$$

must be of the form (12). (Kátai actually gave results of this type for more general classes of multiplicative functions; cf. [16] and the references therein.)

In a similar vein, it was shown in [8] that if $f$ is completely multiplicative and assumes only the values $\pm 1$, then the condition

$$\limsup_{x \to \infty} \frac{(\log\log x)^4}{x} \sum_{n \le x} |f(n+1) - f(n)| = 0$$

implies that $f \equiv 1$. By the same method one could obtain a corresponding result for arbitrary unimodular multiplicative functions.

A different approach was taken in [9] by showing that a unimodular multiplicative function $f$ satisfying (4) and in addition

$$|f(p) - 1| \le c$$

for every prime $p$ must be of the form (12). Here $c$ is an absolute positive constant, an admissible value of which is 0.001. This result implies, for example, that (4) does not hold for the functions $f(n) = e^{i\theta\Omega(n)}$, where $\theta$ is a non-zero real number of modulus $\le c$. In [9] it was also shown that the condition (4) along with

$$\sum_{p} \frac{|1 - f(p)p^{-i\alpha}|^2}{p} < \infty \quad \text{for some} \quad \alpha \in \mathbb{R} \tag{14}$$

on a unimodular multiplicative function $f$ implies (12).

The question whether the condition (4) alone is sufficient thus remains open. In order to approach this problem, one might try to determine the asymptotic behaviour of the averages

$$\frac{1}{x}\sum_{n\leq x} f(n+1)\overline{f(n)} \tag{15}$$

for unimodular multiplicative functions $f$. It is easy to see that (in the case $|f|\equiv 1$) these averages tend to 1 as $x\to\infty$, if and only if (4) is satisfied. In the case (14) is satisfied, one can in fact evaluate (15) up to a term of $o(1)$; see Theorem 2 in [10]. It would be desirable to have a similar result for the case when (14) fails to hold, but this seems to be much more difficult. By comparison, the behaviour of the averages

$$\frac{1}{x}\sum_{n\leq x} f(n) \tag{16}$$

for modular multiplicative functions $f$ is now completely understood after the work of Halász [6]. Unfortunately, Halász' method of treating (16) cannot be adapted to deal with (15), since the associated Dirichlet series do not have an Euler product representation.

The mentioned results on multiplicative functions have applications to the study of group–valued additive functions, i.e., functions $f$ defined on $\mathbb{N}$ with values in an (additively written) Abelian topological group $G$ and satisfying (1) for coprime $n$ and $m$. Conditions like (3) are meaningful in this context, and one might ask whether such conditions characterize some "trivial" subclass of all additive functions with values in $G$, as is the case when $G=\mathbb{R}$. This question has been studied recently by Daróczy and Kátai [2]. Among other things they showed that if $G$ is compact, then any additive function $f:\mathbb{N}\to G$ satisfying (3) must be the restriction to $\mathbb{N}$ of a continuous homomorphism from the multiplicative group of positive real numbers into $G$. The proof depends on Wirsing's result that any unimodular multiplicative function satisfying (3) is of the form (12). This result is applied to the functions $\chi\circ f$, where $\chi$ is a continuous character on $G$. Any improvement of Wirsing's result (for example, by relaxing the condition (3) to (4)) would lead to a corresponding improvement in the theorem of Daróczy and Kátai. For a further discussion of group–valued additive functions we refer to Kátai's survey paper [16].

It should be noted that the study of unimodular multiplicative functions is actually equivalent to that of additive functions with values in the group $\mathbb{R}/\mathbb{Z}$, since the multiplicative group of unimodular complex numbers is isomorphic to the additive group $\mathbb{R}/\mathbb{Z}$. One can therefore restate the above results and conjectures in terms of additive functions with values in $\mathbb{R}/\mathbb{Z}$. For example, the result of Wirsing is equivalent to the assertion that an additive function $f : \mathbb{N} \to \mathbb{R}$ which satisfies

$$\|f(n+1) - f(n)\| = o(1) \quad (n \to \infty)$$

must be of the form

$$f(n) = \lambda \log n + h(n),$$

where $\lambda$ is a constant and $h$ is an integer–valued function. Here $\|u\|$ denotes the distance of $u$ to the nearest integer.

## 5. A General Problem

The characterizations of the logarithm by conditions like (3), (4) or (6) can be qualitatively restated as follows: If an additive function $f$ is not of the form $\lambda$ log, then the differences $f(n+1) - f(n)$ cannot be too small. This formulation suggests the problem to determine, for general additive functions $f$, the behaviour of $f(n+1) - f(n)$, and specifically that of the distribution functions

$$F_x(t) = \frac{1}{[x]} \# \{n \le x : f(n+1) - f(n) \le t\}. \tag{17}$$

This problem has indeed been investigated in the past few years, and one now has a largely satisfactory theory for the distribution of the values $f(n+1) - f(n)$, when $f$ is additive; see Chapter 21 in [3], and [10].

A more fundamental problem would be to study the two–dimensional distribution functions

$$F_x(t_1, t_2) = \frac{1}{[x]} \# \{n \le x : f(n) \le t_1, f(n+1) \le t_2\}. \tag{18}$$

An asymptotic relation for these functions that is uniform in $t_1$ and $t_2$, for example, would imply a corresponding relation for the functions (17). It would also be of interest to compare (18) with the individual distribution functions

$$F_x^{(i)}(t_i) = \frac{1}{[x]} \# \{ n \le x : f(n+i-1) \le t_i \} \quad (i = 1,2),$$

in order to determine the degree of independence between $f(n)$ and $f(n+1)$. The results of Sections 2 and 3 suggest that, unless $f$ is close to a multiple of a logarithm, the functions $f(n)$ and $f(n+1)$ behave much like independent variables.

A number of further generalizations of this problem suggest themselves. For example, one could investigate the behaviour of

$$F_x(t_1,\ldots,t_k) = \frac{1}{[x]} \# \{ n \le x : f_i(n+i-1) \le t_i \quad (i = 1,\ldots,k) \} \tag{19}$$

where $f_1,\ldots,f_k$ are additive functions. Also, one could study the functions (18) with $n$ and $n+1$ replaced by linear forms $an+b$ and $cn+d$.

Up to now, the distribution functions (18) and (19) have been studied only for certain special classes of additive functions $f$ (see, e.g., Kubilius [17]). It would be of interest to develop a general theory for the behaviour of these distribution functions. In light of the recent progress on the characterization problem, I believe this is a feasible task, although it may well be technically quite complicated.

## References


[1] *A.S. Besicovitch*, On additive functions of a positive integer. Studies in Mathematical Analysis and Related Topics, Stanford Univ. Press, Stanford 1962, 38—41.

[2] *Z. Daróczy* and *I. Kátai*. On additive arithmetical functions with values in topological groups I. Publ. Math. Debrecen **33** (1986), 287—291.

[3] *P.D.T.A. Elliott*, Arithmetic functions and integer products. Grundlehren Math. Wiss., Vol. 272, Springer–Verlag, New York 1985.

[4] *P.D.T.A. Elliott*, Functional analysis and additive arithmetic functions. Bull. Amer. Math. Soc. **16** (1987), 179—223.

[5] *P. Erdös* , On the distribution function of additive functions, Ann. Math. **47** (1946), 1—20.

[6] *G. Halász* ,Über die Mittelwerte multiplikativer zahlentheoretischer Funktionen. Acta Math. Sci. Hungar. **19** (1968), 365—403.

[7] *A. Hildebrand* , Characterization of the logarithm as an additive function. Arch. Math. **38** (1982), 535—539.

[8] *A. Hildebrand* , Multiplicative functions at consecutive integers. Math. Proc. Cambr. Phil. Soc. **100** (1986), 229—236.

[9] *A. Hildebrand* , Multiplicative functions at consecutive integers, II. Math. Proc. Cambr. Phil. Soc., **103** (1988), 389—398.

[10] *A. Hildebrand* , An Erdös—Wintner theorem for differences of additive functions. Trans. Amer. Math. Soc., to appear.

[11] *I. Kátai* , Characterization of additive functions by its local behaviour. Ann. Univ. Sci. Budapest. Eötvös Sect. Math. **12** (1969), 35—37.

[12] *I. Kátai* , Some results and problems in the theory of additive functions. Acta. Sci. Math. (Szeged) **30** (1969), 305—311.

[13] *I. Kátai* , On a problem of P. Erdös. J. Number Theory **2** (1970), 1—6.

[14] *I. Kátai* , On additive functions. Publ. Math. Debrecen **25** (1978), 251—257.

[15] *I. Kátai* , Multiplicative functions with regularity properties, III. Acta Math. Hungar. **43** (1984), 259—272.

[16] *I. Kátai* , Characterizations of arithmetical functions, problems and results. To appear in Proc. Number Theory Conference, Quebec (1987).

[17] *J. Kubilius*, Probabilistic methods in the theory of numbers. Trans. Math. Monographs, Vol. 11, Amer. Math. Soc., Providence 1968.

[18] *I.Z. Ruzsa* , Effective results in probabilistic number theory. (Preprint), Budapest 1982.

[19] *E. Wirsing* , A characterization of log $n$ as an additive arithmetic function. Symposia Mathematica, Vol. IV (INDAM, Rome 1968/69), Academic Press, London 1970, 45—57.

[20] *E. Wirsing* , Characterization of the logarithm as an additive function. Sympos. Pure Math., Vol. 20, Amer. Math. Soc., Providence 1971, 275—381.

[21] *E. Wirsing* , Additive and completely additive functions with restricted growth. Recent Progress in Analytic Number Theory, Vol. 2, Academic Press, London 1981, 231—280.

---

Department of Mathematics, University of Illinois, Urbana, IL 61801, USA

# A Note on Dedekind Sums

*Hiroshi Ito* [1]

## 1. Introduction

The connection between the classical Dedekind sum

$$s(a,c) = \frac{1}{c} \sum_{\substack{k \,(\mathrm{mod}\, c) \\ k \not\equiv 0 (\mathrm{mod}\, c)}} \cot\left(\pi \frac{ak}{c}\right) \cot\left(\pi \frac{k}{c}\right)$$

defined for two positive integers $a, c$ with $(2a, c) = 1$ and the Jacobi symbol $\left(\frac{a}{c}\right)$ is well–known, i.e., we have

$$3cs(a,c) \equiv c + 1 - 2\left(\frac{a}{c}\right) (\mathrm{mod}\ 8).$$

In the previous work [3] we have proved a similar congruence between Dedekind sums of Sczech [6] and quadratic residue symbols of imaginary quadratic fields. The main aim of this paper is to give a relation between his Dedekind sums and the cubic residue symbol of the field $\mathbb{Q}(\sqrt{-3})$. In section 2 we prove a general relation between sums of a certain type and power residue symbols of prime degrees by using the so–called Gauss' lemma. Then, in section 3, we apply this result to show that a certain linear combination of Dedekind sums of [6] gives a homomorphism $\psi$ from $\Gamma = \{A \in SL(2, \mathbb{Z}[e^{2\pi i/3}]);\ A \equiv \begin{pmatrix} 1 & \\ & 1 \end{pmatrix} (\mathrm{mod}\ 3)\}$ to the additive group of the ring $\mathbb{Z}$ of rational integers such that

$$\left(\frac{c}{a}\right)_3 = e^{\frac{2\pi i}{3}\psi(A)}, \quad A = \begin{pmatrix} a & b \\ c & d \end{pmatrix} \in \Gamma,$$

where $\left(\frac{c}{a}\right)_3$ denotes the cubic residue symbol of $\mathbb{Q}(\sqrt{-3})$.

---

[1] The author was supported by NSF Grant DMS–8601978.

## 2. Preliminaries

Let $p$ be a prime number and let $K$ be an algebraic number field of finite degree over the rational number field $\mathbb{Q}$ which contains a primitive $p$–th root $\rho$ of unity. Denote by $\mathfrak{o}$ the ring of integers of $K$. If $\varsigma$ is a $p$–th root of unity in $K$, we write $\mathrm{Ind}_\rho$ for the integer $n$ such that $\varsigma = \rho^n$ and $0 \leq n \leq p-1$. Let $\bar{K}$ be an algebraic closure of $K$. Take and fix a prime ideal $\mathfrak{P}$ of the ring of all integers in $\bar{K}$ such that $\mathfrak{P} \cap \mathbb{Z} = p\mathbb{Z}$ and denote by $\mathfrak{O}_\mathfrak{P}$ the ring consisting of all numbers in $\bar{K}$ which are integral at $\mathfrak{P}$.

**Theorem 1.** *Let $\mathfrak{c}$ be an integral ideal of $K$ prime to $p$ and let $f$ and $g$ be maps from $\mathfrak{o}/\mathfrak{c} - \{0\}$ to $\bar{K}$ such that*

$$f(\rho k) = \rho f(k), \quad g(\rho k) = \rho^{-1} g(k) \tag{1}$$

*and*

$$f(k) \equiv g(k) \equiv 1 \pmod{(1-\rho)\mathfrak{O}_\mathfrak{P}} \tag{2}$$

*for every $k \in \mathfrak{o}/\mathfrak{c} - \{0\}$. Then, for every $a \in \mathfrak{o}$ prime to $\mathfrak{c}$,*

$$\sum_{\substack{k \in \mathfrak{o}/\mathfrak{c} \\ k \neq 0}} \{f(ak)g(k) - f(k)g(ak)\} \equiv -2p(1-\rho)\mathrm{Ind}_\rho\left(\frac{a}{\mathfrak{c}}\right)_p \pmod{p(1-\rho)^2\mathfrak{O}_\mathfrak{P}} . \tag{3}$$

*If $p = 2$, we have*

$$\sum_{\substack{k \in \mathfrak{o}/\mathfrak{c} \\ k \neq 0}} f(ak)f(k) \equiv N\mathfrak{c} + 1 - 2\left(\frac{a}{\mathfrak{c}}\right)_2 \pmod{8\mathfrak{O}_\mathfrak{P}}.$$

Here $\left(\frac{a}{\mathfrak{c}}\right)_p$ is the $p$–th power residue symbol of $K$ and $N\mathfrak{c}$ denotes the absolute norm of $\mathfrak{c}$.

**Proof.** Let $R$ be a subset of $\mathfrak{o}/\mathfrak{c}$ such that $R \cap (\rho^n R) = \phi$ $(1 \leq n \leq p-1)$ and $\mathfrak{o}/\mathfrak{c} = \{0\} \cup \left( \bigcup_{n=0}^{p-1} \rho^n R \right)$. By (1),

$$(f(a\rho^n k) - 1)(g(\rho^n k) - 1) = (f(ak) - 1)(g(k) - 1) + (1-\rho^n)f(ak) + (1-\rho^{-n})g(k).$$

Therefore, from (2),

$$\sum_{\substack{k\in \mathfrak{o}/\mathfrak{c}\\ k\neq 0}}(f(ak)-1)(g(k)-1)$$
$$= p\sum_{k\in R}\{(f(ak)-1)(g(k)-1)+f(ak)+g(k)\}$$
$$\equiv p\left\{\sum_{k\in aR} f(k)+\sum_{k\in R} g(k)\right\} \bmod p(1-\rho)^2\mathfrak{D}_{\mathfrak{P}}. \tag{4}$$

In the same way, we have

$$\sum_{\substack{k\in \mathfrak{o}/\mathfrak{c}\\ k\neq 0}}(f(k)-1)(g(ak)-1)$$
$$\equiv p\left\{\sum_{k\in R} f(k)+\sum_{k\in aR} g(k)\right\} \bmod p(1-\rho)^2\mathfrak{D}_{\mathfrak{P}}. \tag{5}$$

Subtracting (5) from (4), we get

$$\sum_{\substack{k\in \mathfrak{o}/\mathfrak{c}\\ k\neq 0}}\{f(ak)g(k)-f(k)g(ak)\}$$
$$\equiv p\sum_{k\in R}\{g(k)-f(k)\}+p\sum_{k\in aR}\{f(k)-g(k)\} \bmod p(1-\rho)^2\mathfrak{D}_{\mathfrak{P}}.$$

Put, for $n = 0, 1, \dots, p-1$,

$$R_n = \{k \in R\,;\, ak \in \rho^n R\}.$$

Then $R_n = \bigcup_{n=0}^{p-1} R_n$ and $aR_n = \bigcup_{n=0}^{p-1} \rho^n R_n$, the unions being disjoint. Therefore, by (1) and (2),

$$\sum_{k\in R} f(k)-\sum_{k\in aR} f(k)=\sum_{n=0}^{p-1}(1-\rho^n)\sum_{k\in R_n} f(k)$$
$$\equiv (1-\rho)\sum_{n=1}^{p-1} n\cdot \#R_n \quad \bmod (1-\rho)^2\mathfrak{D}_{\mathfrak{P}},$$

where $\#R_n$ denotes the cardinality of $R_n$. Similarly,

$$\sum_{k\in R} f(k)-\sum_{k\in aR} g(k)\equiv -(1-\rho)\sum_{n=1}^{p-1} n\cdot \#R_n \quad \bmod (1-\rho)^2\mathfrak{D}_{\mathfrak{P}}.$$

A generalization of Gauss' lemma (*cf.* Reichardt [5]) says

$$\sum_{n=1}^{p-1} n \cdot \#R_n \equiv \mathrm{Ind}_{\mathfrak{p}}\left(\frac{a}{\mathfrak{c}}\right)_p \pmod{p}.$$

This concludes the proof of (3). The assertion concerning the case $p = 2$ can be proved similarly and is treated in [3].

The following remark is useful. For a rational number $\alpha = m/n$ where $m$ and $n$ are coprime integers with $n > 0$, we denote by $(1-\rho)^{\alpha}$ a number $\upsilon$ in $\bar{K}$ such that $\upsilon^n = (1-\rho)^m$. Instead of the condition (2) we can consider the condition that there exist two relational numbers $\alpha$ and $\beta$ with $0 < \alpha \leq 1$, $0 < \beta \leq 1$, and $\alpha + \beta > 1$ such that, for every $k \in \mathfrak{o}/\mathfrak{c} - \{0\}$,

$$f(k) \equiv 1 \pmod{(1-\rho)^{\alpha}\mathfrak{O}_{\mathfrak{P}}},$$

$$g(k) \equiv 1 \pmod{(1-\rho)^{\beta}\mathfrak{O}_{\mathfrak{P}}}.$$

Then we get the same congruence as (3) with modulo $p(1-\rho)^{\gamma}\mathfrak{O}_{\mathfrak{P}}$, where

$$\gamma = \min\{\alpha + \beta,\ 1 + \alpha,\ 1 + \beta\}.$$

## 3. Cubic Residue Symbol of $\mathbb{Q}(\sqrt{-3})$

Let $\wp$ be the Weierstrass $\wp$-function satisfying

$$\wp'^2 = 4\wp^3 - 1$$

and let $L = \mathbb{Z}[\rho]\theta$ ($\rho = e^{2\pi i/3}, \theta > 0$) be the period lattice of $\wp$. We define, for every $z$ in the complex plane $\mathbb{C}$ and $n \in \mathbb{Z}$, $n \geq 0$,

$$E_n(z) = \sum_{\substack{w \in L \\ w+z \neq 0}} (w+z)^{-n} |w+z|^{-s} \Big|_{s=0},$$

where the value at $s = 0$ is to be understood in the sense of analytic continuation. The function $E_n(z)$ is periodic with respect to $L$ and satisfies $E_n(-\rho z) = (-\rho)^{-n} E_n(z)$. It is known (*cf.* [6]) that

$$E_0(z)=\begin{cases}-1, & z\in L\\ 0, & z\notin L\end{cases},$$
$$E_1(z)=\varsigma(z)-\frac{2\pi}{\sqrt{3}\,\theta^2}\bar{z},\qquad z\notin L, \tag{6}$$
$$E_2(z)=\wp(z),\qquad z\notin L$$

with the Weierstrass zeta function $\varsigma(z)$ with respect to $L$. Define the function $E(z)$ by

$$2E(z)=\begin{cases}0, & z\in L\\ \wp(z)-E_1(z)^2, & z\notin L.\end{cases}$$

Put

$$D(a,c;u,v)=\frac{1}{c}\sum_{r\in L/cL}E_1\left(\frac{a(r+u)}{c}+v\right)E_1\left(\frac{k+u}{c}\right)$$

for $a,c\in\mathbb{Z}[\rho]$, $c\neq 0$ and $u,v\in\mathbb{C}$. Let, for $A=\begin{pmatrix}a & b\\ c & d\end{pmatrix}\in SL(2,\mathbb{Z}[\rho])$,

$$\Phi(A)(u,v)=-\overline{\left(\frac{a}{c}\right)}E(u)-\overline{\left(\frac{d}{c}\right)}E(u^*)-\frac{a}{c}E_0(u)E_2(v)$$
$$-\frac{d}{c}E_0(u^*)E_2(v^*)-D(a,c;u,v) \tag{7}$$

if $C\neq 0$ and

$$\Phi(A)(u,v)=-\overline{\left(\frac{b}{d}\right)}E(u)-\frac{b}{d}E_0(u)E_2(v)$$

if $c=0$, where $(u^*,v^*)=(u,v)A$. The values $D(a,c;u,v)$ and $\Phi(A)(u,v)$ depend only on the classes of $u$ and $v$ in $\mathbb{C}/L$. It is shown in [6] that

$$\Phi(AB)(u,v)=\Phi(A)(u,v)+\Phi(B)((u,v)A) \tag{8}$$

for every $A,B\in SL(2,\mathbb{Z}[\rho])$. Put

$$\Psi(A)=\Phi(A)(\mu,\upsilon)+\Phi(A)(-\upsilon,\mu)-\Phi(A)(\mu,-\upsilon)-\Phi(A)(\upsilon,\mu)$$

with

$$\mu=\frac{\theta}{3},\quad \upsilon=\frac{\theta}{\sqrt{-3}}\quad(\sqrt{-3}=\sqrt{3}i).$$

It follows from (8) that $\Psi$ gives a homomorphism from the subgroup $\Gamma$ of $SL(2,\mathbb{Z}[\rho])$ consisting of all matrices congruent to $\begin{pmatrix}1 & \\ & 1\end{pmatrix}$ modulo 3 to the additive group of $\mathbb{C}$.

**Theorem 2.** *The congruence*

$$\Psi(A) \equiv 3\,\mathrm{Ind}_\rho\left[\frac{c}{a}\right]_3 \mod (1-\rho)^3\,\mathbb{Z}[\rho]$$

*holds for every* $A = \begin{pmatrix} a & b \\ c & d \end{pmatrix} \in \Gamma.$

**Proof.** We first observe that

$$\Phi(A)(u\,,v) \in \frac{1}{3}\mathbb{Z}[\rho]$$

for every $A \in SL(2,\mathbb{Z}[\rho])$ and every $u\,,v \in \frac{1}{3}L$. Because of (8) it is sufficient to verify the above for three matrices

$$\begin{pmatrix} 1 & 1 \\ & 1 \end{pmatrix}, \quad \begin{pmatrix} 1 & \rho \\ & 1 \end{pmatrix}, \quad \begin{pmatrix} & 1 \\ -1 & \end{pmatrix}$$

which generate $SL(2,\mathbb{Z}[\rho])$. This is easy to see from the definition of $\Phi(A)(u\,,v)$ and the following 3–division values of our functions:

$$E_1(\mu) = -\frac{1}{\sqrt{3}}, \quad E_2(\mu) = 1, \quad E(\mu) = \frac{1}{3}.$$

In particular $\Psi$ takes values in $\frac{1}{3}\mathbb{Z}[\rho]$. By (7), (8) and by the congruences $a \equiv 1$, $c \equiv 0 \pmod 3$, we see

$$\begin{aligned}
\Psi(A) &= \Psi\left(\begin{pmatrix} & 1 \\ -1 & \end{pmatrix} A\right) \\
&= D(c\,,a;a\mu,-\upsilon) + D(c\,,a;a\upsilon,\mu) - D(c\,,a;a\mu,\upsilon) - D(c\,,a;-a\upsilon,\mu) \\
&= \frac{1}{a}\sum_{r\in L/aL}\Big\{E_1\Big(\frac{cr}{a}-\upsilon\Big)E_1\Big(\frac{r}{a}+\mu\Big) + E_1\Big(\frac{cr}{a}+\mu\Big)E_1\Big(\frac{r}{a}+\upsilon\Big) \\
&\qquad - E_1\Big(\frac{cr}{a}+\upsilon\Big)E_1\Big(\frac{r}{a}+\mu\Big) - E_1\Big(\frac{cr}{a}+\mu\Big)E_1\Big(\frac{r}{a}-\upsilon\Big)\Big\}.
\end{aligned}$$

**Lemma 1.** *For every* $x, y \in \mathbb{C}$ *with* $3x\,,3y \notin L$, *we have*

$$\sum_{n=1}^{3} E_1(\rho^n x + \mu)\,\{E_1(\rho^n y + \upsilon) - E_1(\rho^n y - \upsilon)\} \tag{9}$$

$$= \frac{\sqrt{-3}}{2} \bullet \frac{1+\sqrt{3}^{-1}\,\wp'(x)}{\wp(x)\,\wp(\sqrt{-3}\,x)} \bullet \frac{1}{\wp(y)}.$$

**Proof.** By (6), the addition formula

$$E_1(u)+E_1(v)-E_1(u+v)=-\frac{1}{2}\frac{\wp'(u)-\wp'(v)}{\wp(u)-\wp(v)}$$

holds for every $u, v \in \mathbb{C}$ with $u, v, u \pm v \notin L$. It follows from

$$E_1(\upsilon)=0,\quad \wp(\upsilon)=0,\quad \wp'(\upsilon)=i,\quad \wp(\mu)=1,\quad \wp'(\mu)=-\sqrt{3}$$

that

$$E_1(\rho^n x+\mu)=\rho^{-n}E_1(x)+E_1(\mu)+\frac{1}{2}\frac{\wp'(x)+\sqrt{3}}{\rho^n\wp(x)-1},$$

$$E_1(\rho^n y+\upsilon)-E_1(\rho^n y-\upsilon)=-\rho^{-n}\frac{i}{\wp(y)}.$$

Hence (9) is equal to

$$-\frac{\sqrt{-3}}{2}\bullet\frac{1+\sqrt{3}^{-1}\wp'(x)}{\wp(y)}\sum_{n=1}^{3}\rho^{-n}\frac{1}{\rho^n\wp(x)-1}.$$

The summation over $n$ is equal to

$$\frac{3\wp(x)}{\wp(x)^3-1}=-\frac{1}{\wp(x)\wp(\sqrt{-3}\,x)}.$$

This proves Lemma 1.

Put, for $r\in L$, $r\notin aL$,

$$f(r)=4^{\frac{1}{3}}\wp\left(\frac{r}{a}\right)^{-1}\wp\left(\frac{\sqrt{-3}\,r}{a}\right)^{-1}\left\{1+\sqrt{3}^{-1}\wp'\left(\frac{r}{a}\right)\right\},$$

$$g(r)=4^{-\frac{1}{3}}\wp\left(\frac{r}{a}\right)^{-1}.$$

Because $f(\rho r)=\rho f(r)$ and $g(\rho r)=\rho^{-1}g(r)$, we have, by Lemma 1,

$$\sum_{n=1}^{3}\left\{E_1\left(\frac{\sigma\rho^n}{a}-\upsilon\right)E_1\left(\frac{r\rho^n}{a}+\mu\right)-E_1\left(\frac{\sigma\rho^n}{a}+\upsilon\right)E_1\left(\frac{r\rho^n}{a}+\mu\right)\right\}$$

$$=-\frac{\sqrt{-3}}{2}f(r)g(\sigma)$$

$$=\frac{1}{2\sqrt{-3}}\sum_{n=1}^{3}f(r\rho^n)g(\sigma\rho^n).$$

In the same way,

$$\sum_{n=1}^{3}\left\{E_1\left(\frac{\sigma\rho^n}{a}+\mu\right)E_1\left(\frac{r\rho^n}{a}+\upsilon\right)-E_1\left(\frac{\sigma\rho^n}{a}+\mu\right)E_1\left(\frac{r\rho^n}{a}-\upsilon\right)\right\}$$

$$=-\frac{1}{2\sqrt{-3}}\sum_{n=1}^{3}f(\sigma\rho^n)g(r\rho^n).$$

Therefore,

$$a\Psi(A)=-\frac{1}{2\sqrt{-3}}\sum_{\substack{r\in L/\mathfrak{a}L\\ r\neq 0}}\{f(\sigma)g(r)-f(r)g(\sigma)\}.$$

Let $\mathfrak{P}$ be a prime ideal of the ring of all algebraic integers in $\mathbb{C}$ such that $\mathfrak{P}\cap\mathbb{Z}=3\mathbb{Z}$ and denote by $\mathfrak{O}_{\mathfrak{P}}$ the ring consisting of all algebraic numbers in $\mathbb{C}$ integral at $\mathfrak{P}$.

**Lemma 2.** *If* $\alpha\notin L$ *and* $n\alpha\in L$ *with an integer* $n$ *prime to 3, then we have*

$$4^{\frac{1}{3}}\wp(\alpha)\equiv 1 \mod (1-\rho)\mathfrak{O}_{\mathfrak{P}},$$

$$\wp'(\alpha)\equiv 0 \mod (1-\rho)^{\frac{3}{2}}\mathfrak{O}_{\mathfrak{P}}.$$

**Proof.** If $2\alpha\in L$, the assertions follow from $\mathfrak{p}\left(\frac{\theta}{2}\right)=4^{-\frac{1}{3}}$ and $\mathfrak{p}'(\alpha)=0$. Assume $2\alpha\notin L$. Put

$$t=\frac{12\wp\left(\frac{\theta}{2}\right)}{\wp\left(\frac{\theta}{4}\right)-\wp\left(\frac{\theta}{2}\right)}=4\sqrt{3}$$

and

$$T(z)=\frac{\wp\left(\frac{\theta}{4}\right)-\wp\left(\frac{\theta}{2}\right)}{\wp(z)-\wp\left(\frac{\theta}{2}\right)},$$

$$T_1(z)=\left(\wp\left(\frac{\theta}{4}\right)-\wp\left(\frac{\theta}{2}\right)\right)^{-\frac{1}{2}}\frac{d}{dz}T(z).$$

It is known (Fueter [1]) that $T(\alpha)$ is a root of a polynomial $Z_n(X)$ with coefficients in $\mathbb{Z}[t]$ which has the following property; both the coefficient of the highest term and the constant term of $Z_n(X)$ are $\pm 1$ if $n\equiv 1 \pmod 2$ and $\pm\frac{n}{2}$ if $n\equiv 0 \pmod 2$. Therefore $T(\alpha)$ is a $\mathfrak{P}$-adic unit. The first congruence follows from

$$\wp\left(\frac{\theta}{2}\right)^{-1} \wp(\alpha) = 1 + \frac{12}{tT(\alpha)}.$$

We have

$$\wp'(\alpha) = -\left(\wp\left(\frac{\theta}{4}\right) - \wp\left(\frac{\theta}{2}\right)\right)^{\frac{3}{2}} T(\alpha)^{-2} T_1(\alpha).$$

Because $T_1^2 = T(4T^2 + tT + 4)$ (*cf.* [1]) $T_1(\alpha)$ belongs to $\mathfrak{O}_{\mathfrak{P}}$. We obtain the second congruence from

$$\wp\left(\frac{\theta}{4}\right) - \wp\left(\frac{\theta}{2}\right) = \sqrt{3} \bullet 4^{-\frac{1}{3}}.$$

It follows from this lemma that

$$f(r) \equiv 1 \mod (1-\rho)^{\frac{1}{2}} \mathfrak{O}_{\mathfrak{P}},$$

$$g(r) \equiv 1 \mod (1-\rho) \mathfrak{O}_{\mathfrak{P}}$$

for every $r \in L$, $r \notin aL$. Applying Theorem 1 with $p = 3$, $K = \mathbb{Q}(\rho)$ and $\mathfrak{c} = a\mathbb{Z}[\rho]$ (*cf.* the remark after the proof of it), we see

$$\sum_{\substack{r \in L/aL \\ r \neq 0}} \{f(\sigma)g(r) - f(r)g(\sigma)\}$$

$$\equiv -6(1-\rho) \operatorname{Ind}_\rho \left[\frac{c}{a}\right]_3 \mod 3(1-\rho)^{\frac{3}{2}} \mathfrak{O}_{\mathfrak{P}}.$$

Hence

$$\Psi(A) \equiv 3 \operatorname{ord}_\rho \left[\frac{a}{c}\right]_3 \mod 3(1-\rho)^{\frac{1}{2}} \mathfrak{O}_{\mathfrak{P}}.$$

Theorem 3 follows because $\Psi(A)$ is in $\frac{1}{3}\mathbb{Z}[\rho]$.

We get the homomorphism $\psi : \Gamma \to \mathbb{Z}$ mentioned in section 1 by putting $\psi(A) = -\frac{1}{3}\Psi(A) - \frac{1}{3}\overline{\Psi(A)}$. The homomorphism $\Psi(A)$ can be expressed as a linear combination of $\Phi(A)(u, v)$ with $u, v \in \frac{1}{3}L$ ([2]). Theorem 2 gives another proof for the fact that the map

$$A \to \left(\frac{c}{a}\right)_3 \qquad \left(A = \begin{pmatrix} a & b \\ c & d \end{pmatrix} \in \Gamma\right)$$

is a character of $\Gamma$, which is equivalent to the reciprocity law of the symbol $\left(\frac{c}{a}\right)_3$ (Kubota [4]).

## References

[1] *R. Fueter*, Vorlesungen über die singulären Moduln und die komplexe Multiplikation der elliptischen Functionen, I. Leipzig–Berlin, 1924.

[2] *H. Ito*, On a property of elliptic Dedekind sums. J. Number Theory **27** (1987), 17—21.

[3] *H. Ito*, Dedekind sums and quadratic residue symbols. (Preprint).

[4] *T. Kubota*, Ein arithmetischer Satz über eine Matrizengruppe. J. reine angew. Math. **222** (1966), 55—57.

[5] *H. Reichardt*, Eine Bemerkung zur Theorie des Jacobischen Symbols. Math. Nachr. **19** (1958), 171–175.

[6] *R. Sczech*, Dedekindsummen mit elliptischen Funktionen. Invent. Math. **76** (1984), 523—551.

---

Dept. of Mathematics, Nagoya University, Chikusa–ku, Nagoya 464, Japan.

# A Brief Survey on Distribution Questions for Second Order Linear Recurrences

*Elliot Jacobson*

The purpose of this note is to summarize progress in questions of distribution of the residues of second order linear recurrence sequences. We begin with some notation and definitions.

Let $U_n = AU_{n-1} + BU_{n-2}$, $n \geq 2$, be a linear recurrence with $A, B, U_0, U_1$ integers. Denote $D = A^2 + 4B$, the "discriminant" of $(U_n)$. In the case $A = B = 1$, $U_0 = 0$, $U_1 = 1$, we write $U_n = F_n$ (Fibonacci sequence). For a fixed positive integer $m$, the sequence $(U_n)$ is eventually periodic (mod $m$), and we denote the shortest period by $\lambda = \lambda(m)$. For an integer $b$, we denote by $\nu(m, b)$ the frequency of the residue $b(\bmod m)$ in one shortest period of $U_n(\bmod m)$. The sequence $(U_n)$ is "uniformly distributed" (mod $m$) (notation: u.d.(mod $m$)) if $\nu(m,*)$ is a constant function, that is, if $(U_n)$ attains each residue with the same frequency in any period. In this case we can write $\lambda(m) = mf$ for some positive integer $f$.

The study of recurrence sequences reduced (mod $m$) was formally introduced by Wall [12] where he considered bounding $\lambda(m)$ for the Fibonacci sequence. His main result was:

**Theorem 1.** *Let $p$ be an odd prime. For $(F_n)$ we have:*

*a)* $\lambda(p)$ *is even*

*b)* $\lambda(p)$ *divides* $p-1$ *for* $p \equiv 1, 4 \pmod 5$

*c)* $\lambda(p)$ *divides* $2(p+1)$ *for* $p \equiv 2, 3 \pmod 5$.

The question of uniform distribution was first considered for the Fibonacci sequence. In the early 70's a flurry of activity was culminated by the proof of the following:

**Theorem 2.** (*Kuipers and Shiue* [7], *Niederreiter* [9]) *$(F_n)$ is u.d. modulo m if and only if $m = 5^k$ for some $k > 0$.*

Bumby [1] was the first to completely classify moduli of uniform distribution for arbitrary second order linear recurrences. He established the following:

**Theorem 3.** *$(U_n)$ is u.d (mod m) if and only if the following holds for each prime divisor p of m.*

a) $p \mid D$ *and* $p \nmid B$

b) *For $p \geq 3$ we have* $p \nmid 2U_1 - AU_0$

c) *If $p = 3$ and $9 \mid m$ then* $D \not\equiv 6 \pmod 9$

d) *If $p = 2$ then $U_0$ and $U_1$ are of opposite parity. Moreover, if $4 \mid m$ then $A \equiv 2 \pmod 4$ and $B \equiv 3 \pmod 4$.*

**Example 1.** For $(F_n)$ we have $D = 5$ and $2U_1 - AU_0 = 2$ so Theorem 2 follows from Theorem 3.

**Example 2.** For the "Lucas sequence" $(L_n)$ with data $A = B = 1$, $L_0 = 2$ and $L_1 = 1$ we have $D = 5$, $2L_1 - AL_0 = 0$, so that this sequence is never uniformly distributed for any modulus.

Bumby's classification seemingly completed the study for the second order case until Vélez proved the following theorem, first for $(F_n)$ with Erlebach in 1983 [3], then for general $(U_n)$ in 1987 [11].

**Theorem 4.** *Assume that $(U_n)$ is u.d. $(\bmod\, p^k)$ with $\lambda(p^k) = p^k f$, with p a prime. Then for $s \geq 0$, the sequence of residues*

$$U_{s+kf} \pmod{p^k} \qquad k = 0, 1, \ldots, p^k - 1$$

*form a complete residue system.*

The essential content of this theorem was already apparent in Bumby's work. Its importance lies in that it explicitly describes how the residues are uniformly distributed. This allows an application to non–uniform distribution (non–u.d.) questions, as we now discuss. First observe how much chaos there apparently can be via some examples.

**Example 3.** $F_n$(mod 1583) gives $\lambda$=3168.

**Example 4.** $F_n$(mod 1597) gives $\lambda$=68.

**Example 5.** $F_n$(mod 11) is 0, 1, 1, 2, 3, 5, 8, 2, 10, 1. In this case $\nu(11,*)$ takes on the values 0, 1, 2, 3.

**Example 6.** $F_n$ (mod $5^k$) gives $\varsigma(5^k,*)=4$.

Using Theorem 4 and applying the Chinese Remainder Theorem to certain subscripts the author has established the first distribution result applicable in the non–u.d. situation [5].

**Theorem 5.** *Suppose that* $(U_n)$ *is u.d.* (mod $p^k$) *with* $\lambda(p^k)=p^k f$. *Let* $m \geq 2$ *and denote* $\lambda(m) = Q$. *Assume that* $p \nmid Q$. *Then for every integer b:*

$$\varsigma(p^k m,\ b)=\frac{f}{(f\ ,Q)}\varsigma(m,b).$$

The hypothesis $p \nmid Q$ is justified by considering $F_n$ with $p=5$ and $m=33$. Then $\lambda(33)=40$, and $\nu(33,1)=5$ whereas $\nu(165,1)=3$.

The theorem asserts that $\nu(m,*)$ and $\varsigma(p^k m,*)$ assume the same number of distinct values. In particular by inspecting the periods of $F_n$ modulo 2, 4, 3, and 9 we have (see [4]):

**Theorem 6.** *For the Fibonacci sequence,* $\nu(m,*)$ *assumes exactly two distinct values for any m of the form* $2\cdot 5^k$, $4\cdot 5^k$, $3\cdot 5^k$, $9\cdot 5^k$ *with* $k \geq 0$.

The canonical question this last theorem poses is if the given list is complete. A computer search of moduli $m < 10000$ indicates this is the case. The question of determining the primes for which $\nu(p,*)$ assumes exactly two values dates back to the early 70's. A solution was found by Schinzel at the 1988 CNTA meeting and appears elsewhere in these proceedings. His work easily extends to cover all prime power moduli, and we state this result here for completeness.

**Theorem 7.** *Suppose that for a prime power* $p^k$ *and* $(F_n)$(mod $p^k$) *the function* $\nu(p^k,*)$ *assumes exactly two distinct values. Then* $p^k$ *is one of* 2, 4, 3, *or* 9.

It is worth remarking that by a theorem of Burr [2], except for the prime power moduli 2, 4, $3^k$, $5^k$, and 7, the function $\nu(p^k,*)$ always assumes the value 0. The complete proof of the converse of Theorem 6 is still out of reach.

One other observation is in order. For the Fibonacci sequence, the function $\nu(p,*)$ never "appears" to take on more than 4 distinct values ... as was born out by the computer search for counterexamples to the conjecture above. In fact, this also was proven by Schinzel at the 1988 CNTA meeting. In his proof he went further to characterize exactly what the various values of $\nu$ could be.

**Theorem 8.** *Assume $p \neq 5$. Let $F_q$ be the splitting field over $F_p$ for the polynomial $x^2 - x - 1$. Let $\xi$ be a root of $x^2 - x - 1$ in $F_q$, and let $\delta$ be the multiplicative order of $\xi$ in $F_q^*$. Then :*

a) *If* $\delta \not\equiv 0 \pmod 4$ *then* $\nu(p,*)$ *assumes at most the four values* 0,1,2,3.

b) *If* $\delta \equiv 4 \pmod 8$ *then* $\nu(p,*)$ *assumes at most the three values* 0,2,4.

c) *If* $\delta \equiv 0 \pmod 8$ *then* $\nu(p,*)$ *assumes at most four values, the values coming from exactly one of the lists* 0,1,2,4 *or* 0,2,3,4.

**Example 7.** Let $(U_n)$ be specified by the data $A = 3$, $B = 2$, $U_0 = 0$, and $U_1 = 1$. Then for $(U_n)$mod 89 the function $\nu(89,*)$ takes on the five values 0, 1, 2, 3, 4. A sweeping generalization of Schinzel's theorem therefore seems unlikely.

Bumby stated in his work that results analogous to Theorem 3 should be provable when the recurrence takes values in the integers of a number field. In what follows, let $K$ denote a number field with ring of integers $R$. For an ideal $I$ of $R$ we let $N(I)$ denote its absolute norm. $P, P_i$, etc., will denote prime ideals of $R$. Turnwald [10] was the first to obtain the following theorem.

**Theorem 9.** *Let $I$ be a proper ideal of $R$. then $(U_n)$ is u.d.* (mod $I$) *if and only if the following conditions hold:*

a) *If* $P \mid I$ *then* $N(P) = p$, $P \mid D$, $P \nmid B$.

b) *If* $P_i \mid I$ $(i = 1,2)$ *and* $P_1 \neq P_2$, *then* $N(P_1) \neq N(P_2)$.

c) *If* $P \mid I$ *and* $p \geq 3$ *then* $P \nmid (2U_1 - AU_0)$.

d) *If* $P^2 \mid I$ *and* $p \geq 5$ *then* $e(P/p) = 1$ (*$P$ is unramified*).

e) *If* $P^3 \mid I$ *and* $p = 2$ *or* 3 *then* $e(P/p) = 1$.

f) *If* $P \mid I$ *and* $p = 2$ *then* $U_0 \not\equiv U_1 (\mathrm{mod}\ P)$. *If also* $P^2 \mid I$ *then* $A \not\equiv 0 (\mathrm{mod}\ P^2)$ *and* $B \not\equiv 1 (\mathrm{mod}\ P^2)$.

g) *If* $P^2 \mid I$ *and* $p = 3$ *then* $A^2 \not\equiv B\ (\mathrm{mod}\ P^2)$.

These results were obtained in a different fashion by Vélez and this author [6]. In that paper it was shown that the property given in Theorem 4 above also holds for the Dedekind Domain case. In turn, a reasonable generalization of Theorem 5 also holds. An analogue of Schinzel's result should be possible in the Dedekind Domain case.

The questions posed in studying non-uniform distribution are both difficult and interesting. The area as a whole has barely been touched. The reader should consult the excellent work [8] for a thorough introduction to the theory of uniform distribution and related topics.

## References

[1] *R.T. Bumby*, A distribution property for linear recurrence of the second order. Proc. Amer. Math. Soc. **50** (1975), 101—106.

[2] *S. Burr*, On moduli for which the Fibonacci sequence contains a complete system of residues. Fibonacci Quart. **9** (1971), 497—504.

[3] *L. Erlebach* and *W.Y. Vélez*, Equiprobability in the Fibonacci Sequence. Fibonacci Quart. **21** (1983), 189—191.

[4] *E.T. Jacobson*, Almost uniform distribution of the Fibonacci sequence. (to appear), Fibonacci Quart.

[5] *E.T. Jacobson*, The distribution of residues of two-term recurrence sequences. (preprint).

[6] *E.T. Jacobson* and *W.Y. Vélez*, Uniform and f-uniform distribution of recurrence sequences over Dedekind Domains. (preprint).

[7] *L. Kuipers* and *J. Shiue*, A distribution property of the sequence of Fibonacci numbers. Fibonacci Quart. **10** (1972), 375—376, 392.

[8] *W. Narkiewicz*, Uniform distribution of sequences of integers in residue classes. Lecture Notes in Math., Vol. 1087, Springer—Verlag, New York, 1984.

[9] *H. Niederreiter*, Distribution of Fibonacci numbers mod $5^k$. Fibonacci Quart. **10** (1972), 373—374.

[10] *G. Turnwald*, Uniform distribution of second-order linear recurring sequences. Proc Amer. Math. Soc. **96** (1986), 189—198.

[11] *W.Y. Vélez* , Uniform distribution of two–term recurrence sequences. Trans. Amer. Math. Soc. **301** (1987), 37—45.

[12] *D.D. Wall* , Fibonacci series modulo *m*. Amer. Math. Monthly **67** (1960), 525—532.

---

Department of Mathematics, Ohio University, Athens, OH, USA 45701

# Basis for the Polynomial Time Computable Functions

*J.P. Jones* [1] *and Y. V. Matijasevich*

## 1. Introduction

One approach to complexity of algorithms and classification of classes of number theoretic functions according to computability by certain types of algorithms, is through sets of so called *basis* functions for the class. Here we consider this problem for the class of polynomial time computable functions.

In his classical paper [5], A. Grzegorczyk raised the question of whether recursion can be eliminated from the definition of the classes $E^n$ of computable functions. These Grzegorczyk classes, $E^n$, ($n$ = 1,2,...) were originally defined by starting from an initial set of basic functions and closing under the operations of composition and bounded recursion. Grzegorczyk asked whether it is really necessary to close under recursion. Can the functions in each $E^n$ be generated from the basic functions without recursion?

In order the formulate this question more precisely, one requires the concept of a basis. A *basis* for a class of functions is a finite set of functions sufficient to generate the entire class without recursion. More precisely, $B$ is a basis for $K$ if $B$ is a subset of $K$ and every function in $K$ is obtainable from functions in $B$ by applying only composition.

---

[1] Work supported by Natural Sciences and Engineering Research Council of Canada Research Grant A4525, the N.S.E.R.C. International Scientific Exchange Award Program and the Queen's–Steklov Exchange Program between Canada and the U.S.S.R. Authors also wish to acknowledge valuable conversations with S.R. Buss, University of California, Berkeley, and A.J. Wilkie, Oxford.

By function composition is meant essentially the operation of substitution of one function into another. However one must be slightly careful here because there are different interpretations of what is meant by substitution. Some would not agree that we can obtain $f(x, x)$ from $f(x, y)$ by composition. Such a use of composition (identification of variables is sometimes considered to be an application of one of the so called *explicit transformations*. The explicit transformations are normally the operations of identification of variables, permutation of variables and substitution of a constant for a variable. Here in this paper, we include under composition all such operations and put a small restriction on the latter. As has become a tradition in the subject, we permit replacing a variable by a constant only when the constant is obtainable as a value of some earlier obtained function. Another operation we allow under composition is projection (addition of dummy variables). This is not considered an explicit transformation by some (e.g., according to H. Rose [21]), but in this paper, to decrease the number of trivial functions needed in a basis, this operation will also be allowed under composition. If the reader does not want to include projection in this way, then he would need to add the projection functions, $\Pi^i(x_1, x_2, \ldots, x_n) = x_i$, to the bases constructed here, so to effect all the trivial operations such as passing from the one variable function $f(x)$ to a two variable function $g(x, y) = f(x)$. We are more interested in the mathematical functions (number theoretic functions).

One of the first interesting classes of number theoretic functions to be shown to have a basis was the Grzegorczyk class, $E^3$. This class, $E^3$, at the third level of the Grzegorczyk hierarchy, is the class of *Kalmar elementary functions*, *KF*. By definition *KF* is the smallest class of functions containing zero, addition, multiplication and closed under bounded sums and products. Thus *KF* contains most functions found in elementary number theory. For example, *KF* contains $x^z$, $x!$, the characteristic function of the primes, the Euler $\phi$ function and many other number theoretic functions. The existence of a basis for $KF = E^3$ was first proved by D. Rödding [20]. In the same paper Rödding also proved the existence of a basis for the other Grzegorczyk classes, $E^n$ ($n = 3,4,5,\ldots$). Later S.S. Marchenkov [14] proved that $E^2$ has a basis. The problem of existence of a basis for the two classes $E^0$ and $E^1$ is apparently still open.

Does there exist a basis for the set of all computable functions (all recursive functions)? It is easy to see that the answer is no. This can be proved by a classical diagonal argument. Suppose the class of recursive functions had a finite basis. Then it would be possible to associate with each nonnegative integer $n$, in an effective way, a recursive function $f_n$ such that the list $f_n$ $(n = 0,1,2,\ldots)$, would include every recursive function, Then the diagonal function, $f(x) = f_n(n) + 1$, would be computable but not in the list.

Here in this example we were considering the class of *total* (everywhere defined) recursive functions. It makes a difference. The class of partial recursive functions *does* have a basis. This follows from the existence of a universal Turing machine. The function associated with the universal Turing machine, together with a few other elementary functions gives a basis for the partial recursive functions. So, for this reason, in this paper all functions will be total.

Like the recursive functions, the class of primitive recursive functions also does not have a basis. A simple way to show this (noticed by Alex Wilkie) is to use the fact that the primitive recursive functions are the union of the Grzegorczyk classes $E^n$ $(n = 3,4,5,\ldots)$. Since each class $E^n$ is closed under composition, and each class $E^n$ strictly contains the previous class $E^{n-1}$, the union cannot have a finite basis. This argument can be formulated as a general principle.

**Theorem 1.** *Suppose that* $K$ *is a class of number theoretic functions satisfying the following conditions:* (1) *$K$ is the union of an increasing family of subclasses* $K^n$ $(n = 0,1,2,\ldots)$. (2) *$K$ is not the union of a finite number of the classes* $K^n$. (3) *Each class* $K^n$ *is included in the class* $K^{n+1}$. (4) *Each class* $K^n$ *is closed under composition. Then $K$ does not have a basis.*

In this paper we will consider the basis problem for the class of *polynomial time computable functions.* One reason for considering this class is the widely held view today that the length of computations ought to be in some way restricted. It is considered that if a problem requires exponential time, then it should be considered infeasible, the cost in time and money becoming prohibitive, for large values of the input. Even though computers become faster, still it is the time and space complexity of the algorithm which has the dominant effect on the size of problem which can be considered. For this reason, today *algorithm* has often come to mean a *polynomial time algorithm.*

One could also consider the class PSPACE (*polynomial space* ). However, from the point of view of one interested in the design of efficient algorithms, the time complexity is usually more important than the space complexity. One reason for this is that when a program runs in polynomial time, i.e., in a number of steps which is a polynomial function of the size of the input, then automatically it runs in so called PSPACE. That is, the sizes of all intermediate quantities and hence the size of the output, is a polynomial function of the size of the input. PTIME is contained in PSPACE. We should mention also that in these discussions the *size* of a number is not the number itself but something like its base 2 logarithm. If $x$ is a nonnegative integer, then the size of $x$ is considered to be $\lceil \log_2 (x + 1) \rceil$, i.e., the *length* of $x$ in binary (base 2).

The study of computational complexity and polynomial time computable functions originated in the work A. Cobham [1], J. Edmonds [4], S.A. Cook [2], R.M. Karp [11], and L.A. Levin [13]. In this paper we will use S. Cook's notation *PF* to denote the class of polynomial time computable functions. All computations are restricted to being *deterministic* ones. Each step is determined by the results of the previous steps. There is a related concept of *nondeterministic polynomial time computable functions NPF*, analogous to the class *NP* of sets.

To distinguish functions from sets we will use *P* to denote the class of sets (relations) decidable in deterministic polynomial time. *NP* will denote the class of relations decidable in nondeterministic polynomial time.

For an exact definition of *PF*, there are many equivalent definitions from which to choose. The computational device most often used is the one tape Turing Machine. Here we will use the register machine (Minsky Machine) sometimes also known as the an Unlimited Register Machine. The register machine is also similar to what is called a Random Access Machine (RAM), with small differences.

A *register machine* is a "machine" with a finite number of separately addressable registers $R1, R2, \ldots, Rn$, and a finite program consisting of a finite list of commands considered to be written on labeled lines, *L0,L1,..Ll*. Normally the commands are executed in the sequence given. However, the register machine can transfer to a different location and begin executing commands there.

Four types of commands (and a stop command) are known to be sufficient for a register machine.

$$Li \text{ GOTO} Lk, \quad \text{If } Rj < Rm, \text{GOTO} Li, \quad Rj \leftarrow Rj+1, \quad Rj \leftarrow Rj-1, \qquad (1)$$

Because these commands are sufficient for computation of every recursive function, (see for example Minsky [16]), it is often assumed that the register machine and Turing machine are polynomial time equivalent. However, strictly speaking, this is not the case, If a function is computable in polynomial time on a Turing machine, it is not necessarily polynomial time computable on a register machine. In order to use the register machine for the definition of polynomial time computability, it is necessary to add additional commands to the above four. If only the above four commands are used, then a register machine cannot even add or multiply in polynomial time. (See Jones & Matijasevich [8].) However, if one adds the following two commands, then the Turing Machine and the register machine are polynomial time equivalent:

$$Ri \leftarrow Ri + Rj, \qquad Ri \leftarrow [Ri/2]. \qquad (2)$$

Here $[Ri/2]$ denotes the integer part of $Ri/2$, so these commands (2) permit the register machine to do a right shift in one step. Also they allow addition of the contents of one register to the contents of another register in only one step.

With these commands, a function $f(x)$ of one variable can be defined to be *computable in polynomial time* if there exists a register machine $M$ and a polynomial $P$ such that when started with $x$ in $R1$, $M$ computes $f(x)$ in time $t < P(|x|)$ where $|x| = \lceil \log_2(x+1) \rceil$. Equivalently, one can require the existence of constants $c$ and $d$ such that for all $x$, $t \leq d|x|^c$. (Or one can specify that there is a single constant $c$ such that $t < |x+2|^c$ holds for all $x$. ) Sometimes one doesn't mention any of these constants explicitly. One speaks of a function $f(x)$ computable in time of the *order of n*. This means that $n = |x|$. The constants $c$ and $d$ are implicit.

For a function of several variables, $f(x_1, x_2, \ldots, x_k)$, where $x_1, x_2, \ldots, x_k$ are nonnegative integers, we define polynomial computability as follows. A function $f$ is *computable in polynomial time* if there exists a register machine $M$ and a polynomial $P(x_1, x_2, \ldots, x_k)$, such that $M$, when started with $x_1, x_2, \ldots, x_k$ in $R1$, $R2, \ldots, Rk$ respectively, halts in time $t = P(|x_1|, |x_2|, \ldots, |x_k|)$ steps, with the value $f(x_1, x_2, \ldots, x_k)$ in register $R1$. Alternatively, we can require the existence of constants $c$ and $d$ such that for all $x_1, x_2, \ldots, x_k$, machine $M$ stops in time $t \leq d(|x_1|+|x_2|+\ldots+|x_k|)^c$.

With these definitions the class $PF$ is defined. $PF$ contains many interesting functions of number theory. For example $PF$ contains the function, $r = \text{rem}(x, y)$, (remainder after $x$ is divided by $y$). $PF$ also contains the functions $x + y$, $\gcd(x, y)$, $\text{lcm}(x, y)$, $\text{rem}(x^z, y)$, the *Jacobi symbol* $(x/y)$ and other functions. An example of a function not in $PF$ is $2^x$.

## 2. Existence of a Basis

We show that there exists a basis for $PF$. The first proof of existence of a basis for $PF$ was given by the Russian mathematician A.A. Muchnik [17]. His paper is not well known. It has never been translated into English. Only after inventing our own proof, independently, did we learn about the existence of Muchnik's proof.

We will give a basis for the class of total functions only, (and $PF$ in this paper will refer only to total functions). It will turn out however, that this will also establish the existence of a basis for the partial functions, (for the partial polynomial time computable functions). One can see this as follows: First of all the domain of a partial polynomial time computable function $\varphi$ is a set which must be belong to the polynomial time decidable predicates $P$. Call this set $D$. From the fact that $D \in P$, it follows that every partial polynomial time computable function $\varphi$ can be extended to a total polynomial time computable function $f \in PF$. Hence we can obtain a basis for all the partial functions from a basis for the total functions just by adding some (actually *any*) partial function. For example suppose that $\delta(x) = \lceil 1/x \rceil$, (a partial version of the ceiling function, understood to be undefined at $x = 0$ and for $x > 0$ to have value $\delta(x) = 1$). Let $f$ be any extension of $\varphi$ to a total function in $PF$. Let $\chi$

be the characteristic function of the set $D$ (= the domain of $\varphi$). Then clearly $\varphi(x) = f(x \cdot \delta(\chi(x)))$ because $\delta(0)$ is undefined and $\chi[D'] = \{0\}$. Thus addition of the single function $\delta$ to any basis for the total functions, results in a basis for the partial functions.

**Theorem 2.** *(A.A. Muchnik [17]). There exists a basis for PF.*

**Proof.** There exists no function $U(e,x)$ in $PF$ with the property that every function $f(x)$ in $PF$ is equal to $U(e,x)$ for some $e$. (That is to say, there is no universal function in $PF$ for $PF$.) However, one can find a universal function which belongs to $KF$. Consider the following function:

**Definition.** $Z(e,t,x)$ = *output of machine* $e$ *after* $t$ *steps on input* $x$.

Here the *output* of the $e$th register machine, $M_e$, is considered to be whatever is in register $R1$, after $t$ steps. The $e$th machine is the register machine whose index is $e$ in some encoding. $Z(e,t,x)$ as defined above is computable in time of the order of $t(t + |x| + |e|)$ steps. (See Hartmanis & Stearns [6], M. Minsky [16] Chapter 10 or A.A. Muchnik [17].) In fact Muchnik [17], p. 134, claims an even better estimate for the time (better than quadratic in $t$) to compute $Z(e,t,x)$. Of course $Z(e,t,x)$ is not computable in polynomial time, as a function of $t$. Nevertheless $Z(e,t,x)$ is computable in elementary time, as a function of $e$, $t$ and $x$. So $Z(e,t,x)$ belongs to $KF$, by the principle that whatever is computable in elementary time is elementary, (see e.g., Cutland [3]).

Using the function $Z(e,t,x)$ we can construct a function which is in $PF$ and which behaves very much like a universal function. For any function $f(x)$ in $PF$, we have $f(x) = Z(e,|x|^c,x)$, for some constants $e$ and $c$, because $f(x)$ is computable in time $|x|^c$ for some $c$, (or more precisely in time $\leq d|x|^c$, for some constants $c$ and $d$ though hereafter we will drop the constant $d$ which only complicates the notation).

So the four place function $Z(e,|x|^c,x)$, behaves like a universal function. Unfortunately it is still not in $PF$. As a function of $c$, $Z(e,|x|^c,x)$ belongs to $KF$. It is exponential in $c$. So we will use the function

$$V(e,t,x) = Z(e,|t|,x), \tag{3}$$

and construct $|x|^c$ by composition. The function, $V(e,t,x)$, a kind of polynomially clocked form of $Z(e,t,x)$, does belong to $PF$, by our earlier estimate.

To use $V(e,t,x)$ to compute $f$, we need only find a constant $c$ large enough that $f(x) = V(e,|x|^c,x)$. To obtain the use of this function, through composition, we need to find first a function which is in $PF$ and grows more rapidly than any polynomial. Fortunately there is such a function, the so called *smash* function, $x \# y$, a tradition in the subject of polynomial time computability. (The function $x \# x$ will also do.) The definitions are

$$x \# y = 2^{|x||y|}, \qquad T(x) = x \# x = 2^{|x|^2}. \tag{4}$$

From the inequality $\log_2(x) < |x|$, we find that

$$|x|^{c^2} < |T^{<c>}(x)|, \tag{5}$$

where $T^{<c>}(x)$ denotes *c-fold composition* of the function $T(x)$ with itself. From (5) we have that for any function $f$ in $PF$,

$$f(x) = V(e, T^{<c>}(x), x) = Z(e, |T^{<c>}(x)|, x) \tag{6}$$

for some constants $e$ and $c$. These constants $c$ and $e$ can be obtained by composing the two functions $Z(x) = 0$ and $S(x) = x + 1$. Hence the following four functions are a basis for $PF$.

$$0, \qquad x+1, \qquad x \# y, \qquad V(e,t,x) \tag{7}$$

This proves the existence of a basis for $PF$, i.e., Theorem 2 is proved. Notice that the use of the # function was essential. One cannot use only the multiplication function. Of course one can generate $|x|^c$ by multiplying $|x|$ $c$-times, but something larger is needed because we have only the effective use of $||x||^c$ and $||x||^c$ is only about as large as $c||x||$.

## 3. Size of a Basis

The above proves the existence of a basis of 4 functions for *PF*. Now we ask what is the smallest number of functions sufficient in a basis? We will show next that *one* function is enough.

**Theorem 3.** *There exists a basis for PF consisting of 1 function.*

**Proof.** The one function will be a function of two variables. The first step in the proof is to replace the basis (7), which contains a function of three variables, by a basis consisting of five functions of two variables. It is well known that one can do this in general. Every function of any number of variables can be expressed as the composition of a finite number of functions of two variables. The problem is to show that all the intermediate functions of two variables used, are functions in *PF*. So the first step in the proof is to write the three variable function $V$ so that it is a composition of functions of two variables in *PF*.

As is well known, there exist one–to–one correspondences between the nonnegative integers and pairs of nonnegative integers and such correspondences can be described in an effective way. That is to say, one can find recursive functions $R, L$ and $J$ satisfying equations

$$J(L(z), R(z)) = z, \qquad L(J(x, y)) = x \quad \text{and} \quad R(J(x, y)) = y. \tag{8}$$

A triple of such functions, satisfying (8), is called a *system of pairing functions*. Already in the 19th century, G. Cantor found such a system which he used to prove $\aleph_0 \bullet \aleph_0 = \aleph_0$. In fact the function $J(x, y)$ can be a polynomial with rational coefficients.

$$2J(x, y) = (x + y)^2 + 3y + x, \tag{9}$$

The functions $J$ obviously belongs to *PF*. The inverse functions $R$ and $L$ also belong to *PF*. One way to show that is to derive the following formulas for the inverse functions,

$$R(z) = z - \frac{1}{2}\left[\sqrt{2z + \frac{1}{4}} - \frac{1}{2}\right] \cdot \left[\sqrt{2z + \frac{1}{4}} + \frac{1}{2}\right], \tag{10}$$

$$L(z) = \frac{1}{2}\left[\sqrt{2z + \frac{1}{4}} - \frac{1}{2}\right]^2 + \frac{3}{2}\left[\sqrt{2z + \frac{1}{4}} - \frac{1}{2}\right] - z. \tag{11}$$

Here the square brackets [ ] denote the integer part. Using the pairing functions one can write the three variable function $V(e,t,x)$ as a composition of two functions of two variables. Put $N(e,z) = V(e,L(z),R(z))$. Then $N$ belongs to $PF$, since $V$, $L$ and $R$ belong to $PF$. Also because $V(e,t,x) = N(e,J(t,x))$ and because $J$ is in $PF$, the following five functions are a basis for $PF$.

$$0, \qquad x+1, \qquad J(x,y), \qquad x \# y, \qquad N(x,y). \tag{12}$$

From these five functions of two variables, we can now construct a basis of one function of two variables. Consider the function $H(x,y)$ defined by the following equations:

$$\begin{aligned} H(x,x) &= 5x+1, \\ H(x,5x+1) &= x+1, \\ H(5x+1,5y+2) &= 0, \\ H(5x+1,5y+3) &= J(x,y), \\ H(5x+1,5y+4) &= x \# y, \\ H(5x+1,5y+5) &= N(x,y). \end{aligned} \tag{13}$$

These equations define $H(x,y)$ by cases. They don't apply to every $(x,y)$, strictly speaking, so they only define $H(x,y)$ as a partial function. However, it is trivial to extend $H$ to a total function. We can let $H(x,y)$ be zero in the other cases. So $H(x,y)$ (or some extension) belongs to $PF$ because $N(x,y)$ belongs to $PF$ and $P$ is closed under the Boolean operations. The proof will be complete when we show that the *functions* $0$, $x+1$, $J(x,y)$, $x \# y$ and $N(x,y)$ are obtainable from $H(x,y)$ by composition. But this is evident. One begins with, for example, $H(x,H(x,x)) = x+1$. Then $H(H(x,x),H(x,x)+1) = 0$ and $H(H(x,x),H(y,y)+2) = J(x,y)$, etc. The proof is complete.

Observe that the argument is general. It applies to many classes of functions having a basis. For example, if $K$ is a class of number theoretic functions and $K$ has a basis, and $K$ is closed under composition, then if $K$ contains the functions $0$, $x+1$, $J(x,y)$, $R(z)$, $L(z)$, $x+y$, then $K$ has a basis of one function of two variables.

## 4. Basis for One Variable Functions

How many functions of one variable are necessary in a basis? From functions of one variable one can obtain only functions of one variable, so the question must be reformulated more precisely. We should ask, in order to obtain all $PF$ functions of one variable, with how many functions of one variable must one begin?

We will show that the answer is *two functions*. Actually, two functions are necessary and sufficient. Two functions are necessary because one function $f$ cannot be a basis for all the one variable functions of $PF$. To see this, observe that if $K$ is any class of number theoretic functions of one variable, and if $K$ contains at least the identity fucntion and $K$ has a basis consisting of only one variable, then $K$ is finite. For if $f$ is the one function, then for some value of $c$, we would have $f^c(x) = x$ for every $x$. So for any $r$ and any $m$ we must have $f^{cm+r}(x) = f^r(x)$. Hence $K$ could have cardinality at most $r$. Thus two functions are necessary. The theorem below is therefore best possible.

**Theorem 4.** *There exist two functions $g$ and $h$ in PF, each of one variable, such that every function in PF of one variable is obtainable from these two functions $g$ and $h$, by composition.*

**Proof.** For the proof of Theorem 4, we first replace the use of a pairing function by use of a tripling function and its three inverse functions. Thus we use a function $J(x, y, z)$ and three functions $L$, $M$ and $R$, satisfying the equations

$$J(L(z), M(z), R(z)) = z, \qquad L(J(x, y, z)) = x, \tag{14}$$
$$M(J(x, y, z)) = y, \qquad R(J(x, y, z) = z.$$

Such a system of functions $J$, $L$, $M$ and $R$ can be easily built up from the previous pairing functions given in (8). Re–using the old names, we put $L(z) = L(z)$ and

$$M(z) = L(R(z)), \qquad R(z) = R(R(z)), \qquad J(x, y, z) = J(x, J(y, z)) \tag{15}$$

Then the new three variable version of $J$ and its three inverse functions $L$, $M$ and $R$, belong to $PF$. We use them next to construct four functions of one variable, which are a basis for the one variable functions of $PF$.

$$C(x) = J(0, x, x), \tag{16}$$

$$I(x) = J(L(x)+1,\ M(x), R(x)), \tag{17}$$

$$B(x) = J(L(x), T(M(x)), R(x)), \tag{18}$$

$$E(x) = V(L(x), M(x), R(x)). \tag{19}$$

The idea behind these four functions is that function $C$ encodes $x$, function $I$ increases the index $e$ to the appropriate level, $B$ bumps the clock (to $|x|^c$ ) and $E$ evaluates the resulting encoding.

To see that the four functions, $C, I, B$ and $E$ are a basis for the one variable functions in $PF$, assume that we are given an arbitrary function of one variable, $f(x)$, in $PF$. Suppose too that $f(x)$ is computable in time $|x|^c$, on some register machine with index number $e$. Then equation (6) holds, Hence for all $x$

$$f(x) = E(B^{<c>}(I^{<e>}(C(x)))). \tag{20}$$

The final step in the proof is to compress the 4 functions $C, I, B, E$ into 2 functions. Consider the functions $g$ and $h$ defined by,

$$g(x) = 2x, \qquad h(2x) = 2x+1, \qquad h(8x+1) = C(x), \tag{21}$$

$$h(8x+3) = I(x), \quad h(8x+5) = B(x), \quad h(8x+7) = E(x).$$

These two functions $h$ and $g$ will be the basis for the one variable functions of $PF$. From $g$ and $h$, using only function composition one can obtain the functions $hggg(x) = 8x+1$, $hghgg(x) = 8x+3$, $hgghg(x) = 8x+5$ and $hghghg(x) = 8x+7$. Hence one has the four functions $C(x)$, $I(x)$, $B(x)$ and $E(x)$. Since these four functions are a basis for the one variable functions of $PF$, the two functions $h$ and $g$ are a basis for the one variable functions of $PF$. This completes the proof of Theorem 4.

## 5. Open Problems

Two important problems remain. **Problem 1** is *construction of an explicit basis for PF*. By an *explicit basis* is meant a basis consisting of natural functions. It is not possible to give an exact definition of what is meant by natural functions, so it seems best to give some examples. Consider the following four functions:

$$x \dot{-} y, \qquad [x/y], \qquad \varphi(x, y), \qquad 2^{xy}. \tag{22}$$

These four functions are a basis for the Kalmar elementary functions, *KF*. (For a proof of this see Jones [10].) This theorem is based on an earlier related result of S.S. Marchenkov [3] who defines the function $\varphi(x, y)$ to be the least place in the base $x$ expansion of $y$ where a digit zero occurs. (In order that the function $\varphi(x, y)$ will be total, $\varphi(x, y)$ is defined to be 0 when $x = 0$ and when $x = 1$.) The function $[x/y]$ is defined to be the integer part of $x/y$. Function $x \dot{-} y$ is proper subtraction, ($x \dot{-} y = x - y$ if $y \leq x$, and $x \dot{-} y = 0$ if $x < y$).

As another example of an explicit basis of natural functions, we can mention

$$x + y, \qquad [x/y], \qquad x \dot{-} y, \qquad 2^{x}. \tag{23}$$

These four functions are a basis for the *Kalmar elementary relations*. (The exact meaning of this statement is that all characteristic functions of Kalmar elementary relations are obtainable from functions in (23). Equivalently, one can understand it to mean that every two valued *KF* function is obtainable, or that all *KF* functions with finite range are obtainable.)

For an explicit basis for *PF*, an obvious candidate presents itself. It seems very plausible that the functions

$$x \dot{-} y, \qquad [x/y], \qquad \varphi(x, y), \qquad x \# y \tag{24}$$

constitute a basis for *PF*. Using Lemma 3 of Jones and Matijasevich [7] it is easy to show that these four functions generate most of the known functions in *PF*.

**Open Problem 2.** *Does the class NPF of nondeterministic polynomial time computable functions have a basis?* *NPF* is the analogue for functions of the class *NP* of sets. We have $PF \subseteq NPF \subseteq KF$ i.e., *NPF* lies somewhere between *PF* and *KF*. For a definition of *NPF*, see L.G. Valiant [22] or Jones [9]. A provable characterization of *NPF*, which may be also taken as the definition, is that *NPF* is the class of functions deterministically computable in polynomial time by a Turing machine with a set from $NP \cap co{-}NP$ in the oracle. (See Jones [9].)

The class $NPF$ contains many interesting examples of number theoretic functions. Some examples of functions known to be in $NPF$ are the *Euler* $\phi$ *function (totient),* the function $\tau(n)$ which gives the *number of divisors of n,* the function $\sigma$ for the sum of the divisors of $n$, and the function $\nu(n)$ which gives the *smallest prime divisor of n.* Perhaps some of these natural functions could be used to construct a basis for $NPF$.

In connection with this problem it should also be noted that if $NPF$ has no basis, then $P \neq NP$ follows. This is a consequence of Theorem 2. Also Theorem 1 is interesting in this context. Theorem 1 implies that if $NPF$ is the union of an increasing family of sets closed under composition, then $P \neq NP$. This follows because, by Theorem 2, $PF$ cannot be the union of an increasing hierarchy. Of course $PF$ is the union of the increasing family of functions computable in time $|x|^n$ $(n = 1,2,\ldots)$. However, the classes of this family are not closed under composition. When $f$ and $g$ are computable in time $|x|^n$, computation of the function $f(g(x))$ requires time of the order of $|x|^m$ where $m = n^2$.

## References


[1] *A. Cobham,* The intrinsic computational difficulty of functions. Proc. 1964 International Congress of Mathematicians, in Logic, Methodology and Philosophy of Science, Y. Bar–Hillel, Ed., North–Holland, Amsterdam, 1964, 24—30.

[2] *S.A. Cook* , The complexity of theorem proving procedures. Proc. Third Annual ACM Symposium on the Theory of Computing, May 1971, 151—158.

[3] *N.J. Cutland* , Computability: An Introduction to Recursion Theory. Cambridge University Press, Cambridge, 1980.

[4] *J. Edmonds* , Paths, trees and flowers. Canadian J. Math. **17** (1965), 449—467.

[5] *A. Grzegorczyk* , Some classes of recursive functions. Rozprawy Matematyczene **4** (1953), Institut Matematyczny Polskiej Akad. Nauk. (PAN), Warsawa.

[6] *J. Hartmanis* and *R.E. Stearns* , On the computational complexity of algorithms. Trans. Amer. Math. Soc. **117** (1965), 285—306.

[7] *J.P. Jones* and *Yu.V. Matijasevich* , Exponential diophantine representation of recursively enumerable sets. Proc. Herbrand Symp., Logic Colloquium '81, European Meeting of the Association for Symbolic Logic, July 16—24, 1981, Marseilles, France, Studies in Logic **107**, North–Holland Publishers, Amsterdam, 1982, 159—177. MR 85i 03138.

[8] *J.P. Jones* and *Yu.V. Matijasevich* , Register machine proof of the theorem on exponential diophantine representation of enumerable sets. J. Symbolic Logic **49**, No. 3, (1984), 818—829.

[9] *J.P. Jones* , Elementary properties of the class of nondeterministic polynomial time computable functions. Proc. Semester Math. Problems in Computation Theory, Banach Centre Publications **21** (1988), Warszawa, 277—282.

[10] *J.P. Jones* , Basis for the Kalmar elementary functions. Proc. NATO Advanced Studies Institute on Number Theory and Applications, 1988, Banff, Alberta, NATO ASI Series, Kluwer Academic Publishers, Dordrecht, Netherlands, 1988.

[11] *R.M. Karp*, Reducibility among combinatorial problems. in Complexity of Computer Computations, Ed. R.E. Miller and J.W. Thatcher, Plenum Publishers 1972, 85—103. MR 51 14644.

[12] *N.K. Kosovsky* , Osnovoye Teorii Elementarnyikh Algoritmov. Ucheb. Posobie, Izdatel'stvo L.G.U., 1987, 152 s. Foundations of the Theory of Elementary Algorithms, Leningrad State University, 1987, 152 pp. (Russian — untranslated).

[13] *L.A. Levin* , Universal search problems. Problemy Peredachi Informatsii, vol. **9** (1973), 115—116 (Russian). English translation: Problems of Information Transmission **9** (1973), 265—266.

[14] *S.S. Marchenkov* , Elimination of the scheme of recursion in Grzegorczyk's class $E^2$. Mat. Zametki **5**(1969), 561—568. English translation: Math Notes **5** (1969), 336—340. MR 40 5446.

[15] *S.S. Marchenkov* , On a certain basis with respect to composition for the class of Kalmar elementary functions. Mat. Zametki **27** (1980), 321—332, 492. MR 81e 03039.

[16] *M. Minsky* , Computation: Finite and Infinite Machines. Prentice Hall, Englewood Cliffs, N.J., 1967. MR 50 9050.

[17] *A.A. Muchnik* , On two approaches to the classification of recursive functions. Problems of Mathematical Logic, Complexity of Algorithms and Classes of Computable Functions, Ed. V.A. Kozmidiadi and A.A. Muchnik, Mir, Moscow, 1970, 123—138. (Russian — untranslated).

[18] *C. Parsons* , Hierarchies of primitive recursive functions. Zeitschrift für Math. Logik und Grundlagen der Mathematik **14**(1968), 357—376.

[19] *V.R. Pratt* , Every prime has a succinct certificate. SIAM Jour. Computing, **4** (1975), 214—220.

[20] *D. Rödding* , Über die Eliminierbarkeit von Definitionschemata in der Theorie der Rekursiven Funktionen. Zeitschrift für Math. Logik und Grundlagen der Mathematik, **10** (1964), 315—330.

[21] *H.E. Rose* , Subrecursion, Functions and Hierarchies. Oxford University Press, Oxford, 1984.

[22] *L.G. Valiant* , Relative complexity of checking and evaluating. Information Processing Letters, **5** (1976), 20—23.

---

Department of Mathematics and Statistics, The University of Calgary, Calgary, Alberta T2N 1N4, CANADA.

Steklov Mathematical Institute, Academy of Sciences of the USSR, 27 Fontanka, 191011, Leningrad, USSR.

# Exponential Sums Connected With Quadratic Forms

*M. Jutila*

## 1. Introduction

### 1.1 Transformation of exponential sums

Let $b(n)$ be an arithmetical function, and consider exponential sums of the type

$$\sum_{N}^{N'} b(n) g(n) e(f(n)), \tag{1}$$

where $e(\alpha) = e^{2\pi i \alpha}$, $f$ is a real function, and $g$ is a complex function in the interval $[N, N']$. If $f'(x) \approx h/k$, then the sum (1) is comparable with

$$\sum_{N}^{N'} b(n) g(n) e_k(hn),$$

where $e_k(\alpha) = e(\alpha/k)$. Now, if a summation formula for the sum

$$\sum_{n \le x}{}' \, b(n) e_k(hn) \tag{2}$$

is known, then a summation formula for the sum (1) follows by partial summation. Here $\sum'$ means that if $x$ is an integer, then the last term in the sum is to be halved.

By general principles, a summation formula for the sum (2) exists if the Dirichlet series

$$\sum_{n=1}^{\infty} b(n) e_k(hn) n^{-s} \tag{3}$$

satisfies a functional equation of a suitable form. A classical and typical case is $b(n) = d(n)$ (the divisor function) and $h/k = 1$; the corresponding summation formula is that of Voronoi.

A transformation formula for the sum (1) appears if the integrals occurring in the procedure just described are calculated approximately by the saddle–point method. For details of this method with applications, see [7]–[9] (similar ideas in the case $b(n) = 1$ can be found in [1]). The following cases have been worked out so far:

1) the divisor function $d(n)$,
2) the Fourier coefficients of a holomorphic cusp form for the full modular group,
3) the Fourier coefficients of Maass wave forms.

The cases 1) and 2), in which the functional equation for the respective Dirichlet series (3) is due to T. Estermann [3] and J.R. Wilton [14], were considered in [7]. The case 3) was studied recently by T. Meurman [11]. It is our object to add another class of arithmetical functions to this list, namely

4) $r_Q(n)$, the number of representations of $n$ by the positive quadratic form
$$Q(x, y) = ax^2 + bxy + cy^2. \tag{4}$$

In particular, if $Q(x, y) = x^2 + y^2$, then $r_Q(n)$ is the classical arithmetic function $r(n)$.

## 1.2 Dirichlet series associated with quadratic forms

Though we shall be ultimately concerned with *binary* quadratic forms, this restriction is irrelevant in the study of the series (3). Therefore we let $Q(x)$ be, more generally, a positive definite integral quadratic form in $m \geq 2$ variables. We may write

$$Q(\mathbf{x}) = \frac{1}{2}\mathbf{x}^T A \mathbf{x}, \tag{5}$$

Where $A$ is an integral symmetric $m \times m$ matrix with *even* diagonal elements, and $\mathbf{x} = (x_1, \ldots, x_m)^T$ is the column vector of the indeterminates.

The *adjoint* matrix $A^{\dagger}$ of $A$ is defined by the equation $A^{\dagger}A = DI_m$, where $D = \det(A)$ and $I_m$ is the $m\times m$ identity matrix. The quadratic form attached to $A^{\dagger}$ (as in (5)) is denoted by $Q^{\dagger}(\mathbf{x})$. Note that $Q^{\dagger}$ is positive definite but not necessarily integral (unless $m = 2$). In the following, we shall suppose throughout that $Q^{\dagger}(\mathbf{x})$ is integral.

The Dirichlet series

$$L_Q(s, h/k) = \sum_{n=1}^{\infty} r_Q(n) e_k(hn) n^{-s}$$

was studied by T. Callahan and R.A. Smith [2], who proved its analytic continuability to a meromorphic function having a simple pole at $s = m/2$ with the residue

$$\mathrm{Res}(L_Q(s, h/k), m/2) = (2\pi)^{m/2} D^{-1/2} k^{-m} \Gamma(m/2)^{-1} G_Q(k, h), \tag{6}$$

where

$$G_Q(k, h) = \sum_{\mathbf{x} \bmod k} e_k(hQ(\mathbf{x}))$$

is a Gaussian sum. Here the condition of summation means that each component of $\mathbf{x}$ runs over a complete system of residues (mod $k$). More generally, if $\mathbf{a}$ is an integral vector, we define

$$G_Q(k, h, \mathbf{a}) = \sum_{\mathbf{x} \bmod k} e_k(hQ(\mathbf{x}) + \mathbf{a}\cdot\mathbf{x}),$$

where the dot denotes the inner product.

The function $L_Q(s, h/k)$ satisfies the following functional equation: for $\sigma < 0$,

$$\left(\frac{k\sqrt{D}}{2\pi}\right)^{s} \Gamma(s) L_Q(s, h/k) = \left(\frac{k\sqrt{D}}{2\pi}\right)^{m/2-s} \Gamma(m/2-s) \tilde{L}_Q(m/2-s, h/k), \tag{7}$$

where

$$\tilde{L}_Q(s, h/k) = \sum_{n=1}^{\infty} b_n n^{-s} \tag{8}$$

with

$$b_n = D^{(m-2)/4} k^{-m/2} \sum_{Q^{\dagger}(\mathbf{x})=n} G_Q(k, h, \mathbf{x}). \tag{9}$$

For a proof of this, combine the equations (27) and (29) in [2]; also, this result is explicitly stated in [13], p. 82.

A more symmetric functional equation was obtained in [2] under a suitable condition. As usual, write $\operatorname{ord}_p(n) = \alpha$ if $p^{\alpha}|n$ and $p^{\alpha+1}$ does not divide $n$. Given $D$ and $k$ as above, define

$$\delta_0 = \prod_{\operatorname{ord}_p D \le \operatorname{ord}_p k} p^{\operatorname{ord}_p D}, \tag{10}$$

$$\delta_1 = \prod_{\operatorname{ord}_p D > \operatorname{ord}_p k} p^{\operatorname{ord}_p k}. \tag{11}$$

Then apparently $(D, k) = \delta_0 \delta_1$. The above mentioned condition, assumed in [2], was $\delta_1 = 1$. Write $\delta = (D, k)$,

$$\varepsilon_Q(k, h) = \frac{G_Q(k, h)}{|G_Q(k, h)|},$$

$$N(k, Q) = \operatorname{card}\{\mathbf{x} \bmod k \mid A\,\mathbf{x} \equiv 0 \pmod k\},$$

$$R(k, Q) = \frac{D}{N(k, Q)},$$

$$Z_Q(s, h/k) = \left(\frac{kR(k, Q)^{1/m}}{2\pi}\right)^{s} \Gamma(s) L_Q(s, h/k).$$

Then the symmetric functional equation reads (see [2], Theorem 3):

$$Z_Q(s, h/k) = \varepsilon_Q(k, h) Z_{Q^*}(m/2 - s, -\overline{h(D/\delta)}/k), \tag{12}$$

where $Q^*$ is a certain integral positive definite quadratic form depending on $Q$ and $k$, and the bar indicates the multiplicative inverse (mod $k$), thus $h\bar{h} \equiv 1 \pmod k$.

Modifying the argument in [2], we give a functional equation for $L_Q(s, h/k)$ in the general case (i.e., without the assumption $\delta_1 = 1$). Though our result is not as symmetric as (12), it shares, however, those features of the latter which are of relevance for deriving a summation formula of the Voronoi type for the sum (2) with $b(n) = r_Q(n)$, in the case $m = 2$.

An important step in the proof of (12) was the expression of $G_Q(k, h, \mathbf{a})$ in terms of $G_Q(k, h)$ (see [2], Theorem 1 (i)):

$$G_Q(k, h, \mathbf{a}) = E_\delta(A^\dagger \mathbf{a})\, e_k(-\overline{h(D/\delta)}\delta^{-1} Q^\dagger(\mathbf{a}))G_Q(k, h), \tag{13}$$

where

$$E_\delta(x) = \begin{cases} 1 & \text{if } x \equiv 0 \pmod{\delta} \\ 0 & \text{otherwise.} \end{cases}$$

The assumption $\delta_1 = 1$ is needed in the proof of this relation. In the following theorem, this condition is removed.

**Theorem 1.** *Let $Q(\mathbf{x})$ be a positive definite integral quadratic form in $m \geq 2$ variables, let $A$ with $\det(A) = D$ be the associated matrix, and suppose that the adjoint form $Q^\dagger(\mathbf{x})$ is integral as well. Let $k$ be a positive integer and $h$ an integer with $(h, k) = 1$. Let $\delta_0$ and $\delta_1$ be defined by (10) and (11), and write $\delta = \delta_0\delta_1 = (D, k)$. Then $Q^\dagger(\mathbf{a}) \equiv 0 \pmod{\delta_0}$ whenever $A^\dagger \mathbf{a} \equiv 0 \pmod{\delta_0}$. Write $D_0 = D/\delta_0$, $k_1 = k/\delta_1$, and choose the multiplicative inverse $\bar{D}_0 \pmod{k_1}$ subject to the additional condition*

$$\delta_1 | \bar{D}_0. \tag{14}$$

*Let $g$ be an integer such that*

$$gk_1 \equiv 1 \pmod{\delta_1}. \tag{15}$$

*Then for any $\mathbf{a} \in \mathbb{Z}^m$ we have*

$$\begin{aligned} G_Q(k, h, \mathbf{a}) = {} & E_{\delta_0}(A^\dagger \mathbf{a})\, e_{k_1}(-\bar{h}\bar{D}_0 \delta_0^{-1} Q^\dagger(\mathbf{a})) \\ & \times \sum_{\mathbf{x} \bmod k} e_k(hQ(\mathbf{x}))e_{\delta_1}(\mathbf{x} \cdot g\,\mathbf{a}). \end{aligned} \tag{16}$$

Note that $(D_0, k_1) = (\delta_1, k_1) = 1$, so $\bar{D}_0$ and $g$ exist and the condition (14) can be fulfilled. Since $\bar{D}_0 \delta_0^{-1} Q^\dagger(\mathbf{a})$ is an integer for $E_{\delta_0}(A^\dagger \mathbf{a}) = 1$, it does not matter whether $\bar{h}$ is understood as an inverse $\pmod{k}$ or $\pmod{k_1}$.

The formula (16) reduces to (13) for $\delta_1 = 1$. In the case $\delta_1 > 1$, the sum over $\mathbf{x}$ in (16) is not $G_Q(k, h)$, but the point is that its value depends only on the residue of $\mathbf{a} \pmod{\delta_1}$.

Substituting (16) into (9), and following the argument in [2], Section 6, we obtain a more useful version of the functional equation (7). To formulate the result, let $S = (s_i\,\delta_{ij})$ be the Smith Normal Form of $A^\dagger$, that is to say $S$ is an integral diagonal matrix of the form

$$S = UA^\dagger V\,, \tag{17}$$

where $U, V \in SL(m, \mathbb{Z})$. Define further $b_i = \dfrac{\delta_0}{(\delta_0, s_i)}$, $B = (b_i\,\delta_{ij})$, and

$$A^* = \delta_0^{-1} B V^T A^\dagger V B. \tag{18}$$

**Theorem 2.** *With the assumptions of Theorem 1, the quadratic form $Q^*$ attached to $A^*$ is integral and positive definite, and the following functional equation holds: for $\sigma < 0$,*

$$\left\{\frac{k\sqrt{D_0}}{2\pi}\right\}^s \Gamma(s) L_Q(s, h/k) = D^{(m-2)/4}\,\delta_0^{-m/4}\left\{\frac{k\sqrt{D_0}}{2\pi}\right\}^{m/2-s}$$
$$\times\Gamma(m/2-s)\sum_{n=1}^{\infty}\tilde{r}_{Q^*}(n)\, e_{k_1}(-\bar{h}\bar{D}_0\delta_1^{-1}n)\, n^{s-m/2}, \tag{19}$$

*where*

$$\tilde{r}_{Q^*}(n) = k^{-m/2}\sum_{Q^*(\mathrm{x})=n}\ \sum_{\mathrm{y} \bmod k} e_k(hQ(\mathrm{y}))e_{\delta_1}(\mathrm{y}\cdot gVB\,\mathrm{x}). \tag{20}$$

**Remark 1.** Since (see [13], Lemma 1)

$$\left|G_Q(k, h, \mathbf{x})\right| \le (\delta k)^{m/2},$$

we have

$$\left|\tilde{r}_{Q^*}(n)\right| \le \delta^{m/2} r_{Q^*}(n).$$

**Remark 2.** If $Q$ is the binary form (4), then $D = 4ac - b^2$. If, in particular, $-D$ is a *fundamental discriminant,* then $D$ is square–free up to perhaps the factor 4 or 8, and consequently $\delta_1 = 1, 2$, or 4, depending on $\operatorname{ord}_2 k$. The case $\delta_1 = 2$ occurs e.g., if $Q(x, y) = x^2 + y^2$ (then $D = 4$) and $k \equiv 2 \pmod 4$. The functional equation of $L_Q(s, h/k)$ for this $Q$ was derived by R.A. Smith [12] in all cases; in particular, if $k = 2k_1$ with $k_1$ odd, then the functional equation is

$$L_Q(s, h/k) = 2^s \pi^{2s-2} k^{1-2s} \Gamma^2(1-s) \sin \pi s$$

$$\times \sum_{n=1}^{\infty} \chi(n) r(n) e_{2k}(n\bar{h}(k_1 - 1)) n^{s-1} \quad \text{for } \sigma < 0, \tag{21}$$

where $\chi$ is the non-principal character (mod 4). Another formulation of this is

$$L_Q(s, h/k) = i\,\chi(\bar{h}) 2^s \pi^{2s-2} k^{1-2s} \Gamma^2(1-s) \sin \pi s$$

$$\times \sum_{n \equiv 1 \,(\mathrm{mod}\ 4)} r(n) e_{2k}(-n\bar{h}) n^{s-1}, \tag{22}$$

where $\bar{h}$ is defined (mod $k$). Indeed, $\chi(n) r(n) = 0$ unless $n \equiv 1 \,(\mathrm{mod}\ 4)$, and therefore the series on the right of (21) is

$$\sum_{n \equiv 1 \,(\mathrm{mod}\ 4)} r(n) e_{2k}(-n\bar{h}) e(\bar{h}/4) n^{s-1} = i\,\chi(\bar{h}) \sum_{n \equiv 1 \,(\mathrm{mod}\ 4)} r(n) e_{2k}(-n\bar{h}) n^{s-1}.$$

It turned out in [12] that the cases $k \not\equiv 2 \,(\mathrm{mod}\ 4)$ and $k \equiv 2 \,(\mathrm{mod}\ 4)$ (i.e., the cases $\delta_1 = 1$ and $\delta_1 = 2$) are essentially different, so that the extra complications in Theorems 1 and 2 above, compared with the corresponding results in [2], seem inevitable. In Section 4, we are going to show how (19) reduces to (22) when $Q(x, y) = x^2 + y^2$ and $k \equiv 2 \,(\mathrm{mod}\ 4)$.

**Remark 3.** The assumption that $Q^\dagger$ be integral is not a serious restriction in Theorems 1 and 2. Namely, we may replace $Q(x)$ by $2Q(x)$ if necessary. Then $2Q(x)^\dagger$ is integral, $G_Q(k, h, \mathbf{a})$ equals $G_{2Q}(k, h/2, \mathbf{a})$ or $2^{-m} G_{2Q}(2k, h, 2\mathbf{a})$ according as $h$ is even or odd, and likewise $L_Q(s, h/k)$ equals $2^s L_{2Q}(s, (h/2)/k)$ or $2^s L_{2Q}(s, h/2k)$.

### 1.3 Summation formulae

Turning to the summation formula involving $r_Q(n) e_k(nh)$, we specify $m = 2$, because then we obtain a formula for the sum

$$\sum_{n \le x}{}' r_Q(n) e_k(nh). \tag{23}$$

For $m > 2$, it would be possible to prove analogous results for Riesz means (as in [13]), but we leave this topic aside here.

For $m = 2$, the form $Q^{\dagger}$ is integral along with $Q$, so the assumption concerning $Q^{\dagger}$ in Theorem 1 and 2 is redundant. The functional equation (19) can be written as follows:

$$L_Q(s, h/k) = \delta_0^{-1/2}\left(\frac{k\sqrt{D_0}}{2\pi}\right)^{1-2s} \Gamma(1-s)\Gamma(s)^{-1} \times \sum_{n=1}^{\infty} \tilde{r}_{Q^*}(n) e_{k_1}(-\overline{hD_0}\,\delta_1^{-1} n) n^{s-1}. \tag{24}$$

This should be compared with the functional equation (Lemma 1.2 in [7])

$$\varphi(s, h/k) = (-1)^{\kappa/2} (k/2\pi)^{\kappa-2s} \Gamma(\kappa - s)\Gamma(s)^{-1} \varphi(\kappa - s, -\bar{h}/k)$$

for the function

$$\varphi(s, h/k) = \sum_{n=1}^{\infty} a(n) e_k(nh) n^{-s},$$

where $a(n)$ is the $n$th Fourier coefficient of a cusp form of weight $\kappa$. The corresponding summation formulae (Theorems 1.6 and 1.7 in [7]) read

$$\sum_{n \le x}{}' a(n) e_k(nh) = (-x)^{\kappa/2} \sum_{n=1}^{\infty} a(n) n^{-\kappa/2} e_k(-n\bar{h}) J_\kappa\left(4\pi\frac{\sqrt{nx}}{k}\right) \tag{25}$$

and

$$\sum_{a \le n \le b}{}' a(n) e_k(nh) f(n) = 2\pi k^{-1} (-1)^{\kappa/2} \sum_{n=1}^{\infty} a(n) e_k(-n\bar{h}) n^{-(\kappa-1)/2} \times \int_a^b x^{(\kappa-1)/2} J_{\kappa-1}\left(4\pi\frac{\sqrt{nx}}{k}\right) f(x)\,dx, \tag{26}$$

where $J_\nu(x)$ stands for a Bessel function, in the standard notation. In (26), the function $f$ should be sufficiently nice, say $f \in C^1[a, b]$, and the summation convention is that if $a$ or $b$ is an integer, then the corresponding term is halved.

Appealing to the analogy between the functional equations for $L_Q(s, h/k)$ and $\varphi(s, h/k)$, we may now construct analogues for the identities (25) and (26). As a matter of fact, the series in (25) corresponds to the *error term* in the asymptotic formula for the sum (23); the leading terms are the residues of the function $L_Q(s, h/k)x^s/s$ at $s = 1$ and $s = 0$. The residue of $L_Q(s, h/k)$ at $s = 1$ was given in (6). The resulting formula is

$$\sum_{n \le x}{}' r_Q(n) e_k(nh) = 2\pi D^{-1/2} k^{-2} G_Q(k, h) x + L_Q(0, h/k) + \delta_0^{-1/2} x^{1/2}$$

$$\times \sum_{n=1}^{\infty} \tilde{r}_{Q^*}(n) n^{-1/2} e_{k_1}(-\overline{hD_0}\, \delta_1^{-1} n) J_1\left(4\pi \frac{\sqrt{nx/D_0}}{k}\right), \tag{27}$$

and as an analogue of (26), we have

$$\sum_{a \le n \le b}{}' r_Q(n) e_k(nh) f(n) = 2\pi D^{-1/2} k^{-2} G_Q(k, h) \int_a^b f(x)\, dx$$

$$+ 2\pi(k\sqrt{D})^{-1} \sum_{n=1}^{\infty} \tilde{r}_{Q^*}(n) e_{k_1}(-\overline{hD_0}\, \delta_1^{-1} n) \int_a^b J_0\left(4\pi \frac{\sqrt{nx/D_0}}{k}\right) f(x)\, dx. \tag{28}$$

In the special case $h/k = 1$, the formula (27) has been pointed out by G.H.Hardy [4].

### 1.4 Applications

The summation formula (28) serves as a starting point for the derivation of transformation formulae for exponential sums of the type

$$\sum_{N}^{N'} r_Q(n) g(n) e(f(n))$$

along the lines of [7], Chapter III. As applications, one obtains estimates for the *Epstein zeta-function* $L_Q(s) = L_Q(s, 1)$. For instance, an analogue of the estimate $\zeta(1/2 + it) \ll t^{1/6}$ is

$$L_Q(1/2 + it) \ll t^{1/3+\varepsilon}. \tag{29}$$

As analogues of the mean value theorems of D.R. Heath–Brown [5] and H. Iwaniec [6] for the zeta–function, we have

$$\int_0^T |L_Q(1/2+it)|^6 dt \ll T^{2+\varepsilon}, \tag{30}$$

$$\int_T^{T+T^{2/3}} |L_Q(1/2+it)|^2 dt \ll T^{2/3+\varepsilon}. \tag{31}$$

The implied constants in (29) through (31) depend on $D$. By a closer analysis, this dependence could be made explicit. The methods of the proofs of these estimates are adaptations of those in [7] (Sections 4.2 and 4.4) and in [8].

If $L_Q(s)$ is a product of $\zeta(s)$ and an $L$-function, then (29) is known, but in general the estimate

$$L_Q(1/2+it) \ll t^{5/12} \log t$$

due to I. Kalniņš [10] seems to be the sharpest to be found in the literature, though modern methods in the theory of double exponential sums would no doubt give somewhat better results.

The error term $\Delta_Q(x)$ in the asymptotic formula for the sum function of $r_Q(n)$ measures the error made when the number of lattice points in an elliptical region defined in terms of the quadratic form $Q$ is estimated by the area of this region. The difference $\Delta_Q(x+U)-\Delta_Q(x)$ is an analogous lattice point error for the zone between two ellipses. This may be compared with $\Delta(x+U)-\Delta(x)$, where $\Delta(x)$ is the error term in Dirichlet's divisor problem. In [9] we proved that

$$\int_X^{X+H} (\Delta(x+U)-\Delta(x))^2 dx \ll (HU + X^{2/3} U^{4/3}) X^{\varepsilon}$$

for $1 \le H, U \le X$, and the same result can now be obtained for $\Delta_Q(x)$ as well if the implied constant is allowed to depend on $D$.

## 2. Proof of Theorem 1

To begin with, we verify the first assertion of the theorem. Suppose that $A^{\dagger}\mathbf{a} \equiv 0 \pmod{\delta_0}$. Then

$$D\mathbf{a}^T A^{\dagger}\mathbf{a} = 2Q(A^{\dagger}\mathbf{a}) \equiv 0 \pmod{2\delta_0^2}.$$

But $(D/\delta_0, \delta_0) = 1$, so $(D/\delta_0, 2\delta_0) = 1$ if $\delta_0$ is even. Then

$$2Q^\dagger(\mathbf{a}) = \mathbf{a}^T A^\dagger \mathbf{a} \equiv 0 \pmod{2\delta_0}. \tag{32}$$

Also, this congruence holds $(\bmod\ \delta_0)$ and $(\bmod\ 2)$ if $\delta_0$ is odd. Thus (32) is valid in any case, and it follows that $\delta_0 | Q^\dagger(\mathbf{a})$.

The proof of the formula (16) will be based on the following lemma (Lemma 1 in [2]); the argument does not differ much from that in [2].

**Lemma 1.** *Let $A$ be an integral symmetric $m \times m$ matrix with $\det(A) = D$ such that the associated quadratic forms $Q$ and $Q^\dagger$ are integral and positive definite. Let $q$ be a positive integer, write $\delta = (D, q)$, and suppose that*

$$(D/\delta, q) = 1. \tag{33}$$

*Then the congruence $A\mathbf{x} + \mathbf{a} \equiv 0 \pmod{q}$ is solvable in $\mathbb{Z}^m$ if and only if $A^\dagger \mathbf{a} \equiv 0 \pmod{\delta}$. If the congruence is solvable, then*

$$-\overline{(D/\delta)}\delta^{-1} A^\dagger \mathbf{a} \tag{34}$$

*is a solution. (The bar indicates a multiplicative inverse $(\bmod\ q)$).*

Now let $\gamma$ be an arbitrary vector in $\mathbb{Z}^m$. On replacing $\mathbf{x}$ by $\mathbf{x} + \bar{h}\gamma$ (where $\bar{h}$ is taken $(\bmod\ k)$), we obtain

$$G_Q(k, h, \mathbf{a}) = e_k(\bar{h}Q(\gamma) + \bar{h}\,\mathbf{a} \cdot \gamma) G_Q(k, h, A\gamma + \mathbf{a}). \tag{35}$$

First suppose that $A^\dagger \mathbf{a} \equiv 0 \pmod{\delta_0}$. We may apply the lemma with $q = k_1$. Indeed, $(D_0, k_1) = 1$, so that the condition (33) is satisfied. By the lemma, there exists a vector $\gamma$ such that

$$A\gamma + \mathbf{a} = 0 \pmod{k_1}, \tag{36}$$

and we may take $\gamma = -\bar{D}_0 \delta_0^{-1} A^\dagger \mathbf{a}$ by (34). We specify here $\bar{D}_0$ as required in the theorem; thus $\delta_1 | D_0$. Then, if $g$ is as in the theorem, we have

$$A\gamma + \mathbf{a} \equiv g k_1 \mathbf{a} \pmod{k}. \tag{37}$$

In the first place, this holds $(\bmod k_1)$ by (36). Also, since $\gamma$ is divisible by $\delta_1$, both sides of (37) are congruent to $\mathbf{a} (\bmod \delta_1)$. Hence, all in all, this congruence holds $(\bmod k)$.

Next, we calculate

$$\bar{h}Q(\gamma) + \bar{h}\,\mathbf{a}\cdot\gamma = \frac{1}{2}\overline{hD_0^2\delta_0^{-2}}\,\mathbf{a}^T A^\dagger A A^\dagger \mathbf{a} - \overline{hD_0}\,\delta_0^{-1}\mathbf{a}\cdot A^\dagger\mathbf{a}$$
$$= \overline{hD_0^2\delta_0^{-2}}\,DQ^\dagger(\mathbf{a}) - 2\overline{hD_0}\,\delta_0^{-1}Q^\dagger(\mathbf{a})$$
$$\equiv -\overline{hD_0}\,\delta_0^{-1}Q^\dagger(\mathbf{a}) (\bmod k_1).$$

This also holds $(\bmod \delta_1)$ since, owing to the assumption (14), both sides of the congruence are divisible by $\delta_1$. Hence the congruence holds $(\bmod k)$, and together with (35) and (37), this proves the validity of the formula (16) in the case $A^\dagger\mathbf{a} \equiv 0$ $(\bmod \delta_0)$.

To complete the proof of the theorem, we still have to show that $G_Q(k, h, \mathbf{a}) = 0$ if $A^\dagger\mathbf{a} \not\equiv 0 (\bmod \delta_0)$. This condition implies, by the lemma, that the congruence $A\mathbf{x} + \mathbf{a} \equiv 0 (\bmod k_1)$ is not solvable. Define $k_2 = k/\delta = k_1/\delta_0$, and replace $\gamma$ by $k_2A^\dagger\gamma$ in (35). Then we obtain

$$G_Q(k, h, \mathbf{a}) = e_k(\bar{h}k_2A^\dagger\mathbf{a}\cdot\gamma)G_Q(k, h, \mathbf{a}).$$

If $A^\dagger\mathbf{a}\cdot\gamma \equiv 0 (\bmod \delta_0)$ for all $\gamma$, then $A^\dagger\mathbf{a} \equiv 0 (\bmod \delta_0)$, contrary to the assumption. Hence there exists a vector $\gamma$ such that $A^\dagger\mathbf{a}\cdot\gamma \not\equiv 0 (\bmod \delta_0)$. Then $\bar{h}k_2A^\dagger\mathbf{a}\cdot\gamma \not\equiv 0 (\bmod k)$, and hence necessarily $G_Q(k, h, \mathbf{a}) = 0$.

## 3. Proof of Theorem 2

In view of the functional equation (7), the actual problem is to analyse the sum

$$\sum_{Q^\dagger(\mathbf{x})=n} G_Q(k, h, \mathbf{x}). \tag{38}$$

We may suppose here that $A^\dagger\mathbf{x} \equiv 0 (\bmod \delta_0)$, for otherwise $G_Q(k, h, \mathbf{x}) = 0$ by Theorem 1. Then $\delta_0 | n$ by the same theorem, so let us therefore write $n\delta_0$ in place of

$n$. Further, since the mapping $x \to Vx$ is a bijection of $\mathbb{Z}^m$, we may replace $x$ by $Vx$ in (38). Then, by (7)–(9), we have

$$\left(\frac{k\sqrt{D}}{2\pi}\right)^{s} \Gamma(s) L_Q(s, h/k) = \left(\frac{k\sqrt{D}}{2\pi}\right)^{m/2-s} \Gamma(m/2-s) \times D^{(m-2)/4} k^{-m/2} \delta_0^{s-m/2} \times \sum_{n=1}^{\infty} \sum_{\substack{Q^{\dagger}(V\mathbf{x})/\delta_0 = n \\ A^{\dagger}V\mathbf{x} \equiv 0 \ (\mathrm{mod}\ \delta_0)}} G_Q(k, h, V\mathbf{x}) n^{s-m/2}. \tag{39}$$

By (17), the condition $A^{\dagger}V\mathbf{x} \equiv 0 \ (\mathrm{mod}\ \delta_0)$ is equivalent to $S\mathbf{x} \equiv 0 \ (\mathrm{mod}\ \delta_0)$. This, in turn, is satisfied precisely by the vectors $B\mathbf{y}$ with $\mathbf{y} \in \mathbb{Z}^m$. Hence, the congruence condition in (39) can be omitted if $\mathbf{x}$ is replaced by $B\mathbf{x}$. Now $A^{\dagger}VB\mathbf{x} \equiv 0 \ (\mathrm{mod}\ \delta_0)$, for all $x$, and therefore $Q^{\dagger}(VB\mathbf{x}) \equiv 0 \ (\mathrm{mod}\ \delta_0)$, by Theorem 1. Consequently,

$$Q^{*}(\mathbf{x}) = \delta_0^{-1} Q^{\dagger}(VB\mathbf{x})$$

is an integral quadratic form, and the condition of summation with respect to $\mathbf{x}$ can be written simply as $Q^{*}(\mathbf{x}) = n$. By (16), we have

$$\sum_{Q^{*}(\mathbf{x})=n} G_Q(k, h, VB\mathbf{x}) = e_{k_1}(-\overline{hD_0}\,\delta_1^{-1} n) \sum_{Q^{*}(\mathbf{x})=n} \sum_{\mathbf{y} \bmod k} e_k(hQ(\mathbf{y})) e_{\delta_1}(\mathbf{y} \cdot gVB\mathbf{x}) .$$

The assertion (19) now follows when this expression is substituted into (39).

## 4. An Example: Proof of the Functional Equation (22)

Let $Q(x, y) = x^2 + y^2$ and $k \equiv 2 \ (\mathrm{mod}\ 4)$. Then $A = A^{\dagger} = 2I_2$, $D = 4$, $\delta_0 = 1$, $\delta_1 = 2$, $k_1 = k/2$, $D_0 = 4$, and $g = 1$. Further, in (17), $S = 2I_2$ and $U = V = I_2$. Thus $A^{*} = 2I_2$ by (18), so $Q^{*} = Q$.

The following facts about ordinary Gaussian sums

$$G(q, p) = \sum_{x \bmod q} e_q(px^2)$$

will be needed:

$$G(q, p) = 0 \quad \text{for } q \equiv 2 \pmod 4,$$
$$G^2(q, p) = \chi(q)q \quad \text{for odd } q.$$

Consider the coefficients $\tilde{r}_{Q^*}(n)$ defined in (20). If $\begin{pmatrix} x_1 \\ x_2 \end{pmatrix} \not\equiv \begin{pmatrix} 1 \\ 1 \end{pmatrix} \pmod 2$, then the sum over the vectors $\mathbf{y}$ vanishes because a vanishing Gaussian sum $G(k, h)$ can be factored out.

If $x_i \equiv 1 \pmod 2$ for $= 1,2$, then $n = x_1^2 + x_2^2 \equiv 2 \pmod 4$, and the number of such representations equals the number $r(n)$ of *all* representations of $n$ as the sum of two squares. Thus $\tilde{r}_{Q^*}(n) = 0$ unless $n \equiv 2 \pmod 4$, in which case

$$\begin{aligned}
\tilde{r}_{Q^*}(n) &= k^{-1}\left(\sum_{z=1}^{k} e_k(hz^2)(-1)^z\right)^2 r(n) \\
&= k^{-1}\left(2\sum_{\substack{z=1\\2|z}}^{k} e_k(hz^2) - \sum_{z=1}^{k} e_k(hz^2)\right)^2 r(n) \\
&= k^{-1}(2G(k_1, 2h) - G(k, h))^2 r(n) \\
&= 2\chi(k_1) r(n).
\end{aligned}$$

Write here $4n + 2$ in place of $n$. The function $\frac{1}{4} r(n)$ is multiplicative, so $r(4n + 2) = r(2n + 1)$ since $r(2) = 4$.

By these calculations, the functional equation (19) takes the form

$$\begin{aligned}
(k/\pi)^s \Gamma(s) L_Q(s, h/k) = \chi(k_1)(k/\pi)^{1-s} \Gamma(1-s) 2^s \\
\times \sum_{\substack{n=1\\ n \equiv 1 \bmod 4}}^{\infty} r(n) e_{k_1}(-\overline{h4} n) n^{s-1},
\end{aligned} \tag{40}$$

where $h\bar{h} \equiv 1 \pmod{k_1}$, $4\bar{4} \equiv 1 \pmod{k_1}$. Let

$$\frac{\overline{h4}}{k_1} - \frac{\bar{h}}{4k_1} \equiv w \pmod 1.$$

Writing $4\bar{4} = 1 + rk_1$, we obtain $w \equiv \frac{r\bar{h}}{4} \pmod 1.$ Thus

$$e_{k_1}(-\overline{h4}n) = e_{2k}(-\bar{h}n)e_4(-r\bar{h}).$$

But $rk_1 \equiv -1 \pmod 4$, whence $r \equiv -k_1 \pmod 4$, and

$$e_4(-r\bar{h}) = e_4(\bar{h}k_1) = i\,\chi(\bar{h}k_1).$$

Then, since $\Gamma(s)^{-1} = \pi^{-1}\Gamma(1-s)\sin\pi s$, the functional equation (22) follows from (40).

## References


[1] *E. Bombieri* and *H. Iwaniec,* On the order of $\zeta(1/2+it)$. Ann. Scuola Norm. Sup. Pisa Cl. Sci. **13** (1986), 449—472.

[2] *T. Callahan* and *R.A. Smith,* $L$-functions of a quadratic form. Trans. Amer. Math. Soc. **217** (1976), 297—309.

[3] *T. Estermann,* On the representation of a number as the sum of two products. Proc. London Math. Soc. (2) **31** (1930), 123—133.

[4] *G.H. Hardy,* The lattice points of a circle. Proc. Royal Soc. Ser. A **107** (1925), 623—635.

[5] *D.R. Heath-Brown,* The twelfth power moment of the Riemann zeta–function. Quart. J. Math. Oxford **29** (1978), 443—462.

[6] *H. Iwaniec,* Fourier Coefficients of Cusp Forms and the Riemann Zeta–function. Séminaire de Théorie des Nombres, Univ. Bordeaux 1979/80, exposé no. **18**, 36 pp.

[7] *M. Jutila,* Lectures on a Method in the Theory of Exponential Sums, Tata Institute of Fundamental Research, Lectures on Mathematics and Physics **80**; published for the Tata Institute for Fundamental Research by Springer–Verlag, Berlin–Heidelberg–New York–Tokyo, 1987.

[8] *M. Jutila,* The fourth power moment of the Riemann zeta-function over a short interval. Proc. Coll. János Bolyai, Colloquium on Number Theory (Budapest 1987), North–Holland Publ. Comp., Amsterdam (to appear).

[9] *M. Jutila,* Mean value estimates for exponential sums. Journées Arithmétiques Ulm 1987 (to appear).

[10] *I. Kalninš,* On the Epstein zeta–function. (In Russian) Latvijas PSR Zinātnu Ser. 1965 no. **5**, 66—75.

[11] *T. Meurman,* On exponential sums involving the Fourier coefficients of Maass wave forms. J. reine angew. Math. **384** (1988), 192—207.

[12] *R.A. Smith,* The circle problem in an arithmetic progression. Canad. Math. Bull. **11** (1968), 175—184.

[13] *R.A.* Smith, The average order of a class of arithmetic functions over arithmetic progressions with applications to quadratic forms. J. reine angew. Math. **317** (1980), 74—87.

[14] *J.R. Wilton,* A note on Ramanujan's arithmetical function $\tau(n)$. Proc. Cambridge Phil. Soc. **25** (1929), 121—129.

---

Department of Mathematics, University of Turku, SF–20500 Turku, FINLAND.

# On Quartic Fields with the Symmetric Group

*Gérard Kientega and Pierre Barrucand*

## 1. Introduction

Let $K_4$ be a quartic field over $\mathbb{Q}$ and let $K$ be the algebraic closure with $G = \mathrm{Gal}(K/\mathbb{Q})$ the Galois group of $K/\mathbb{Q}$.

Put $K_4 = \mathbb{Q}(x)$, where $x$ admits the following minimal polynomial

$$f(X) = X^4 + a_2X^2 + a_1X + a_0 = (X - x_1)(X - x_2)(X - x_3)(X - x_4)$$

over $\mathbb{Q}$, with $x_1, x_2, x_3$ and $x_4 \in K$ the distinct roots of $f(x)$ in $K$. We recall that $f(x)$ is

$$g(X) = X^3 + 2a_2X^2 + (a_2^2 - 4a_0)X - a_1^2 = (X - \theta_1)(X - \theta_2)(X - \theta_3)$$

and $\theta_1, \theta_2, \theta_3$ lie in some subfield of $K$. We recall that $f(X)$ and $g(X)$ have the same discriminant; which we shall denote by $\Delta$; and that the roots of $f(X)$ are given by the following formulas

$$2x_1 = \sqrt{\theta_1} + \sqrt{\theta_2} + \sqrt{\theta_3}$$

$$2x_2 = \sqrt{\theta_1} - \sqrt{\theta_2} - \sqrt{\theta_3}$$

$$2x_3 = -\sqrt{\theta_1} + \sqrt{\theta_2} - \sqrt{\theta_3}$$

$$2x_4 = -\sqrt{\theta_1} - \sqrt{\theta_2} + \sqrt{\theta_3}$$

If $g(X)$ is irreducible then $G = S_4$ the symmetric group on four letters, if $\Delta$ is not square. Otherwise $G = A_4$, the subgroup of even permutations of $S_4$. We are mainly interested in the case $G = S_4$.

First we express the zeta function of each subfield of $K$ as a product of Artin $L$–functions associated with irreducible characters of $G$ and give local factors of those $L$–functions at non–ramified primes. In the second part we study ramification in $K$; the discriminants of $K_4$ and the two non–normal sextic subfields of $K$ are computed and higher ramification is studied for the case of wild ramification. Finally, we summarize the results in tables which complement and correct those of Dribin [2] and [3].

## 2. Induced Characters and Local Factors of $L$–Functions

The subgroups of $K$ are: (*cf.* [2])

| Degree | Subfield | Group |
|---|---|---|
| 2 | $K_2 = Q(\sqrt{\Delta})$ | $A_4 = \{1, (12)(34), (13)(24), (14)(23), (123), (132)(124), (142), (134), (143), (234), (243)\}$ |
| 3 | $K_3 = Q(\theta_1)$ | $D_4 = \{1, (12)(34), (13)(24), (14)(23), (12), (34), (1324), (1423)\}$ |
| 4 | $K_4 = Q(x_1)$ | $S_3' = \{1, (23), (24), (34), (234), (243)\}$ |
| 6 | $\bar{K}_3 = Q(\theta_1, \theta_2, \theta_3)$ | $B_4 = \{1, (12)(34), (13)(24), (14)(23)\}$ |
|  | $K_6 = Q(\sqrt{\theta_1})$ | $B_4' = \{1, (12)(34), (12), (34)\}$ |
|  | $\bar{K}_6 = Q(\sqrt{\Delta\,\theta_1})$ | $C_4 = \{1, (12)(34), (1324), (1423)\}$ |
| 8 | $K_8 = K_4 K_2$ | $C_3 = \{1, (234), (243)\}$ |
| 12 | $K_{12} = \bar{K}_3(\sqrt{\theta_1})$ | $C_2 = \{1, (12)(34)\}$ |
|  | $\bar{K}_{12} = Q(x_1 x_2)$ | $T_2 = \{1, (34)\}$ |

$S_4$ contains five conjugate classes. Let $\chi_i$, $i \in \{0,1,\ldots,4\}$ be the corresponding characters where $\chi_0$ is trivial, $\chi_1$ is of order 1 and $\chi_2$ is of order 2. $S_4$ admits a four dimensional natural representation which is a direct sum of a one dimensional representation and a 3–dimensional irreducible representation $\rho_3$. Then $\chi_3$ is the character of $\rho_3$ and $\chi_4 = \chi_3\chi_1$. Let $L_i(s) = L(s, \chi_i, K/\mathbb{Q})$. We have $L_0(s) = \zeta(s)$ the Riemann zeta function. By elementary Galois Theory and formalism of Artin $L-$functions it is easy to compute the following table:

| Subfield | Induced Character | $J$ function |
|---|---|---|
| $K_2$ | $\chi_0 + \chi_1$ | $\zeta(s)L_1(s)$ |
| $K_3$ | $\chi_0 + \chi_2$ | $\zeta(s)L_2(s)$ |
| $K_4$ | $\chi_0 + \chi_3$ | $\zeta(s)L_3(s)$ |
| $\overline{K}_3$ | $\chi_0 + \chi_1 + 2\chi_2$ | $\zeta(s)L_1(s)L_2^2(s)$ |
| $K_6$ | $\chi_0 + \chi_2 + \chi_3$ | $\zeta(s)L_2(s)L_3(s)$ |
| $K'_6$ | $\chi_0 + \chi_2 + \chi_4$ | $\zeta(s)L_2(s)L_4(s)$ |
| $K_8$ | $\chi_0 + \chi_1 + \chi_3 + \chi_4$ | $\zeta(s)L_1(s)L_3(s)L_4(s)$ |
| $K_{12}$ | $\chi_0 + \chi_1 + 2\chi_2 + \chi_3 + \chi_4$ | $\zeta(s)L_1(s)L_2^2(s)L_3(s)L_4(s)$ |
| $K'_{12}$ | $\chi_0 + \chi_2 + 2\chi_3 + \chi_4$ | $\zeta(s)L_2(s)L_3^2(s)L_4(s)$ |
| $K$ | $\chi_0 + \chi_1 + 2\chi_2 + 3\chi_3 + 3\chi_4$ | $\zeta(s)L_1(s)L_2^2(s)L_3^3(s)L_4^3(s)$ |

**Definition:** *Let $k$ be a number field and suppose that the rational prime number $p$ has the decomposition $\mathfrak{p}_1^{e_1} \ldots \mathfrak{p}_r^{e_r}$ with $f_i$, $i = 1, \ldots, r$ the residue class degree of $\mathfrak{p}_i$. Then we say that $p$ belongs to the class $(f_1^{e_1} \ldots f_r^{e_r})$.*

With this definition, any method of decomposition yields this table for non-ramified primes.

| $K_3$ | $K_4$ | $K_6$ | $K'_6$ |
|---|---|---|---|
| 1 1 1 | 1 1 1 1 | 1 1 1 1 1 1 | 1 1 1 1 1 1 |
| 1 1 1 | 2 2 | 1 1 2 2 | 1 1 2 2 |
| 3 | 1 3 | 3 3 | 3 3 |
| 1 2 | 1 1 2 | 1 1 2 2 | 2 2 2 |
| 1 2 | 4 | 2 4 | 1 1 4 |

Now it is a straight forward computation to use those two tables to compute the local factor for the non-ramified primes.

| $K_4$ | $L_1(s)$ | $L_2(s)$ | $L_3(s)$ | $L_4(s)$ |
|---|---|---|---|---|
| 1 1 1 1 | $\frac{1}{1-p^{-s}}$ | $\frac{1}{(1-p^{-s})^2}$ | $\frac{1}{(1-p^{-s})^3}$ | $\frac{1}{(1-p^{-s})^3}$ |
| 2 2 | $\frac{1}{1-p^{-s}}$ | $\frac{1}{(1-p^{-s})}$ | $\frac{1}{(1-p^{-2s})(1+p^{-s})}$ | $\frac{1}{(1-p^{-s})^2(1+p^{-s})}$ |
| 1 3 | $\frac{1}{1-p^{-s}}$ | $\frac{1}{1+p^{-s}+p^{-2s}}$ | $\frac{1}{1-p^{-3s}}$ | $\frac{1}{1-p^{-3s}}$ |
| 1 1 2 | $\frac{1}{1+p^{-s}}$ | $\frac{1}{1-p^{-2s}}$ | $\frac{1}{(1-p^{-s})(1-p^{-2s})}$ | $\frac{1}{(1+p^{-s})(1-p^{-2s})}$ |

## 3. Ramification

Let $K_2 = \mathbb{Q}(\sqrt{d})$, $d$ square free and $d_2, d_3 = d_2 f^2$, $d_4, d_6, d_{\tilde{6}}$ be the absolute discriminants of $K_2, K_3, K_4, K_6$, and $\tilde{K}_6$. We shall use the known results about $d_3$ [4].

Recall that $K_6 = \mathbb{Q}(\sqrt{\theta})$, $\tilde{K}_6 = \mathbb{Q}(\sqrt{\Delta\theta}) = \mathbb{Q}(\sqrt{d\,\theta})$. Let

$$(\theta) = \prod_{i=1}^{t} \mathfrak{p}_i^{e_i}$$

be the decomposition of the principal ideal $(\theta)$ in $K_3$. Let $p_i$ denote the unique positive rational prime lying below $\mathfrak{p}_i$. Define

$$h_0 = \prod_{\substack{p_i \mid d \\ e_i \equiv 1 (\mathrm{mod}\ 2)}} p_i$$

$$h_1 = \prod_{\substack{p_i \nmid d \\ e_i \equiv 1 (\mathrm{mod}\ 2)}} p_i$$

Notation: For $k$ a number field, we denote by $N$ the absolute norm $N_{k/\mathbb{Q}} : k \to \mathbb{Q}$.

**Theorem 1.** *The discriminants* $d_4, d_6$, *and* $\tilde{d}_6$ *are given in the following formulas:*

$$d_4 = d_3 h_0^2 h_1^2 4^k = d_2 f^2 h_0^2 h_1^2 4^k$$
$$d_6 = d_3 d_4 = d_2 f^4 h_0^2 h_1^2 4^k$$
$$\tilde{d}_6 = d_2^3 f^4 h_1^2 4^{k'},$$

*where* $k,\ k'$ *are natural numbers such that* $0 \le k,\ k' \le 3$.

**Proof.** Let $\delta_6 = d_{K_6/K_3}$, $\tilde{\delta}_6 = d_{\tilde{K}_6/K_3}$, be the discriminants of $K_6/K_3$ and $\tilde{K}_6/K_3$. We have the following lemmas:

**Lemma 1.** $N(\delta_6) = h_0^2 h_1^2 4^k$, $0 \le k \le 3$.

**Lemma 2.** $N(\tilde{\delta}_6) = h_0^2 4^{k'}$, $0 \le k' \le 3$.

Clearly it suffices to prove the above lemmas. Let us prove the second lemma which is the more difficult. The first is easy and uses the same idea.

Put

$$d = \prod_{i=1}^{t_1} p_i .$$

If $p_i \ne 3$, then it is not totally ramified in $K_3$; write

$$(p_i) = q_i^2 q_i$$

such that

$$(d) = \prod_{i=1}^{t_1} q_i^2 q'_i$$

if 3 is not a divisor of the $\gcd(d, f)$. Otherwise

$$(d) = q_{i_1}^3 \prod_{i \ne i_1} q_i^2 q'_i .$$

If

$$(\theta) = \mathcal{A}^2 \prod_{\ell=1}^{t} q_\ell$$

where ?? is the entire ideal of $K_3$ and $q_\ell$ $(\ell = 1,\ldots,t)$ are prime ideals of $K_3$, then

$$q_\ell \cap Z = (p)$$

if and only if $p | h_0 h_1$. Suppose that $\gcd(d, f)$ is not a multiple of 3. Then

$$(d\,\theta) = ??^2 \prod_{i=1}^{t_1} q_i^2 q'_i \prod_{\ell=1}^{t} q_\ell$$

For $p | h_1$, there exists $\ell$ and $\ell'$ such that $q_\ell$ and $q_{\ell'}$ divide $(d\,\theta)$ to an odd power and lie over $(p)$. This implies that $q_\ell$ and $q_{\ell'}$ divide $\delta_6$ and taking the norm we find that $p^2 | N(\delta_6)$. If $p \neq 2$ then $p^2 \| N(\delta_6)$. (If $p$ is a rational prime and $n$ an entire rational number, $p^\alpha \| n$, means $p^\alpha | n$ and $n$ is not a multiple of $p^{\alpha+1}$.)

Now it is clear from $\mathbb{Q}(\sqrt{d_6}) = \mathbb{Q}(\sqrt{d})$ that

$$d h_1^2 | N(\delta_6)$$

If $p \neq 2$ and $p | N(\delta_6)$ then $p | d h_1^2$ and we have

$$N(\delta_6) = d h_1^2 4^K .$$

for some natural number $K$.

It remains to show that for all $d$,

$$N(\delta_6) = d_2 h_1^2 4^{k'}, \quad 0 \leq k' \leq 3.$$

This is easy if $d \equiv 1 \pmod 4$, or $d \equiv 2 \pmod 4$. Now for the case $d \equiv 3 \pmod 4$, we observe that in $K_3$

$$(2) = q_1^2 q'_1$$

so it suffices to prove that one of the two ideals must ramify in $K_6$. If one assumes the contrary, the study of the biquadratic extension $K_{12}/K_3$ would yield a contradiction.

**Theorem 2.** a) *Let $p > 2$ be a rational prime. We have*

$$p|h_1 \text{ if and only if } (p) = (\mathfrak{p}_1, \mathfrak{p}'_1)^2 \text{ or } (p) = \mathfrak{p}_2^2$$

$$p|h_0 \text{ if and only if } (p) = \mathfrak{p}^4$$

$$p|f \text{ if and only if } (p) = \mathfrak{p}_1^3 \mathfrak{p}'_1$$

$p|d_2$ *and* $p$ *is not a divisor of* $h_0$ $f$ *if and only if* $((p) = \mathfrak{p}_1^2 p'_1 p''_1$ or $(p) = \mathfrak{p}_1^2 \mathfrak{p}_2)$.

b) *If a rational prime greater than 2 remains inert or totally ramifies in $K_3$, then it cannot divide the product $h_0 h_1$.*

**Proof.** This is an easy consequence of Theorem 1 and classical results from local fields (*cf.* [5] Proposition 13, Chapter III).

**Remark 1.** Those results don't give information for $p = 2$. This case is complicated and cannot be studied in full; we have made tables which give the results for $p = 2$. As far as we know Dribin was the first to make tables for quartic fields, but they are not complete and they contain some errors. In particular, when the decomposition group is $S_4$ or $D_4$ (the eighth order dihedral group) his conclusions are not exact, (*cf.* [1], page 48) and the following result.

**Theorem 3.** *Let $L/\mathbb{Q}_2$ be a $D_4$ extension of the 2–adic field $\mathbb{Q}_2$. Denote by $L_4$ and $L'_4$ the two non-normal quartic subfields of $L$, $L_2$ and $L'_2$ their quadratic subfields. Let $H = \mathrm{Gal}(L/L_2)$, $H' = \mathrm{Gal}(L/L'_2)$. Finally, let $L''_2$ be the last quadratic subfield of $L$, $C_4 = \mathrm{Gal}(L/L''_2)$, $C_2$ the centre of $D_4$, $d_2$, $d'_2$, $d''_2$, $d_4$ and $d'_4$ the discriminants of $L_2$, $L'_2$, $L''_2$, $L_4$ and $L'_4$. We have*

1) $2^2 \| d_2$, $2^3 \| d'_2$ *then* $2^9 \| d_4$, $2^{10} \| d'_4$ *and* $G_0 = G_1 = D_4$, $G_2 = G_3 = H'$, $G_4 = G_5 = C_2$, $G_6 = \{1\}$.
2) $2^3 \| d_2$, $2^2 \| d'_2$ *then* $2^{10} \| d_4$, $2^9 \| d'_4$ *and* $G_0 = G_1 = D_4$, $G_2 = G_3 = H$, $G_4 = G_5 = C_2$, $G_6 = \{1\}$.

3) $2^3 \| d_2$, $2^3 \| d'_2$ *then* $2^{11} \| d_4$, $2^{11} \| d'_4$ *and* $G_0 = G_1 = D_4$, $G_2 = G_3 = C_4$, $G_4 = \ldots = G_7 = C_2$, $G_8 = \{1\}$.

**Proof.** Let $\upsilon_2$ be the normalized valuation of $\mathbb{Q}_2$. Let $H_i = G_i \cap H$. From the fact $G_0 = G_1$, we have $H_0 = H_1$ and if $\varphi$ is the Herbrand function we have

$$\varphi_{L/L_2}(0) = 0, \quad \varphi_{L/L_2}(1) = 1$$

$$\varphi_{L/L_2}(2) = \begin{cases} 2 & \text{if } H_1 = H_2 \\ \frac{3}{2} & \text{if } H_1 \neq H_2 \end{cases}$$

$$\varphi_{L/L_2}(3) = \begin{cases} 3 & \text{if } H_1 = H_2 \\ 2 & \text{if } H_1 \neq H_2 \end{cases}$$

$$\varphi_{L/L_2}(i) > 2 \text{ if } i > 3.$$

because $G_i = G_{i+1}$ if $i$ is even (*cf.* [5] page 79, ex. 3). For the case 1) $(G/H)_0 = (G/H)_1 \cong C_2$, $(G/H)_2 = 1$. Applying the Herbrand Theorem, we must have $G_2 = H$ so $G_2 = G_3 = H$. Using the same idea for $L_4|L_2$, we find the result for that case. For the case 3) $\upsilon_2(d_2) = \upsilon_2(d'_2) = 3$ implies $G_2 = G_3 = C_4$ and we must have $G_3 \neq G_4 = \ldots = G_7 = C_2$, $G_8 = \{1\}$.

Conclusions about discriminants are immediate.

## 4. The Alternating Group $A_4$

If $\Delta$ is a square, then $G = A_4$, $K_3$ is cyclic, $K_6 = \tilde{K}_6$. The results on discriminants remains true, letting $d_2 = 1$. In particular the only prime which may be totally ramified in $K_4$ is 2. If $L(s)$ is the $L$-function associated with one of the two conjugate characters of the cyclic group of order three then

$$\zeta_{K_4}(s) = \zeta(s) L(s) \bar{L}(s)$$

$$\zeta_{K_4}(s) = \zeta(s) L_3(s)$$

$$\zeta_{K_6}(s) = \zeta(s) L(s) \bar{L}(s) L_3(s)$$

where $L_3(s)$ is the $L$-function associated with the character of degree 3 of $A_4$.

## 5. The Table for Ramification

The following table gives the decomposition and ramification groups for $p \neq 2$. We see that in some cases we can give precise information. When $p|d_3$ and $p$ does not divide $f$ the last line tells us that $p \equiv 1 \pmod 4$. The three other lines don't enable us to say anything like that so it is reasonable to think that the two cases $p \equiv 1 \pmod 4$ and $p \equiv 3 \pmod 4$ are possible.

| $\frac{d_3}{p}$ | $p$ | | $K_3$ | $K_4$ | $K_6$ | $\tilde{K}_6$ | $K$ | $G_{-1}$ | $G_i$ $(i = 0,1,\ldots)$ |
|---|---|---|---|---|---|---|---|---|---|
| +1 | | | 111 | $1^21^2$ | $1^21^211$ | $1^21^211$ | $1^2$ | $C_2$ | $C_2$ |
| | | | 111 | $2^2$ | $1^21^22$ | $1^21^22$ | $2^2$ | $B_4$ | $C_2$ |
| -1 | | | 12 | $1^21^2$ | $112^2$ | $2^22$ | $2^2$ | $B'_4$ | $C_2$ |
| | | | 12 | $2^2$ | $2^22^2$ | $112^2$ | $2^2$ | $C_4$ | $C_2$ |
| 0 | $p \mid f$ | $p \equiv 1(3)$ | $1^3$ | $1^31$ | $1^31^3$ | $1^31^3$ | $1^3$ | $C_3$ | $C_3$ |
| | | $p \equiv -1(3)$ | $1^3$ | $1^31$ | $1^31^3$ | $2^3$ | $2^3$ | $S_3$ | $C_3$ |
| | | $p=3$ $\left(\frac{d_2}{3}\right)=+1$ | $1^3$ | $1^31$ | $1^31^3$ | $1^31^3$ | $1^3$ | $C_3$ | $C_3C_3$ |
| | | $p=3$ $\left(\frac{d_2}{3}\right)=-1$ | $1^3$ | $1^31$ | $1^31^3$ | $2^3$ | $2^3$ | $S_3$ | $C_3C_3$ |
| | | $p=3$ $\left(\frac{d_2}{3}\right)=0$ | $1^3$ | $1^31$ | $1^31^3$ | $1^6$ | $1^6$ | $S_3$ | $S_3C_3$ |
| | | | $1^3$ | $1^31$ | $1^31^3$ | $1^6$ | $1^6$ | $S_3$ | $S_3C_3C_3C_3$ |
| | $p \nmid f$ | | $1^21$ | $1^22$ | $2^211$ | $2^21^2$ | $2^2$ | $B'_4$ | $T_2$ |
| | | | $1^21$ | $1^211$ | $1^21^211$ | $1^21^21^2$ | $1^2$ | $T_2$ | $T_2$ |
| | | | $1^21$ | $1^4$ | $1^41^2$ | $1^42$ | $2^4$ | $D_4$ | $C_4$ |
| | | $p \equiv 1(4)$ | $1^21$ | $1^4$ | $1^41^2$ | $1^411$ | $1^4$ | $C_4$ | $C_4$ |

## References

[1] *Buhler*, I–Cosahedral Galois Representations. Springer–Verlag 1978.

[2] *D.M. Dribin*, Quartic fields with the symmetric group. Annals of Math, vol. 38, n° 3, 1937, 739—749.

[3] *D.M. Dribin*, Normal extensions of quartic fields with the symmetric groups. Annals of Math, vol. 39, n° 2, 1938, 341—349.

[4] *Hasse*, Arithmetische Theorie der kubischen Zahlkorper auf klassenkorpe theoretische Grundlage. Math. Zeitsch., **31** (1930), 565—587.

[5] *Serre*, Corps Locaux. Hermann Paris (1968).

[6] *Serre*, Representations Linéaires des Groupes Finis. Hermann Paris (1971).

---

Institut de Mathématique et de Science Physique, Université B.P. 7021—OUAGADOUGOU, Burkina—Faso.

C.N.R.S., Université Pierre et Marie Curie, UER 47, 4, Place Jussieu—75005 Paris, FRANCE.

# Sets With Non–Squarefree Sums

*C.B. Lacampagne, C.A. Nicol and J.L. Selfridge*

## 1. Good Sets and Non–Squarefree Integers

Notice that the set $\{4,8,\ldots,4n\}$ has all sums of pairs divisible by 4 and hence non–squarefree. Let us use the adjective *good*.

**Definition.** *A set or an infinite sequence* $A$ *is said to be* **good** *if for each distinct pair* $a,b \in A$, $a+b$ *is non–squarefree.*

The set $\{2,6,\ldots,4n-2\}$ is also a good set. Our first result shows that this set is best possible.

**Theorem 1.** *If* $n > 4$, *then any set of positive integers* $\{a_1,\ldots,a_n\}$ *with* $a_1 < a_2 < \ldots < a_n < 4n-2$ *has a pair of distinct elements whose sum is squarefree.*

**Proof.** Following Filaseta [1], we show that there are not enough non–squarefree positive integers up to the maximum pair sum. We use the inequality

$$NSF_0(x) < \frac{x}{2}\left(1-\frac{8}{\pi^2}\right)+\frac{\sqrt{x}}{2}+\frac{1}{8} \qquad (1)$$

for a bound on the number of non–squarefree odd positive integers up to $x$. (Theorem 2 proves this bound.) We have three cases: The set $A$ contains both odd and even integers, only even integers, or only odd integers.

<u>Case 1.</u> The set contains at least one even and at least one odd integer.

Let $A = \{a_1, a_2, \ldots, a_n\}$, $B = \{b \in A : b \text{ even}\}$ and $C = \{c \in A : c \text{ odd}\}$. Then $\#(B) + \#(C) = n$, and $B + C$ has at least $n-1$ distinct elements (since if $b_1 < \ldots < b_r$ and $c_1 < \ldots < c_s$ then $b_1+c_1 < \ldots < b_1+c_s < b_2+c_s < \ldots < b_r+c_s$ ).

Note that all of the elements of $B + C$ are odd and less than or equal to $8n - 7$. In order to insure that $A$ will contain a squarefree sum, we need only show that the number of non–squarefree odd positive integers up $8n - 7$ is less than $n - 1$, the least possible number of distinct sums. That is, we will show that for large $x$

$$NSF_0(x) < \frac{x}{2}(1 - \frac{8}{\pi^2}) + \frac{\sqrt{x}}{2} + \frac{1}{8} < n - 1,$$

where $x = 8n - 7$, or

$$\frac{x}{2}(1 - \frac{8}{\pi^2}) + \frac{\sqrt{x}}{2} < \frac{x - 2}{8}$$

Thus

$$\sqrt{x} > \frac{2 + \sqrt{\frac{64}{\pi^2} - 2}}{\frac{32}{\pi^2} - 3},$$

or

$$x > 288.9,$$

and this is true for $n \geq 37$. For $n < 37$, the 27 non–squarefree odd positive integers up to 289 are easily written down and the theorem checked for $n > 2$.

Similar results are true in Case 2 for all $n$ and in Case 3 for $n > 2$, $n \neq 4$. In Case 2 and Case 3 we can find all good sets even when $a_n$ is allowed to be as large as $9n - 1$. This is a "natural boundary", since the set $\{9, 18, \ldots, 9n\}$ is a set with both odd and even integers having no squarefree sum. Since $a_n$ may be larger than $4n$, we must of course exclude from Case 2 any set $A$ of even integers which are all congruent modulo 4.

Case 2. $A$ contains only even integers.

As in the previous case, we write $A = B \cup C$ where $B = \{b \in A : b \equiv 0 \bmod 4\}$ and $C = \{c \in A : c \equiv 2 \bmod 4\}$. Note that we exclude the case where either $B$ or $C$ is empty. Since $\#(B) + \#(C) = n$, as in Case 1, $B + C$ has at least $n - 1$ elements. Here $b + c \equiv 2 \bmod 4$, for all sums $b + c$. Since $a_n < 9n$, $b + c \leq 18n - 4$.

Denote the $n - 1$ distinct sums $\{2t_1, \ldots, 2t_{n-1}\}$ with $2t_{n-1} \leq 18n - 4$. Since the $t$'s are odd, we use inequality (1), and as before we make the right side of this inequality less than $n-1$ to insure that $2t_i$ is squarefree for some $i$ where $9(n-1) = x - 7$.

Thus we need to show that for large $x$

$$\frac{x}{2}\left(1-\frac{8}{\pi^2}\right)+\frac{\sqrt{x}}{2}+\frac{1}{8}<\frac{x-7}{9}$$

or

$$\sqrt{x} > \frac{9+\sqrt{\frac{4680}{\pi^2}-374}}{2\left(\frac{72}{\pi^2}-7\right)}$$

or

$$x > 1037.2$$

which is true for $n \geq 116$. A computer search verified that there were insufficiently many non–squarefree numbers up to the maximum sum for $n = 5$ or $6 < n < 116$. It is easy to show that there are no good sets with $a_n < 9n$ when $n = 3,4$ or 6. When $n = 2$ there are just four good sets: $\{8,10\}$, $\{6,12\}$, $\{4,14\}$, and $\{2,16\}$.

<u>Case 3.</u> $A$ contains only odd integers.

Write $A_n = B \cup C$ where $B = \{b \in A: b \equiv 1 \bmod 4\}$ and $C = \{c \in A: c \equiv 3 \bmod 4\}$. Let $D$ be $B$ or $C$, whichever has the larger cardinality. Thus $\#(D) \geq \frac{n}{2}$ if $n$ is even and $\#(D) \geq \frac{n+1}{2}$ if $n$ is odd. If $\#(D) = k$, we note that there are at least $2k-3$ distinct sums of the form $d_i + d_j$, $i \neq j$ because if $D = \{d_1, \ldots, d_k\}$ where $d_1 < \ldots < d_k$, then $d_1 + d_2 < \ldots < d_1 + d_k < d_2 + d_k < \ldots < d_{k-1} + d_k$. Hence there are at least $n-3$ distinct sums $d_i + d_j$, $i \neq j$, if $n$ is even and $n-2$ if $n$ is odd. Furthermore, $d_i + d_j \leq 18n - 6$ and $d_i + d_j \equiv 2 \bmod 4$. We are now essentially back to Case 2 with either $n-3$ or $n-2$ distinct integers, and, as in Case 2, those integers congruent to two modulo 4 are replaced by a sequence of odd integers all of which are less than or equal to $9n-3$. We again use the inequality (1) and require that the right member be less than $n-3$ to insure a squarefree pair. Thus

$$\frac{x}{2}\left(1-\frac{8}{\pi^2}\right)+\frac{\sqrt{x}}{2}+\frac{1}{8}<\frac{x-24}{9}$$

where $x = 9n - 3$. Since $n > 4$, $x > 41$, and this inequality becomes

$$\sqrt{x} > \frac{9 + \sqrt{\frac{14472}{\pi^2} - 1326}}{2(\frac{72}{\pi^2} - 7)}$$

or $x > 1247.3$ and $n \geq 139$. Again, a computer search verified Case 3 for $10 < n < 139$ except for $n = 20$. When $n > 5$ easy special arguments rule out the six remaining values of $n$. When $n = 5$, $C = \{7,11,43\}$, and there are ten good sets. When $n = 4$, there is a pair with both elements congruent to 1 mod 4, and the other pair with both elements congruent to 3 mod 4, with each pair summing to 18, 50 or 54. It is easy to find the 42 good sets. When $n = 3$, there are four pairs summing to 18 and six or seven choices for the remaining number for a total of 26 good sets. When $n = 2$, there are 24 good sets with $a_2 < 18$. when $n = 1$, the four sets are trivially good.

**Remark 1.** The theorem is also true for $n = 3$. For $n = 4$ the only exception is $\{5,7,11,13\}$. For $n = 2$, the exceptions are $\{1,3\}$, $\{3,5\}$, and $\{4,5\}$. For $n = 1$, $\{1\}$ is the trivial exception.

**Remark 2.** Notice that if we remove the condition that the summands be distinct, the theorem is true for $n \geq 1$.

**Theorem 2.**

$$NSF_0(x) < \frac{x}{2}(1 - \frac{8}{\pi^2}) + \frac{\sqrt{x}}{2} + \frac{1}{8} \tag{1}$$

*where $NSF_0(x)$ denotes the number of non–squarefree odd positive integers less than or equal to $x$.*

**Proof.** We begin with the observation that the number of odd multiples of $d^2 \leq x$ is $\left[\frac{x}{2d^2} + \frac{1}{2}\right]$ where $x \geq 0$ and [ ] denotes the integer part.

Let $S_0(x)$ denote the number of squarefree odd integers less than or equal to $x$. Then

$$S_0(x) = \left[\frac{x+1}{2}\right] + \sum_{\substack{1 < d \leq \sqrt{x} \\ d \text{ odd}}} \mu(d)\left[\frac{x}{2d^2} + \frac{1}{2}\right],$$

so that

$$NSF_0(x) = \left[\frac{x+1}{2}\right] - S_0(x) = -\sum_{\substack{1<d\le\sqrt{x}\\ d \text{ odd}}} \mu(d)\left[\frac{x}{2d^2}+\frac{1}{2}\right].$$

Next we wish to show that

$$-\mu(d)\left[\frac{x}{2d^2}+\frac{1}{2}\right] \le \frac{\mu(d)x}{2d^2}+\frac{1}{2}.$$

But this is easily checked for $\mu(d)=0$, $\mu(d)=-1$ and $\mu(d)=1$. Hence we have

$$NSF_0(x) \le -\frac{x}{2}\sum_{\substack{1<d\le\sqrt{x}\\ d \text{ odd}}} \frac{\mu(d)}{d^2} + \frac{1}{2}\sum_{\substack{1<d\le\sqrt{x}\\ d \text{ odd}}} 1.$$

Note that

$$\sum_{\substack{1<d\le\sqrt{x}\\ d \text{ odd}}} \frac{\mu(d)}{d^2} = \sum_{\substack{d\ge 1\\ d \text{ odd}}} \frac{\mu(d)}{d^2} - \sum_{\substack{d>\sqrt{x}\\ d \text{ odd}}} \frac{\mu(d)}{d^2} - \frac{\mu(1)}{1}.$$

It is well known that

$$\sum_{d=1}^{\infty} \frac{\mu(d)}{d^2} = \frac{6}{\pi^2}$$

(Niven and Zuckerman, [2]). Since $\mu(4k)=0$, there are two classes of summands: $d$ odd and $d$ twice an odd. If $d$ is odd then $\frac{\mu(2d)}{(2d)^2} = -\frac{1}{4}\frac{\mu(d)}{d^2}$. Hence

$$\sum_{d\ge 1} \frac{\mu(d)}{d^2} = \sum_{\substack{d\ge 1\\ d \text{ odd}}} \frac{\mu(d)}{d^2} + \sum_{\substack{d\ge 1\\ d \text{ odd}}} \frac{\mu(2d)}{(2d)^2} = (1-\frac{1}{4})\sum_{\substack{d\ge 1\\ d \text{ odd}}} \frac{\mu(d)}{d^2}.$$

Since

$$\sum_{d=1}^{\infty} \frac{\mu(d)}{d^2} = \frac{6}{\pi^2},$$

we have

$$\sum_{\substack{d\ge 1\\ d \text{ odd}}} \frac{\mu(d)}{d^2} = \frac{4}{3}\left(\frac{6}{\pi^2}\right) = \frac{8}{\pi^2}.$$

Our inequality now becomes

$$NSF_0(x) \le \frac{x}{2}(1 - \frac{8}{\pi^2}) + \frac{x}{2} \sum_{\substack{d > \sqrt{x} \\ d \text{ odd}}} \frac{\mu(d)}{d^2} + \frac{1}{2} \sum_{\substack{1 < d \le \sqrt{x} \\ d \text{ odd}}} 1.$$

First

$$\frac{1}{2} \sum_{\substack{1 < d \le \sqrt{x} \\ d \text{ odd}}} 1 \le \frac{1}{4}(\sqrt{x} - 1).$$

Next

$$\sum_{\substack{d > \sqrt{x} \\ d \text{ odd}}} \frac{\mu(d)}{d^2} < \sum_{\substack{d > \sqrt{x} \\ d \text{ odd}}} \frac{1}{d^2} = \sum_{m > \frac{\sqrt{x} - 1}{2}} \frac{1}{(2m + 1)^2}$$

$$< \int_{\frac{\sqrt{x} - 2}{2}}^{\infty} \frac{dt}{(2t + 1)^2} = \frac{1}{2(\sqrt{x} - 1)},$$

where we erect our trapezoid under the curve from $t - \frac{1}{2}$ to $t + \frac{1}{2}$ with average height $\frac{1}{(2t + 1)^2}$. Thus we have

$$NSF_0(x) < \frac{x}{2}(1 - \frac{8}{\pi^2}) + \frac{1}{4}(\sqrt{x} - 1) + \frac{x}{4(\sqrt{x} - 1)}.$$

Combining the two final terms and noting that $\frac{1}{4(\sqrt{x} - 1)} \le \frac{1}{8}$ when $NSF_0(x) > 0$, we get

$$NSF_0(x) < \frac{x}{2}(1 - \frac{8}{\pi^2}) + \frac{\sqrt{x}}{2} + \frac{1}{8}. \tag{1}$$

■

## 2. Density

By Theorem 1, set $A$ can be good only if $a_n \ge 4n - 2$. It is natural to say that the *density* of the set $\{4, 8, \ldots, 4n\}$ is $\frac{1}{4}$ and that the density of the densest good set $\{2, 6, \ldots, 4n - 2\}$ approaches $\frac{1}{4}$.

**Definition.** *The* ***density*** *of a finite set of distinct positive integers,* $\delta(A)$, *is the ratio of the number of its elements to its largest element. The* ***asymptotic density*** *of an infinite increasing sequence of positive integers,* $\delta(A)$, *is*

$$\delta(A) = \lim_{n \to \infty} \frac{n}{a_n}.$$

**Theorem 3.** (Filaseta [1]) *Let $A$ be infinite with not all $a \equiv 0 \bmod 4$ and not all $a \equiv 2 \bmod 4$ be a good set. Then the maximum density taken over all such good sets $A$ is bounded by*

$$\frac{1}{9} \le \max \delta(A) \le 1 - \frac{8}{\pi^2} \approx \frac{1}{5.28}.$$

The upper bound holds for any infinite sequence. The lower bound is achieved for the sequence $\{9,18,\ldots,9n,\ldots\}$. Our proofs of Cases 2 and 3 of Theorem 1 show that any finite good set $A$ with density greater that $\frac{1}{9}$ contains even and odd integers if $n > 5$.

We have constructed our proof of Theorem 1 along the lines of Filaseta's proof of Theorem 3.

## 3. Complete Sets

Note that in the sets $\{4,8,\ldots,4n\}$ and $\{2,6,\ldots,4n-2\}$, no number less than $a_n$ could be added to the set with the resulting set remaining good. This is not true for the set $\{9,18,\ldots,9n\}$ for $n = 2$, since $\{7,9,18\}$ is also good. It is natural to say that the set $\{9,18\}$ is *incomplete*.

**Definition.** *A good set $A$ with $a_1 < \ldots < a_n$ is said to be* **complete** *if there is no positive integer less than $a_n$ that can be added to the set with the resulting set remaining good. An infinite increasing sequence of positive integers is complete for $n > N$ if each of its initial subsets of more than $N$ elements is complete.*

**Theorem 4.** *If $a_1 < \ldots < a_n$ is a good set and $a_n > nP_n^2$ where $P_n = \prod_{i=1}^{n} p_i$ and $p_i$ is the i-th prime, then the set is not complete.*

**Proof.** Since $a_n > nP_n^2$, there exists a pair $a_{i-1}$, $a_i$ such that $a_i - a_{i-1} > P_n^2$ where $i = 1,\ldots,n$ and $a_0 = 0$. (If $a_i - a_{i-1} \le P_n^2$ for all $i = 1,\ldots,n$, then $a_n \le nP_n^2$.) By the Chinese Remainder Theorem, there exists an integer $a > 0$ such that the system of congruences, $a + a_1 \equiv 0 \bmod 4$, $a + a_2 \equiv 0 \bmod 9$, ..., $a + a_n \equiv 0 \bmod p_n^2$ is solvable, and any two such solutions are congruent $\bmod P_n^2$. Since

$a_i - a_{i-1} > P_n^2$, there will be an integer $k$ such that $a_{i-1} < a + kP_n^2 < a_i$. Thus the set $A$ is incomplete.

Trivially, the complete sets of size 1 are $\{1\}$, $\{2\}$, and $\{4\}$. It is easy to show that each of the sets of size two listed below is complete and that all other good pairs with $a_2 \le 25$ are incomplete.

| | | | |
|---|---|---|---|
| {1,3} | {2,6}* | | |
| {3,5} | {1,7}** | | |
| {4,5} | {3,6}** | {2,7}** | {1,8}** |
| {5,7} | {3,9}** | | |
| {4,8} | | | |
| {7,9} | | | |
| {8,10} | {6,12}** | {4,14}** | {2,16}** |
| {12,13} | {10,15}** | | |
| {6,19} | {4,21}** | | |
| {9,19} | | | |
| {9,23} | | | |
| {19,25} | | | |

Here the * set {2,6} is a multiple of {1,3}, and each of the ** sets have 2 added to one element and subtracted from the other element of a complete set to its left.

We now prove that the above list of 23 complete pairs is complete.

**Theorem 5.** *Every good set* $\{a_1, a_2\}$ *with* $a_2 > 25$ *is incomplete.*

**Proof.** Let $A = \{a_1, a_2\}$. Let $b \ne a_i$ be such that $1 \le b \le 36$. Suppose $a_2 > 36$. There are three cases.

<u>Case 1.</u> $a_1$ is not a multiple of 9.

There exists a $b$ such that $b + a_2 \equiv 0 \bmod 4$ and $b + a_2 \equiv 0 \bmod 9$. Thus $b$ may be included in the set $A$ and the resulting set is good.

<u>Case 2.</u> $9 | a_1$ and $a_2$ is not a multiple of 9.

There exists a $b$ such that $b + a_1 \equiv 0 \bmod 4$ and $b + a_2 \equiv 0 \bmod 9$. Then $b \not\equiv 0 \bmod 9$, and again $b$ may be included in the set $A$ and the resulting set is good.

<u>Case 3.</u> $9 | a_1$ and $9 | a_2$.

There exists a $b$ which is an available multiple of 9 less than or equal to 18.

Thus in all three cases, $b$ may be included in the set $A$ and the resulting set is good. Hence the set $A$ is not complete.

Now suppose $25 < a_2 \leq 36$. If $4 | a_2 - a_1$, use $24 - a_1$ for $b$ when $a_1 < 24$, $a_1 \neq 12$, else use $36 - a_1$. Else if $4 | a_1 + a_2$, then $a_1$ and $a_2$ are odd and we use $50 - a_2$ for $b$, since $4 | a_1 + 50 - a_2$. Else if $a_1 + a_2 = 54$, $a_1$ is even and we use $a_1 - 4$. Else if $a_1 + a_2 = 50$, use $a_1 + 4$ for $b$ when $a_2 > 26$ else use 1. If $a_1 + a_2 = 63$ use 45 minus the even $a_i$. In the remaining cases, use $25 - a_1$ if $a_2$ is even and use $27 - a_1$ if $a_2$ is odd.

Thus the set $A$ is not complete. So all complete pairs have $a_2 \leq 25$. These are listed above.

When $n > 7$, most sets with relatively small $a_n$ belong to infinite sequences exemplified by those discussed later. When $n \leq 7$, we list below the complete sets with small $a_n$ which seem unusual. We omit those mentioned in the proof of Theorem 1 and those easily obtained from others.

$\{11,13,14\},\{8,10,17\},\{1,24,26\},\{3,24,25\},\{20,24,25\}$
$\{1,8,17,19\},\{23,2,22,26\},\{5,22,23,27\},\{11,16,29,34\},\{16,20,29,34\},\{10,15,17,35\}$
$\{7,11,25,38,43\},\{2,7,25,38,43\},\{7,20,25,29,43\}$
$\{7,11,37,38,43,61\},\{7,11,21,29,43,69\}$
$\{2,7,25,43,47,73,74\},\{43,2,6,38,74,78,82\},\{13,14,31,50,67,85,86\}$

In Table 1, for each $n \leq 16$ we list the least value of $a_n$ for a good set (omitting the three complete sets where $a_n = 4n - 2, 4n$, or $9n$). The corresponding sets for $n = 5$ and $n \geq 8$ use a range of non–squarefree numbers from a residue class mod 4, and this range is indicated. For example, when $n = 8$, we subtract 23 from 25, 45, 49, 81, 117, 121, and 125 to get the set.

**Table 1. Minimum largest element for a good set of $n$ elements**

| $n$ | $a_n$ |
|---|---|
| 1 | 1 |
| 2 | 3 |
| 3 | 11 |
| 4 | 13 |
| 5 | 42 = 49 – 9 + 2 |
| 6 | 61 |
| 7 | 74 |
| 8 | 102 = 125 – 25 + 2 |
| 9 | 118 = 125 – 9 + 2 |
| 10 | 146 = 153 – 9 + 2 = 169 - 25 + 2 |
| | = 189 – 45 + 2 = 261 – 117 + 2 |
| 11 | 162 = 169 – 9 + 2 |
| 12 | 182 = 189 – 9 + 2 = 297 – 117 + 2 |
| 13 | 202 = 350 – 150 + 2 |
| 14 | 218 = 333 – 117 + 2 |
| 15 | 246 = 361 – 117 + 2 |
| 16 | 254 = 369 – 117 + 2 |

For $n > 16$, the least $a_n$ is $18n - 26$ for $n$ even and $18n - 16$ for $n$ odd. We stopped the computation at $n = 20$, but we expect that this holds for all $n > 16$.

## 4. Fairly Dense Infinite Sequences

Except for the case when all the $a_i$ are even and congruent modulo 4 or all divisible by 9, the maximum density we are able to find for large $n$ is $\frac{1}{18}$.

The following infinite sequences seem to be complete, with the exceptions indicated:

$\{4,8,\ldots,4n,\ldots\}$
$\{2,6,\ldots,4n-2,\ldots\}$
$\{9,18,\ldots,9n,\ldots\}$, $n > 2$
$\{8,10,44,46,\ldots,18n-26,\ldots\}$ ($a_i \equiv 8,10 \bmod 36$ and $n$ even)
$\{2,16,38,52,74,\ldots,18n-16,\ldots\}$ ($a_i \equiv 2,16 \bmod 36$ and $n$ odd), $n \neq 3$
$\{7,2,18,38,42,\ldots,r_{n-1}-7,\ldots\}$, where $r_n$ is the $n$-th NSF $\equiv 1 \bmod 4$
$\{25,2,38,50,74,\ldots,s_{n-1}-25,\ldots\}$, where $s_n$ is the $n$-th NSF $\equiv 3 \bmod 4$, $n > 3$
$\{16,2,34,38,74,\ldots,t_{n-1}-16,\ldots\}$, where $t_n$ is the $n$-th NSF $\equiv 2 \bmod 4$

The first two sequences are complete by Theorem 1. To show that the third sequence is complete, notice that if it were not complete for a given $n$ then there would have to be $n$ non–squarefree numbers all differing by 9 but not divisible by 9. We leave it to the reader to complete the proof. Initial subsets of the remaining sequences are complete for small values of $n$. We believe these sequences are complete, for suppose there were an initial subset of one of the sequences that was not complete. Then there would be a smallest such subset which implies that the element to be added would have to be inserted between $a_{n-1}$ and $a_n$, and this would require $n-2$ available non–squarefree sums below $2n$ meeting a variety of constraints. However, theorems similar to Theorem 1 would seem to be needed to prove that these sequences are complete.

The asymptotic density of the first two infinite sequences is $\frac{1}{4}$, of the next is $\frac{1}{9}$, of the next two is $\frac{1}{18}$, and of the last three is $\frac{1}{2}\left(\frac{6}{\pi^2}-\frac{1}{2}\right)\approx\frac{1}{18.53}$.

Sequences (usually having small asymptotic densities) may be generated by a greedy algorithm; that is, given the complete set $\{a_1, a_2, \ldots, a_k\}$, find the next integer greater than $a_k$ such that the new set $\{a_1, a_2, \ldots, a_k, a_{k+1}\}$, is good and hence also complete. (Note that if we did not demand that the next available integer be used, the Chinese Remainder Theorem would quickly find a good (but probably incomplete) set with one more element.)

Starting with 1, a computer search using the greedy algorithm and non–squarefree sums up to $1.2 \times 10^4$ yields

$$\{1,3,15,17,147,557,735,3615,4335,8117\}.$$

If we start with 2, we simply get the infinite sequence $\{2,6,\ldots,4n-2,\ldots\}$. This follows immediately from Theorem 1. However, starting with 4, instead of $\{4,8,\ldots, 4n,\ldots\}$ we get $\{4,5,20,40,44,\ldots\}$. When we start with 5, we get $\{5,7,11,13,43,85,\ldots\}$ which starts dense before becoming exponentially thin. Starting with 6, we again get density $\frac{1}{4}$.

We computed all non–squarefree numbers 1, 2, and 3 mod 4 up to $10^6$. Parts of these computations were used in the proof of Theorem 1. Just as the numbers divisible by 4 dominate the classes of non–squarefree numbers, so the numbers divisible by 9 dominate the non–squarefree numbers which are 1, 2, or 3 mod 4. And for each multiple of 9 which is 3 mod 4, there is a smaller multiple of 9 which is 2 mod 4 and an even smaller multiple which is 1 mod 4. So it is not surprising that the count of 1 mod 4 non–squarefree numbers usually led the counts of 2 and 3 mod 4 non–squarefree numbers.

Table 2 below shows the number of non–squarefree integers in given ranges. Notice that the count of 1 mod 4 non–squarefree integers usually leads but the difference between the counts of 1, 2, and 3 mod 4 in the ranges listed in the table is never more than 28. We thank Richard Blecksmith for his help with these computations and Michael Filaseta for helpful discussions, especially for his improvement of our original version of Theorem 2.

## References

[1] *M. Filaseta,* Sets with elements summing to squarefree numbers. C.R. Math. Rep. Acad. Sci. Canada **IX** (1987), 243—246.

[2] *I. Niven* and *H.S. Zuckerman,* An Introduction to the Theory of Numbers. New York, John Wiley and Sons, 1980.

---

Department of Mathematical Sciences, Northern Illinois University, DeKalb, IL 60115–2888, USA

Department of Mathematics, University of South Carolina, Columbia, SC 29208, USA

Department of Mathematical Sciences, Northern Illinois University, DeKalb, IL 60115–2888, USA

**Table 2. Number of Non–Squarefree Integers to $N$**

| $N$ | 1 mod 4 | 2 mod 4 | 3 mod 4 |
|---|---|---|---|
| 100 | 5 | 5 | 4 |
| 200 | 11 | 9 | 8 |
| 300 | 16 | 14 | 12 |
| 400 | 20 | 19 | 18 |
| 500 | 24 | 23 | 22 |
| 600 | 29 | 28 | 27 |
| 700 | 35 | 32 | 30 |
| 800 | 38 | 38 | 35 |
| 900 | 45 | 42 | 41 |
| 1000 | 50 | 46 | 46 |
| 2000 | 97 | 96 | 92 |
| 3000 | 145 | 140 | 141 |
| 4000 | 191 | 189 | 187 |
| 5000 | 241 | 235 | 232 |
| 6000 | 285 | 286 | 283 |
| 7000 | 336 | 329 | 330 |
| 8000 | 380 | 378 | 377 |
| 9000 | 428 | 426 | 423 |
| 10000 | 474 | 473 | 470 |
| 20000 | 951 | 944 | 945 |
| 30000 | 1424 | 1419 | 1415 |
| 40000 | 1901 | 1896 | 1893 |
| 50000 | 2368 | 2366 | 2365 |
| 60000 | 2849 | 2839 | 2839 |
| 70000 | 3319 | 3315 | 3310 |
| 80000 | 3791 | 3794 | 3788 |
| 90000 | 4264 | 4259 | 4264 |
| 100000 | 4740 | 4733 | 4733 |
| 200000 | 9480 | 9473 | 9466 |
| 300000 | 14210 | 14207 | 14205 |
| 400000 | 18954 | 18946 | 18935 |
| 500000 | 23688 | 23682 | 23672 |
| 600000 | 28413 | 28415 | 28418 |
| 700000 | 33159 | 33137 | 33146 |
| 800000 | 37892 | 37889 | 37877 |
| 900000 | 42622 | 42627 | 42620 |
| 1000000 | 47364 | 47360 | 47350 |

# Class Numbers of Simplest Quartic Fields

*Andrew J. Lazarus*

## 1. Introduction

In [12], Shanks was led to study the cyclic cubic fields defined by the roots of

$$X^3 - tX^2 - (t+3)X - 1, \qquad t \in \mathbb{Z} \tag{1}$$

(where $t^2 + 3t + 9$ is prime) because they had particularly small regulator—any two units which are the roots of the polynomial (1) are a fundamental system—and, therefore, generally had large class number. He referred to these fields as the simplest cubics. In [8], Lettl found an explicit, effectively computable lower bound for the class number of these fields and calculated a complete list of simplest cubic fields with class number less than sixteen.

Shanks also observed that a similar family of quadratic fields is defined by

$$X^2 - tX - 1, \qquad t \in \mathbb{Z}^+. \tag{2}$$

This polynomial generates the field $\mathbb{Q}\left[\sqrt{t^2+4}\right]$, and its larger root is the fundamental unit. The class number one problem for a larger class of quadratic extensions, the Richaud–Degert type, has recently been solved by Mollin, up to either one exception or assumption of the Generalized Riemann Hypothesis [9], [10], [11].

We examine the quartic analog, defined by the polynomials

$$f_t(X) = X^4 - tX^3 - 6X^2 + tX + 1, \qquad t \in \mathbb{Z}^+,\ t \neq 3. \tag{3}$$

Here $t$ may be specified greater than zero since $f_t$ and $f_{-t}$ generate the same extension; $f_0$ and $f_3$ are reducible. We exhibit effective lower bound for the class number in terms of $t$, when $t^2 + 16$ has no odd square free factor. In particular, only finitely many such fields have class number one.

Working from Hasse's seminal paper [5] on cyclic quartics, or from the quartic formula, we obtain the following expression for the roots:

$$\varepsilon_{1,2,3,4} = \pm \frac{\sqrt[4]{t^2+16}\sqrt{\sqrt{t^2+16} \pm t}}{2\sqrt{2}} \pm \frac{\sqrt{t^2+16}}{4} + \frac{t}{4} \tag{4}$$

where the second and third ambiguous signs agree.

The four roots, obviously all real, are permuted by the linear fractional transformation

$$\sigma : \varepsilon \longmapsto \frac{\varepsilon - 1}{\varepsilon + 1} \tag{5}$$

Hence simplest quartic extensions are cyclic.

Cornell and Washington [1] show that simplest fields may arise from torsion elements in $PSL(2,\mathbb{Q})$ acting as linear fractional transformations. In the quartic case, $\begin{pmatrix} 1 & -1 \\ 1 & 1 \end{pmatrix}$ has order 4. There are also elements in this group of order 2 and 3, corresponding to the simplest fields defined in (2) and (1), and 6, corresponding to the sextic fields of Marie Gras [3].

Define $\varepsilon$ to be the largest root of $f_t(X)$, $K = K_t$ to be $\mathbb{Q}[\varepsilon]$, $k$ to be the quadratic subfield, and $\sigma$ to be the generator of $\mathrm{Gal}(K/\mathbb{Q}) \cong \mathbb{Z}/4\mathbb{Z}$. We will drop the subscript $t$ whenever there is no risk of confusion.

Consider

$$\varepsilon^{1+\sigma} = \frac{t\sqrt{t^2+16}}{2\sqrt{4\sqrt{t^2+16} + t^2 + 16}} + \frac{\sqrt{t^2+16}}{2} - 1 \in K$$

This element is clearly in $\mathbb{Q}\left[\sqrt{4\sqrt{t^2+16} + t^2 + 16}\right]$, a quartic number field which is therefore $K$. Since $K/\mathbb{Q}$ is cyclic, the unique quadratic subfield $k$ is $k = \mathbb{Q}\left[\sqrt{t^2+16}\right]$.

It is convenient to exclude from our family $t$ such that $t^2 + 16$ is divisible by an *odd* square. Gras [2] shows that the form $t^2 + 16$ represents infinitely many square–free integers, hence odd–square–free parts, so our family remains infinite.

The conductors and discriminants of $K$ and $k$ depend upon the 2–adic valuation of $t$. When $2|t$ we may remove its powers from the radicands as follows:

$$\text{CASE 1:} \quad t \equiv 1 \bmod 2 \qquad K = \mathbb{Q}\left[\sqrt{4\sqrt{t^2+16} + t^2 + 16}\right]$$

$$\text{CASE 2:} \quad t \equiv 2 \bmod 4 \qquad K = \mathbb{Q}\left[\sqrt{2\sqrt{\left(\frac{t}{2}\right)^2 + 4} + \left(\frac{t}{2}\right)^2 + 4}\right]$$

$$\left.\begin{array}{l}\text{CASE 3:} \quad t \equiv 4 \bmod 8 \\ \text{CASE 4:} \quad t \equiv 0 \bmod 8\end{array}\right\} \quad K = \mathbb{Q}\left[\sqrt{2\sqrt{\left(\frac{t}{4}\right)^2 + 1} + \left(\frac{t}{4}\right)^2 + 1}\right]$$

Hudson and Williams, *et al.*, [4] show that representation of cyclic quartics in the form $\mathbb{Q}\left[\sqrt{a(d + b\sqrt{d})}\right]$ is unique under certain conditions. (This $d$ is the radicand of the quadratic field $k$, not the discriminant of $K$.) The description of $K$ above satisfies these conditions because $t^2 + 16$ is now assumed to be odd–square–free. They also calculate the conductors of $K$ and $k$, and the discriminants using the Conductor–Discriminant formula. Define $f$ to be the conductor of $K$, $m$ to be the conductor of $k$, and $D$ to be the discriminant of $K$.

| CASE | $f$ | $m = D_k$ | $D$ |
|---|---|---|---|
| 1 | $t^2 + 16$ | $t^2 + 16$ | $(t^2 + 16)^3$ |
| 2 | $t^2 + 16$ | $\frac{t^2 + 16}{4}$ | $\frac{(t^2 + 16)^3}{4}$ |
| 3 | $\frac{t^2 + 16}{2}$ | $\frac{t^2 + 16}{4}$ | $\frac{(t^2 + 16)^3}{16}$ |
| 4 | $\frac{t^2 + 16}{2}$ | $\frac{t^2 + 16}{16}$ | $\frac{(t^2 + 16)^3}{64}$ |

(6)

Note that $f$ is odd only in Case 1 and $m$ is even only in Case 3. Gras also mentions these types, but she categorizes them in terms of the $\mathbb{Z}[i]$–factorization of $t^2 + 16$ instead of analyzing $t$.

## 2. Class Number Formula

The analytic class number formula of Dirichlet states in general

$$h_K = \frac{w\sqrt{|D|}\prod_{\chi\neq 1} L(1,\chi)}{2^{r_1}(2\pi)^{r_2} R}$$

where

$h_K$ = the class number of $K$

$w$ = number of roots of unity in $K$

$D$ = discriminant of $K$

$r_1$ = number of embeddings $K \hookrightarrow \mathbb{R}$

$r_2$ = the number of pairs of non–real embeddings $K \hookrightarrow \mathbb{C}$

$R$ = the regulator of $K$

and the product is over the non–trivial primitive characters of the Galois group $\mathrm{Gal}(K/\mathbb{Q})$.

For our family this formula simplifies to

$$h = \frac{\sqrt{D}\prod_{\chi\neq 1} L(1,\chi)}{8R} \tag{7}$$

Our goal is to bound effectively each of the terms in the formula (7) for $h$, thereby bounding $h$ itself.

## 3. Units and the Regulator

Let $E_K$ and $E_k$ be the units of $K$ and $k$ respectively. For any abelian number field $K$ there is an isomorphism between $\mathrm{Gal}(K/\mathbb{Q})$ and the group of characters $X = \{\chi : \mathrm{Gal}(K/\mathbb{Q}) \to \mathbb{C}^*\}$. By Kronecker–Weber, we may embed $K$ in some cyclotomic field $\mathbb{Q}(\zeta_f)$, where the minimal such $f$ is the conductor. For a cyclic extension, there exists by Galois correspondence a character $\chi = \chi_K$ of $\mathrm{Gal}(\mathbb{Q}(\zeta_f)/\mathbb{Q})$ whose kernel is the subgroup $\mathrm{Gal}(\mathbb{Q}(\zeta_f)/K)$. We say $K$ *belongs* to $\chi$. Note that $\chi$ is determined only up to complex conjugation.

With this we can define $\chi$*–relative units,* a term owing to Leopoldt [7]. Let $K$ belong to $\chi$; then the $\chi$–relative units

$$E_\chi = \{\varepsilon \in E_K : N_L^K(\varepsilon) = \pm 1\}$$

for all fields $L, \mathbb{Q} \subsetneq L \subsetneq K$. It is easy to see that $E_\chi$ is a subgroup of $E_K$.

Now to specialize to the quartic family. Let $\chi$ generate $X \cong \mathbb{Z}/4\mathbb{Z}$. Similarly, let $\sigma$ generate $\mathrm{Gal}(K/\mathbb{Q})$. The only non–trivial subfield of $K$ is $k$. We have

$$E_\chi = \{\varepsilon : \varepsilon^{1+\sigma^2} = \pm 1\}.$$

Obviously $E_\chi \cap E_k = \{\pm 1\}$.

The Dirichlet Unit Theorem shows that $E_K$ has rank 3 as a $\mathbb{Z}$–module. By definition we have the following exact sequence:

$$0 \longrightarrow E_\chi \longrightarrow E_K \xrightarrow{N_k^K/\pm 1} E_k/\pm 1.$$

Since the group $E_k/\pm 1$ has $\mathbb{Z}$–rank 1, $E_\chi$ has $\mathbb{Z}$–rank at least 2. Since $E_k \subset E_K$ has image of rank 1, we conclude that $E_\chi$ has $\mathbb{Z}$–rank exactly 2.

We see from this discussion that $Q_K = [E_K : E_\chi E_k]$ is finite. Let $\eta$ be the fundamental unit of $k$. Consider, for any unit $\nu \in E_K$, the relative norm $N_k^K(\nu) = \nu^{1+\sigma^2} = \pm\eta^r$, $r \in \mathbb{Z}$. If $r$ is always even, then $\nu/\eta^{r/2}$ exists and is a $\chi$–relative unit, so $Q = 1$. Conversely, $Q = 1$ implies that $r$ is always even.

On the other hand, suppose there exists $\varepsilon$ such that $\varepsilon^{1+\sigma^2} = \pm\eta^{2r+1}$. Dividing $\varepsilon$ by $\eta^r$ we may assume $r = 0$. Then $\varepsilon^{1+\sigma} \in E_\chi$, $\varepsilon^{1+\sigma^2} \in E_k$, and $\varepsilon^2/\eta$ is a $\chi$–relative unit. It follows that any $\nu \in E_K$ may be written as $\pm\varepsilon^r \nu_0$ where $r = 0$ or 1 and $\nu_0 \in E_\chi E_k$. Therefore $Q = 2$.

We have just re–proven part of the following elementary result, which was known to Hasse [5]:

**Theorem 1.** *The following are equivalent:*

1. $Q = 2$
2. *There exists* $\varepsilon$ *such that* $E_K = \langle -1, \varepsilon, \varepsilon^\sigma, \varepsilon^{\sigma^2} \rangle$.
3. *There exists* $\varepsilon$ *such that* $E_k = \langle -1, \varepsilon^{1+\sigma^2} \rangle$.
4. *There exists* $\varepsilon$ *such that* $E_\chi = \langle -1, \varepsilon^{1+\sigma} \rangle$.

Each $\varepsilon$ appearing above is the same, up to sign and conjugation by $\pm\sigma$.

Bouvier and Payan [0] show that when the conductor $f$ of a cyclic quartic (not necessarily simplest) is prime, necessarily $Q = 2$. Perusal of the tables [2] work leads to the following conjecture:

**Conjecture.** *If the conductor* $f$ *of a simplest quartic field* $K$ *is divisible by more than one prime (including* 2*), then* $Q = 1$.

The tables show this conjecture is false when 'simplest' is weakened to 'cyclic' or when $t^2 + 16$ has odd square factors.

For our immediate purposes the value of $Q$ is not important, because any subgroup of $E_K$ which has finite index gives a lower bound on the class number $h$; a fundamental basis for $E_K$ gives the smallest regulator and best bound.

Nevertheless, we are not that far from a fundamental basis. Considering the map $\sigma$ in (5), we see that $\varepsilon_i^{1+\sigma^2} = -1$, $i = 1,2,3,4$, so the roots of the polynomial $f_t(X)$ are $\chi$–relative units. Hasse proves that the subgroup $\left\langle \varepsilon_1, \varepsilon_2 = \varepsilon_1^\sigma \right\rangle = E_\chi$, so these roots are fundamental relative units.

In cases 2, 3, and 4, $k$ is a simplest quadratic field and the fundamental unit $\eta$ may be computed easily. This allows us to write down $R$ in terms of $t$ explicitly. In Case 1, $\eta$ is not well understood at this time, but an upper bound will suffice.

**Proposition 2.** *When $t$ is even, $\eta$ is given by*

$$\eta = \begin{cases} \dfrac{(t/2) + \sqrt{(t/2)^2 + 4}}{2} & t \equiv 2 \bmod 4 \\ \dfrac{1+\sqrt{5}}{2} & t = 8 \\ (t/4) + \sqrt{(t/4)^2 + 1} & \textit{otherwise}. \end{cases}$$

**Proof.** Not that the first and third expressions are in fact identical, but the given representation motivates the proof. We know that a fundamental unit of $k$ has minimal coefficients. In Case 2 the coefficients are 1, obviously minimal. In Cases 3 and 4, $(t/4)+\sqrt{(t/4)^2+1}$ is a unit and a calculation shows that

$$\frac{A + \sqrt{(t/4)^2 + 1}}{2}$$

can be a unit only in the special case $t = 8$, $A = 1$, $m = 5$. ■

**Proposition 3.** *(Schur) For any quadratic field $k$ of conductor $m$, $\eta < m^{\sqrt{m}}$.*

**Proof.** See [6], Chapter 12.13. ■

**Remark 1.** In fact, [6] also has the slightly stronger result $\eta < (e\sqrt{m})^{\sqrt{m}}$.

Of course, the quadratic regulator $R_k = \log \eta$. Note that $\log|\varepsilon_3| = -\log \varepsilon_1$ and $\log \eta^{\sigma} = -\log \eta$. Form the regulator

$$R' = \begin{vmatrix} \log \varepsilon_1 & \log \varepsilon_2 & -\log \varepsilon_1 \\ \log \varepsilon_2 & -\log \varepsilon_1 & -\log \varepsilon_2 \\ \log \eta & -\log \eta & \log \eta \end{vmatrix} = 2 \log \eta \,(\log^2 \varepsilon_1 + \log^2 \varepsilon_2). \tag{8}$$

We see that $R' > 0$ and $R'/R = Q_K$. Less trivially, from Proposition 2 and equation (4) we have, using $O$-notation, $R = R' = O(\log^3 t)$ when $t$ is even. When $t$ is odd, $R' = O(t \log^3 t)$.

Define the relative regulator $R^{-} = R'/R_k$.

## 4. Stark's Bound on $\prod_{\chi \neq 1} L(1,\chi)$

Stark [13] showed that for certain fields, including a simplest cyclic quartic, it is possible to bound $\prod_{\chi \neq 1} L(1,\chi)$ below in an effective way. We begin by recapitulating his results. Theorem 4 is a specialization of [13] Theorem 1′ to abelian $K/\mathbb{Q}$. Stark suggests the following values for constants which appear in the theorem:

$$c_1 = 0$$

$$c_2 = 2/\log 3$$

$$c_3 = \exp\left(\frac{21}{8} + \frac{c_1}{2} - \frac{c_2\Gamma'(\frac{1}{2})}{8\Gamma(\frac{1}{2})}\right)$$

$$c_4 = 2c_3 = 43.16211498072652\ldots$$

$$c_8 = \pi/6$$

**Theorem 4.** *(Stark) If $K$ is an abelian number field of degree $n$ and discriminant $D$, then*

$$\prod_{\chi \neq 1} L(1,\chi) > c_4^{-1} \min[(4\log|D|)^{-1}, (c_8|D|^{1/n})^{-1}]$$

A brief calculation shows that for our family the latter term in this minimum is the lesser for all $t > 57$. We can then conclude that for $t > 57$

$$\prod_{\chi \neq 1} L(1,\chi) > (c_4 c_8 |D|^{1/4})^{-1} \tag{9}$$

## 5. Class Number Bound

Define

$$A_0 = (8c_4 c_8)^{-1} = 0.00553\ldots$$

**Theorem 5.** *For $K$ a simplest quartic field with $t^2+16$ odd–square–free, the class number $h_K$ is bounded:*

$$h_K > \begin{cases} \dfrac{3Q_K A_0 D^{1/12}}{R^{-} \log D} & t \text{ odd} \\ \dfrac{Q_K A_0 D^{1/4}}{R'} & t \text{ even} \end{cases}$$

**Proof.** By substitution of (9) and Proposition 3, in the $t$ odd case, into the class number formula (7). ■

We can rewrite this result in terms of $t$ instead of $D$ using (4), (6), (8), and Proposition 3. When constants depend of the case, i.e., the 2–adic valuation of $t$, tuple notation is used.

**Corollary 6.** *For $K$ as above, there exists a constant $C$ independent of the field such that for $t \gg 0$,*

$$h_K > \begin{cases} C \dfrac{t^{1/2}}{\log^3 t} & t \text{ odd} \\ C \dfrac{t^{3/2}}{\log^3 t} & t \text{ even} \end{cases}$$

*In particular, we may take $C = (0.0024, 0.00193, 0.00137, 0.00096)$ in cases 1 to 4, respectively.*

**Proof.** There exist constants $A_i$ such that for sufficiently large $t$ we have

$$D > A_1 t^6$$

$$\prod_{\chi \neq 1} L(1,\chi) > A_2 t^{-3/2}$$

$$R < \begin{cases} A_3 t \log^3 t & t \text{ odd} \\ A_3 \log^3 t & t \text{ even} \end{cases}$$

The corollary follows by substitution. The given tuple $C$ was obtained by setting $A_1 = (1, \frac{1}{4}, \frac{1}{16}, \frac{1}{64})$ and finding $A_3$ for $t = 57$. It is evident from (8) that such an approximation for $A_3$ becomes better with increasing $t$. In Case 1, Remark 1 after Proposition 3 was used to yield a 2-fold improvement. ■

**Remark 2.** To ensure $h_K > 1$ in Theorem 5 it suffices to take $t \geq (1.8 \cdot 10^{14}, 4204, 5700, 7712)$. The magnitude of $t$ in the first case corresponds to fields whose discriminants have 85 decimal digits.

Stark observed that although the quadratic fields are closest in degree to the rational numbers, they are the fields for which the class number one problem is most intractable. The quadratic fields in Case 1 are a family for which possibly infinitely many have class number one. We have shown here that this is not possible for a certain simple extension of these fields.

## References


[0] *L. Bouvier* and *J.-J. Payan,* Modules sur certains anneaux de Dedekind. Crelle **274/275** (1975), 278—286.

[1] *G. Cornell* and *L.C. Washington,* Class numbers of cyclotomic fields. J. Number Theory, **21**:3 (1985), 260—274.

[2] *M.-N. Gras,* Table numérique du nombre de classes et des unités des extensions cycliques de degré 4 de $\mathbb{Q}$. Publ. math. fasc. 2, Fac. Sci. Besançon (1977/1978).

[3] *M.-N. Gras,* Special units in real cyclic sextic fields. Math. Comp. **48**:177 (1987), 179—182.

[4] *K. Hardy, R.H. Hudson, D. Richman, K.S. Williams,* and *N.M. Holtz,* Calculation of the Class Numbers of Imaginary Cyclic Quartic Fields. Mathematical Lecture Note Series 7, Carleton University and Université d'Ottawa (July 1986).

[5] *H. Hasse,* Arithmetische Bestimmung von Grundeinheit und Klassenzahl in zyklischen kubischen und biquadratischen Zahlkörpern. In Mathematische Abhandlungen. Walter deGruyter, Berlin (1975).

[6] *L.K. Hua,* Introduction to Number Theory. Springer-Verlag, Berlin, English edition (1982).

[7] *H.W. Leopoldt,* Über Einheitengruppe und Klassenzahl reeler abelscher Zahlkörper. Abh. Deutsche Akad. Wiss. Berlin 2, (1953), 1—48.

[8] *G. Lettl,* A lower bound for the class number of certain cubic number fields. Math. Comp. **46**:174 (1986), 659—666.

[9] *R.A. Mollin,* Class number one criterion for real quadratic fields I. Proc. Japan Acad., **63**:Ser. A (1987), 121—125.

[10] *R.A. Mollin,* Class number one criterion for real quadratic fields II. Proc. Japan Acad., **63**:Ser. A (1987), 162—164.

[11] *R.A. Mollin* and *H.C. Williams,* On prime–valued polynomials and class numbers of real quadratic fields. Nagoya Math. J. **112** (1988), 143—151.

[12] *D. Shanks,* The simplest cubic fields. Math. Comp. **28**:128 (1974), 1137—1152.

[13] *H.M. Stark,* Some effective cases of the Brauer–Siegel Theorem. Invent. Math. **23** (1974), 135—152.

---

Department of Mathematics, University of California, Berkeley, CA 94720.

# On Improving Ramachandra's Unit Index

*Claude Levesque*

## 1. Introduction

Fix an integer $n \not\equiv 2 \pmod 4$ with $n = \prod_{i=1}^{s} p_i^{e_i}$ (a factorization involving $s$ different primes), and let $\zeta_n$ be a primitive $n$-th root of unity. The cyclotomic field $\mathbb{Q}(\zeta_n)$ of degree $\phi(n)$ (Euler $\phi$-function) over the rationals $\mathbb{Q}$ contains the maximal real subfield $\mathbb{Q}(\zeta_n)^+ = \mathbb{Q}(\zeta_n + \zeta_n^{-1})$ of degree $\frac{\phi(n)}{2}$ over $\mathbb{Q}$ with $\mathrm{Gal}(\mathbb{Q}(\zeta_n)^+/\mathbb{Q}) = (\mathbb{Z}/n\mathbb{Z})^\times/\{1,-1\}$, where $(\mathbb{Z}/n\mathbb{Z})^\times$ is the set of invertible integers modulo $n$.

$$
\begin{array}{lcl}
\mathbb{Q}(\zeta_n) & \longleftrightarrow & \{1\} \\
\Big| \, 2 & & \\
\mathbb{Q}(\zeta_n)^+ & \longleftrightarrow & \{1,-1\} \\
\Big| \, \frac{\phi(n)}{2} & & \\
\mathbb{Q} & \longleftrightarrow & \mathrm{Gal}(\mathbb{Q}(\zeta_n)/\mathbb{Q}) = (\mathbb{Z}/n\mathbb{Z})^\times
\end{array}
$$

**Figure 1.**

For a subset $I$ of $S = \{1, \ldots, s\}$, define

$$n_I := \prod_{i \in I} p_i^{e_i}.$$

Then for $1 \le a < n$ and $(a, n) = 1$, K. Ramachandra [2] introduced the units

$$\xi_a := \zeta_n^{d_a} \prod_{I \subset S} \frac{1-\zeta_n^{a n_I}}{1-\zeta_n^{n_I}} \quad \text{with } d_a = \frac{1-a}{2} \sum_{I \subset S} n_I , \tag{1}$$

and proved that

$$\mathcal{S}_1 = \{\xi_a : 1 < a < \tfrac{n}{2} \text{ and } (a,n)=1\} \tag{2}$$

is a maximal independent system of units of $\mathbb{Q}(\zeta_n)^+$. More precisely, if the class number and the unit group of $\mathbb{Q}(\zeta_n)^+$ are respectively denoted by $h_n^+$ and $E_n^+$, and if $C_n'$ is the group generated by $-1$ and the units of $\mathcal{S}_1$, then K. Ramachandra [2] obtained the unit index formula

$$\left[E_n^+ : C_n'\right] = h_n^+ \left[ \prod_{\chi \text{ even } \neq 1} \left[ \prod_{\substack{i \in S \\ p_i \nmid f_\chi}} (\phi(p_i^{e_i}) + 1 - \chi(p_i)) \right]\right] \neq 0 \tag{3}$$

where $\chi$, of conductor $f_\chi$, runs through all the non–trivial even characters of $(\mathbb{Z}/n\mathbb{Z})^\times$.

A nice feature about $\mathcal{S}_1$, is that for each $n$ one can explicitly exhibit in a *canonical* way a maximal independent system of units, viz. the $r$ units of $\mathcal{S}_1$ where $r := \frac{\phi(n)}{2} - 1$ is the rank of the unit group of $\mathbb{Q}(\zeta_n)^+$ (and of $\mathbb{Q}(\zeta_n)$).

Now let $V_n$ be the multiplicative group generated by

$$\{\zeta_n, -\zeta_n, 1-\zeta_n^a : 1 < a \leq n-1\}$$

and let $E_n$ be the group of units of $\mathbb{Q}(\zeta_n)$; define

$$C_n := V_n \cap E_n$$

to be the group of *cyclotomic units* of $\mathbb{Q}(\zeta_n)$. So the group of cyclotomic units of $\mathbb{Q}(\zeta_n)^+$ will be $C_n^+ := V_n \cap E_n^+$. Then W. Sinnott proved in [3] that

$$[E_n^+ : C_n^+] = 2^b h_n^+ \quad \text{with} \quad b = \begin{cases} 0 & \text{if } s = 1, \\ 2^{s-2} + 1 - s & \text{if } s \geq 2. \end{cases} \tag{4}$$

What is interesting about (4) is that one usually has a smaller unit index, since the fudge factor $2^b$ is in general much smaller than $[E_n^+ : C_n'] / h_n^+$.

As is well known, the system of units

$$\mathcal{S}_0 = \left\{ \zeta_n^{\frac{1-a}{2}} \cdot \frac{1-\zeta_n^a}{1-\zeta_n} : 1 < a < n/2 \text{ and } (a,n) = 1 \right\} \tag{5}$$

is never independent for $s \geq 4$ (see exercise 8.8 of [4]), though we know from Kummer that for $s = 1$, $\mathcal{S}_0$ is independent and generates with $-1$ a subgroup of index $h_n^+$ in $E_n^+$. In general, for a given $n$ with $n \not\equiv 2 \pmod 4$, K. Feng and D.Y. Pei [1] proved that $\mathcal{S}_0$ generates with $-1$ a group with rank equal to the cardinality of the set

$$\{\chi \text{ even } \neq 1 : \prod_{i \in S} (1 - \chi(p_i)) \neq 0\}.$$

They also showed that this rank is equal to

$$\frac{\phi(n)}{2} + \sum_{k=1}^{s} (-1)^k \sum_{\substack{I \subseteq S \\ \#I = k}} m_I$$

where $m_I$ is defined by

$$m_I := \left[ \left( \mathbb{Z}/\frac{n}{n_I}\mathbb{Z} \right)^{\times} : \left\langle -1, \{p_i : i \in I\} \right\rangle \right],$$

i.e., $m_I$ is the index of the subgroup generated by $-1$ and the primes of $\{p_i : i \in I\}$ in the group of invertible elements modulo $\frac{n}{n_I}$. This led them to conclude that $\mathcal{S}_0$ is an independent (maximal) system of units if and only if for each $i \in S$

$$\left( \mathbb{Z}/\frac{n}{p_i^{e_i}}\mathbb{Z} \right)^{\times} = \left\langle -1, p_i \right\rangle.$$

This last criterion allowed K. Feng and D.Y. Pei [1] to describe all integers $n$ such that $\mathcal{S}_0$ is independent, in which case the index of the group generated by $-1$ and $\mathcal{S}_0$ in $E_n^+$ is

$$h_n^+ \left[ \prod_{\chi \text{ even } \neq 1} \left[ \prod_{i \in S} (1 - \chi(p_i)) \right] \right]. \tag{6}$$

See Theorem 4 of [1] for this list of integers.

Therefore at this point we are facing two problems:

(a) to find a canonical basis of the units of $C_n^+$;

(b) to exhibit a canonical independent system of units leading to a unit index smaller than that of (3).

We will concentrate on (b) since (a) appears to be hard. In Section 2, we will obtain our main theorem and in Section 3 we will show how Ramachandra's unit index is lowered.

## 2. Modification of Ramachandra's Units

As before, let $n \not\equiv 2 \pmod 4$ and let $n = \prod_{i \in S} p_i^{e_i}$ be its prime factorization.

We plan to prove our main result.

**Theorem 1.** *Let $\mathfrak{D}$ be a fixed subset of the set $\{d : 1 \leq d < n, d \mid n\}$ of proper divisors of $n$. For $1 < a < n/2$ and $(a, n) = 1$, define*

$$\lambda_a := \zeta_n^{b_a} \prod_{d \in \mathfrak{D}} \frac{1 - \zeta_n^{ad}}{1 - \zeta_n^{d}} \quad \text{with} \quad b_a = \left(\frac{1-a}{2}\right) \sum_{d \in \mathfrak{D}} d . \tag{7}$$

*If $\mathcal{C}_n$ denotes the group generated by $-1$ and the set*

$$\mathcal{S} = \{\lambda_a : 1 < a < n/2, (a, n) = 1\},$$

*then*

$$[E_n^+ : \mathcal{C}_n^+] = h_n^+ \left[ \prod_{\chi \text{ even} \neq 1} \left[ \sum_{\substack{d \in \mathfrak{D} \\ f_\chi \mid \frac{n}{d}}} \left( \frac{\phi(n)}{\phi(n/d)} \prod_{p \mid \frac{n}{d}} (1 - \chi(p)) \right) \right] \right], \tag{8}$$

*the index being considered as infinite when the right hand side is* 0.

In fact, in this paper, the only good idea we claim to have is the definition of $\lambda_a$, since the proof goes along the same lines as those of [2] and [4]: see the excellent, elegant and inspiring proof given by L.C. Washington in his book ([4], pp. 147–150).

In the course of the proof we will use the following results.

**Lemma 1.** *Let* $d|n$. *If* $f_\chi$ *does not divide* $\frac{n}{d}$, *then*

$$\sum_{\substack{a=1\\(a,n)=1}}^{n} \chi(a)\log|1-\zeta_n^{ad}| = 0.$$

**Proof.** See page 148 of [4].

**Lemma 2.** *Let* $d|n$ *and let* $f_\chi|\frac{n}{d}$. *Then*

$$\sum_{\substack{a=1\\(a,n)=1}}^{n} \chi(a)\log|1-\zeta_n^{ad}| = \frac{\phi(n)}{\phi(n/d)}\left(\sum_{\substack{b=1\\(b,n/d)=1}}^{n/d} \chi(b)\log|1-\zeta_{n/d}^{b}|\right).$$

**Proof.** As in page 148 of [4], we write

$$a = b + c\cdot\frac{n}{d} \text{ with } 1 \le b < \frac{n}{d} \text{ and } 0 \le c < d.$$

On the one hand, if $(a,n)=1$ then $(b,\frac{n}{d})=1$. On the other hand, suppose $(b,\frac{n}{d})=1$; there are $\phi(n)$ possible values for $a$; since there are $\phi(\frac{n}{d})$ possible candidates for $b$, there will be $\phi(n)/\phi(\frac{n}{d})$ choices for $c$ such that $(b+c\frac{n}{d},n)=1$. Moreover, $\chi(a)$ depends only on $b$ since $f_\chi|\frac{n}{d}$. The conclusion follows since $\zeta_n^{ad} = \zeta_n^{bd+cn} = \zeta_n^{bd}$.

**Lemma 3.** *Let* $m|n$. *If* $f_\chi|m$, *then*

$$\sum_{\substack{b=1\\(b,m)=1}}^{m} \chi(b)\log\left|1-\zeta_m^{b}\right| = \left[\prod_{p|m}(1-\chi(p))\right]\sum_{a=1}^{m} \chi(a)\log\left|1-\zeta_m^{a}\right|.$$

**Proof.** See page 148 of [4]; here as usual, $p$ stands for a prime.

**Lemma 4.** *Suppose* $F, g, t$ *are positive integers with* $f_\chi|F$ *and* $g|F$. *Then*

$$\sum_{\substack{a=1\\(a,g)=1}}^{Ft} \chi(a)\log|1-\zeta_{Ft}^{a}| = \sum_{\substack{b=1\\(b,g)=1}}^{F} \chi(b)\log|1-\zeta_F^{b}|.$$

**Proof.** See page 148 of [4].

**Proposition 1.** *Let* $f = f_\chi$ *and let*

$$\tau(\chi) := \sum_{a=1}^{f} \chi(a) \varepsilon^{\frac{2\pi i}{f} a}.$$

*If* $\chi$ *is even* $\neq 1$, *then*

$$\sum_{a=1}^{f} \chi(a) \log\left|1 - \zeta_f^a\right| = -\tau(\chi) L(1, \bar{\chi}),$$

*where*

$$L(s, \chi) := \sum_{k=1}^{\infty} \frac{\chi(k)}{k^s}, \quad \Re(s) > 1.$$

*Moreover*

$$\pm \prod_{\chi \text{ even } \neq 1} \frac{-\tau(\chi) L(1, \bar{\chi})}{2} = h_n^+ R_n^+,$$

*where* $R_n^+$ *is the regulator of* $\mathbb{Q}(\zeta_n)^+$.

**Proof.** See pages 37 and 35 of [4].

**Proof of Theorem 1.** As in page 147 of [4], the regulator $R(\{\lambda_a\})$ of the units of $\mathcal{S}$ is

$$R(\{\lambda_a\}) = \pm \prod_{\chi \text{ even } \neq 1} \left[ \frac{1}{2} \sum_{d \in \mathcal{D}} \left( \sum_{\substack{a=1 \\ (a,n)=1}}^{n} \chi(a) \log\left|1 - \zeta_n^{ad}\right| \right) \right].$$

By Lemma 1, we can restrict ourselves to the case $f_\chi \mid \frac{n}{d}$. Therefore, by Lemmas 2 and 3 respectively

$$\sum_{\substack{a=1 \\ (a,n)=1}}^{n} \chi(a) \log\left|1 - \zeta_n^{ad}\right| = \frac{\phi(n)}{\phi(n/d)} \left( \sum_{\substack{b=1 \\ (b, n/d)=1}}^{n/d} \chi(b) \log\left|1 - \zeta_{n/d}^{b}\right| \right)$$

$$= \frac{\phi(n)}{\phi(n/d)} \left( \prod_{p \mid \frac{n}{d}} (1 - \chi(p)) \right) \sum_{a=1}^{n/d} \chi(a) \log\left|1 - \zeta_{n/d}^{a}\right|.$$

Writing $\frac{n}{d} = f_\chi \cdot \frac{n}{df_\chi}$ and using Lemma 4 with $F = f_\chi$, $t = \frac{n}{df_\chi}$, $g = 1$, we obtain

$$\sum_{a=1}^{n/d} \chi(a)\log\left|1-\zeta_{n/d}^{a}\right| = \sum_{b=1}^{f_\chi} \chi(b)\log\left|1-\zeta_{f_\chi}^{b}\right|$$

$$= -\tau(\chi)L(1,\bar{\chi}),$$

the last equality being given by Proposition 1.

Therefore, we conclude that

$$R(\{\lambda_a\}) = \pm \prod_{\chi \text{ even } \neq 1} \left[\left(\frac{-\tau(\chi)L(1,\bar{\chi})}{2}\right) \sum_{\substack{d\in\mathfrak{D} \\ f_\chi | \frac{n}{d}}} \left[\frac{\phi(n)}{\phi(n/d)}\left(\prod_{p|\frac{n}{d}}(1-\chi(p))\right)\right]\right]$$

$$= h_n^{+} R_n^{+}\left[\prod_{\chi \text{ even } \neq 1}\left[\sum_{\substack{d\in\mathfrak{D} \\ f_\chi | \frac{n}{d}}}\left(\frac{\phi(n)}{\phi(n/d)}\prod_{p|\frac{n}{d}}(1-\chi(p))\right)\right]\right].$$

This concludes the proof.

We will see in the following section the kind of improvements some choices of $\mathfrak{D}$ lead to.

## 3. Unit Indices Smaller than Ramachandra's One

The notations being as before, a first improvement is given by

**Theorem 2.** *Let*

$$T := \{i \in S : \exists\, \chi \text{ even} \neq 1 \text{ with } \chi(p_i) = 1\}.$$

*If* $\mathfrak{D} = \{n_I : I \subseteq T\}$ *then* $\mathcal{S}$ *is a maximal independent system of units and*

$$[E_n^{+} : \mathcal{C}_n] = h_n^{+}\left[\prod_{\chi \text{ even } \neq 1}\left[\left(\prod_{\substack{i\in T \\ p_i \nmid f_\chi}}(\phi(p_i^{e_i}) + 1 - \chi(p_i))\right)\left(\prod_{j\in S\setminus T}(1-\chi(p_j))\right)\right]\right].$$

**Note.** It is useful to define $\dfrac{1-\zeta_n^{an_I}}{1-\zeta_n^{n_I}}$ as being 1 whenever $I = S$. Moreover, an empty product is equal to 1 by definition, and an empty sum is 0.

**Proof.** It remains only to show that if $H = H(\chi)$ is defined by

$$H := \sum_{\substack{I \subseteq T \\ f_\chi \mid \frac{n}{n_I}}} \left[ \phi(n_I) \prod_{i \in S \setminus I} (1-\chi(p_i)) \right],$$

then

$$H = \left[ \prod_{\substack{i \in T \\ p_i \nmid f_\chi}} (\phi(p_i^{e_i}) + 1 - \chi(p_i)) \right] \left[ \prod_{i \in S \setminus T} (1-\chi(p_i)) \right].$$

Let us use the fact that for any given set $A = \{1, \ldots, s\}$ of indices, the equality

$$\prod_{i \in A} (Y_i + a_i) = \sum_{J \subseteq A} \left[ \left( \prod_{j \in J} Y_i \right) \left( \prod_{i \in A \setminus J} a_i \right) \right]$$

holds true. So along the lines of [4], we deduce

$$\begin{aligned} H &= \sum_{\substack{I \subseteq T \\ f_\chi \mid \frac{n}{n_I}}} \left[ \left( \prod_{i \in I} \phi(p_i^{e_i}) \right) \prod_{i \in S \setminus I} (1-\chi(p_i)) \right] \\ &= \sum_{\substack{I \subseteq T \\ f_\chi \mid \frac{n}{n_I}}} \left[ \left( \prod_{i \in I} \phi(p_i^{e_i}) \right) \left( \prod_{t \in T \setminus I} (1-\chi(p_t)) \right) \prod_{j \in S \setminus T} (1-\chi(p_j)) \right] \\ &= \left[ \prod_{j \in S \setminus T} (1-\chi(p_j)) \right] \left[ \sum_{\substack{I \subseteq T \\ f_\chi \mid \frac{n}{n_I}}} \left[ \left( \prod_{i \in I} \phi(p_i^{e_i}) \right) \prod_{t \in T \setminus I} (1-\chi(p_t)) \right] \right]. \end{aligned}$$

Defining $A = A(\chi)$ by

$$A := \{ i \in T : p_i \nmid f_\chi \},$$

we have

$$\prod_{t\in T\setminus I}(1-\chi(p_t))=\prod_{t\in A\setminus I}(1-\chi(p_t)),$$

since $1-\chi(p_t)=1$ whenever $p_t \nmid f_\chi$. Therefore

$$H=\left[\prod_{j\in S\setminus T}(1-\chi(p_j))\right]\left[\sum_{I\subseteq A}\left[\left(\prod_{i\in I}\phi(p_i^{e_i})\right)\prod_{t\in A\setminus I}(1-\chi(p_t))\right]\right]$$

$$=\left[\prod_{j\in S\setminus T}(1-\chi(p_j))\right]\left[\prod_{\substack{i\in T\\ p_i\nmid f_\chi}}(\phi(p_i^{e_i})+1-\chi(p_i))\right].$$

Here

$$\prod_{\substack{i\in T\\ p_i\nmid f_\chi}}(\phi(p_i^{e_i})+1-\chi(p_i))\neq 0$$

since the real part of $\phi(p_i^{e_i})+1-\chi(p_i)$ is positive, and

$$\prod_{j\in S\setminus T}(1-\chi(p_j))\neq 0$$

since by construction $\chi(p_j)\neq 1$ for $j\notin T$. This concludes the proof. Note that this new index is smaller than or equal to that of K. Ramachandra, as is easily seen.

**Examples.** Write the unit index of formula (8) as

$$[E_n^+ : C_n^+] = h_n^+ \cdot i(\mathfrak{D}).$$

(1) Let $n=5\cdot 7=35$. Then we have the following values for $i(\mathfrak{D})$:

| $\mathfrak{D}$ | {1} | {1,7} | {1,5} | {1,5,7} |
|---|---|---|---|---|
| $i(\mathfrak{D})$ | 6 | 24 | 62 | 248 |

**Table 1**

(2) Let $n=5\cdot 7\cdot 11=385$. There are three non–trivial even characters $\chi$ such that $\chi(p)=1$ for a $p\in\{5,7,11\}$; one of those has conductor 5 and the other two have conductor 35; in the three cases, the only prime $p$ involved is 11. Let us consider the cases where $\mathfrak{D}$ is a subset of $B=\{1,5,7,11,35,55,77\}$ containing $\{1,11\}$. Table 2 gives the 32 possibilities for the fudge factor $i(\mathfrak{D})$, and one can witness the growth

rate of $i(\mathfrak{D})$ up to the last value corresponding to Ramachandra's formula. Most calculations were done by hand, though some crucial ones were verified with the help of MAPLE (version 4.0) running on a SUN 3.

| $\mathfrak{D}$ | $i(\mathfrak{D})$ |
|---|---|
| {1,11} | 187740432600000 |
| {1,11,77} | 750961730400000 |
| {1,11,55} | 1618143728600000 |
| {1,11,55,77} | 6472574914400000 |
| {1,11,35} | 3080152143025944000 |
| {1,11,35,77} | 12320608572103776000 |
| {1,11,35,55} | 26547977994652184000 |
| {1,11,35,55,77} | 106191911978608736000 |
| {1,11,7} | 31208822068433494222815000000 |
| {1,11,7,77} | 124835288273733976891260000000 |
| {1,11,7,55} | 268990323542212497825215000000 |
| {1,11,7,55,77} | 1075961294168849991300860000000 |
| {1,11,7,35} | 2732517886867303657222023000000 |
| {1,11,7,35,77} | 10930071547469214628888092000000 |
| {1,11,7,35,55} | 23551701786808664855104103000000 |
| {1,11,7,35,55,77} | 94206807147234659420416412000000 |
| {1,11,5} | 213098638811178745761765579720000 |
| {1,11,5,77} | 852394555244714983047062318880000 |
| {1,11,5,55} | 1457057525403797626493796824520000 |
| {1,11,5,55,77} | 5828230101615190505975187298080000 |
| {1,11,5,35} | 47059655637348761683447356489864000 |
| {1,11,5,35,77} | 188238622549395046733789425959456000 |
| {1,11,5,35,55} | 321769419888538879636572886911624000 |
| {1,11,5,35,55,77} | 1287077679554155518546291547646496000 |
| {1,11,5,7} | 932816623182427794643249939432034843565000000 |
| {1,11,5,7,77} | 3731266492729711178572999757728139374260000000 |
| {1,11,5,7,55} | 6378114324015173793224078773264145373165000000 |
| {1,11,5,7,55,77} | 25512457296060695172896315093056581492660000000 |
| {1,11,5,7,35} | 35424215282667104894253543021485465065893000000 |
| {1,11,5,7,35,77} | 141696861130668419577014172085941860263572000000 |
| {1,11,5,7,35,55} | 242212337662415378903826464141931297623013000000 |
| {1,11,5,7,35,55,77} | 968849350649661515615305856567725190492052000000 |

Table 2.

Let us now have a close look at formula (8). If we want to avoid that the product in (8) be 0, it is necessary to include 1 in $\mathcal{D}$, since by the Conductor–Discriminant Formula ([4], p. 27), there are non–trivial even characters $\chi$ with conductor $n$. What else should we include in $\mathcal{D}$ in order to always get a non–zero

$$i(\mathcal{D}) := \prod_{\chi \text{ even } \neq 1} \left[ \sum_{\substack{d \in \mathcal{D} \\ f_\chi | \frac{n}{d}}} \left( \frac{\phi(n)}{\phi(n/d)} \prod_{p | \frac{n}{d}} (1 - \chi(p)) \right) \right] ? \tag{9}$$

The set $\mathcal{D}$ under investigation should at least have the property that the terms of the sum

$$\sum_{\substack{d \in \mathcal{D} \\ f_\chi | \frac{n}{d}}} \left( \frac{\phi(n)}{\phi(n/d)} \prod_{p | \frac{n}{d}} (1 - \chi(p)) \right) \tag{10}$$

are not all 0. Therefore *each time there exists a prime* $p$ *such that* $\chi(p) = 1$ *for a character* $\chi$ *of conductor* $f_\chi$, *it is necessary to count on a divisor* $d$ *of* $n$ *such that* $f_\chi | \frac{n}{d}$ *and such that* $\chi(q) \neq 1$ *for any prime divisor* $q$ *of* $\frac{n}{d}$.

Fix $f$ an integer which is a divisor of $n$, and define

$$S_f := \{ i \in S : \exists \chi \text{ even} \neq 1 \text{ of conductor } f \text{ with } \chi(p_i) = 1 \}$$

and

$$n_f := \prod_{i \in S_f} p_i^{e_i}.$$

If $S_f = \emptyset$ then by convention $n_f = 1$; for instance $S_1 = \emptyset$ and $n_1 = 1$. Moreover, $S_f \subset S$ (strict inclusion), because for any character $\chi$ of conductor $f_\chi$, we have $\chi(q) = 0$ $(\neq 1)$ for any divisor $q$ of $f_\chi$. Therefore each time the product in $i(\mathcal{D})$ hits a non–trivial even character $\chi$ of conductor $f_\chi$ such that

$$\chi(p_i) = 1 \text{ for each } i \in B := \{i_1, \ldots, i_t\}$$

and

$$\chi(p_j) \neq 1 \text{ for each } j \in S \setminus B,$$

then

$$p_i \nmid f_\chi \text{ for each } i \in B,$$

whence

$$f_\chi \Big| \frac{n}{n_{f_\chi}} \text{ and } \prod_{p \mid \frac{n}{n_{f_\chi}}} (1-\chi(p)) \neq 0.$$

So when $\mathcal{D}$ contains $\{n_f : f \mid n\}$ (a set to which $n_1 = 1$ belongs), then the summation in (10) corresponding to a given character $\chi$ of conductor $g$ contains the term

$$\frac{\phi(n)}{\phi(n/n_g)}\left[\prod_{p \mid \frac{n}{n_g}} (1-\chi(p))\right] = \phi(n_g)\left[\prod_{i \in S \setminus S_g} (1-\chi(p_i))\right]$$

which is non–zero by definition of $S_g$. This looks like an interesting construction but unfortunately we cannot prove that the whole summation in (10) will always be different from 0.

One can also show that any set $\mathcal{D}$ which contains

$$\left\{ n_f \cdot m_f : f \mid n \text{ and } m_f \text{ is a fixed divisor of } \frac{n}{f \cdot n_f} \right\}$$

will share the same property as the one stated in the last paragraph.

Before moving to the next section and motivated by the data of Table 2, we raise the following questions:

(i) If $\mathcal{D}^* = \mathcal{D} \cup \{d^*\}$ with $d^* \mid n$, $d^* \notin \mathcal{D}$, is $i(\mathcal{D}) < i(\mathcal{D}^*)$?

(ii) If $d_1 \in \mathcal{D}_1$ and $d_2 \in \mathcal{D}_2$ are such that $\mathcal{D}_1 \setminus \{d_1\} = \mathcal{D}_2 \setminus \{d_2\}$ and if $d_1 < d_2$, is $i(\mathcal{D}_2) < i(\mathcal{D}_1)$?

Unfortunately we again have no answer.

## 4. The Case of a Totally Real Abelian Number Field

As in page 152 of [4], we can state without proof the equivalent of Theorem 1 for a totally real abelian number field $K$ of degree $r$ over $\mathbb{Q}$. Denote the unit group and the class number of $K$ by $E_K$ and $h_K$ respectively. Assume $n$ to be the conductor of $K$, i.e., $n$ is the smallest integer such that $K \subseteq \mathbb{Q}(\zeta_n)^+$. Therefore if

$$G := \mathrm{Gal}(\mathbb{Q}(\zeta_n)^+/\mathbb{Q}) = \{\sigma_a : 1 \le a < \tfrac{n}{2} \text{ and } (a,n) = 1\},$$

$$H := \mathrm{Gal}(\mathbb{Q}(\zeta_n)^+/K),$$

then

$$G/H = \mathrm{Gal}(K/\mathbb{Q}).$$

Let $R \subseteq G$ be a set of coset representatives for $G/H$ and let $R' \subset R$ be a set of representatives for $G/H - \{H\}$. Let us denote by $N$ the norm from $\mathbb{Q}(\zeta_n)^+$ to $K$. Then it is left to the reader to prove as in page 152 of [4] the next result.

**Theorem 3.** *Let $\mathfrak{D}$ be a fixed subset of the set $\{d : 1 \le d < n, d \mid n\}$ of proper divisors of $n = \prod_{i \in S} p_i^{e_i}$. For $1 \le a < \frac{n}{2}$ and $(a,n) = 1$, define*

$$\lambda_a := \zeta_n^{b_a} \frac{\alpha^{\sigma_a}}{\alpha} \tag{11}$$

*where*

$$\alpha = \prod_{d \in \mathfrak{D}} (1 - \zeta_n^d) \text{ and } b_a = \frac{1-a}{2} \sum_{d \in \mathfrak{D}} d.$$

*If $\mathcal{C}_K$ denotes the group generated by $-1$ and the set*

$$S := \{N\lambda_a : \sigma_a \in R'\},$$

*then*

$$[E_K : \mathcal{C}_K] = h_K \left[ \prod_{\substack{\chi \in \hat{G} \\ \chi(H)=1 \\ \chi \ne 1}} \left[ \sum_{\substack{d \in \mathfrak{D} \\ f_\chi \mid \frac{n}{d}}} \left( \frac{\phi(n)}{\phi(n/d)} \prod_{p \mid \frac{n}{d}} (1 - \chi(p)) \right) \right] \right]. \tag{12}$$

*where the product is taken over all non–trivial characters of $G$ with $\chi(H) = 1$, the index being infinite if the product is 0. Moreover if*

$$T := \{i \in S : \exists \text{ non-trivial } \chi \in \hat{G} \text{ with } \chi(H) = 1 \text{ and } \chi(p_i) = 1\}$$

*and if* $\mathfrak{D} = \{n_I : I \subseteq T\}$, *then* $\mathfrak{S}$ *is a maximal independent system of units and*

$$[E_K : \mathcal{C}_K] = h_K \left[ \prod_{\substack{\chi \in \hat{G} \\ \chi(H)=1 \\ \chi \neq 1}} \left[ \left( \prod_{\substack{i \in T \\ p_i \nmid f_\chi}} (\phi(p_i^{e_i}) + 1 - \chi(p_i)) \right) \left( \prod_{j \in S \setminus T} (1 - \chi(p_j)) \right) \right] \right].$$

**Acknowledgement.** The author is grateful to professor Richard Foot for some stimulating discussions.

## References

[1] *K. Feng,* The rank of group of cyclotomic units in abelian fields. J. Number Theory **14** (1982), 315—326.

[2] *K. Ramachandra,* On the units of cyclotomic fields. Acta Arith. **XII** (1966), 165—173.

[3] *W. Sinnott,* On the Stickelberger ideal and the circular units of a cyclotomic field. Ann. of Math. (2) **108** (1978), 107—134.

[4] *L.C. Washington,* Introduction to Cyclotomic Fields. Graduate Texts no. 83, Springer–Verlag (1982), xi+384 pages.

---

Département de Mathématiques et de Statistique, Université Laval, Québec, P.Q., CANADA G1K 7P4

# Rational Functions, Diagonals, Automata and Arithmetic

*Leonard Lipshitz*[1] *and Alfred J. van der Poorten*[2]

**Dedicated to the memory of Kurt Mahler**

## 1. Notions and Notation

By $R$ we denote an integral domain; then $R[[x]]$ is the ring of formal power series over $R$ in the variables $x = (x_1, \dots, x_n)$. We write a series $f(x) \in R[[x]]$ as $f(x) = \sum a_{i_1 \dots i_n} x_1^{i_1} \dots x_n^{i_n} = \sum a_\nu x^\nu$ where $\nu = (\nu_1, \dots, \nu_n)$ is a multi–index and $x^\nu = x_1^{\nu_1} \dots x_n^{\nu_n}$. In the sequel $R$ is usually the finite field $\mathbb{F}_p$ of $p$ elements or the ring of $p$–adic integers $\mathbb{Z}_p$. One says that $f(x) \in R[[x]]$ is *algebraic* if it is algebraic over the quotient field of the polynomial ring $R[x]$. If $f(x) \in \mathbb{Z}_p[[x]]$ we say that $f$ is *algebraic* mod $p^s$ if there is an algebraic $g(x) \in \mathbb{Z}_p[[x]]$ with $f \equiv g \bmod p^s$.

Given $f(x, y) = \sum a_{ij} x^i y^j$ it is natural to refer to the series $I_{xy} f = \sum a_{ii} x^i$ as its *diagonal*. In general, if $f(x) = \sum a_{i_1 \dots i_n} x_1^{i_1} \dots x_n^{i_n}$ we define its diagonal $I_{12} f$ by $I_{12} f = \sum a_{i_1 i_1 i_3 \dots i_n} x_1^{i_1} x_3^{i_3} \dots x_n^{i_n}$; generally, for $k \neq l$, the other $I_{kl}$ are defined correspondingly. By a *diagonal* we mean any composition of the $I_{kl}$ s. The (complete) diagonal is $If = \sum a_{ii \dots i} x_1^i$. We shall also need the *off–diagonals* $J_{k\ell} f = \sum_{i_k > i_\ell} a_{i_1 \dots i_n} x_1^{i_1} \dots x_n^{i_n}$.

The *Hadamard product* of series $f(x) = \sum a_\nu x^\nu$ and $g(x) = \sum b_\nu x^\nu$ is (their "child's product") $f * g(x) = \sum a_\nu b_\nu x^\nu$.

---

[1] Supported in part by a grant from the NSF.
[2] Work supported in part by the Australian Research Council.

The operations of diagonal and Hadamard product are connected by:

$$f * g(x) = I_{1,n+1} \cdots I_{n,2n} f(x_1, \ldots, x_n) g(x_{n+1}, \ldots, x_{2n}); \tag{1}$$

$$I_{12} f = f * \left( \frac{1}{1 - x_1 x_2} \prod_{j=3}^{n} \frac{1}{1 - x_j} \right). \tag{2}$$

Over $\mathbb{C}$, the diagonal is also given by the integral

$$I_{12} f(t, x_3, \ldots, x_n) = \frac{1}{2\pi i} \oint_{x_1 x_2 = t} f(x_1, \ldots, x_n) \frac{dx_1 \wedge dx_2}{dt}$$

$$= \frac{1}{2\pi i} \oint_{|y| = \varepsilon} f\left(\frac{t}{y}, y, x_3, \ldots, x_n\right) \frac{dy}{y}. \tag{3}$$

Noting (1) and (2), one sees that if a ring of power series is closed under one of the two operations of taking diagonals or Hadamard products then it is also closed under the other operation. From (3) it follows (*cf.* [19]) that the diagonal of a rational function $f(x, y)$ of two variables is an algebraic function. Indeed,

$$(If)(t) = \frac{1}{2\pi i} \oint_{|y| = \varepsilon} f\left(\frac{t}{y}, y\right) \frac{dy}{y} = \frac{1}{2\pi i} \oint_{|y| = \varepsilon} \frac{P(t, y)}{Q(t, y)} dy.$$

Writing $Q(t, y) = \prod (y - y_i(t))$, where the $y_i(t)$ are algebraic, and evaluating the integral by residues verifies the claim. In fact, every algebraic power series of one variable is the diagonal of a rational power series of two variables ([19]) and, indeed, every algebraic power series in $n$ variables is the diagonal of a rational power series in $2n$ variables ([16]). However, as we see in section 2, in characteristic zero diagonals of rational functions in more than two variables do not generally yield algebraic functions.

A *finite automaton* (*cf.* [32]) is a 'machine' with a finite number of states $\mathcal{S}$, a finite input alphabet $\mathcal{I}$, and a finite output alphabet $\mathcal{O}$. In the immediate sequel we will take $\mathcal{I} = \{0, 1, \ldots, p-1\}$ and will presume there to be $n$ input tapes. The output $\mathcal{O}$ will consist of elements of $\mathbb{F}_p$. At each stage of the computation the automaton reads one digit from each tape and accordingly alters its internal state, reporting its new state as an element of $\mathbb{F}_p$. Thus, formally, a finite automaton is a transition function

$\tau : \mathbb{Z}^n \times \mathcal{S} \to \mathcal{S}$ and an output function $o : \mathcal{S} \to \mathbb{F}_p$. Some distinguished state is the initial state.

Such an automaton generates its characteristic function $f = \sum a_\nu x^\nu$ by the rule that $a_\nu$ is the output after the words $\nu = (\nu_1, \ldots, \nu_n)$, each expressed in $p$–adic notation (that is, in base $p$ ), have been fully read from the input tapes. The principle of *irrelevance of symbols* is plain: the elements of the output alphabet just serve as markers—if $f = \sum a_\nu x^\nu$ is generated by a finite automaton then, for each output symbol $i$, so are the series $f^{(i)}(x) = \sum_{a_\nu = i} x^\nu$; and conversely. Thus our choice of output alphabet is of no matter. In section 3 we consider automata with output alphabet $\mathbb{Z}/p^s$. In the above spirit, notice that if for each $k < s$ the series $\sum a_\nu(k) x^\nu$ is generated by a finite automaton then so is the series

$$\sum (a_\nu(0) + a_\nu(1)p + \ldots + a_\nu(s-1)p^{s-1}) x^\nu;$$

and, once again, conversely.

A basic property of finite automata is their cyclic nature: Since it has only finitely many internal states, after a sufficiently long input the automaton will be in the same state as it was after a shorter portion of that input. (Incidentally, it follows that a finite automaton can add numbers in base $p$, but that no finite automaton can square arbitrarily large numbers.)

We digress, both to detail the foregoing and to provide an (apparently) alternative description for finite automata. Consider, for example, the sequence $(r_h)$

0001001000 0111010001 0010111000 1000010010 0001110111.....

which, for the viewer's convenience and as a blow in the battle against the binary, we have split into groups of 10 symbols.

The 'pattern', which is to say a formation rule, becomes plain, by our viewing the symbols in pairs as binary integers. We then see that the resulting sequence is self reproducing under the uniform (or *regular* ) 2–substitution $\theta$ on the symbols $\{0,1,2,3\}$:

$$0 \mapsto 01, \qquad 1 \mapsto 02, \qquad 2 \mapsto 31, \qquad 3 \mapsto 32.$$

Finally the output map $o$ replacing 3 by 1, 2 by 1 and 1 by 0 yields the given sequence:

```
0001001000 0111010001 0010111000 1000010010 0001110111....
 0 1 0 2 0  1 3 1 0 1  0 2 3 2 0  2 0 1 0 2  0 1 3 1 3
0102013101 0232020102 0131323101 3101020131 0102320232....
0001001000 0111010001 0010111000 1000010010 0001110111....
```

The 'what is going on here' is somewhat disguised by our notation. The intermediate symbols {0,1,2,3} represent the four states $\{s_0, s_1, s_2, s_3\}$ of a binary automaton ($p = 2$ in our description above). The substitution $\theta$ provides the transition map $\tau$ as in the transition table:

| | 0 | 1 |
|---|---|---|
| $s_0$ | $s_0$ | $s_1$ |
| $s_1$ | $s_0$ | $s_2$ |
| $s_2$ | $s_3$ | $s_1$ |
| $s_3$ | $s_3$ | $s_2$ |

The output map $o$ is as described above.

The sequence $(r_h)$ is *generated* by the automaton in all of the following senses: It is (an image of) a sequence invariant under the substitution $\theta$ (that is, under the transition map). That invariant sequence is $\lim_{h\to\infty} \theta^h(0)$, where

$$\theta(0) = 01,\ \theta^2(0) = \theta(01) = \theta(0)\theta(1) = 0102, \ldots, \theta^h(0) = \theta(\theta^{h-1}(0)\theta^{h-1}(1)) = \ldots$$

Finally, the automaton induces a map $h \mapsto r_h$ on the nonnegative integers in the following way (as is obviously equivalent to the foregoing): We have just one input tape containing the digits of $h$ written in base 2; $s_0$ is the initial state. The automaton reads the digits of $h$ successively (form left to right, disregarding irrelevant leading zeros since these leave the automaton in state $s_0$). The final state reached is transformed by the output map and yields $r_h$.

We will, in the sequel, use the sequence $(r_h)$ as a convenient and interesting example of an *automatic sequence*. Cobham [13] shows that the interconnections displayed by the example are general.

We have already remarked that, more generally, a multiplicity of input tapes causes a finite automaton to induce a map $\nu \mapsto a_\nu$ on $n$–tuples of nonnegative integers.

The following "breaking up" procedure will be fundamental to several of our arguments: If $y(x) \in \mathbb{F}_p[[x]]$ and $S = \{0,1,\ldots,p-1\}^n$ then, for $\alpha \in S$, there are unique $y_\alpha(x) \in \mathbb{F}_p[[x]]$ such that $y(x) = \sum_{\alpha \in S} x^\alpha y_\alpha^p(x)$. In different words: $\{x^\alpha = x_1^{\alpha_1} \ldots x_n^{\alpha_n} : \alpha \in S\}$ is a basis for $\mathbb{F}_p[[x]]$ over $(\mathbb{F}_p[[x]])^p$.

We choose to work over $\mathbb{F}_p$ for clarity and definiteness, but with only minor modifications, the results cited remain valid if $\mathbb{F}_p$ is replaced by an arbitrary finite field **F** and $\mathbb{Z}_p$ by the complete discrete valuation ring with prime $p$ and residue field **F**.

## 2. Algebraic Power Series in Characteristic $p$, Diagonals and Automata

If $y(x) \in \mathbb{F}_p[[x]]$ is algebraic then $y$ satisfies an equation of the shape

$$\sum_{i=r}^{s} f_i(x) y^{p^i} = 0,$$

with $r, s \in \mathbb{N}$, the $f_i \in \mathbb{F}_p[x]$ and $f_r \neq 0$. In fact, we may take $r = 0$, for if $r > 0$ then writing $f_i = \sum_\alpha x^\alpha f_{i\alpha}^p$ as above we get that

$$\sum_\alpha x^\alpha \left( \sum_{i=r}^{s} f_{i\alpha}(x) y^{p^{i-1}} \right)^p = 0.$$

Hence, $\sum_{i=r-1}^{s-1} f_{i+1,\alpha}(x) y^{p^i} = 0$ and some $f_{r,\alpha} \neq 0$.

Thus if $y \in \mathbb{F}_p[[x]]$ is algebraic, $y$ satisfies an equation of the shape

$$f(x) y = \sum_{i=1}^{s} f_i(x) y^{p^i} = L(y^p, \ldots, y^{p^s}),$$

where $L$ is linear with coefficients polynomials in $x$. After multiplying by $f^{p-1}$, breaking up $y$ and the coefficients of $L$ as above and then taking $p$–th roots, we get the equations

$$f(x)y_{\alpha_1} = L_{\alpha_1}(y, y^p, \ldots, y^{p^{s-1}}). \tag{4}$$

Now multiplying by $f^{p-1}$ and substituting for $f(x)y$ on the right hand side of (4) yields

$$f^p y_{\alpha_1} = L_{\alpha_1}(f^{p-2}L(y^p, \ldots, y^{p^s}), f^{p-1}y^p, \ldots, f^{p-1}y^{p^{s-1}}),$$

which is linear in $y^p, \ldots, y^{p^s}$. This brings us back, more or less, to the start and shows that iterating the process described leads to equations of the shape

$$f^p y_{\alpha_1 \ldots \alpha_e} = L_{\alpha_1 \ldots \alpha_e}(y, y^p, \ldots, y^{p^{s-1}}).$$

If, during this procedure, we keep track of the (multi–) degree in $x$ of $L_{\alpha_1 \ldots \alpha_e}$ we see that that degree remains bounded. Hence, since $\mathbb{F}_p$ is finite, there are only a finite number of distinct $y_{\alpha_1 \ldots \alpha_e}$ and we have:

**Theorem 1.** ([9], [16]) *If* $y = \sum a_\nu x^\nu$ *is algebraic then*

(F) *there is an* $e$ *such that for every* $(\alpha_1, \ldots, \alpha_e) \in S^e$ *there is an* $e' < e$ *and a* $(\beta_1, \ldots, \beta_{e'}) \in S^{e'}$ *such that*

$$y_{\alpha_1 \ldots \alpha_e} = y_{\beta_1 \ldots \beta_{e'}};$$

*equivalently,*

(A) *there is an* $e$ *such that for all* $j = (j_1, \ldots, j_n)$ *with the* $j_i < p^e$ *there is an* $e' < e$ *and a* $j' = (j'_1, \ldots, j'_n)$ *with the* $j'_i < p^{e'}$ *such that*

$$a_{p^e \nu + j} = a_{p^{e'} \nu + j'} \text{ for all } \nu.$$

Conversely, if $y$ satisfies (F) then taking $y_{\beta_1 \ldots \beta_{e'}}$ and breaking it up $e - e'$ times, we see that the $y_{\alpha_1 \ldots \alpha_e}$ satisfy a system of the form

$$y_{\alpha_1 \ldots \alpha_e} = \sum x^\gamma y_{\beta_1 \ldots \beta_e}^{p^{e-e'}}. \tag{5}$$

Viewing this system as a system of equations in the $y_{\delta_1 \ldots \delta_e}$ we see (because, happily, the derivative of a $p-$th power vanishes) that its Jacobian is 1. By the Lemma below it follows that the $y_{\alpha_1 \ldots \alpha_e}$, and hence $y$, must be algebraic.

**Lemma 1.** ([25], p. 286) *If* $\mathbf{K} \subset \mathbf{L}$ *are fields and* $y_1, \ldots, y_N \in \mathbf{L}$ *satisfy a system of* $N$ *polynomial equations* $F_i(Y_1, \ldots, Y_N) = 0$ *over* $\mathbf{K}$ *with* $J(y_1, \ldots, y_N) \neq 0$, *where* $J = \det(\partial F_i / \partial Y_i)$ *is the Jacobian of the system, then the* $y_i$ *are all algebraic over* $\mathbf{K}$.

To see this, observe that if the conclusion were false there would be a nontrivial derivation $D$ on $\mathbf{K}(y_1, \ldots y_N)$ which is trivial on $\mathbf{K}$. But by applying $D$ to the equations $F_i(y_1, \ldots y_N) = 0$ we obtain

$$\left(\frac{\partial F_i}{\partial Y_j}(y_1, \ldots, y_N)\right)(Dy_i) = 0,$$

which forces $Dy_i = 0$ for all $i$.

If we interpret the equations (5), in the unknowns $y_{\delta_1 \ldots \delta_e}$, over $\mathbb{Z}_p$ instead of over $\mathbb{F}_p$, they have Jacobian $\equiv 1 \bmod p$. Hence, by the Implicit Function Theorem for $\mathbb{Z}_p[[x]]$ (which is just Hensel's Lemma; an elementary proof is given at [16], p. 50) these equations have a unique solution $\tilde{y}_{\alpha_1 \ldots \alpha_e} \in \mathbb{Z}_p[[x]]$ with $\overline{\tilde{y}_{\alpha_1 \ldots \alpha_e}} = y_{\alpha_1 \ldots \alpha_e}$, where the $\overline{\phantom{x}}$ denotes reduction mod $p$. By Lemma 1 the $\tilde{y}_{\alpha_1 \ldots \alpha_e}$ are also algebraic. Thus we see that every algebraic power series in $\mathbb{F}_p[[x]]$ is the reduction of an algebraic power series in $\mathbb{Z}_p[[x]]$. In section 4 below we mention a further, rather extraordinary, "lifting theorem" whereby every algebraic power series in $\mathbb{F}_p[[x]]$ is shown to lift to a series in $\mathbb{Z}[[x]]$ which is a solution of a system of functional equations.

Suppose that the series $\sum a_\nu x^\nu$ is generated by a finite automaton $\mathcal{M}$, as explained in section 1. Choose $e$ so large that every state that $\mathcal{M}$ enters in the course of any computation is entered in a computation of length less than $e$. Then the $a_\nu$ satisfy version (A) of Theorem 1.

Conversely, if the $a_\nu$ satisfy (A) one can construct a finite automaton $\mathcal{M}$ that generates $\sum a_\nu x^\nu$. $\mathcal{M}$ will be equipped with a table detailing the identifications (A) and an output list of the values of the $a_j$ for $j=(j_1,\ldots,j_n)$ with the $j_i < p^e$. $\mathcal{M}$ computes as follows: It reads $e$ digits from each tape. Then it uses the table (A) to replace those $e$ digits by $e'$ digits. It reads a further $e-e'$ digits and iterates. At each stage it outputs the appropriate value from its output list. So we have:

**Theorem 2.** ([9], [16]) $\sum a_\nu x^\nu \in \mathbb{F}_p[[x]]$ *is algebraic if and only if it is generated by a finite automaton.*

This has the immediate:

**Corollary 1.** ([19], [15], [16]) Let $f, g \in \mathbb{F}_p[[x]]$ *be algebraic. Then*

(i) *Every diagonal of $f$ is algebraic, as is every off–diagonal.*

(ii) *The Hadamard product $f * g$ is algebraic.*

(iii) *Each characteristic series* $f^{(i)} = \sum_{a_\nu = i} x^\nu$ *is algebraic.*

We emphasize that the situation in characteristic 0 is very different. Neither diagonals nor (equivalently, see section 1) Hadamard products preserve algebraicity. For the latter, the standard example is

$$(1-4x_1)^{-1/2} = \sum \binom{2h}{h} x_1^h \,, \quad \text{but} \sum \binom{2h}{h}^2 x_1^h$$

is not algebraic. We note that the first remark is just the useful identity

$$\binom{2h}{h} = (-1)^h \binom{-\frac{1}{2}}{h}.$$

Congenial facts such as this are of interest in constructions useful to logicians; see, for example [23]. Then, with a little work and some first–year introductory calculus[1] the latter series is given by the integral

$$\frac{2}{\pi}\int_0^{\pi/2} \frac{dt}{\sqrt{(1-16x_1\sin^2 t)}}.$$

This is a complete elliptic integral well known not to represent an algebraic function.

[1] Nowadays, probably at best second–year, if one is lucky.

## 3. Algebraic Power Series mod $p^s$

**Theorem 3.** *If $f \in \mathbb{Z}_p[[x]]$ is algebraic and $I$ is any diagonal then $If$ is algebraic mod $p^s$ for all $s \in \mathbb{Z}$.*

**Proof.** It is enough to show the Theorem for $I = I_{12}$. As above, $\bar{\ }$ denotes reduction mod $p$. Then $\bar{f} \in \mathbb{F}_p[[x]]$ is algebraic and hence so is $I_{12}\bar{f}$. As shown previously there is an algebraic $g \in \mathbb{Z}_p[[x_1, x_3, \ldots, x_n]]$ such that $\bar{g} = I_{12}\bar{f}$. Similarly, there are algebraic $h_1, h_2 \in \mathbb{Z}_p[[x]]$ such that $\bar{h}_1 = J_{12}\bar{f}$ and $\bar{h}_2 = J_{21}\bar{f}$, whilst $I_{12}h_1 = I_{21}h_2 = 0$. (Before lifting to characteristic 0 write $J_{12}\bar{f} = x_1 k(x_1, x_1x_2, x_3, \ldots, x_n)$ and lift $k(x_1, z, x_3, \ldots, x_n)$.) Set $\tilde{g} = g(x_1x_2, x_3, \ldots, x_n)$ and note that $I_{12}\tilde{g} = g$. Now, by induction on $s$, $I_{12}\frac{1}{p}(f - \tilde{g} - h_1 - h_2)$ is algebraic mod $p^{s-1}$.

**Corollary.** *If $f, g \in \mathbb{Z}_p[[x]]$ are algebraic then their Hadamard product $f * g$ is algebraic mod $p^s$, for all $s$.*

Next we generalize Theorem 1:

**Theorem 4.** ([15], [16]) *If $f(x) = \sum a_\nu x^\nu \in \mathbb{Z}_p[[x]]$ is algebraic then for every $s$ there is an $e$ such that for every $j = (j_1, \ldots, j_n)$ with the $j_i < p^e$ there is an $e' < e$ and a $j' = (j'_1, \ldots, j'_n)$ with the $j'_i < p^{e'}$ such that*

$$a_{\nu p^e + j} \equiv a_{\nu p^{e'} + j'} \mod p^s \text{ for all } \nu$$

(and hence the $a_\nu$ mod $p^s$ can be generated by a finite automaton).

**Proof.** For $i \in \{1, \ldots, p-1\}$ each characteristic series $\bar{f}^{(i)}$ is algebraic. Lift $\bar{f}^{(i)}$ to $f^{(i)} \in \mathbb{Z}_p[[x]]$ with $f^{(i)}$ algebraic and $\overline{f^{(i)}} = \bar{f}^{(i)}$. Then $g_i = *^{p^s} f^{(i)}$, the $p^s$-fold Hadamard product of $f^{(i)}$ is algebraic mod $p^s$. Of course each coefficient $a_{\nu i}$ of $g_i = \sum a_{\nu i} x^\nu$ congruent to either 0 or 1 mod $p^s$. Hence $g = \frac{1}{p}(f - \sum i g_i)$ is algebraic mod $p^{s-1}$ and the result follows by induction.

## 4. Automata, Functional Equations and Arithmetic

We return to the example 2–automatic sequence $(r_h)$ of section 1 to notice the pattern:

```
0001001000 0111010001 0010111000 1000010010 0001110111....
0 0 0 1 0  0 1 0 0 0  0 1 1 1 0  1 0 0 0 1  0 0 1 0 1 ....
 0 1 0 0 0  1 1 1 0 1  0 0 1 0 0  0 0 1 0 0  0 1 1 1 1....
```

The second row is $(r_{2h})$. Remarkably, it coincides with the original sequence, illustrating the $r_h = r_{2h}$. The third row is $(r_{2h+1})$. With careful attention, we see that $r_{4h} = r_{4h+1}$ but $r_{4h+2} \neq r_{4h+3}$. Setting $P(X) = \sum (-1)^{r_h} X^h$, these observations amount to the functional equation

$$P(X) = P(X^2) + XP(-X^2). \tag{6}$$

Noticing that for an arbitrary sequence $(i_h)$, with $i_h \in \{0,1\}$, one has

$$2\sum i_h X^h = (1 - X)^{-1} - \sum (-1)^{i_h} X^h,$$

we find that the function $R(X) = \sum r_h X^h$ satisfies

$$2X(1 - X^4)R(X^4) + (1 - X^4)(1 - X)R(X^2) - (1 - X^4)R(X) + X^3 = 0. \tag{7}$$

To discover this functional equation directly we recall the generated sequence

```
0102013101 0232020102 0131323101 3101020131 0102320232...
```

and the defining substitutions

$$0 \mapsto 01, \qquad 1 \mapsto 02, \qquad 2 \mapsto 31, \qquad 3 \mapsto 32.$$

It is quite as convenient to deal with the general case (as in [11]): Accordingly, let $a_0, a_1, \ldots, a_{m-1}$ be a given alphabet and suppose we are given a substitution

$$a_0 \mapsto w_0, \qquad a_1 \mapsto w_1, \quad \ldots, \quad a_{m-1} \mapsto w_{m-1}$$

with words $w_i$, in the $a_j$, each of length $t \geq 2$. Denote by $s_0 s_1 s_2 \ldots$ a sequence fixed by the substitution and consider its generating function $\sum s_h X^h$. Denote the characteristic function of each symbol $a_i$ by $f_i(X) = \sum_{s_h = a_i} X^h = \sum_{h \geq 0} u_{ih} X^h$, so that $\sum_{h \geq 0} s_h X^h = \sum_{i=0}^{m-1} a_i f_i(X)$. Note that $s_{th} s_{th+1} \ldots$ $\ldots s_{t(h+1)-1}$ depends only on $s_h$. Accordingly, set $v_{ijk} = 1$ if $a_i$ is the $(k+1)$–st symbol of the word $w_j$ (and 0 otherwise), so that $u_{i,th+k} = \sum_{j=0}^{m-1} v_{ijk} u_{jh}$. In other words,

$$f_i(X) = \sum_{s=0}^{\infty} u_{is} X^s = \sum_{h=0}^{\infty} \sum_{k=0}^{t-1} u_{i,th+k} X^{th+k}$$
$$= \sum_{j=0}^{m-1} \left( \sum_{k=0}^{t-1} v_{ijk} X^k \right) \sum_{h=0}^{\infty} u_{jh} X^{th} = \sum_{j=0}^{m-1} p_{ij}(X) f_j(X^t).$$

If we denote by $A(X)$ the $m \times m$ matrix $A(X) = (p_{ij}(X))$ and by $f(X)$ the column vector $f(X) = (f_0(X), f_1(X), \ldots, f_{m-1}(X))'$, then we have the matrix functional equation

$$f(X) = A(X) f(X^t).$$

Moreover, it is plain that every linear combination of the $f_i(X)$ over the field of rational functions satisfies an equation of the shape $\sum_{i=0}^{m} c_i(X) g(X^{t^i}) = 0$, with polynomial coefficients $c_i(X)$.

Special cases of such functional equations were studied by Kurt Mahler in the late twenties; see [28]. It is therefore fitting to refer to these systems of equations as Mahler systems and to their solutions as Mahler functions.

Let $f(x_1) \in \mathbb{Q}[[x_1]]$ be algebraic. Because of Eisenstein's Theorem $f$ has a reduction $\bar{f}$ mod $p$ for almost all primes $p$, and that reduction is, of course, an algebraic element of $\mathbb{F}_p[[x_1]]$. As shown in section 2 such an algebraic power series lifts to an algebraic power series in $\mathbb{Z}_p[[x_1]]$ and its reductions mod $p^s$ have coefficients generated by an automaton which reads its input in base $p$. Treating its coefficients, which are of the shape $a_h(0) + a_h(1)p + \ldots + a_h(s-1)p^{s-1}$, as elements of $\mathbb{Z}$ the new series satisfies a Mahler $p-$functional equation by the argument above. The theorem of Pólya–Carlson (the most convenient reference is [33], part VIII, Chap. 3, §5) tells us that the new series is either rational, in which case its sequence of coefficients is (eventually) periodic, or it represents a transcendental function with the unit circle as natural boundary. (In the light of the last paragraph of section 1 we also refer the reader to [3], Chap. 5.)

Thus the reductions of algebraic power series are either rational or, when viewed in characteristic zero, are transcendental functions satisfying Mahler functional equations.

Our example $R(X) = \sum r_h X^h$, when viewed as an element of $\mathbb{F}_2[[X]]$, satisfies the algebraic equation

$$(1+X)^5 R^2 + (1+X)^4 R + X^3 = 0.$$

When viewed as an element of $\mathbb{C}[[X]]$ the series is a transcendental function and the algebraic equation lifts to the Mahler 2–functional equation (7)

$$2X(1-X^4)R(X^4) + (1-X^4)(1-X)R(X^2) - (1-X^4)R(X) + X^3 = 0.$$

In much this spirit, there is a theorem of Cobham:

**Theorem 5.** ([12]) *If $f(x_1) = \sum_{i \in S} x_1^i$ is generated by both an $s$– and a $t$–automaton and $s$ and $t$ are multiplicatively independent integers (equivalently,* $\log s / \log t$ *is irrational), then the set $S$ is a finite union of arithmetic progressions (equivalently, the given function is rational).*

Recall that by "irrelevance of symbols" there is no loss of generality in supposing given algebraic power series over finite fields to have just coefficients 0 and 1. We have that, if $\gcd(p,q) = 1$, a power series algebraic as an element of both $\mathbb{F}_p[[x]]$ and $\mathbb{F}_q[[x]]$, is rational. Incorporating the previous remarks we see that a power series is either rational, or is algebraic in at most one characteristic and transcendental in all others.

The argument in [12] seems easy locally, but is difficult globally. It remains of interest to find a more direct proof (*cf.* [38]).

The definition of finite $b$–automaton entails that a sequence of integers generated by the automaton exhibits some 'digit pattern' in the base $b$ representation of the integers. For example, the sequence $(r_h)$ counts the number of occurrences (mod 2) of the pair 11 in the binary expansion of h:

| 0 | 1 | 10 | 11 | 100 | 101 | 110 | 111 | 1000 | 1001 | 1010 | 1011 | 1100 |
|---|---|---|---|---|---|---|---|---|---|---|---|---|
| 0 | 0 | 0 | 1 | 0 | 0 | 1 | 0 | 0 | 0 | 0 | 1 | 1 |

| 1101 | 1110 | 1111 | 10000 | 10001 | 10010 | 10011... |
|---|---|---|---|---|---|---|
| 1 | 0 | 1 | 0 | 0 | 0 | 1 ... |

In these terms, Cobham's Theorem states that a sequence of integers has 'digit pattern' in multiplicatively independent bases if and only if it is (eventually) periodic.

In characteristic zero it makes sense to speak of values taken by a series. In an attempt to decrease the awe of the transcendental we sketch the proof of the theorem:

**Theorem 6.** ([29]) *Let $t \geq 2$ be an integer. If $s_0 s_1 s_2 \ldots$ is a t–automatic sequence and $f(X) = \sum s_h X^h$ is a transcendental Mahler t–function then, for integers $b \geq 2$, the numbers $f(b^{-1})$ are transcendental.*

To understand transcendence theory one need only know the fundamental lemma of the subject:

$$\textit{If } n \in \mathbb{Z} \textit{ and } |n| < 1 \textit{ then } n = 0.$$

Notice, firstly, that to show that a number $\alpha$ is transcendental amounts to demonstrating that no nontrivial expression $a_0 + a_1\alpha + \ldots + a_m\alpha^{m-1}$, with the $a_i$ in $\mathbb{Z}$, can vanish. Thus one seeks to establish the linear independence of numbers $1, \alpha, \ldots, \alpha^{m-1}$ over $\mathbb{Z}$ for all $m$. Similarly, to prove that numbers $\alpha_1, \ldots, \alpha_n$ are algebraically independent one considers all monomials $\alpha^{\mu} = \alpha_1^{\mu_1} \ldots \alpha_n^{\mu_n}$, and proves the linear independence over $\mathbb{Z}$ of arbitrary finite collections of the $\alpha^{\mu}$.

We have a single transcendental function $f(z)$ satisfying a functional equation of the shape

$$c_n(x) f(z^{t^n}) + \ldots + c_1(z) f(z^t) + c_0(z) f(z) = c(z),$$

with $c(z)$ and the $c_i(z) \in \mathbb{Z}[z]$ and wish to show that $\alpha = f(b^{-1})$ is transcendental. Set $f_i(z) = f(z^{t^i})$. We take a finite set of monomials $f^{\mu} = f_0^{\mu_0} \ldots f_n^{\mu_n}$, linearly independent and stable under the transformation $z \mapsto z^t$ over $\mathbb{Q}(z)$. For the present application the column vector $g$, say, of these monomials may be supposed to satisfy a matrix functional equation of the shape $g(z) = A(z) g(z^t)$, with an $m \times m$ matrix $A$ of polynomials with integer coefficients.

We apply this as follows: Suppose (contrary to what we want to show) that there is, over $\mathbb{Z}$, a nontrivial linear relation

$$a_1 g_1(b^{-1}) + \ldots + a_m g_m(b^{-1}) := a \cdot g(b^{-1}) = 0.$$

Then, by the iterated form $g(z) = A^{(k)}(z)g(z^{t^k})$ of the functional equation, we have for $k = 0, 1, \ldots$

$$0 = a \cdot A^{(k)}(b^{-1})g(b^{-t^k}) := a(k) \cdot g(b^{-t^k}).$$

This allows us to study the functions $g_i$ at points conveniently close to the origin.

To prove the transcendence result, we construct a nonzero integer which, given the relation just mentioned, has absolute value less than 1. Our construction requires some notions from the theory of simultaneous polynomial approximation of functions that derive from the work of Mahler ([30]).

In the sequel $\rho = (\rho_1, \ldots, \rho_m)$ denotes an $m$-tuple of integers with sum $\sigma$. By ord we denote the order of vanishing at $z = 0$. We say that the vector of polynomials $(p_1, \ldots, p_m)$ with $\deg p_j \le \rho_j - 1$, approximates the vector $g$ at $\rho$ if

$$\operatorname{ord}\Big(r(z) = \sum p_j(z) g_j(z)\Big) \ge \sigma - 1.$$

If no approximations at $\rho$ have $\operatorname{ord} r(z) > \sigma - 1$ then the vector $g$ is said to be *normal at* $\rho$. We claim that if $g$ is normal at $\rho$ then there exists a (unique) $m \times m$ matrix $P_\rho = (p_{ij})$ of polynomials with the following properties:

(i) The $p_{ii}$ are monic and $\deg p_{ii} = \rho_i$; whilst, off the diagonal, $\deg p_{ij} \le \rho_i - 1$.

(ii) $\operatorname{ord}(r_i(z) = \sum p_{ij}(z) g_j(z)) \ge \sigma$.

(iii) $\det P_\rho = z^\sigma$.

This is not hard to see. The polynomials $p_{ii}$ must have exact degree $\rho_i$, since otherwise we contradict normality at $\rho$. Plainly $\det P_\rho = z^\sigma +$ lower order terms, whilst, by multiplying by a $g_i$, and applying (ii), we see that $\operatorname{ord}(\det P_\rho) \ge \sigma$.

Thus $\det P_\rho$ vanishes only at $z = 0$. Set $B = b^{t^k}$. There is then no loss of generality in supposing that the vector $a(k)$ and the last $m - 1$ rows of $P_\rho(B^{-1})$ are linearly independent over $\mathbb{C}$. Now consider the $m \times m$ determinant $\Delta$ whose rows are these $m$ vectors, noting that $\Delta$ is a nonzero rational number. There is some integer $Q_\rho$, depending only on $\rho$ and not on $k$, so that all the $Q_\rho p_{ij}$ have integer coefficients and there is a constant $C > 0$, independent of $\rho$ and $k$, so that $B^C a(k)$ is an integer vector. Set $\rho_{\min} = \min \rho_i$. Then, on expanding the determinant $\Delta$ by its first row, we see that

$$B^C Q_\rho^{m-1} B^{\sigma-\rho_{\min}} \Delta \tag{8}$$

is an integer.

Now multiply $\Delta$ by $g_1(B^{-1})$, say. After simple manipulation, $g_1(B^{-1})\Delta$ assumes the shape

$$\begin{vmatrix} 0 & a_2(k) & \cdots & a_m(k) \\ r_2(B^{-1}) & p_{22}(B^{-1}) & \cdots & p_{2m}(B^{-1}) \\ \vdots & \vdots & \ddots & \vdots \\ r_m(B^{-1}) & p_{m2}(B^{-1}) & \cdots & p_{mm}(B^{-1}) \end{vmatrix}.$$

Expanding by the first column shows that $\Delta$ is surprisingly small; only the factor $B^{-\sigma}$ depends jointly on the parameters $k$ and $\rho$.

By Theorem 1 of [29] we may suppose that the vector $g$ is normal at a sequence of parameter points $\rho$ with $\rho_{\min} \to \infty$. Thence, choose $\rho$ so that $\rho_{\min}$ is much larger than $C$ and subsequently select $k$ so that $B$ dominates the remaining quantities (which depend only on $\rho$ and not on $k$). It follows that the nonzero integer in (8) has absolute value less than 1. This contradiction proves the transcendence result.

We have shown that if $s_0 s_1 s_2 \ldots$ is a sequence generated by a finite automaton then a number $\sum s_i b^{-i} = s_0 . s_1 s_2 \ldots$ (with the "decimal"point indicating expansion in base $b$) is either rational or transcendental. In different words: *The digits of an irrational algebraic number cannot be generated by a finite automaton.*

There is transcendence theory in characteristic $p > 0$. For example ([31]), consider, over $\mathbb{F}_p$, the formal power series

$$\sum_{h \geq 0} \binom{\lambda}{h} X^h = (1+X)^\lambda = (1+X)^{\sum_{i\geq 0} \lambda_i p^i} = \prod_{i \geq 0} (1+X^{p^i})^{\lambda_i} = \sum_{h \geq 0} \prod_{i \geq 0} \binom{\lambda_i}{h_i} X^h;$$

with $\lambda = \sum \lambda_i p^i \in \mathbb{Z}_p$ and $h = \sum h_i p^i$ expanded in base $p$. It is natural to guess that $(1+X)^\lambda$ yields a formal power series algebraic over $\mathbb{F}_p(X)$ if and only if

$\lambda$ is rational. Indeed, if the series is algebraic then the sequence $\left(\binom{\lambda}{h}\right)$ is $p$-automatic, and that entails that the sequence $\left(\binom{\lambda}{p^h}\right)$ is periodic. But, as we saw above, $\left(\binom{\lambda}{p^h}\right) = \lambda_h$, so the sequence $(\lambda_h)$ is periodic, which is to say that $\lambda$ is rational as we had predicted. More generally, similar, though more sophisticated arguments [2] show that if $f \in \mathbb{F}_p[[X]]$, and is algebraic with constant coefficient 1, then the formal power series $f^{\lambda_1},\dots,f^{\lambda_s}$ are algebraically independent over $\mathbb{F}_p(X)$ if and only if the $p$-adic integers $1,\lambda_1,\dots,\lambda_s$ are linearly independent over $\mathbb{Z}$.

The classical theory of diophantine approximation in fields of positive characteristic is described by Geijsel [20].

The growing literature on the subject of finite automata is readily reached by iteration of references. Our example automatic sequence $(r_h)$, the Rudin–Shapiro sequence, is a hero of the story "FOLDS!" [14]. A more recent survey is that of Allouche [1].

## 5. Diagonals of Rational Functions

In this section we consider the class of power series in $\mathbb{Q}[[x]]$ which occur as the diagonals of rational functions $f(x) = P(x)/Q(x)$ with $P, Q$ polynomials and $Q(0,\dots,0) \neq 0$.

Many combinatorial generating functions are diagonals of rational functions (*cf.* [34]). Moreover, functions of number–theoretic interest arise in this way, as, for example, with Apéry's function ([37])

$$A(x_1) = \sum_{h=0}^{\infty}\sum_{k=0}^{\infty}\binom{h}{k}^2\binom{h+k}{k}^2 x_1^h,$$

which occurs in the proof of the irrationality of $\zeta(3)$, and which is the diagonal

$$A(x_1) = I\,\{(1-x_1)[(1-x_2)(1-x_3)(1-x_4)(1-x_5) - x_1x_2x_3]\}^{-1}.$$

In section 1 we mentioned that every algebraic power series is the diagonal of a rational power series, whilst we saw in section 3 that every such diagonal is algebraic mod $p^s$ for all $s$ and almost all $p$ (since if $f(x)$ is a rational power series over $\mathbb{Q}$ then $f(x) \in \mathbb{Z}_p[[x]]$ for almost all $p$).

The complete diagonals of rational power series have many other interesting properties:

**Theorem 7.** (*cf.* [8]) *If* $f(x_1)$ *is the diagonal of a rational power series over* $\mathbb{Q}$ *then*

(i) $f$ *has positive radius of convergence* $r_p$ *at every place* $p$ *of* $\mathbb{Q}$ *and* $r_p = 1$ *for almost all* $p$.

(ii) *for almost all* $p$ *the function* $f$ *is bounded on the disc* $D_p(1^-) = (t \in \mathbb{C}_p : |t| < 1\}$, *where* $\mathbb{C}_p$ *is the completion of the algebraic closure of the p–adic rationals* $\mathbb{Q}_p$ *and, for all* $p$, $\sup\{|f(t)| : t \in D_p(1^-)\} = 1$.

(iii) $f$ *satisfies a linear differential equation over* $\mathbb{Q}[x_1]$; *and*

(iv) *this equation is a Picard–Fuchs equation.*

Whilst (i) and (ii) are reasonably straightforward, (iii) and (iv) are more difficult and are proved by Deligne using resolution of singularities (see [6], the footnote on p. 5). Dwork [18] has given a proof which avoids resolution. An elementary proof of (iii) and its generalizations appears in [26].

Picard–Fuchs equations are equations that "come from geometry"—they are satisfied by those analytic functions (of the parameter) which are the periods of differential forms along cycles in pencils of curves. (A number of concrete examples are worked out in [36].) The singular points of these equations are regular with rational exponents.

One readily notices that $f(x_1) = \sum a_i x_1^i$ satisfying a linear equation with coefficients from $\mathbb{Q}[x_1]$ is equivalent to the Taylor coefficients $a_i$ satisfying a linear recurrence relation

$$p_0(h)a_{h+k} = p_1(h)a_{h+k-1} + \ldots + p_k(h)a_h,$$

where the $p_j$ are polynomials; see [35], and [27] for generalizations to several variables. Algebraic functions over $\mathbb{Q}$, of course, satisfy differential equations of the sort just mentioned. The linear recurrence can be useful in computing their Taylor coefficients (*cf.* [10]).

There is a great deal of folklore on the subject matter of Theorem 7. There are conjectures of Bombieri and Dwork in [17] to the effect that if the solutions of a linear differential equation, over $\mathbb{Q}[x_1]$, have a "large" radii of convergence in $\mathbb{C}_p$ for almost all $p$, then they are functions that "come from geometry". This has been verified for the Apéry function ([17]). Christol [8] conjectures that if $f(x_1)$ satisfies the first three properties cited in Theorem 7 then $f$ is the diagonal of a rational function; a special case is proved in [8].

The results of section 3 constitute a vast range of congruences for the Taylor coefficients of diagonals of rational functions. Such, and other, congruences have been obtained for the coefficients of the Apéry function (for example [21], by elementary arguments, and [4], [36], [5]). It would be interesting to see if these congruences can be obtained by the methods of section 3, which are, of course, effective.

We conclude by mentioning Grothendieck's Conjecture: *If a linear homogeneous differential equation with coefficients from* $\mathbb{Q}[x_1]$, *and of order* $n$, *has, for almost all* $p$, $n$ *independent solutions in* $\mathbb{F}_p[[x_1]]$ *then all its solutions are algebraic.* This has been proved in a number of special cases (for example, for Picard–Fuchs equations) by Katz (see [24]). Some results have also been obtained by elementary methods ([22]).

## References


[1] *J–P Allouche,* Automates finis en théorie des nombres. Expositiones Math., **5** (1987), 239—266.

[2] *J–P Allouche, M.Mendès France* and *A.J. van der Poorten,* Indépendance algébrique de certaines séries formelles. Bull. Soc. Math. France **116** (1988).

[3] *Y. Amice,* Les nombres $p$–adique. Presses Universitaires de France (1975).

[4] *F. Beukers,* Some congruences for the Apéry numbers. J. Number Theory **21** (1985), 144—155.

[5] *F. Beukers,* Another congruence for the Apéry numbers. J. Number Theory **25** (1987), 201—210.

[6] *G. Christol,* Diagonales de fractions rationelles et équations différentielles. Groupe d'étude d'analyse ultramétrique (Amice, Christol, Robba), Inst. Henri Poincaré, Paris, 1984/85 n° 18.

[7] *G. Christol,* Diagonales de fractions et équations de Picard–Fuchs. Groupe d'étude d'analyse ultramétrique (Amice, Christol, Robba), Inst. Henri Poincaré, Paris. 1984/85 n° 13.

[8] *G. Christol,* Fonctions hypergéométriques borneés. Groupe d'étude d'analyse ultramétrique (Amice, Christol, Robba), Inst. Henri Poincaré, Paris, 1986/87 n° 10.

[9] *G. Christol, T. Kamae, M. Mendès France and G. Rauzy,* Suites algébriques, automates et substitutions. Bull. Soc. Math. France, **108** (1980), 401—419.

[10] *D.V. Chudnovsky* and *G.V. Chudnovsky,* On expansion of algebraic functions in power and Puiseux series. IBM Research Report **RC 11365**, IBM Research Centre, Yorktown Heights, New York, (1985).

[11] *A. Cobham,* On the Hartmanis–Stearns problem for a class of tag machines. Technical report **RC 2178**, IBM Research Centre, Yorktown Heights, New York, 1968.

[12] *A. Cobham,* On the base dependence of sets of numbers recognizable by finite automata. Math. Systems Theory, **3** (1969), 186—192.

[13] *A. Cobham,* Uniform tag sequences. Math. Systems Theory **6** (1972), 164—192.

[14] *M. Dekking, M. Mendès France* and *A. van der Poorten,* FOLDS! The Mathematical Intelligencer, **4** (1982), 130—138; II: Symmetry disturbed, *ibid.* 173—181; III: More morphisms, *ibid.* 190—195.

[15] *P. Deligne,* Intégration sur un cycle évanescent, Invent Math. **76** (1984), 129—143.

[16] *J. Denef* and *L. Lipshitz,* Algebraic power series and diagonals. J. Number Theory **26** (1987), 46—67.

[17] *B. Dwork,* On Apéry's differential operator. Groupe d'étude d'analyse ultramétrique (Amice, Christol, Robba), Inst. Henri Poincaré, Paris, 1979/81 $n^0$ 25.

[18] *B. Dwork,* Differential systems associated with families of singular hypersurfaces. Preprint.

[19] *H. Furstenberg,* Algebraic functions over finite fields. J. Alg. **7** (1967), 271—277.

[20] *J.M. Geijsel,* Transcendence in fields of positive characteristic. PhD Thesis, Amsterdam (1978) = Mathematical Centrum Tracts (Amsterdam) **91** (1979).

[21] *I. Gessel,* Some congruences for the Apéry numbers. J. Number Theory **14** (1982), 362—368.

[22] *T. Honda,* Algebraic differential equations. Symposia Mathematica **XXIV** Academic Press (1981), 169—204.

[23] *J.P. Jones* and *Yu. V. Matijasevic,* Exponential diophantine representation of recursively enumerable sets. Proc. Herbrand Symp. (Logic Colloquium '81, European Meeting of the Association of Symbolic Logic, July 16–24, 1981, Marseille, France), Studies in Logic **107**, North–Holland Publishers, Amsterdam (1982), 159—177.

[24] *N. Katz,* Algebraic solutions of differential equations. Invent. Math. **18** (1972), 1—118.

[25] *S. Lang,* Algebra. Addison–Wesley (1965).

[26] *L. Lipshitz,* The diagonal of a $D$-finite power series is $D$-finite. J. Alg. **113** (1988), 373—378.

[27] *L. Lipshitz,* $D$-finite power series, J. Alg. (to appear).

[28] *J.H. Loxton* and *A.J. van der Poorten,* Transcendence and algebraic independence by a method of Mahler. In: Transcendence Theory—Advances and Applications. *A. Baker* and *D.W. Masser,* eds. (Academic Press, London and New York, 1977), Chapter 15, 211—226.

[29] *J.H. Loxton* and *A.J. van der Poorten,* Arithmetic properties of automata: regular sequences. J. für Math. (to appear).

[30] *K. Mahler,* Perfect systems. Compositio Math. **19** (1968), 95—166.

[31] *M. Mendès France* and *A.J. van der Poorten,* Automata and the arithmetic of formal power series. Acta Arith. **46** (1986), 211—214.

[32] *M. L. Minsky,* Computation: Finite and Infinite Machines. Prentice–Hall (1967).

[33] *G. Polya* and *G. Szegö,* Problems and Theorems in Analysis. Springer–Verlag (translation of 4th edition, 1976).

[34] *R.P. Stanley,* Generating functions. In *Gian–Carlo Rota,* ed., Studies in Combinatorics, MAA Studies in Mathematics, **17** (1978), 100—141.

[35] *R.P. Stanley,* Differentiably finite power series. European J. Comb. **1** (1980), 175—188.

[36] *J. Stienstra* and *F. Beukers,* On the Picard–Fuchs equation and the formal Brauer group of certain elliptic $K$3–surfaces. Math. Ann. **271** (1985), 269—304.

[37] *A.J. van der Poorten,* A proof that Euler missed...Apéry's proof of the irrationality of $\zeta(3)$; An informal report. The Mathematical Intelligencer, **1** (1979), 195—203.

[38] *A.J. van der Poorten,* Remarks on automata, functional equations and transcendence. Sém. théorie des nombres de Bordeaux, année 1986–1987, exp. n$^{o}$ 27.

---

Department of Mathematics, Purdue University , West Lafayette, IN 47907, USA

School of Mathematics, Physics, Computing and Electronics, Macquarie University, NSW 2109, Australia

# Discriminants and Permutation Groups[1]

*J. Martinet*

Let $K$ be a field. One can attach to any finite separable extension $E/K$ a pair $(G,\Omega)$, where $\Omega$ is a finite set with $|\Omega| = [E:K]$, acted on (on the left) by a finite group $G$: One takes for $G$ the Galois group of a Galois closure $F/K$ of $E/K$, defined up to isomorphism by $E/K$, and for $\Omega$ the set of conjugates in an algebraic closure of $K$ of a primitive element of $E/K$.

Given $v: K \to \mathbb{C}$ (or $v: K \to \overline{\mathbb{Q}}_p$), the conjugacy class $c_w$ of the Frobenius of some $w|v$ in $F$ is well defined.

**Problem.** Given $(G,\Omega)$, to construct, for $K = \mathbb{Q}$, an $E$ which corresponds to $(G,\Omega)$, with a prescribed infinite Frobenius and with minimal discriminant.

[The condition on the Frobenius is a more restrictive one than the mere signature of $E/\mathbb{Q}$, see below.]

The answer to this problem is not known in general: The inverse problem of Galois theory is a special case of our problem! We believe however that the construction of $E/K$ is always possible. Note that this is not true for a finite Frobenius: Counter–examples are provided by the "Grunwald–Wang phenomenon", cf. [2], ch. 10. (In all known counter–examples, $v$ is a place above 2 and the Frobenius is of even order.)

We intend to discuss methods and results on the above problem. In §1 (group actions) and §2 (signature and higher invariants), we consider

[1] Laboratoire Associé au C.N.R.S. no. 040226

arbitrary finite separable extensions. We then restrict ourselves to number fields, and discuss briefly geometric methods (§3) and class field thoery (§4). We then describe results, for degree at most 4 (§5), for extensions of prime degree (§6) and finally for sixth and eighth degree extensions (§7,§8). An appendix gives tables of 1 and 2 dimensional invariants and the number of conjugacy classes of a given signature for all transitive and faithful groups of degree up to seven.

Soicher ([26]) gives an example of an extension of $\mathbb{Q}$ for each permutation group of degree up to seven, but his study does not involve a search for minimal discriminants, and the signature is not considered.

This paper corresponds to talks delivered in 1988 in Bordeaux (May 13th), in Oberwolfach (May 25th) and at the Bowdoin College Conference (Brunswick, Maine, U.S.A., July 13th).

## 1. Group Actions

Recall that $\Omega$ is a non–empty finite set with $n$ elements. The action of $G$ on $\Omega$ satisfies the rules.

(i) $1_G \cdot x = x$;

(ii) $s\cdot(t\cdot x) = (st)\cdot x$ ($\Omega$ is a $G$–set). Hence, there is a canonical homomorphism $\psi : G \to S(\Omega)$ (the symmetric group of $\Omega$) such that, for all $s \in G$, $\psi(s) = (x \mapsto s\cdot x)$; conversely, such a homomorphism defines a group action on $\Omega$.

### 1a. Isomorphisms

An *isomorphism* between a $G$–set $\Omega$ and a $G'$–set $\Omega'$ is a pair $(\varphi, f)$ where $\varphi : G \to G'$ is an isomorphism and $f : \Omega \to \Omega'$ is one–to–one, such that one has $\varphi(s)\cdot f(x) = f(s\cdot x)$ for all $s \in G$ and $x \in \Omega$. If $(\varphi, f)$ is an isomorphism, so is $(t'\varphi t'^{-1}, t'f)$ for any $t' \in G'$. When $G' = G$, $\varphi \in \mathrm{Aut}(G)$, and, if $\varphi \in \mathrm{Int}(G)$, one can replace $(\varphi, f)$ by another pair $(\varphi', f')$ with $\varphi' = Id$. This leads to the following definition:

Two permutations of the same group $G$ on two sets $\Omega$ and $\Omega'$ are *weakly equivalent* if they are isomorphic, and are *strongly equivalent* if one can take $\varphi = Id$.

We say that $G$ acts *transitively* (on the set $\Omega$) if there is only one orbit of $G$ in $\Omega$, that $G$ acts *faithfully* if $1_G$ is the only element which acts trivially.

The classification of $G$–sets for a given group $G$ reduces immediately to the case when $G$ is both transitive and faithful. (We simply say that "$G$ is of degree $n$".)

Let $G_x = \{s \in G \mid s \cdot x = x\}$. Then, if $G$ is transitive, the $G$–set $\Omega$ is strongly equivalent to $G/G_x$, where $G$ acts on a quotient space $G/H$ by $s\cdot(tH) = (st)H$. Then, $G$ is faithful on $G/H$ if and only if $\bigcap_{t\in G} tHt^{-} = \{1\}$. Moreover, $G/H$ and $G/H'$ are weakly (resp. strongly equivalent) if and only if $H'$ is the image of $H$ under an automorphism of $G$ (resp. $H'$ and $H$ are conjugate).

## 1b. Étale Algebras

Let $K$ be a field, let $K_s$ be a separable closure of $K$ and let $G_K = Gal(K_s/K)$. Let $E = \prod_{i=1}^{r} E_i$ be an étale $K$–algebra (i.e., the $E_i$'s are separable extensions of $K$). There is a one–to–one correspondence

| étale algebras (up to isomorphism) | $\longmapsto$ | $G_K$–sets |
|---|---|---|
| given by $E$ | $\longmapsto$ | $\mathrm{Hom}_K(E, K_s)$. |

Here is a strongly equivalent definition: If $E_i \simeq K[t]/(f_i)$ (we then say that $E$ is defined by the separable polynomial $f = \prod f_i$, even if the $f_i$'s are not all distinct), we can take for $\Omega$ the disjoint union of the sets $\Omega_i$'s of roots of $f_i$ in $K_s$.

Clearly, "$G_K$ is transitive" $\Longleftrightarrow$ "$E$ is a field".

Note that $G_K$ acts through a finite faithful quotient, namely the Galois group of the Galois closure in $K_s$ of the field generated by the embeddings in $K_s$ of the $E_i$'s.

### 1c. Primitive Groups

(i) We say that a non-empty subset $\Omega'$ of the $G$-set $\Omega$ is an *imprimitivity set* if, for every $s \in G$, one has either $s\Omega' = \Omega'$ or $\Omega \cap \Omega' = \emptyset$.

(ii) We say that $\Omega$ (or $G$) is *primitive* if $G$ is transitive on $\Omega$, and if the only imprimitivity sets are $\Omega$ itself and the subsets with one element.

(iii) We say that $G$ is *k-transitive* if $G$ is transitive on the ordered sets of $k$ elements of $G$.

**Remark**

(i) "2-transitive" $\Longrightarrow$ "primitive".

(ii) With the notation of 1b, $G_K$ is primitive if and only if $E$ is a field and $E/K$ does not contain any trivial subextension.

### 1d. Characters

The permutation representation defined by the $G$-set $\Omega$ is $R_\Omega = \oplus\, \mathbb{C} e_x$, where $G$ acts by $s \cdot e_x = e_{sx}$ ; the corresponding *augmentation representation* is $U_\Omega = R_\Omega / D$, where $D$ is the line $\mathbb{C}(\sum e_x)$.

The corresponding characters are denoted by $r_\Omega = r$ and $u_\Omega = u$ respectively; one has $r = u + 1$.

**Proposition.**

*(i) $G$ is transitive if and only if $\langle u,1 \rangle = 0$ (orthogonality).*

*(ii) $G$ is 2-transitive if and only if $u$ is irreducible.*

This proposition is easily proved by calculating the values of $r$, namely $r(s) = |\{x \in \Omega \mid s \cdot x = x\}|$ for all $s$ in $G$. Similarly, one shows:

**Proposition.**

*For any imprimitivity subset* $\Omega'$ *of* $\Omega$, *one has* $r_\Omega = Ind_H^G(r_{\Omega'})$, *where* $H = \{s \in G \mid s\Omega' = \Omega'\}$. *In particular,* $r_\Omega = Ind_G^G\, 1_{G_x}$.

Let us come back to the situation of 1b, with a transitive operation of $G_K$, and suppose that we are given a Dedekind domain with quotient field $K$. Then, the discriminant $\delta_{F/E}$ is defined for any $E,F$ finite over $K$, and the Artin conductor $\mathcal{F}(\mathcal{X})$ is defined for any character $\mathcal{X}$ of $G_K$ with open kernel. The above proposition together with the induction formula of Artin conductors, cf. [23], ch. VI, immediately shows:

**Corollary.** $\delta_{E/K} = \mathcal{F}(r) = \mathcal{F}(u)$.

Various relations between relative discriminants and conductors inside a given Galois extension can be obtained by using induction and decomposition into irreducible characters. Nevertheless, it would be interesting to be more precise. For instance, what can be said for an extension with Galois group $A_5$ or $S_5$ about the conductors of the irreducible characters of degrees 4 and 5? Moreover, some precise results need special information on ramification groups; for instance, for $E/K$ dihedral of prime degree $\ell$, the formula $\delta_{E/K} = \delta_{L/K}^{(\ell-1)/2} \cdot \mathcal{F}^{(\ell-1)}$ with $[L:K] = 2$ for some ideal $\mathcal{F}$ of $K$ is not completely obvious: The induction formula only gives $N_{L/K}(\mathcal{F})^{(\ell-1)/2}$ for some ideal $\mathcal{F}$ of $L$, and we need some more information on ramification groups.

### 1e. Infinite Frobenius

We consider a group $G$ and an element $\sigma$ of $G$ with $\sigma^2 = 1_G$. Put $r_1(H,\sigma) = |\{s \in G/H \mid \sigma \in sHs^{-1}\}|$. Then, $r_1 \equiv \bmod 2$, for $\sigma \in sHs^{-1}$ and $s\sigma \in sHs^{-1}$ imply $s \in H$. We thus attach to the pair $(H,\sigma)$ a signature $(r_1,r_2)$ such that $r_1 + 2r_2 = n$; since this signature depends solely on the conjugacy class of $H$, a signature, for a given $\sigma$, is attached to any strong equivalence class of $G$-set, and depends only on the conjugacy class of $\sigma$.

For an extension $E/K$, when $G = G_K$ and $H$ fixes $E$ in $G_K$, we obtain the signature of $E$ above a given real embedding $v$ of $K$ by taking for $\sigma$ the Frobenius of some extension $w$ of $v$ to $E$. ($\sigma$ is defined by $\sigma w = w$, and its conjugacy class depends only on $v$.)

Now, when we classify the permutation groups, we only see the weak equivalence classes. To take into account the signature, we must consider for each conjugacy class $c$ of order 1 or 2 the subgroup of $Aut(G)/Int(G)$ which carries $c$ on a class with the same signature. We then obtain, for each signature, an integer, namely the number of orbits under $Aut(G)$ of elements $\sigma$ of order 1 or 2 whose signature is the given one. This number can be 0 (e.g. for $H$ normal in $G$ if $0 < r_1 < n$), and is exactly 1 when $r_1 = n$. These numbers are calculated in the appendix for $n \leq 7$.

### 1f. Examples

(i) The symmetric group $S_n$ for $n \geq 1$ (resp. the alternating group $A_n$ for $n \geq 3$) is $n$ (resp. $n-2$) – transitive; there is one Frobenius for each signature for $S_n$, and also for $A_n$ if one restricts oneself to the signatures with even $r_2$.

(ii) The dihedral group $D_n = \langle \sigma, \tau \mid \sigma^n = \tau^2 = 1, \tau\sigma\tau^{-1} = \sigma^{-1} \rangle$ of order $2n$ is 1–transitive for $n \geq 3$; the possible signatures are $(n,0)$ and $(1,(n-1)/2)$ for $n$ odd, $(n,0)$, $(2,(n-2)/2)$ and $(0,n/2)$ for $n$ even. There is one orbit of Frobenius except when $r_1 = 0$, where there are two orbits. They correspond to the two possible signatures of the subextension of degree 2 (resp. $n/2$) of $E$ for $n \equiv 0 \bmod 4$ (resp. $n \equiv 2 \bmod 4$).

(iii) (Burnside) For $\ell$ prime, groups of degree $\ell$ are either solvable, and then contained in $\mathrm{Aff}(\mathbb{F}_\ell)$, or non–solvable, and then 2–transitive.

(iv) Transitive groups of degree $\leq 4$ are $C_2, C_3, D_3 \simeq S_3, C_4, C_2 \times C_2, D_4, A_4, S_4$. (Here, $C_n$ denotes the cyclic group of order $n$.)

## 2. Signature and Higher Invariants

We keep the previous notation: $G, \Omega, K, K_s, G_K, E, n, \ldots$ .

**2a. The Signature**

This has nothing to do with 1e: The *signature* of the $G$-set $\Omega$ is the homomorphism $\epsilon : G \longrightarrow \{-1,+1\}$ such that $\epsilon(s)$ is the signature of the permutation $x \longmapsto s\cdot x$ of $\Omega$. We say that $(G,\Omega)$ is even (+) if Ker $\epsilon = G$, odd (–) otherwise.

The set of quadratic subextensions of $K_s/K$ (including $K$ itself) can be given a natural group structure by making use of the compositum of 2 quadratic fields. One obtains an Abelian group $\mathcal{Q}(K)$ of exponent 2, which is canonically isomorphic to $\mathrm{Hom}_{\mathrm{cont}}(G_K,\{\pm1\})$. Thus, $\epsilon$ defines for each finite extension $E/K$ an element of $\mathcal{Q}(K)$, and hence a discriminant $d_{E/K} \in K^*/K^{*2}$ if char $(K) \neq 2$, and an addive one in $K/\wp(K)$ if char $(K) = 2$. (Here, $\wp$ is the Artin–Schreier map $x \longmapsto x^2 - x$; we simply write $d_E$ when $K = \mathbb{Q}$.)

With the notation of 1b, the signature $\epsilon : G_K \longrightarrow \{\pm1\}$ defines a quadratic extension $E_q$ of $K$.

**2b. The Norm.**

Let $K_0 \subset K$, with $[K:K_0] < \infty$. Then, the norm $N_{K/K_0}$ for char $(K) \neq 2$ (or the trace $Tr_{K/K_0}$ for char $(K) = 2$) defines a homomorphism $\mathcal{N}_{K/K_0}$: $\mathcal{Q}(K) \longrightarrow \mathcal{Q}(K_0)$, called the *norm*. Thanks to the canonical isomorphism $\mathrm{Hom}_{\mathrm{cont}}(G_K,\{\pm1\}) \xrightarrow{\sim} H^1(G_{K_0},\{\pm1\})$, the corestriction Cor : $\mathrm{H}^1(\mathrm{G}_K,\{\pm1\}) \longrightarrow \mathrm{H}^1(\mathrm{G}_{K_0},\{\pm1\})$ gives a unified definition of the norm.

**Application for n = 4:**

(i) Extensions $E/K$ of degree 4 containing a given quadratic extension $L/K$ $(L \neq K)$ are classified by $\mathcal{N}_{L/K}(E/L)$ (type $C_2 \times C_2$ if $\mathcal{N}(E/L) = K$, $C_4$ if $\mathcal{N}(E/L) = L$, and $D_4$ otherwise).

(ii) Primitive extensions of degree 4 are classified by a cubic extension $K_3/K$ (cyclic in the $A_4$ case, dihedral otherwise), and, for a given $K_3$, completely determined by an element of Ker $\mathcal{N}_{K_3/K}$, unique only up to conjugacy in the $A_4$ case.

**2c. Example : Dihedral Extensions** ($n \geq 3$)

In the diagrams of fields, we indicate the number of conjugates when > 1 and the degree when > 2. We consider an extension $E/K$ of degree $n$ with Galois closure $F/K$ such that $G$ $(= Gal(F/K))$ is isomorphic to $D_n$ and define $L \subset F$ by $[L:K] = 2$ and $N/L$ cyclic. The signature depends on $n$ mod 4. The diagrams in Figure 1 are complete when $n$ is a prime, or twice a prime, or equal to 8.

**2d. Example : Primitive Quartic Extensions**

For $G = S_4$, the $G$-set $G/H$ with $H = \langle (12),(34) \rangle$ (resp. $H = \langle 1234 \rangle$) is is even (resp. odd), and defines the 6–th degree permutations $S_4^+$ (resp. $S_4^-$) of $S_4$ ; the corresponding fields are denoted by $L^+$ (respectively $L^-$) in Figure 2; there is one field $L^+$, one field $L^-$ and one field among $M$, $M'$, $M''$ which contain a given conjugate of $K_3$ ; the product over $K_3$ is $L^+ \cdot L^- = K_6$.

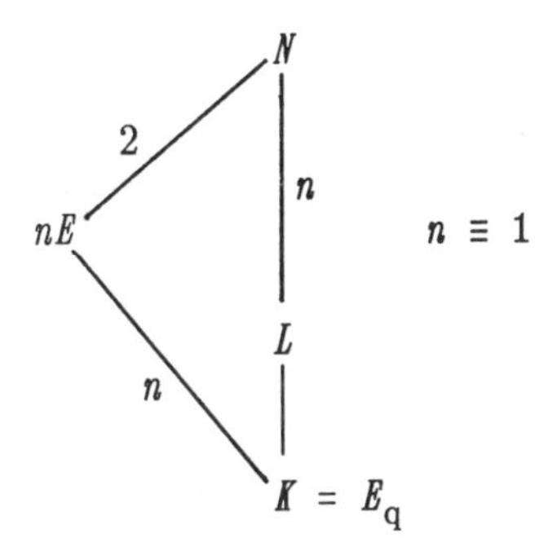


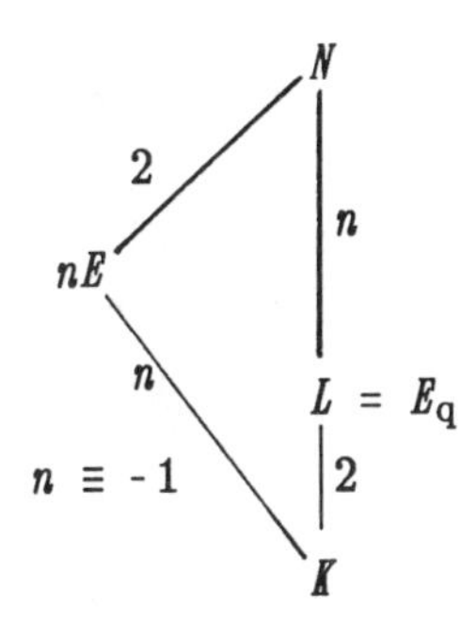


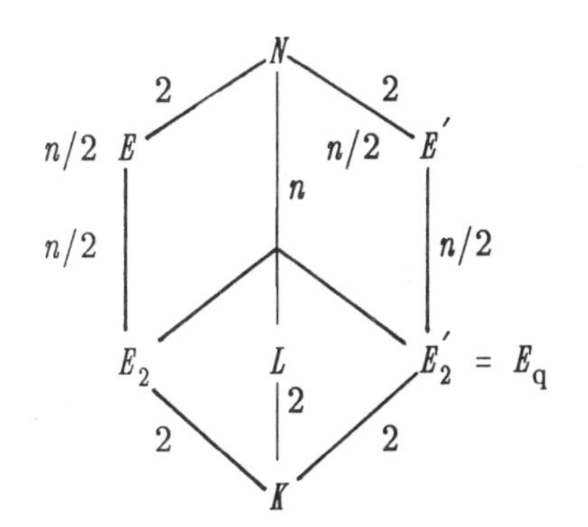


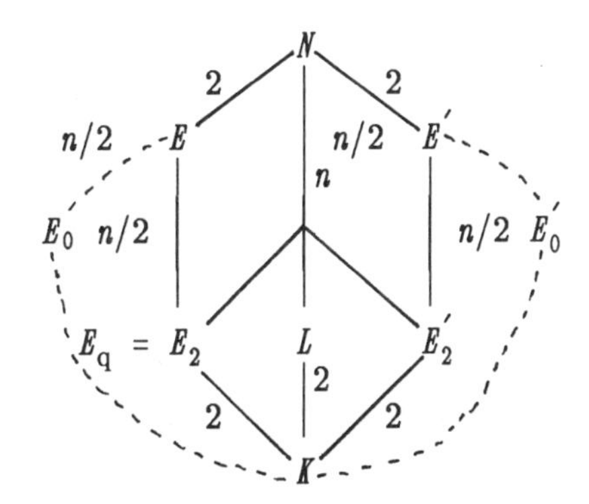


**Figure 1 : Dihedral Extensions**

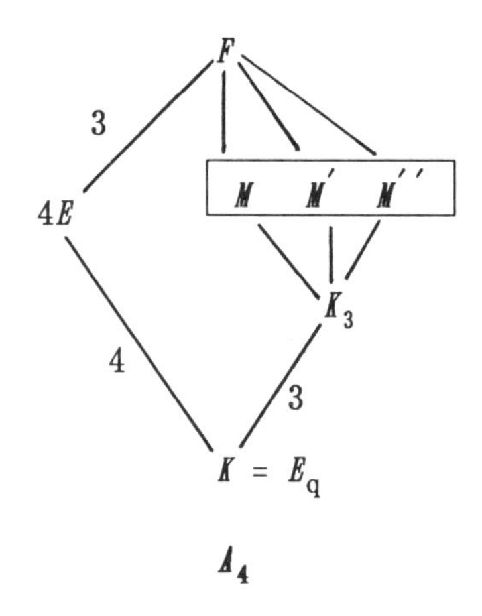


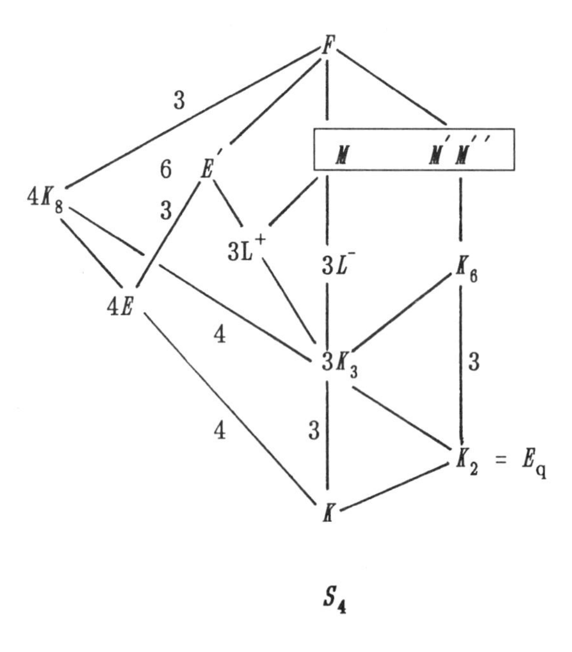


**Figure 2 : Primitive Quartic Extensions**

Since there is exactly one field $L^+$ containing a given conjugate of $K_3$, the group structure of Ker $\mathcal{N}_{K_3/K}$ defines a group structure on the set of $S_4$ fields contained in $K_s$, a fact equally noticed by Serre. (One must accept the field $K_3$ itself, or the algebra $K \times K_3$ of degree 4.) This remark allows us to construct the $S_4$ fields over $\mathbb{Q}$ associated to a given $K_3$, at least when $K_3$ has an odd class number, as products of fields in which only a few primes ramify outside those which are ramified in $K_3/\mathbb{Q}$: This is the same kind of construction that the one which gives the usual quadratic fields, taking as elementary fields the fields with discriminants $-4$, $+8$ and $(-1)^{(p-1)/2}\,p$, $p$ an odd prime. Amazingly, there is no composition law in the $A_4$ case, for there is no canonical choice in Ker $\mathcal{N}_{K_3/K}$ to define $E/K$.

## 2e. 2–Dimensional Invariants

The definition of the signature in 1a could have been given by means of the group $H^1(G,\{\pm1\})$. It is thus natural to look at $H^2(G,\{\pm1\})$, which yields to interesting embedding problems when $G = G_K$. (For dimensions $\geq 3$, cf. B. Kahn, [16].) It was proved by Schur ([22]) that $H^2(S_n,\{\pm1\})$ is of order 4 for $n \geq 4$ (its order is 1 for $n = 1$, and 2 for $n = 2,3$). We shall now construct three groups $S_n'$, $\hat{S}_n$, $\tilde{S}_n$, which are 2–coverings of $S_n$, and which define distinct non–trivial extensions of $S_n$ for $n \geq 4$.

The following three diagrams define successively $S_n'$, a 2–covering $\overline{A}_n$ of $A_n$ and $\hat{S}_n$:

$$\begin{array}{ccc} S_n' & \longrightarrow & C_4 \\ \downarrow & & \downarrow \\ S_n & \longrightarrow & \{\pm1\} \end{array} \qquad \begin{array}{ccc} \overline{A}_n & \longrightarrow & \mathrm{Spin}_{n-1}(\mathbb{R}) \\ \downarrow & & \downarrow \\ A_n & \longrightarrow & SO_{n-1}(\mathbb{R}) \end{array} \qquad \begin{array}{ccc} \hat{S}_n & \longrightarrow & \overline{A}_{n+2} \\ \downarrow & & \downarrow \\ S_n & \longrightarrow & A_{n+2} \end{array}$$

and $\tilde{S}_n$ is then $S_n' \cdot \hat{S}_n$ (the Baer multiplication, corresponding to the product in $H^2(S_n, \{\pm1\})$). The extensions $S_n'$, $\hat{S}_n$ and $\tilde{S}_n$ of $S_n$ can be characterized by the following lifting property: An element of order 2 of

$S_n$, product of $t$ transpositions, has a lifting of order 2 in $S_n'$ (resp. $\hat{S}_n$, resp. $\tilde{S}_n$) if and only if $t \equiv 0$ or $2$ (resp. $t \equiv 0$ or $-1$, resp. $t \equiv 0$ or $1$) modulo 4.

Now, given a $G$-set $\Omega$, the corresponding embedding $G \to S_n$ (defined up to an inner automorphism of $S_n$) defines by restriction group extensions $G_n'$, $\hat{G}_n$, and $\tilde{G}_n$, which depend, up to isomorphism, only on the strong equivalence class of $(G,\Omega)$ (but they do depend on the operation of $G$ on $\Omega$, not only on $G$! ).

One has $G_n' = G_n \times C_n$ (and $\hat{G}_n = \tilde{G}_2$) when $(G,\Omega)$ is even; in particular, $A_n' = A_n \times C_2$ and $\hat{A}_n = \tilde{A}_n = \bar{A}_n$.

Given $K$ with char $(K) \neq 2$, and an operation of $G_K$ on $\Omega$ (with $|\Omega| = n$), there corresponds to $S_n'$, $\hat{S}_n$ and $\tilde{S}_n$ embedding problems, to which we associate elements $e'$, $\hat{e}$, $\tilde{e} \in H^2(G_K, \pm 1) \simeq Br_2(K)$. Serre ([26]) has given a way of calculating $\tilde{e}$ by means of the invariants of the trace form $x \mapsto T_{r_{E/K}}(x^2)$. Explicit embeddings were found by Witt ([27]), 1936) in some special cases. Recent work of Ms. Crespo ([6]) gives such explicit embeddings for the groups $A_n$ and $S_n$; this answers for these groups a question of Serre.

**2f. Examples**

(i) The pull back in $S^3 \to SO_3(\mathbb{R})$ of $C_n$ (respectively $D_n$, $A_4$, $S_4$, $A_5$) is $C_{2n}$ (resp. $H_n$, $\hat{A}_4 = \tilde{A}_4$, $\hat{S}_4$, $\hat{A}_5 = \tilde{A}_5$). (The quaternion group $H_n$ is $\langle \sigma,\tau\,;\, \sigma^n = \tau^2,\ \tau^4 = 1,\ \tau\sigma\tau^{-1} = \sigma^{-1} \rangle$.)

(ii) $\tilde{S}_4 \to S_4$ is isomorphic to $G\ell(2,3) \to PG\ell(2,3)$, and $\hat{A}_4 \to A_4$ to $SL(2,3) \to PSL(2,3)$.

(iii) For $D_n$ $(n \geq 3)$, viewed as a group of degree $n$, $D_n'$, $\hat{D}_n$ and $\tilde{D}_n$ depend on $n$ modulo 8.

## 3. Geometric Methods

From now on, $K$ is a number field. We want to find all extension $E/K$ (up to isomorphism) with discriminant $|d_E| \leq B$ for some bound $B$, and with prescribed ramification in $E/K$ of the infinite primes of $E$; for $K = \mathbb{Q}$, the last condition simply amounts to the knowledge of the signature of $E$. Taking $B$ large enough, one can hope to find all the possible permutation groups with the given signature.

### 3a. Smallest Discriminants

The absolute smallest discriminants are known for $n \leq 7$ and $n = 8$, $r_1 = 0$ (and probably for $n = 8$, $r_2 = 0$). Here are the results for increasing $r_1$:

$n = 2 : d = -3,5$.
$n = 3 : d = -23,49$ (Furtwängler, 1896).
$n = 4 : d = 117,-275,725$ (Mayer, 1929).
$n = 5 : d = 1609,-4511,14641$ (Hunter, 1957).
$n = 6 : d = -9747,28037,-92729,300125$ (Pohst, 1982; also Kaur, 1970 for $r_1 = 6$).
$n = 7 : d = -184607,612233,-2306599,20134393$ (Diaz y Diaz, 1982, 1984, 1988; Pohst, 1977).
$n = 8$, $r_1 = 0 : d = 1257728$ (Diaz y Diaz, 1987). (For $n = r_1 = 8$, a cyclic extension of $\mathbb{Q}(\sqrt{2})$ quoted by Lenstra has discriminant $d = 2^{12} \cdot 41^3 = 282300416$; no primitive field (Pohst and al.) and no extension of quartic field exists with a smaller discriminant.)[2]

It should be remarked that Diaz y Diaz results, as well as the work quoted above for $n = r_1 = 8$, take into account the behaviour of the finite primes of small norm; finite primes do not appear in geometry of numbers.

[2] It is the minimal discriminant for $n = r_1 = f$, cf. F. Diaz y Diaz, J. Martinet, M. Pohst, to appear.

### 3b. Inequalities

Embed $E$ in the $n$-dimensional space $V = \mathbb{R} \otimes E$, and put on $V$ the Euclidean structure which comes from the norm $\|x\| = \sum_{\sigma: E \to \mathbb{C}} |\sigma x|^2$ on $E$. Applying an argument "à la Minkowski" for the orthogonal projection of $\mathbb{Z}_E$ (the lattice of integers of $E$) on the subspace orthogonal to $K$ allows one to find a $\theta \in \mathbb{Z}_E$, $\theta \notin K$, with not too big conjugates with respect to $|d_E|$. Let $K = \mathbb{Q}$. One finds only finitely many possible polynomials for $\theta$. If $\mathbb{Q}(\theta) \neq E$, one can use successive minima, or relative computations (after having calculated bounds for the possible discriminants of subfields of $E$), relying on the following theorem (cf. [20], th. 2.8):

**Theorem.** *Let $E/K$ be an extension of number fields, of degree $n$. Let $p = [K{:}\mathbb{Q}]$ and $q = [E{:}\mathbb{Q}]$ $(= np)$. Then, there exists $\theta \in \mathbb{Z}_E$, $\theta \notin K$, such that*

$$\sum_{\sigma: E \to \mathbb{C}} |\sigma\theta|^2 \leq \sum_{\tau: k \to \mathbb{C}} |\tau Tr_{E/K}(\theta)|^2 + \gamma_{q-p} \left| \frac{d_E}{n^p \cdot d_K} \right|^{1/(q-p)},$$

*and $\theta$ can be chosen arbitrarily modulo $\mathbb{Z}_K$. (Here, $\gamma_d$ is the Hermite constant for the dimension $d$, i.e. $\gamma_d^{-d/2} = \Gamma_d$ is the lattice constant of the unit ball in $\mathbb{R}^d$.)*

### 3c. Relative Computations

It will be probably very difficult to find the minimal discriminants beyond degree 8 with the above methods. Interesting partial results can be obtained by restricting oneself to extensions of, say, quadratic or cubic fields. In this direction, a recent joint work with A.-M. Bergé and M. Olivier provides long tables of sextic fields containing a quadratic subfield ([3]). Since integral bases are no more available, discriminants are calculated by making use of local methods.

## 4. Class Fields

### 4a. Abelian Extensions

Let $K$ be a number field, and let $\mathcal{M} = \mathcal{M}_0\mathcal{M}_\infty$ be a "module" of $K$, formal product of an integral ideal $\mathcal{M}_0$ and of an again formal product $\mathcal{M}_\infty$ of real places of $K$. To the module $\mathcal{M}$, class field theory associates a canonical Abelian extension $K^{\mathcal{M}}/K$ (the "ray class field" modulo $\mathcal{M}$), in which only the prime divisors of $\mathcal{M}$ can ramify (thus, $K^{(1)}$ = Hilbert $(K)$); for $K = \mathbb{Q}$, and $m > 0$, one has $\mathbb{Q}^{(m_\infty)} = \mathbb{Q}(e^{2in/m})$ and $\mathbb{Q}^{(m)} = \mathbb{Q}(2\cos(2\pi/m))$.

If $E/K$ is an Abelian extension with Galois group $G$, one can define the conductor $\mathcal{F}$ of $E/K$, related to discriminant and conductors by Hasse's rules $\mathcal{F} = \ell.\text{c.m.}\ (\mathcal{F}(\mathcal{X}))$ and $\delta_{E/K} = \prod \mathcal{F}(\mathcal{X})$. The fundamental result of class field theory is that $E$ is contained in $K^{\mathcal{F}}$, and that, for each module $\mathcal{M}$, one can determine Gal$(K^{\mathcal{M}}/K)$ and $\delta_{K^{\mathcal{M}}/K}$ from data of $K$ itself, namely $C\ell_K$ (the class group) and $E_K$ (the unit group). We thus have a description of Abelian extensions of $K$ with a given discriminant which does not make use of polynomials.

To find fields $E$ with a given signature and with discriminant up to a given bound, and which are moreover Abelian over some field $K$ with a given signature, is now possible, provided one has long tables of fields of the signature of $K$ with class groups and units. Tables exist for quadratic fields, for cubic fields (Angell ([1]) for negative discriminants, Ennola and Turenen ([10]) for positive discriminants), for real cyclic fields of degree 3, 4 (M.–N. Gras, [13], [14]) and 6 (Mäki, [19]), for imaginary cyclic quartic fields (Hardy and al., [15]). Work in progress by Buchmann, Ford, Pohst, v. Schmettow will provide soon tables for quartic fields. (The totally real case is now done, cf. [3a,3b]).

**4b. Solvable Extensions**

The extension $E/K$ is solvable if and only if there exists a tower $K_0 = K \subset K_1 \subset \ldots \subset K_r$ of finite Abelian extensions $K_{i+1}/K_i$ with $E \subset K_r$. One can construct cubic extensions by taking $r = 2$, $[K_1 : K_0] = 2$ and $[K_2 : K_1] = 3$. Similarly, $A_4$ and $S_4$ fields correspond to degrees (3,4) and (2,3,4) respectively (or (3,2,2) and (2,3,2,2)), but, in practice, they are characterized by $K_1'/K_0$ cubic and $K_2'/K_1'$ quadratic, cf. 2d. Explicit calculations of discriminants are done by means of various Artin conductors.

Note that, when climbing in the tower $K_1 \subset K_2 \subset \ldots$ , we must at each step find units and class groups when no table is available. Since these informations are not provided by class field theory, we must find polynomials which define the extensions $K_{i+1}/K_i$. Kummer theory is often convenient, especially for quadratic extensions, since it then does not require the adjonctions of roots of unity to the base field.

## 4e. Embedding Problems

We just quote a refinement of class field and Kummer methods, by looking at the following example: $E/K$ Galois of degree 8, with quaternion Galois group ($G \simeq H_2$). The diagram of fields below is complete.

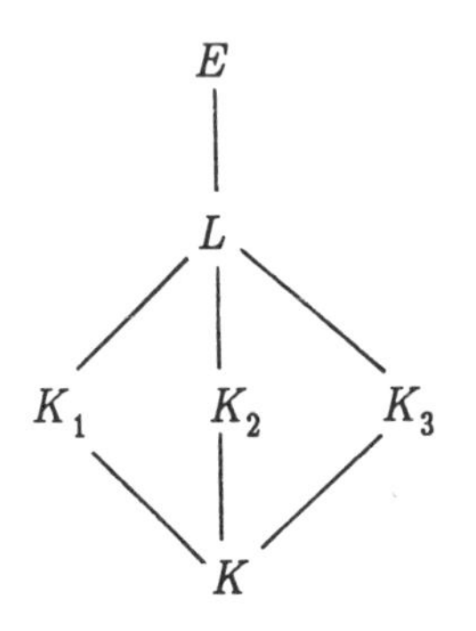


Write $K_i = K(\sqrt{x_i})$. Then, $E$ exists above $L$ if and only if one has the equivalence
$T_{r_{L/K}}(x^2) \sim_K \sum x_i^2$ (Witt).

Thus, we first select the possible $L$'s by the above theorem, and apply then class field theory or Kummer theory to find $E/L$. Quaternion extensions of degrees 8, 12, 20 have been dealt with by S.–H. Kwon ([17]).

## 5. Fields of Degree $\leq 4$

All permutation groups of degree $\leq 4$ except $A_4$ are covered by the "classical" tables (Delone–Fadeev, [7], for $n = 3$; Godwin, [11], circa 1956, for $n = 4$). The minimal discriminants for $A_4$ fields are easily found by class field theory: the fields $M$ (notation of 2d) are $\mathbb{Q}(\sqrt{2\cos(2\pi/7)})$ for $r_1 = 0$ and an unramified quadratic extension of the cyclic field of conductor 163 for $r_1 = 4$. The results are summarized in Table 1 below.

**Table 1: Fields of Degree 3 and 4**

| | | | | |
|---|---|---|---|---|
| $C_3$ | — | 49 | $2\cos\frac{2\pi}{7}$ | $\sim X^3 + X^2 - 2X - 1$ |
| $D_3$ | $-23_*$ | 148 | $X^3 - X - 1$ | $X^3 - 4X - 2$ |

| | | | | | | |
|---|---|---|---|---|---|---|
| $C_4$ | 125 | — | 1125 | $\xi_5$ | - | $2\cos\frac{2\pi}{15}$ |
| $C_2 \times C_2$ | 144 | — | 1600 | $\xi_{12}$ | - | $\sqrt{5} + \sqrt{2}$ |
| $D_4$ | 117*<br>1025* | $-275_*$ | $725_*$ | $-1 + 2\sqrt{-3}$ | | $-11 + 4\sqrt{5}$ |
| $A_4$ | 3136 | — | $26569_*$ | $3 + 2\sqrt{5}$ | | $7 + 2\sqrt{5}$ |
| $S_4$ | $229_*$ | $-283_*$ | $1957_*$ | | | |

$A_4 : X^4 - 2X^3 + 2X^2 + 2, \quad X^4 - 6X^3 + 5X^2 + 7X + 1$

$S_4 : X^4 - X + 1, \quad X^4 - X - 1, \quad X^4 - 4X^2 - X + 1$

*Related to Hilbert class fields, in the ordinary or in the narrow sense. The fields to be considered are the quadratic fields of discriminants –23 ($D3$), –39, +205, –55, +145 ($D4$), the cyclic cubic field of conductor 163 ($A_4$), and the non–cyclic cubic fields of discriminants +229, –283 and +1957 ($S_4$).

## 6. Groups of Prime degree $\ell > 3$

We consider the case of a solvable group $G$. Then, up to isomorphism, $G$ is classified by its order $|G| = \ell q$, $q | \ell-1$, and one has $r_1 = \ell$ for $q$ odd, and $r_1 = \ell$ or 1 for $q$ even. Minimal discriminants are known for $\ell = 5$, $q = 1,2,4$ and $\ell = 7$, $q = 1,2,3$, cf. Kwon and Martinet, [18]. For instance, in the case $\ell = 5$, $q = 4$, $r_1 = 1$, the minimal discriminant is

$2^4 \cdot 13^3 = 35152$; the polynomial $f_1(X) = X^5 - 2X^4 - 4X^3 - 96X^2 - 352X - 568$ given in [18], of discriminant $2^4 \cdot 13^2 \cdot (2^4 \cdot 10429)^2$, has been transformed by Diaz y Diaz into $f_2(X) = X^5 + X^4 + 4X^3 + 2X^2 + X + 1$ of discriminant exactly $2^4 \cdot 13^2$. (For $r_1 = 5$, Diaz y Diaz found the polynomial $f_3(X) = X^5 - 2X^4 - 15X^3 + 48X^2 - 43X + 12$, with $df_3 = 2^4 \cdot 53^3 = 2{,}382{,}032$, the discriminant of the field.)

**Table 2: Fields of Degree 5 and 7**

Table 2 below gives the known results for the degrees 5 and 7.

| | | | | |
|---|---|---|---|---|
| $C_5$ | —— | —— | 14641 | $11^4$ |
| $D_5$ | 2209 | —— | 160801 | $47^2, 401^2$ |
| $Aff_5$ | 35152 | —— | 2382032 | |
| $A_5$ | 18496* | —— | ? | $(2^3 \cdot 17)^2$ |
| $S_5$ | 1609 | -4511 | 24217 | |

* Buhler.

| | | | | |
|---|---|---|---|---|
| $C_7$ | —— | —— | —— | 594,823,321 |
| $D_7$ | -357911 | —— | —— | 192,100,033 |
| $C_7 \cdot C_3$ | —— | —— | —— | 1,817,487,424 |
| $S_7$ | -184607 | 612233 | -2,306,599 | 20,134,393 |

The minima are not known for $Aff(7)$, $PSL(3,2)$ and $A_7$.

## 7. Groups of Degree 6

The classification of sixth degree permutation groups involves 4 series defined by imprimitivity questions: For an extension $E/K$, Type I (resp. II, III, IV) correspond to the existence of both a quadratic and a cubic subextension (resp. only a quadratic one, resp. only a cubic one, resp. no non-trivial subextension). Groups of the $4^{\text{th}}$ kind are not solvable, and isomorphic to $PSL(2,5) \simeq A_5$, $PGL(2,5) \simeq S_5$, $A_6$ or $S_6$. The others are obviously solvable. Here are the 3 lists (we use semi–direct products which are easy to guess for the second list):

Type I: $C_6$, $D_3 = S_3$, $D_6 = S_3 \times C_2$

Type II:

$$G_{18} = (C_3 \times C_3) \cdot C_2 (= C_3 \times S_3),\ G^+_{36} = (C^3 \times C_3) \cdot C_4,$$

$$G^-_{36} = (C_3 \times C_3) \cdot (C_2 \times C_2)\ (= S_3 \times S_3),$$

$$G_{72} = (C_3 \times C_3) \cdot D_4$$

Type III: $A_4$, $A_4 \times C_2$, $S_4^+$, $S_4^-$, $S_4 \times C_2$.

Long tables of cubic extensions of quadratic fields have been computed in a joint work with A.–M. Bergé and M. Olivier ([3]). As a consequence, minimal discriminants are known for all groups of Type I and II, cf. Table 3 below. Work in progress[3] will produce long tables of sextic fields of Type III, and hence the minimal discriminants (which could be however computed by hand; for instance, in the $S_4^+$ case, one finds $229^2$ for $r_1 = 0$, $2^6 \cdot 23^2$ for $r_1 = 2$ and $2^6 \cdot 229^2$ for $r_1 = 4$.)

For Type IV extensions, the minimal discriminants are known only in the $S_6$ case. It would be interesting to study closely the connection between degrees 5 and 6 to handle the $S_5$ case. However, $A_5$ and $A_6$ extensions are probably out of our computational possibilities.

[3] Completed (M. Olivier, to appear).

**Table 3: Sextic Fields of Type I, II.**

| Gal \ Sgn | (0,3) | (2,2) | (4,1) | (6,0) |
|---|---|---|---|---|
| $C_6$ | –16807 | —— | —— | 300125 |
| $D_3$ | –12167 | —— | —— | 810448 |
| $D_6$ | –14283<br>–309123 | 66125 | —— | 2738000 |
| $G_{18}$ | –9747 | —— | —— | 722000 |
| $G_{36}^{+}$ | —— | 525625 | —— | 55130625 |
| $G_{36}^{-}$ | –309123 | 242000 | —— | 27848000 |
| $G_{72}$ | –11691 | 30125 | –104875 | 485125 |

To finish with §7, we give two diagrams illustrating the 2 direct products which appear in Type II permutations.

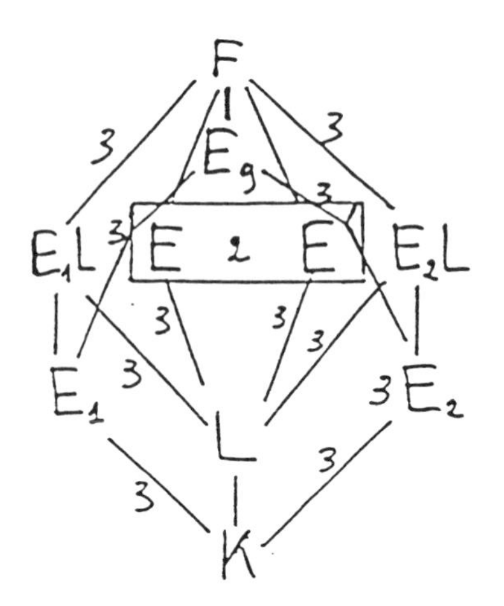


To obtain the discriminant –9747 (resp. +722000), take for $E_1$, $E_2$ fields with discriminants $-3\cdot 19^2$, $19^2$ (resp. $5\cdot(2^2\cdot 19)^2$, $19^2$

Figure 3a: The Group $G_{18}$

To obtain the discriminant $-309123$ (resp. $+242000$; resp. $+27848000$), take for $E_1$, $E_2$ the fields with discriminants $-107$, $3 \cdot 107$ (resp. $-4 \cdot 11^2$, $-20 \cdot 11^2$; resp. $2^3 \cdot 59^2$, $(2^3 \cdot 5)59^2$.)

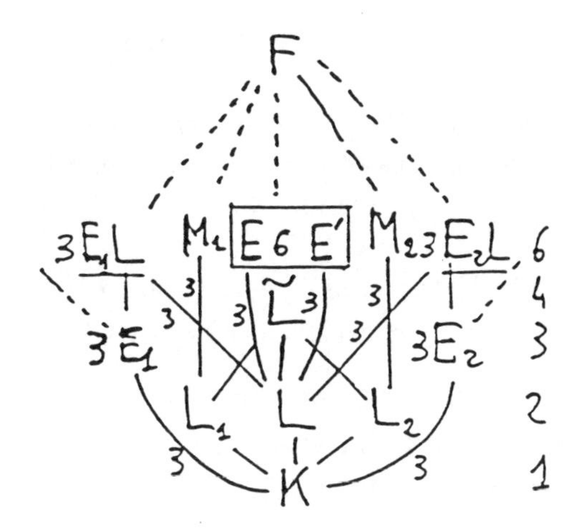


**Figure 3b: The Group $G^+_{36}$ .**

## 8. A glance at octic fields

Here the classification involves 50 groups, and 5 types according to primitivity and solvability: Type I to IV are solvable, with subextensions of degrees 2 and 4 for Type I (29 groups), of degree 2 only for Type II and 4 only for Type III (7 groups in each case); Type IV extensions are primitive (2 groups); we are left with 5 non–solvable groups. Only a few examples have been dealt with, including the 5 Galois extensions, all of Type I, as well as $A_4 \times C_2, S_4, S_4 \times C_2, \dots$ . Type II extensions are primitive quartic extensions of a quadratic field; they are closely related to sextic extensions of Type II (cf. §7 and §2d). Examples of Type III extensions are provided by the groups $\tilde{A}_4 \simeq SL(2,3)$ and $\tilde{S}_4 \simeq G\ell(2,3)$. An example of Type IV extension is provided by a generalization of $A_4$ : just make $C_7$ act on $C_2 \times C_2 \times C_2$ as a cyclic permutation of the elements of order 2. Type V groups involve $PSL(2,7)$ or $A_8$ ; for instance, one can replace in the preceeding example $C_7$ by $PSL(3,2) \simeq PSL(2,7)$.

The totally real octic field referred to in 3a is of Type I; it is the field $\mathbb{Q}(\sqrt{M})$, where $M = (m + (1+\sqrt{2})\sqrt{m})/2$, with $m = 7 + 2\sqrt{2}$; it is the ray class field modulo a prime above 41 over $\mathbb{Q}(\sqrt{2})$.

## Appendix

The following table (Table 4) gives for $2 \leq n \leq 7$ the list of permutation groups of degree $n$ together with the following data:

– the signature (+ or –).

– the trivial 2–coverings (+ if all are trivial, – if none is trivial, the trivial one (′, ~, or ^) if exactly one is trivial).

– $[(n+1)/2]$ columns, one for each value of $r_1$ which is a priori possible from 0 or 1 to $n$, in which is indicated the number of classes or order 1 or 2 for the given signature, modulo the automorphisms which preserve this signature.

It relies on results found in [5].

Table 4.

$n = 2$

| $G$ \ $r_1$ | | | 0 | 2 |
|---|---|---|---|---|
| $C_2$ | $-$ | $\sim$ | 1 | 1 |

$n = 3$

| $G$ \ $r_1$ | | | 1 | 3 |
|---|---|---|---|---|
| $C^3$ | $+$ | $+$ | 0 | 1 |
| $D^3 \simeq S_3$ | $-$ | $\sim$ | 1 | 1 |

$n = 4$

| $G$ \ $r_1$ | | | 0 | 2 | 4 |
|---|---|---|---|---|---|
| $C_2 \times C_2$ | $+$ | $'$ | 1 | 0 | 1 |
| $C_4$ | $-$ | $'$ | 1 | 0 | 1 |
| $D_4$ | $-$ | $-$ | 2 | 1 | 1 |
| $A_4$ | $+$ | $'$ | 1 | 0 | 1 |
| $S_4$ | $-$ | $-$ | 1 | 1 | 1 |

$n = 5$

| $G$ \ $r_1$ | | | 1 | 3 | 5 |
|---|---|---|---|---|---|
| $C_5$ | $+$ | $+$ | 0 | 0 | 1 |
| $D_5$ | $+$ | $'$ | 1 | 0 | 1 |
| $Aff(5)$ | $-$ | $'$ | 1 | 0 | 1 |
| $A_5$ | $+$ | $'$ | 1 | 0 | 1 |
| $S_5$ | $-$ | $-$ | 1 | 1 | 1 |

$n = 7$

| $G$ \ $r_1$ | | | 1 | 3 | 5 | 7 |
|---|---|---|---|---|---|---|
| $C_7$ | $+$ | $+$ | 0 | 0 | 0 | 1 |
| $D_7$ | $-$ | $\wedge$ | 1 | 0 | 0 | 1 |
| $C_7 \cdot C_3$ | $+$ | $+$ | 0 | 0 | 0 | 1 |
| $Aff(7)$ | $-$ | $\wedge$ | 1 | 0 | 0 | 1 |
| $PSL(3,2)$ | $+$ | $'$ | 0 | 1 | 0 | 1 |
| $A_7$ | $+$ | $'$ | 0 | 1 | 0 | 1 |
| $S_7$ | $-$ | $-$ | 1 | 1 | 1 | 1 |

**Table 4.**

$n = 6$

| $G$ \ $r_1$ | | | 0 | 2 | 4 | 6 |
|---|---|---|---|---|---|---|
| $C_6$ | – | ∧ | 1 | 0 | 0 | 1 |
| $D_3 \simeq S_3$ | – | ∧ | 1 | 0 | 0 | 1 |
| $D_6 \simeq S_3 \times C_2$ | – | – | 2 | 1 | 0 | 1 |
| $G_{18}$ | – | ∧ | 1 | 0 | 0 | 1 |
| $G_{36}^+$ | + | ′ | 0 | 1 | 0 | 1 |
| $G_{36}^-$ | – | – | 1 | 1 | 0 | 1 |
| $G_{72}$ | – | – | 1 | 1 | 1 | 1 |
| $A_4$ | + | ′ | 0 | 1 | 0 | 1 |
| $A_4 \times C_2$ | – | – | 1 | 1 | 1 | 1 |
| $S_4^+$ | + | ′ | 0 | 2 | 0 | 1 |
| $S_4^-$ | – | – | 1 | 1 | 0 | 1 |
| $S_4 \times C_2$ | – | – | 2 | 2 | 1 | 1 |
| $A_5$ | + | ′ | 0 | 1 | 0 | 1 |
| $S_5$ | – | – | 1 | 1 | 0 | 1 |
| $A_6$ | + | ′ | 0 | 1 | 0 | 1 |
| $S_6$ | – | – | 1 | 1 | 1 | 1 |

## References

§§1,2 – Group Theory: [5], [21], [22], [24]; §2e: [6], [25], [27].

§3a – References are in [20], except [8] and [9].

§4 – Class Field Theory: [2]; Tables: [1], [3b], [10], [13], [14], [15], [19].

§5 – [7], [11], [12].

§6 – [4], [5], [18].

§7 – [3], [5].


[1] *I.O. Angell*, A table of complex cubic fields. Bull. London Math. Soc. **5** (1957), 37–38.

[2] *E. Artin* and *J. Tate*, Class Field Theory. Harvard, 1954.

[3] *A.–M. Bergé*, *J. Martinet* and *M. Olivier*, The computation of sextic fields with a quadratic subfield. Math Comp., to appear.

[3a] *J.A. Buchmann* and *D. Ford*, On the computation of totally real quartic fields of small discriminants. Math. Comp., to appear.

[3b] *J. Buchmann*, *M. Pohst* and *I.v. Schmettow*, On the computation of unit groups and class groups of totally real quartic fields. Math. Comp., to appear.

[4] *J.P. Buhler*, Icosahedral Galois Representations. Springer Lecture Notes 654, 1978.

[5] *G. Butler* and *J. McKay*, The transitive groups of degree up to eleven. Comm. Alg. **11** (1983), 863–911.

[6] *T. Crespo*, (a) Explicit construction of $\tilde{A}_n$ type fields; (b) Explicit construction of $2S_n$ Galois extensions. Preprints, Barcelona.

[7] *B.N. Delone* and *D.K. Fadeev*, The Theory of Irrationalities of the Third Degree. A.M.S., Providence, 1964, Ed. originale, Moscou, 1940 (en Russe).

*F. Diaz y Diaz*, cf. [20]

[8] *F. Diaz y Diaz*, Petits discriminants des corps de nombres totalement imaginaires de degré 8. J. Number T. **25** (1987), 34–52.

[9] *F. Diaz y Diaz*, Discrimant minimal et petits discriminants des corps de nombres de degré 7 avec 5 places réeles. J. London Math. Soc. **38** (1988), 33–46.

[10] *V. Ennola* and *R. Turunen*, On totally real cubic fields. Math. Comp. **44** (1985), 495–518.

[11] *H.J. Godwin*, (a) Real quartic fields with small discriminant. J. London Math. Soc. **31** (1956), 478–485; (b) On totally complex quartic fields with small discriminants. Proc. Cambridge Phil. Soc. **53** (1957), 1–4; (c) On quartic fields of signature one with small discriminant. Quart. J. Math. Oxford **8** (1957), 214–222.

[12] *H.J. Godwin*, On quartic fields of signature one, II. Math. Comp. **42** (1984), 707–711; corrigenda, ibid. **43** (1984), 621.

[13] *M.–N. Gras*, Méthodes et algorithmes pour le calcul numérique du nombre de classes et des unités des extensions cubiques cycliques de $\mathbb{Q}$. J. Crelle **277** (1975), 89–116.

[14] *M.–N. Gras*, Classes et unités des extensions cycliques réelles de degré 4 de $\mathbb{Q}$. Ann. Inst. Fourier **29** (1979), 107–124, and Publ. Math. Besançon, Th. Nombres, fasc. 2, 1977/1978.

[15] *K. Hardy*, *R.H. Hudson*, *D. Richman*, *K.S. Williams* and *N.M. Holtz*, Calculation of the Class Numbers of Imaginary Cyclic Quartic Fields. Math. Comp. 49 (1987), 615–620.

*J. Hunter*, cf. [20].

[16] *B. Kahn*, Classes de Stiefel–Whitney de formes quadratiques et de représentations galoisiennes réelles. Invent. Math. **78** (1984), 223–256.

*G. Kaur*, cf. [20].

[17] *S.–H. Kwon*, Sur les discriminants minimaux des corps quaternioniens. Preprint, Bordeaux.

[18] *S.–H. Kwon* and *J. Martinet*, Sur les corps résolubles de degré premier. J. Crelle 375/376 (1987), 12–23.

[19] *S. Mäki*, The Determination of Units in Real Cyclic Sextic Fields. Springer Lecture Notes 797, 1980.

[20] *J. Martinet*, Méthodes géométriques dans la recherche des petits discriminants. Progress in Mathematics, Vol. 59, Birhhäuser, 1985, 147–179.

*J. Mayer*, cf. [20]

[21] *D.S. Passman*, Permutation groups. Benjamin, New York, 1968.

*M. Pohst*, cf. [20]

[22] *I. Schur*, Über die Darstellung der symmetrischen und alternierenden Gruppe durch gebrochene lineare Subtitutionen. J. Crelle **139** (1911), 155–250, Ges. Abh. I, 346–441.

[23] *J.–P. Serre*, Corps Locaux. 3$^{\text{ieme}}$ éd., Hermann, Paris, 1968.

[24] *J.-P. Serre*, Représentations linéaires des groupes finis. $2^{ieme}$ édition, Hermann, Paris, 1971.

[25] *J.-P. Serre*, L'invariant de Witt de la forme $Tr(x^2)$. Comm. Math. Helv. **59** (1984), 651–676, oeuvres III, 675–700.

[26] *L. Soicher*, The Computation of Galois groups. Thesis, Concordia University, Montréal.

[27] *E. Witt*, Konstruktion von galoisschen Körpern der Charakteristik p zu vorgegebene Gruppe der Ordung $p^f$. J. Crelle **174** (1936), 237–245.

---

Centre de Recherche en Mathematiques de Bordeaux,
351, cours de la Liberation 33405 Talence Cedex

# The Riemann Hypothesis From a Logician's Point of View

*Yuri V. Matijasevich*

The aim of this paper is to survey some results and ideas connected with the Riemann hypothesis (RH) which were developed by specialists in mathematical logic.

As a first example we consider the problem of deciding, given a natural number $N$, whether or not it is prime. For technical reasons it is preferable to state the problem in the form

$$\text{is } N \text{ composite?} \tag{1}$$

A mathematician with no interest in computational number theory would call it a trivial problem, but modern mathematicians and logicians are interested in measures of the difficulty of mathematical problems of this kind. While a complexity measure was for a long time intuitively clear for those who were interested in computation, and can for example be found in a 1969 paper by D.H. Lehmer [5]. Computational complexity developed a rigorous general theory in papers by S.A. Cook [1] and R.M. Karp [3]. There, two classes of problems, $P$ and $NP$, were introduced. Every problem in $NP$ consists of infinitely many individual subproblems each of which requires an answer YES or NO. Problem (1) can serve as an example of a problem in $NP$.

The class $P$ is a subclass of $NP$. For a problem to belong to $P$, it should admit an algorithmic solution with no more than $T(M)$ steps, where $M$ is the volume of information specifying a particular subproblem, and $T$ is a polynomial. Note that, because of the tradition of representing numbers in binary notation, as a function of $N$, the number of steps should be no more than $T(\log N)$. To make the definition of $P$ precise we need to choose the algorithmic steps which are to be counted, but it turns out that as long as we do not pay attention to the degree of $T$ a large variety of algorithms can be used when defining the same class $P$.

The definition of $NP$ is a bit more tricky. It is similar to the definition of $P$ but the algorithm can make "guesses" in the course of its work. In our example it could "guess" two factors $X$ and $Y$ of $N$, and check (in polynomial time!) the equality $N = XY$ before giving the answer YES (however, no checking is required if the answer is NO).

Thus problem (1) does belong to $NP$. The stronger assertion "problem (1) belongs to $P$" was left open in [3], and is still open. All algorithms for testing primality known today require at least $N^{\alpha}$ steps instead of requiring at most $(\log N)^{\beta}$ steps.

There are many other problems which evidently belong to $NP$ but we cannot say today whether they belong to $P$ or not. On the other hand, no problem is known to belong to $NP \setminus P$ and the (supposed) inequality

$$P \neq NP \tag{2}$$

is still the object of tantalizing efforts of many researchers.

What problems are candidates to lie in $NP \setminus P$ if (2) is true? It turns out that we can easily find them in abundance. A problem from $NP$ is called ***NP*-complete** if any other problem can be reduced to it (in polynomial time, of course—we omit other technical details of such a reduction). Thus if any $NP$-complete problem lies in $P$, then all problems from $NP$ lie in $P$ also, i.e., $P = NP$. Paper [3] contained a handful of examples of $NP$-complete problems, nowadays their quantity is measured in the thousands. Nevertheless we still do not know whether

(1) is an $NP$-complete problem. (3)

Our belief in it was shaken by G. L. Miller [9] who, under the assumption of

the extended Riemann hypothesis (4)

proved that (1) belongs to $P$, i.e., primality or non-primality of N can be recognized in $M(\log N)$ steps, where $M$ is a polynomial.

We can consider Miller's algorithm as one more contribution to the very large number of conditional results depending on RH. However we can look at it in a different way too. Namely, we know from Miller's theorem that of the three statements (2) – (4), at least one of them is not true. Thus to refute RH it would suffice to prove (2) and (3). While (2) seems to have no number-theoretic flavour, (3) would serve as a bridge between number theory and logic.

Let us now consider another classification studied by logicians. This time it will be a classification of mathematical statements about natural numbers. Such a statement can contain parameters. The top level in the hierarchy will consist of **decidable** statements. A statement $\mathcal{D}$ with parameters $a_1, a_2, \dots, a_k$ is called decidable if for given numerical values of the parameters we can, at least in principle, decide in a finite number of steps whether $\mathcal{D}(a_1, a_2, \dots, a_k)$ is true or not. The statement "$a$ is a prime number" can serve as an example of a decidable statement.

One level below decidable statements one finds two more classes of statements that can be obtained from decidable statements by bounding some of the variables, either by existential quantifiers or by universal quantifiers. If for some values of the parameters $a_1, a_2, \dots, a_k$ the statement

$$\exists\, x_1, x_2, \dots, x_n \;\; \mathcal{D}(a_1, a_2, \dots, a_k, x_1, x_2, \dots, x_n) \tag{5}$$

where $\mathcal{D}(a_1, a_2, \dots, a_k, x_1, x_2, \dots, x_n)$ is a decidable statement with $k+n$ parameters is true, then we can determine the validity of (5) by finding corresponding values of $x_1, x_2, \dots, x_n$ and verifying $\mathcal{D}$.

If for some values of the parameters $a_1, a_2, \dots, a_k$ the statement

$$\forall\, x_1, x_2, \dots, x_n \;\; \mathcal{D}(a_1, a_2, \dots, a_k, x_1, x_2, \dots, x_n) \tag{6}$$

is false then we again can determine that fact. However, if (5) is false or (6) is true, we have, in general, no method to discover it.

On the third level there are also two classes of statements obtainable by adding existential or universal quantifiers. We now have no means for determining the validity or the falsehood of the statements

$$\exists\, y_1, y_2, \dots, y_m \;\; \forall\, x_1, x_2, \dots, x_n \;\; \mathcal{D}(\bar{a}, \bar{x}, \bar{y})$$

or

$$\forall\, y_1, y_2, \ldots, y_m\ \exists\, x_1, x_2, \ldots, x_n\ \mathcal{D}(\bar{a}, \bar{x}, \bar{y})$$

even if $D$ is decidable.

The hierarchy goes down infinitely. Note that it is the number of changes of quantifiers which is significant, not the number of quantifiers themselves.

So, where in this hierarchy does RH lie? We need to make this question more precise, RH can be reformulated in many ways which are mathematically equivalent but which result in statements in different classes in the considered hierarchy. More precisely, the question is as follows: in what classes can we find statements today (without parameters) which are equivalent to RH?

The traditional mathematician usually thinks in terms of only the top level class, but, to place RH there means exactly to prove or disprove it. Logicians are more flexible, they have a richer language. Actually, the question of the position of RH in the above hierarchy was of interest to logicians for at least half a century. In 1939 A.M. Turing [11] proved that

$$\text{RH} \Leftrightarrow \forall n\, \exists m\ T(n, m)$$

where $T(n,m)$ is a decidable relation between natural numbers $n$ and $m$. Twenty years later G. Kreisel [4] improved Turing's representation by showing that

$$\text{RH} \Leftrightarrow \forall n\ K(n) \tag{7}$$

for some decidable property $K(n)$ of natural numbers. (Actually Kreisel obtained much more general results about analytic functions.)

Kreisel's proof is too technical to be reproduced here. Instead, we will provide some heuristic evidence in favour of the possibility of such a reformulation of RH.

The first idea that comes to mind is to simply define $K(n)$ as the statement that the first $n$ zeros of the zeta function do lie on the critical line. In fact, recently this $K(n)$ has been verified, with the aid of computers, up to a very large bound, which will surely be exceeded by the time of publication of this paper.

Unfortunately, methods actually used for the location of the zeros depend heavily on the (fortunate) fact that all these zeros turned out to be simple. However, these methods would fail in the case of a multiple zero even lying on the critical line.

The other idea (the one actually exploited by Kreisel) is as follows: We can represent the half-strip $0 < \mathrm{Re}\, s < 0.5$ as the union of countably many closed rectangles. Then one can define $K(n)$ as the statement that the $n^{\text{th}}$ rectangle contains no zeros. To decide whether $K(n)$ is true or not, one can evaluate the integral

$$\frac{1}{2\pi i} \int \frac{\zeta'(s)}{\zeta(s)} ds \tag{8}$$

where the integral is taken over the boundary of the $n^{th}$ rectangle. Since (8) is known to be an integer, it suffices to calculate it with an error not greater than, say 1/3 which can be done by numerical methods.

Unfortunately we cannot exclude, a priori, the possibility of a zero lying exactly on the boundary of the rectangle and, in such a case, calculation of (8) would fail. To overcome this difficulty we can shift the position of the rectangle a bit. For details see Kreisel's papers.

We have outlined two constructions of decidable $K(n)$ satisfying (7). In both cases the equivalence to RH was evident, although the decidability was not. Now we wish to exhibit an obviously decidable example of a statement $K(n)$ which also satisfies (7), but that equivalence is not so evident.

**Theorem.** *The Riemann hypothesis is equivalent to the assertion that for all* $n$

$$\left( \sum_{m \le \delta(n)} \frac{1}{m} - \frac{n^2}{2} \right)^2 < 36 n^3 \tag{9}$$

*where*

$$\delta(n) = \prod_{m < n} \prod_{l \le m} \eta(l),$$

*and*

$$\eta(l) = \begin{cases} p, & \text{if } l = p^w, \ p \text{ is prime} \\ 1, & \text{otherwise.} \end{cases}$$

The inequality (9) is nothing but the restatement, in integers, of the inequality

$$\left|\psi_1(n) - n\right| < 6n\sqrt{n}$$

where

$$\psi_1(n) = \int_1^n \psi(t)\,dt = \ln \delta(n)$$

$\psi(t)$ is the Tchebyshev psi function. The idea of using the latter inequality is due to H.N. Shapiro. For a detailed proof see [2]. A. Schinzel and M. Jutila observed that similar representation could be obtained from corresponding inequalities for the Tchebyshev theta function (the numerical value of the constant in the $O$-symbol can be taken, for example, from [10]).

It is obvious that a proof of RH would have many important consequences, but could we make any use of the much weaker result (7)? The answer is: yes, we could.

First, we mention the following result which was obtained by logicians as a by product of their investigations into Hilbert's tenth problem (see [2]): every statement of the form (7) with a decidable predicate $\mathcal{D}$ is equivalent to a statement of the form

$$\forall\, x_1, x_2, \ldots, x_m \;\; P(a_1, a_2, \ldots, a_k, x_1, x_2, \ldots, x_n) \neq 0$$

where $P$ is a polynomial with integer coefficients. Thus (7) implies the existence of a particular polynomial $R$ with integer coefficients such that

$$\text{RH} \Leftrightarrow \forall\, x_1, x_2, \ldots, x_m \;\; R\left(x_1, x_2, \ldots, x_m\right) \neq 0 \qquad (10)$$

Such an $R$ could be exhibited explicitly but would be rather cumbersome. Now we define $K(n)$ as

$$\forall\, x_1 < n,\, x_2 < n, \ldots,\, x_m < n \;\; R\left(x_1, x_2, \ldots, x_m\right) \neq 0$$

to get one more restatement of RH in the form (7). Number theorists did not suspect the possibility of being able to express the RH in terms of a single Diophantine equation.

The restatement of RH in the form (7) is useful because it provides a clear answer to the following question often asked by non-logicians: Should not logicians try to prove that

RH is independent from the usual set theoretic axiomatic systems as they did with, for example, the continuum hypothesis?

Statement (10) shows that unless an axiomatic theory $T$ is extremely weak, to prove RH to be independent from $T$ would imply a proof of RH to be valid. Indeed, suppose RH is not true, then its refutation could consist of calculating the polynomial $R$ for the proper values of $x_1, x_2,..., x_m$. Thus the theory $T$ would have to be so weak that either it does not allow arithmetic operations on integers or the equivalence (10) cannot even be proved in $T$.

Finally let us consider the heuristic meaning of (7). According to (7), we can split RH into countably many subhypotheses and we can prove each individual subhypothesis (provided, of course, that RH is true). That is, we can prove $K$ (1), $K$ (2), and so on, and thus proving RH is just a matter of proper generalization! The success of such an approach depends crucially upon the nature of this decidable $R$. The proofs of $K$ (1), $K$ (2),... should be informative and give us insight. Unfortunately, we cannot say so about the above outlined existing constructions of $K$. In the opinion of this author we should look for new formulations of RH in the form (7) which would allow proofs of $K$ (1), $K$ (2),... which gives us genuine insight. In the remainder of this paper we will describe one of the author's attempts to do so (see also [6], [7]).

The first problem to be tackled is the definition of the zeta function itself. Canonically it is defined by a Dirichlet series in the half-plane Re $s$ >1 while all the zeros lie in the opposite half-plane Re $s$ <1. The zeta function can be defined on the whole plane by its Laurent expansion at the pole:

$$\zeta(s) = \frac{1}{s-1} + \sum_{n=0}^{\infty} \frac{(-1)^n \gamma_n (s-1)^n}{n!} \tag{11}$$

Of course, to use (11) as a definition of $\zeta$, we need to define the $\gamma$'s in an independent way. Fortunately, expressions for the $\gamma$'s had been found by T.J. Stieltjes (and were rediscovered many times afterwards, see for example [12]). Namely,

$$\gamma_n = \mathrm{T}\left\{ \frac{(\ln(t))^n}{t} \right\}, \qquad n = 1,2,\ldots$$

where $\mathrm{T}f$ is the limit of the error of the trapezoid quadrature formula:

$$\mathrm{T}f = \lim_{M \to \infty} \left\{ \sum_{m=1}^{M-1} \frac{f(m) + f(m+1)}{2} - \int_1^M f(t)\,dt \right\} \tag{12}$$

As for $\gamma_0$, it is known to be equal to the Euler constant $\gamma = 0.577215...$ $= 0.5 + \mathrm{T}\{1/t\}$.

To verify for particular values of $s$ that $\zeta(s) = 0$ according to (11) one needs to know *all* the $\gamma$'s. Our first idea was to try to define each $K(n)$ depending only on *finitely many* of the $\gamma$'s, and in fact this can be done. For technical reasons it is more natural to enumerate the subhypothesis with two indices, i.e., instead of (7) we will have

$$\mathrm{RH} \Leftrightarrow \forall n \, \forall m \; M(n, m)$$

where $M(n, m)$ is a decidable relation to be defined (in [7] the author outlined another restatement of RH of similar style but resulting naturally in a one variable representation like (7)).

Let

$$\theta_{n,m} = \sum_{k_1 < \ldots < k_n} \prod_{l=1}^{n} [\rho_{k_l}(1 - \rho_{k_l})]^{-m} \tag{13}$$

where $\rho_1, \rho_2, \ldots$ is the list of all non-trivial zeros of the zeta function.

Now we define

$$M(n, m) \Leftrightarrow \theta_{n,m} > 0$$

The implication

$$\mathrm{RH} \Rightarrow \forall n \; \forall m \; \theta_{n,m} > 0$$

is evident because in this case $1 - \rho_{k_l} = \bar{\rho}_{k_l}$ and hence, all the quantities $\rho_{k_l}(1 - \rho_{k_l})$ in (13) are positive.

To see intuitively why the converse implication holds consider the case $n = 1$ and define for simplicity

$$\theta_m = \theta_{1,m} = \sum \frac{1}{[\rho_k(1 - \rho_k)]^m} \tag{14}$$

We may assume that the zeros are enumerated in such a way that

$$\left|\rho_k(1-\rho_k)\right| \le \left|\rho_{k+1}(1-\rho_{k+1})\right| \tag{15}$$

and in case of equality in (15) that

$$\left|\operatorname{Re}\rho_k - 0.5\right| \le \left|\operatorname{Re}\,\rho_{k+1} - 0.5\right|$$

Let $k_0$ be the least value of $k$ for which the strict inequality in (15) holds. Then the sum (14) will, for large $m$, be dominated by the contribution of the first $k_0$ summands (the precise notion of such a dominance is rather involved and is not given here). As a result of this domination, the $\theta_m$ cannot be positive for all $m$ unless $\operatorname{Re}\rho_1 = 0.5$.

We can now proceed by induction. The sum (13) is dominated for large $m$ by several first terms with the absolute value equal to

$$\prod_{k=1}^{n}\left[\rho_k(1-\rho_k)\right]^{-m}$$

If we know that $\operatorname{Re}\rho_k = 0.5$ for $k = 1,2,\ldots,n-1$, then $\theta_{n,m}$ cannot be positive for all $m$ unless $\operatorname{Re}\rho_n = 0.5$.

Each of the $\theta_{n,m}$ can be expressed in closed form in terms of finitely many $\gamma$'s and some other more tractable constants (see, for example, [8]). In particular,

$$\theta_1 = \gamma - \ln(4\pi) + 2 \tag{16}$$

We can easily check that $\theta_1 = 0.046\ldots > 0$ but let us take a closer look at (16). Here we have three apparent irrationalities, seemingly very different in their nature: $\gamma$, $\ln\pi$, and $\ln 2$. What drives them to make (16) positive? What do they have in common? Besides the raw numerical calculation we know there is a deep underlying reason for

$$\theta_1 > 0 \tag{17}$$

due to its connection (14) with the $\rho$'s. Can we find something intermediate, namely an unconditional analytical "explanation" of (17)?

We have been searching for a long time for such an "explanation." We gave one in [7]. However, the explanation in [7] failed to generalize (17) and so in this paper we will present another "explanation" which is more conducive to generalization.

Let us represent $\theta_1$ as $\theta_1 = \sigma_1 - \tau_1$ where $\sigma_1 = \gamma - \ln(2\pi) + 1$ and $\tau_1 = \ln 2 - 1$. Clearly,

$$\tau_1 = -\frac{1}{2} + \frac{1}{3} - \frac{1}{4} + \ldots = -\frac{1}{2\cdot 3} - \frac{1}{4\cdot 5} - \ldots = \sum_{\substack{r=-\infty \\ r \equiv 0 \,(\mathrm{mod}\ 2)}}^{-2} \frac{1}{r(1-r)}$$

In other words, $\sigma_1$ is a sum similar to (14) with $m = 1$, but with summation over all the trivial zeros. Hence, $\tau_1$ is also a sum of the same type but with summation over all the zeros. It is not difficult to check that

$$\sigma_1 = 2\mathrm{T}\psi \tag{18}$$

where $\tau$ is defined by (12) and $\psi(t) = \dfrac{\Gamma'(t)}{\Gamma(t)}$ is now the logarithmic derivative of the gamma function. The $\gamma$ summand in $\sigma_1$ comes from the sum in (12) while $\ln(2\pi)$ is due to the integral which is, of course, $\ln\Gamma(M)$ and contains $\ln(2\pi)$ in the Stirling expansion.

Now we need to find an expression for $\tau_1$, moreover, it should be similar to (18) to imply (17). Fortunately,

$$\tau_1 = 2\mathrm{T}\mathcal{A}\psi \tag{19}$$

where $\mathcal{A}$ is an averaging operator:

$$(\mathcal{A}f)(t) = \int_{-0.5}^{0.5} f(t+x)\,dx \tag{20}$$

On other words, the contributors of the trivial zeros to $\tau_1$, the sum over all the zeros, can be obtained just by averaging $\psi$ in (18).

Respectively,

$$\theta_1 = 2\mathrm{T}(\mathcal{E} - \mathcal{A})\psi$$

where $\mathcal{E}$ is the identity operator.

It is easy to derive from (12) and (20) that

$$T(\mathcal{E} - \mathcal{A}) f = \int_{0.5}^{\infty} W(t)\, f''''(t)\, dt$$

with non-positive weight $W(t)$ and hence (17) is a corollary of the well-known property of $\psi$, that $\psi''''(t) < 0$ for $t > 0$. This is our "explanation" for the positiveness of $\theta_1$.

Let us now turn to $\theta_2$. It can be shown that

$$\theta_2 = 2\gamma^2 + 4\gamma_1 + 2\gamma - 2\ln(4\pi) + \frac{\pi^2}{4} + 2$$

We can also split $\theta_2$ into $\theta_2 = \sigma_2 - \tau_2$ where

$$\sigma_2 = 2\gamma^2 + 4\gamma_1 + 2\gamma - 2\ln(2\pi),$$

and

$$\tau_2 = 2\ln 2 - \frac{\pi^2}{4}$$

Again $\sigma_2$ is the counterpart to $\theta_2$ with summation over all the zeros, while $\tau_2$ accounts for the trivial ones. Now we have the following analog of (18):

$$\sigma_2 = 4T\varphi \tag{21}$$

where $\varphi(t) = \dfrac{(t\,\varphi(t) + \gamma)}{(t - 1)}$.

However, a direct analogy with (19) fails:

$$\tau_2 \neq 4T\mathcal{A}\varphi$$

It is not difficult to see the "reason" for this failure: $\theta_2$ is much less than $\theta_1$ and the operator $\mathcal{A}$ is too crude an approximation of $\mathcal{E}$:

$$((\mathcal{A} - \mathcal{E})f)(t) = \frac{f''(t + \epsilon_1)}{24} \tag{22}$$

where $|\epsilon_1| \le 0.5$.

Double averaging is even worse:

$$((\mathcal{A}^2 - \mathcal{E})f)(t) = \frac{f''(t + \epsilon_2)}{12} \tag{23}$$

However, (22) and (23) imply that the operator $\mathcal{A}_2 = 2\mathcal{A}^2 - \mathcal{A} - \frac{\mathcal{D}}{8}$ is a better approximation to $\mathcal{E}$ where

$$(\mathcal{D}f)(t) = (\mathcal{A}f'')(t) = f'(t+0.5) - f'(t-0.5) = f''(t+\epsilon_3)$$

and in fact it gives us the desired formula for $\tau_2$:

$$\tau_2 = 4\mathrm{T}\mathcal{A}_2\varphi \tag{24}$$

So, we see the same phenomenon: the contribution of the trivial zeros to $\sigma_2$ being obtained for from a T-representation for $\theta_2$ by a proper averaging.

The pairs of formulas, (18)/(19) and (21)/(24) look very promising for further generalization, and one could expect that they are the beginning of an infinite series of equations of the type

$$\sigma_n = \mathrm{T}\varphi_n$$
$$\tau_n = \mathrm{T}\mathcal{A}_n\varphi_n$$
$$\theta_n = \sigma_n - \tau_n$$

However, we have been unable to find a suitable $\varphi_3$ and $\mathcal{A}_3$. Nevertheless, the pairs (18)/(19) and (21)/(24) do look attractive by themselves. They give us one more relation between the zeta and gamma functions. The formulas were found by trial and error and are essentially numerical identities. It would be interesting to discover if they reflect some intrinsic analytical relation between the zeta and gamma functions

To conclude, I wish to give my own suggestion, as a logician, for future progress on the Riemann Hypothesis. We should take advantage of the possibility to express the RH in a simple, logical form such as $\forall n\ K(n)$ with decidable $K(n)$, and try to find informative proofs for $K(1), K(2), \ldots$ for different choices of $K$. Then we can hope to gain enough insight to prove $\forall n\ K(n)$, i.e., RH.

## Appendix: An Alternative to the Euler–MacLaurin Summation Formula

In my search for formulas for $\sigma_n$ and $\tau_n$ I used a computer extensively to calculate $\mathrm{T}f$, for various $f$, with sufficient accuracy. A straight forward way to compute $\mathrm{T}f$ is to use the Euler–MacLauren summation formula. However, this formula is not very suitable for software implementation since it includes, besides $f$ and

$F(x) = \int f(t)\,dt$, a number of derivatives of $f$. So one has either to program each of the derivatives separately, by hand, or to use some kind of computer algebra package to produce the derivatives automatically. Below we present another method for calculating multi-precision values of $\mathrm{T}f$ which require calculations of $f$ and $F$.

We can start with any quadrature formula of the form

$$\sum_{k=0}^{N} A_k\, g(k) - \frac{1}{N}\int_0^N g(t)\,dt = c g^{<w>}(x) \tag{25}$$

where $g^{<w>}$ is the $w$ th derivative of $g$ and $x \in [0,N]$. To be able to have a small $c$ we do not require (25) to be a precise quadrature formula of degree $N$ but the equality

$$A_0 + \ldots + A_N = 1$$

is necessary.

We now define

$$\mathrm{T}_M f = \sum_{m=M}^{\infty}\left(\sum_{k=0}^{N} A_k f(m+k) - \frac{1}{N}\int_m^{m+N} f(t)\,dt\right) \tag{26}$$

(definition (12) is the same as (26) with $M = N = 1$, $A_0 = A_1 = 0.5$.) It is easy to verify that

$$\mathrm{T}f = 0.5 f(1) + \sum_{k=2}^{M-1} f(k) + \sum_{k=0}^{N-1} B_k f(M+k) + F(1) - \frac{1}{N}\sum_{k=0}^{N-1} F(M+k) + \mathrm{T}_M f$$

where

$$B_k = \sum_{l=k+1}^{N} A_l \qquad k = 0,\ldots,N-1$$

Neglecting the last summands results in the error

$$\mathrm{T}_M f = c \sum_{m=M}^{\infty} f^{<w>}(x_m) \approx c f^{<w-1>}(x)$$

**Acknowledgement.** The author wishes to thank Professor R.A. Mollin and J.P. Jones for inviting him to the First Conference of the Canadian Number Theory Association.

## References

[1] *S.A. Cook* , The Complexity of Theorem–Proving Procedures. Conference Records of the Third ACM Symp. on the Theory of Computing (1970), 151—158.

[2] *M. Davis* , *Yu.V. Matijasevich* and *J. Robinson* , Hilbert's Tenth Problem. Diophantine Equations: Positive Aspects of a Negative Solution. Mathematical Developments Arising From the Hilbert Problems. Proc. Sympos. Pure Math., **28** (1976), AMS, Providence, RI, 323—378.

[3] *R.M. Karp* , Reducibility Among Combinatorial Problems. Complexity of Computer Computations, R.E. Miller, J.W. Thatcher, Editors, Plenum, New York, (1972), 85—103.

[4] *G. Kreisel* , Mathematical Significance of Consistency Proofs. J. Symbolic Logic **23** (1958), No. 2, 155—182.

[5] *D.H. Lehmer* , Computer Technology Applied to the Theory of Numbers. MAA Studies in Mathematics **6** (1969), Studies in Number Theory, W.J. Leveque, Editor, 117—151.

[6] *Yu.V. Matijasevich* , Yet Another Machine Experiment in Support of the Riemann Hypothesis. (Russian), Kibernetika (Kiev), n6 (1982), 10, 22, (English translation in: Cybernetics **18** (1982), n6, 705—707(1983)).

[7] *Yu.V. Matijasevich* , An Analytical Representation For the Sum of Values, Inverse to Non–Trivial Zeros of the Riemann Function. (Russian) Trudy Mat. Inst. Steklov **163** (1984), 181—182, MR86d—11069.

[8] *Y.A. Matsuoka* , A Sequence Associated With the Zeros of Riemann Zeta Functions. Tsukuba J. Math. **10** (1986), n2, 249—254.

[9] *G.L. Miller* , Riemann's Hypothesis and Tests For Primality. J. Comput. System Sci. **13** (1976), 300—317.

[10] *L. Schönfeld*, Sharper Bounds for the Chebyshef Functions $\theta(x)$ and $\psi(x)$. Math. Comp., **30** (1976), n134, 337—360, n136, 900.

[11] *A.M. Turing* , Systems of Logic Based On Ordinals. Proc. London Math. Soc. (2) **45**(1939), 161—228.

[12] *S.C. van Veen* , Review 2232, Math. Reviews **29** (1965), n3, 429—430.

---

Steklov Mathematical Institute, Academy of Sciences of the USSR, 27 Fontanka, 191011, Leningrad, USSR.

# Triangular Farey Arrays

*Michael E. Mays*

## Introduction

The Farey series of order $n$, $F_n$, is usually given as the ascending series of irreducible fractions between 0 and 1 whose denominators do not exceed $n$. A standard reference is [3], with more historical material available in [1]. For example, $F_6$ is

$$0/1,\ 1/6,\ 1/5,\ 1/4,\ 1/3,\ 2/5,\ 1/2,\ 3/5,\ 2/3,\ 3/4,\ 4/5,\ 5/6,\ 1/1.$$

Two basic properties of Farey series are

(1) If $h/k$ and $h'/k'$ are two successive terms of $F_n$, then $kh' - hk' = 1$, and

(2) If $h/k$, $h''/k''$, and $h'/k'$ are three successive terms of $F_n$, then $h''/k'' = (h+h')/(k+k')$.

The fraction $(h+h')/(k+k')$ is called the mediant of $h/k$ and $h'/k'$. "Mediant" is an appropriate term because if $h/k < h'/k'$, then $h/k < h''/k'' < h'/k'$. A mediant may be inserted between any pair of fractions, and mediants between the resulting pairs, and so on, but the resulting fractions need not be irreducible. However, $F_n$ may be viewed as arising from $F_{n-1}$ by inserting the appropriate $\phi(n)$ mediants, each of which is irreducible. Hence, if the initial fractions are 0/1 and 1/1 then not only are the mediants inserted all irreducible, but also every rational number between 0 and 1 eventually arises exactly once as a mediant.

We can disregard the restriction on the size of the denominators to insert mediants in a different order. Beginning at step 0 with 0/1 and 1/1, first insert the mediant 1/2 at step 1, then the mediants between 0/1 and 1/2 and between 1/2 and 1/1 at step 2, then the 4 mediants in each of the new subintervals (instead of just the two having denominator 4) at step 3, and so on. This produces a sequence of families of rational numbers between 0 and 1 as well, in which the family after step $n$ contains $2^n + 1$ numbers, and for which properties (1) and (2) still hold. All rationals between 0 and 1 are still eventually inserted, but rationals with the same denominator need not be inserted in the same step.

Gould [2] considered a generalization of this process, defining a mediant of three fractions by forming from $a/b$, $c/d$, and $e/f$ the (reduced) fraction

$$F(a/b,\ c/d,\ e/f) = (a+c+e)/(b+d+f). \tag{3}$$

As there are first 1, then 2, then 4, then 8 mediants to insert between initial fractions $a/b$ and $c/d$ in the process described above, in this case an initial set of three fractions provides first 1, then 3, then 9, then 27 mediants. Powers of 2 are replaced by powers of 3. Gould provided calculations for particular starting values and raised questions about which fractions can appear from given initial fractions. He also asked about the average of the fractions introduced at step $n$.

This paper is concerned with a related, but different, situation. We also start with points at three vertices of a triangle and build successive mediants, but the mediants are formed by considering points only two at a time, and placed on the line connecting the two points rather than in the interior of the triangle. In the first part of the paper we consider the points to be ordered pairs (fractions) and relate this construction to Gould's ideas. In particular, we obtain results about irreducibility and limiting values in the "central triangles." In the second part of the paper we represent the points as ordered triples and develop a more general theory which parallels that of Farey series.

## The Central Triangles

We will use ordered pair notation in this section to emphasize that if reducible fractions arise, the cancellation of common factors will not be done. Reducing fractions introduces a complication in that since a cancelled fraction doesn't have as much weight as one that doesn't reduce (with cancellations affecting future terms), it is hard to predict how the values evolve.

To approach the center of the triangle whose vertices are $(a_0, b_0)$, $(c_0, d_0)$, and $(e_0, f_0)$, we should always insert new mediants by picking the central triangle to subdivide. This gives a sequence of values as shown in Figure 1.

**Figure 1. Stepping Towards the Center Via Central Triangles**

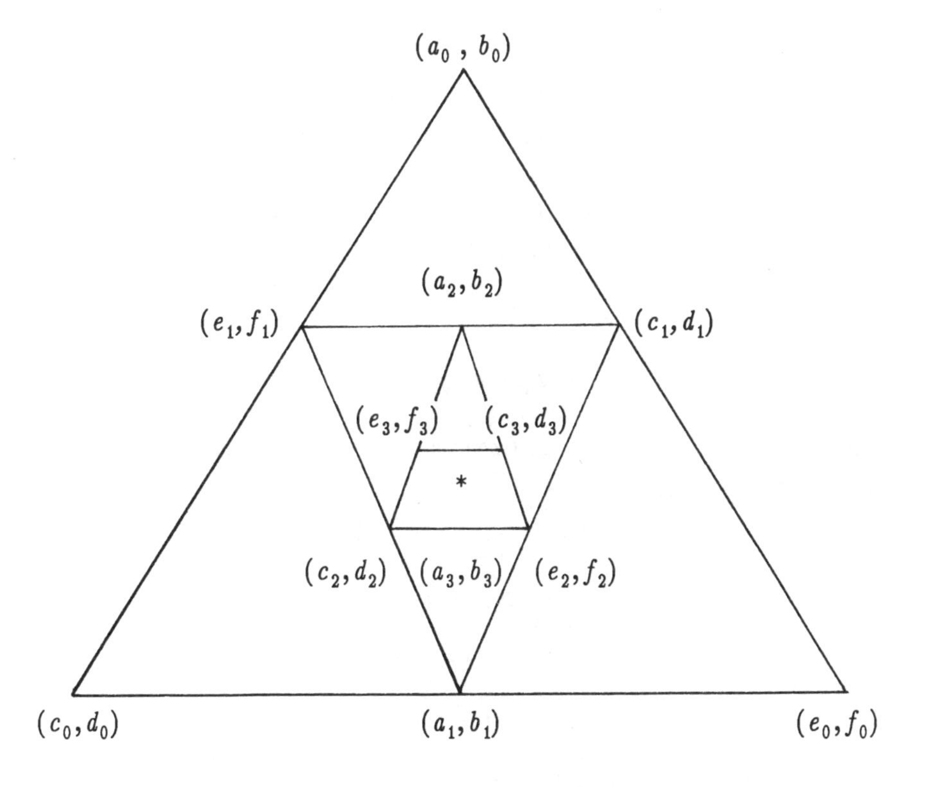

This construction is then related to that of Gould by the following result.

**Theorem 1.** *Let* $(a_0, b_0)$, $(c_0, d_0)$, *and* $(e_0, f_0)$ *be arbitrary ordered pairs of numbers. Define three sequences of ordered pairs recursively by*

$$\begin{aligned}(a_{n+1}, b_{n+1}) &= (c_n + e_n, d_n + f_n)\\ (c_{n+1}, d_{n+1}) &= (a_n + e_n, b_n + f_n)\\ (e_{n+1}, f_{n+1}) &= (a_n + c_n, b_n + d_n).\end{aligned}$$

*Then* $\lim_{n\to\infty} (a_n/b_n) = \lim_{n\to\infty} (c_n/d_n) = \lim_{n\to\infty} (e_n/f_n) =$

$$(a_0 + c_0 + e_0)/(b_0 + d_0 + f_0).$$

**Proof.** We exploit the symmetry among $\{(a_n, b_n)\}$, $\{(c_n, d_n)\}$, and $\{(e_n, f_n)\}$ as well as that between first and second coordinates in each case. Notice that the first few values of $a_n$, written in terms of $a_0$, $c_0$, and $e$, are

$$\begin{aligned}a_0 &= a_0\\ a_1 &= c_0 + e_0\\ a_2 &= 2a_0 + c_0 + e_0\\ a_3 &= 2a_0 + 3c_0 + 3e_0\\ a_4 &= 6a_0 + 5c_0 + 5e_0\\ a_5 &= 10a_0 + 11c_0 + 11e_0\end{aligned}$$

and so on. Write in general

$$a_n = x_n a_0 + y_n c_0 + z_n e_0,$$

and it is easy to use the recurrences to establish that, after $x_0 = 1$, $y_0 = z_0 = 0$, we have

$$\begin{aligned}x_{n+1} &= 2(x_n + (-1)^{n+1})\\ y_{n+1} &= 2y_n + (-1)^n \qquad (4)\\ z_{n+1} &= 2z_n + (-1)^n.\end{aligned}$$

Thus in particular $y_n = z_n$ for any $n$, all three of $x_n$, $y_n$, and $z_n$ approach infinity as $n$ increases, and $x_n = y_n + (-1)^{n+1}$. These last two facts are enough to conclude that $\lim_{n\to\infty} x_n/y_n = 1$.

By symmetry, the same coefficients work to give $b_n$ in terms of $b_0$, $d_0$, and $f_0$, and permutations of the coefficients arise in the formulas for $c_n$, $d_n$, $e_n$, and $f_n$.

Now write

$$\lim_{n\to\infty} a_n/b_n = \lim_{n\to\infty} (x_n a_0 + y_n c_0 + z_n e_0)/(x_n b_0 + y_n d_0 + z_n f_0) =$$

$$\lim_{n\to\infty} ((x_n/y_n)a_0 + c_0 + e_0)/((x_n/y_n)b_0 + d_0 + f_0) =$$

$$(a_0 + c_0 + e_0)/(b_0 + d_0 + f_0). \qquad \square$$

**Irreducibility**

Despite the appeal of reducing fractions, there are desirable properties that hold only if the fractions are left unreduced. For example, we would have for

$$F(a/b,\ c/d,\ e/f) = (a+c+e)/(b+d+f) = g/h$$

that

$$F(F(a/b,\ c/d,\ g/h),\ F(c/d,\ e/f,\ g/h),\ F(a/b,\ e/f,\ g/h))$$
$$= F(a/b,\ c/d,\ e/f).$$

Some choices of initial fractions will guarantee that all fractions in the central triangles remain irreducible. This happens for initial fractions 1/3, 2/5, and 7/12. Other behaviors that can arise are that of the initial fractions 3/8, 4/11, and 5/13, in which one sequence of vertices consists of fractions with ever increasing common factors, and that of the initial fractions 5/7, 6/23, and 11/40, in which one sequence of vertices has periodic common factors of 5 (period 12) and 11 (period 15). We account for these different behaviors in the necessary and sufficient condition for irreducibility below. We first note that, of the sequences $\{x_n\}$, $\{y_n\}$, and $\{z_n\}$ defined earlier, only one is necessary. We can build all three recurrences in (4) from the recurrence which $\{y_n\}$ satisfies: $y_0 = 0$, $y_1 = 1$, $y_{i+1} = y_i + 2y_{i-1}$ for $i \geq 1$, and the observation that $z_n = y_n$ and $x_n = 2y_{n-1}$.

**Theorem 2.** *Given the initial fractions* $a_0/b_0$ , $c_0/d_0$ *and* $e_0/f_0$ , *build* $a_1/b_1 = (c_0 + e_0)/(d_0 + f_0)$, $u/v = (a_0 + a_1)/(b_0 + b_1)$, *and* $a_{i+1}/b_{i+1} = (2y_{i-1}a_0 + y_i a_1)/(2y_{i-1}b_0 + y_i b_1)$, *where* $y_0 = 0$, $y_1 = 1$, *and* $y_{i+1} = y_i + 2y_{i-1}$ *for* $i \geq 1$. *Write*

$$a_{i+1}/b_{i+1} = ua_{i+1}/(va_{i+1} + r_{i+1}). \tag{5}$$

*Then* $r_{i+1} = (-1)^i(a_1b_0 - a_0b_1)$, *and the only possible common prime factors* $p$ *of* $a_{i+1}$ *and* $b_{i+1}$ *are those primes which divide* $u(a_1b_0 - a_0b_1)$. *Furthermore, the sequences* $\{a_i\}$ *and* $\{b_i\}$ *modulo* $p$ *are periodic, so only finitely many terms need to be compared for a common zero to decide if a common* $p$ *factor exists.*

**Proof.** From (5), $va_{i+1} + r_{i+1} = ub_{i+1}$, which says

$$(b_0 + b_1)(2y_{i-1}a_0 + y_i a_1) + r_{i+1} = (a_0 + a_1)(2y_{i-1}b_0 + y_i b_1).$$

Hence $r_{i+1} = (2y_{i-1} - y_i)(a_1b_0 - a_0b_1)$. But $y_i = 2y_{i-1} - (-1)^i$ so this reduces to $(-1)^i(a_1b_0 - a_0b_1)$.

Now suppose the fraction $a_{i+1}/b_{i+1}$ is reducible. A prime $p$ dividing top and bottom of the right hand side of (5), $ua_{i+1}/(va_{i+1} + r_{i+1})$, in particular either divides $u$ or $a_{i+1}$. There are only finitely many choices if $p$ is to divide $u$. If $p$ divides $a_{i+1}$ then since $p$ divides the denominator $p$ must divide $r_{i+1} = \pm(a_1b_0 - a_0b_1)$ as well, so again there are only finitely many choices.

The periodicity follows because $a_{i+1}$ depends only on $y_i$ and $y_{i-1}$, $y_{i+1}$ depends only on $y_i$ and $y_{i-1}$, and there are at most $p^2$ distinct pairs $(y_i, y_{i-1})$ available before pairs must repeat modulo $p$. □

**Corollary 3.** *Given the initial fractions* $a_0/b_0$ , $c_0/d_0$ , *and* $e_0/f_0$, *define* $u = a_0 + c_0 + e_0$ , $r = (c_0 + e_0)b_0 - a_0(d_0 + f_0)$, $s = (a_0 + e_0)d_0 - c_0(b_0 + f_0)$, *and* $t = (a_0 + c_0)f_0 - e_0(b_0 + d_0)$. *Then any reducible fractions in the central triangles must have a common factor divisor of rstu.*

**Proof.** Apply Theorem 2 to each sequence $\{a_i/b_i\}$, $\{c_i/d_i\}$, and $\{e_i/f_i\}$. □

We remark that $r+s+t=0$, which is an aid to calculation.

Now we can account for the different behaviors of the examples earlier. A parity argument lets us make an initial screening and thereafter ignore powers of 2. For 1/3, 2/5, and 7/12, $u=1$, $r=10$, $s=10$, $t=-20$, and there is no common zero modulo 5. For 3/8, 4/11, and 5/13, $s=0$ (4/11 is the mediant of 3/8 and 5/13). For 5/7, 6/23, and 11/40, $u=11$, $r=-196$, $s=-86$, and $t=110$, and the 5 and 11 factors of $t$ give ommon zeros in the sequences for $\{e_i\}$ and $\{f_i\}$ modulo 5 and modulo 11. These give fractions divisible top and bottom by 5 and by 11, which periodically coincide to give fractions with common factor 55.

**Theorem 4.** *There are infinitely many ways to choose initial fractions so that all the fractions arising in the central triangles are irreducible.*

**Proof.** As long as we start with vertices which are odd/odd, odd/even, and even/odd, this pattern is preserved in all subtriangles. Thus we can assume that none of the fractions in the central triangles can be of the form even/even. In fact, any subtriangle must have three vertices which are odd/odd, odd/even, and even/odd. This means that we can guarantee irreducibility by Corollary 3 if the initial fractions can be chosen so that $r$, $s$, $t$, and $u$ are all powers of 2. There are many ways to do this. For example, if

$$\begin{aligned} a_0/b_0 &= (2^k-1)2^k, \\ c_0/d_0 &= (2^k+2)/(2^k+3), \text{ and} \\ e_0/f_0 &= (2^{k+1}-1)/(2^{k+1}+1), \end{aligned}$$

then the limiting value is $2^k/(2^k+1)$, $r=-4$, $s=8$, and $t=-4$. It is clear that different $k$ produce different triples of fractions here. □

The algebra is somewhat simplified as we watch the triangle with initial vertices $(2^k+1)/(2^k+2)$, $(2^k-2)/(2^k-1)$, and 1/1 converge to $2^k/(2^k+1)$, and even more simplified as we approach 1/3 starting from 0/1, $(2^k-1)/(3\cdot 2^k-2)$, and 1/1, or 2/3 starting from 0/1, $(2^{k+1}-1)/(3\cdot 2^k-2)$, and 1/1. These last examples are interesting in that, unlike the first two, $r$, $s$, and $t$ are not constant.

**Theorem 5.** *No choice of initial fractions allows all mediants in all subtriangles to be irreducible.*

**Proof.** It is sufficient to consider numerators and denominators modulo 3. As soon as a fraction of the form 0/0 (mod 3) arises, irreducibility is violated. No two vertices can be the same because the second mediant would be of the form 0/0: $a/b$---0/0---$2a/2b$---0/0---$a/b$. From the 8 allowable forms $a/b$, then, there are ${}_8C_3 = 56$ triples to build, and a case by case analysis shows that 0/0 arises in each of them. □

## A Triangle of Triples

Sometimes properties of Farey series are developed from the starting values 0/1 and 1/0 rather than 0/1 and 1/1. In this case 1/1 is the first mediant, and there is a symmetry between proper fractions inserted to the left of 1/1 and their reciprocals to the right. In fact, it is the convention of Hurwitz [H1] to extend this so that the fractions in Farey series assume both positive and negative values, beginning with –1/0, 0/1, and 1/0. In [H2] Hurwitz considers the "Farey polygons" arising when these series are wrapped around a circle with 1/0 and –1/0 identified as the point $\infty$ , and 0/1, 1/1, and $\infty$ at the vertices of an inscribed equilateral triangle. This is used in an algorithm for reducing binary quadratic forms.

We introduce points as ordered triples to label the vertices of a triangular Farey array (TFA) in a symmetric way and avoid the ambiguity of starting with arbitrary fractions. The initial labelling is (1,0,0) at the top, (0,0,1) at the lower left, and (0,1,0) at the lower right. Assume the initial triangle has sides of unit length. Mediants are gotten by adding coordinates: the mediant of $(a,b,c)$ and $(d,e,f)$ is given as $(a+d,\ b+e,\ c+f)$, and is placed at the midpoint of the line joining $(a,b,c)$ and $(d,e,f)$. The first few triangles are shown in Figure 2. We follow the convention at the end of the first section, and at each stage insert mediants between all adjacent points from the step before. Adjacent points in step $n$ are points which are $1/2^n$ units apart. This produces a series of arrays $T_0, T_1, T_2, \ldots$ in which each triangle in $T_i$ is replaced by four subtriangles in $T_{i+1}$. In Theorem 6 we will want to distinguish between the three outer subtriangles in in $T_{i+1}$ that share a vertex with the original triangle in $T_i$ and the central subtriangle (which is upside down compared with the triangle in $T_i$). Note the appearance of numerators and denominators of Farey fractions along each side in the non-zero coordinates.

**Figure 2. The First Triangular Farey Arrays**

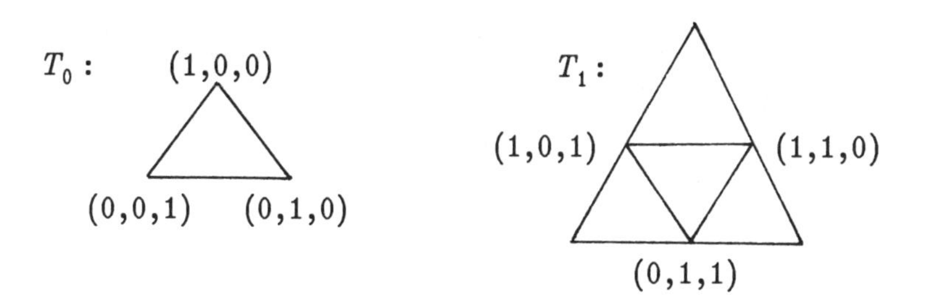


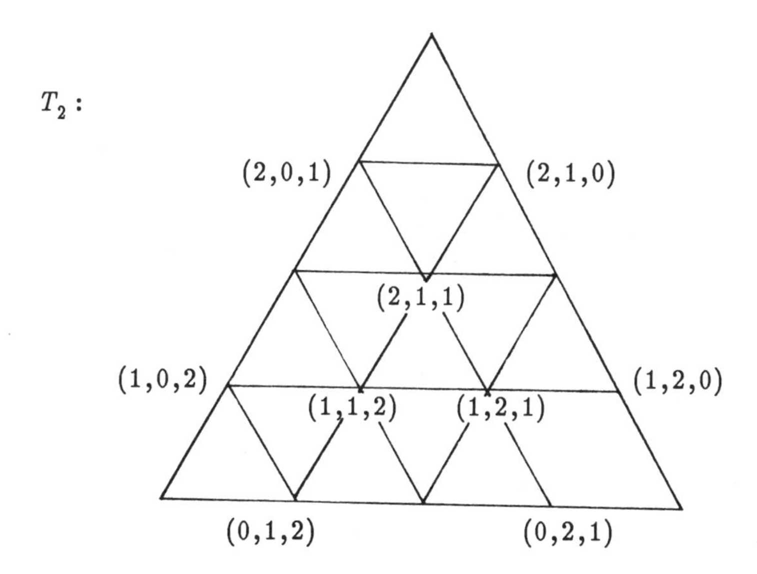


It is worthwhile when later arrays are drawn to exploit the six-fold symmetry of the points plotted: If $(a,b,c)$ appears in the triangle then every permutation of $(a,b,c)$ also appears, representing a point in the triangle which is a reflection of the original point about one or more altitudes. This is illustrated in Figure 3, and the triangle highlighted in the lower left corner with the property that every point $(a,b,c)$ in it has $a \leq b \leq c$ we call the fundamental triangle. This triangle is given in more detail in Figure 4, with a subtriangle expanded.

**Figure 3. Sixfold Symmetry and the Fundamental Triangle**

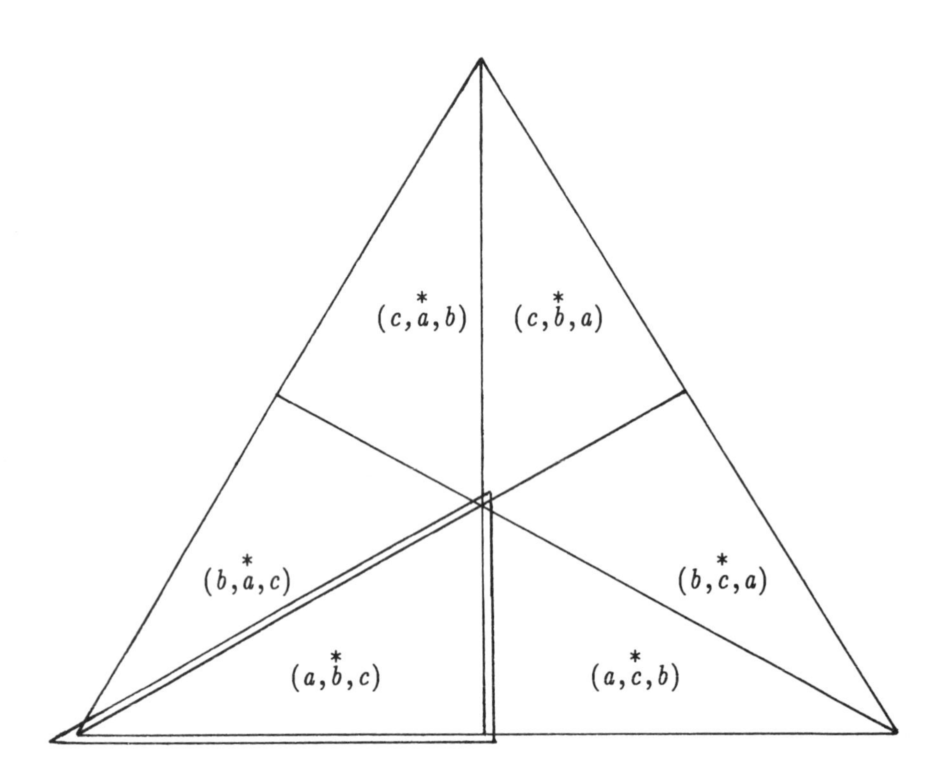

**Figure 4. The Fundamental Triangle and A Magified View**

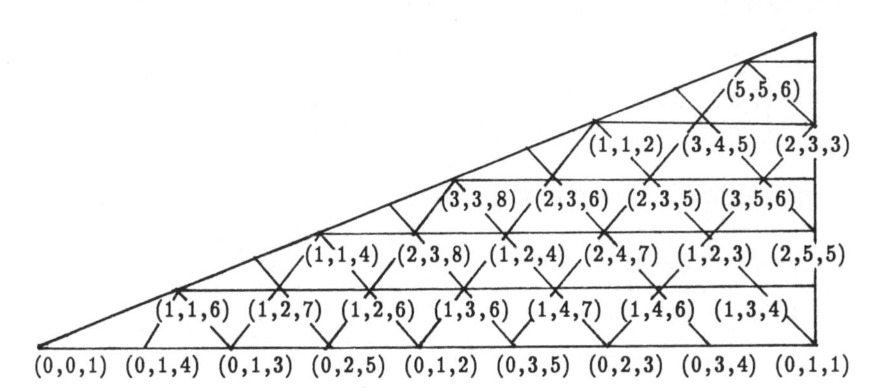


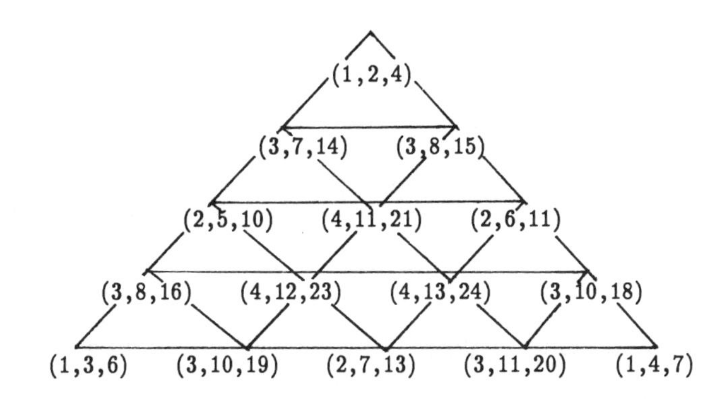


We begin by giving a theorem for TFAs analogous to property (1).

**Theorem 6.** *If (a,b,c), (d,e,f), and (g,h,i) are vertices of a subtriangle in* $T_k$ *read clockwise, then* $\begin{vmatrix} a & b & c \\ d & e & f \\ g & h & i \end{vmatrix} = 2^j$ *for that j,* $0 \leq j \leq k$*, which counts how many times the central triangle was chosen in the k intermediate steps in reaching that subtriangle in* $T_k$ *from the original triangle* $T_0$.

**Proof.** The original triangle, with vertices (1,0,0), (0,0,1), and (0,1,0), has $\begin{vmatrix} 1 & 0 & 0 \\ 0 & 1 & 0 \\ 0 & 0 & 1 \end{vmatrix} = 1$. If we consider an arbitrary triangle with vertices *(a,b,c)*, *(d,e,f)*, *(g,h,i)* in $T_n$ and consider the triangles formed within it in $T_{n+1}$

when the mediants are inserted, it is a consequence of a basic property of determinants that for the three outer triangles

$$\begin{vmatrix} a & b & c \\ d & e & f \\ g & h & i \end{vmatrix} = \begin{vmatrix} a & b & c \\ a+d & b+e & c+f \\ a+g & b+h & c+i \end{vmatrix} = \begin{vmatrix} a+d & b+e & c+f \\ d & e & f \\ d+g & e+h & f+i \end{vmatrix} = \begin{vmatrix} a+g & b+h & c+i \\ d+g & e+h & f+i \\ g & h & i \end{vmatrix}$$

However, stepping to the inner triangle gives

$$\begin{vmatrix} d+g & e+h & f+i \\ a+g & b+h & c+i \\ a+d & b+e & c+f \end{vmatrix} = \begin{vmatrix} d & e & f \\ a+g & b+h & c+i \\ a+d & b+e & c+f \end{vmatrix} + \begin{vmatrix} g & h & i \\ a+g & b+h & c+i \\ a+d & b+e & c+f \end{vmatrix} =$$

$$\begin{vmatrix} d & e & f \\ a+g & b+h & c+i \\ a & b & c \end{vmatrix} + \begin{vmatrix} g & h & i \\ a & b & c \\ a+d & b+e & c+f \end{vmatrix} = \begin{vmatrix} d & e & f \\ g & h & i \\ a & b & c \end{vmatrix} + \begin{vmatrix} g & h & i \\ a & b & c \\ d & e & f \end{vmatrix} = 2 \cdot \begin{vmatrix} a & b & c \\ d & e & f \\ g & h & i \end{vmatrix} .$$

□

This gives an analog of the irreducibility of Farey fractions as well.

**Corollary 7.** *If $(a,b,c)$ is a point in $T_n$, then the greatest common divisor of $a$, $b$, and $c$ is 1.*

**Proof.** A parity argument gives that not all of $a$, $b$, and $c$ are even. Every point $(a,b,c)$ consists of two even numbers and an odd number or two odd numbers and an even. If an odd prime $p$ existed with $a = pa'$, $b = pb'$, and $c = pc'$, then for $(d,e,f)$ and $(g,h,i)$ adjacent points to $(a,b,c)$ in $T_n$,

$$\begin{vmatrix} a & b & c \\ d & e & f \\ g & h & i \end{vmatrix} = p \begin{vmatrix} a' & b' & c' \\ d & e & f \\ g & h & i \end{vmatrix},$$

which contradicts Theorem 6. □

Now we give some instances of how variants of the mediant construction account for particular values in the fundamental triangle.

**Theorem 8.** *The (extended) hypotenuse of the fundamental triangle in $T_n$ consists of triples $(a,a,c)$ obtained by inserting weighted mediants in pairs.*

*First, between* (0,0,1) *and* (1,1,0) *is inserted the weighted mediant* (1,1,2) *in* $T_2$. *Then, if between* $(a,a,c)$ *and* $(d,d,f)$ *was inserted* $(g,g,i)$ *in* $T_n$, $(2a+g,\ 2a+g,\ 2c+i)$ *is inserted between* $(a,a,c)$ *and* $(g,g,i)$ *and* $(g+2d,\ g+2d,\ i+2f)$ *is inserted between* $(g,g,i)$ *and* $(d,d,f)$ *in* $T_{n+1}$.

**Proof.** Consider the initial rhombus with vertices labelled (0,0,1), (1,0,1), (1,1,0), and (0,1,0). The long diagonal is the extended hypotenuse of the fundamental triangle and the center of the rhombus is at (1,1,2), the first mediant. Successive weighted mediants are inserted as the rhombus is divided into subrhombi, with a weighted mediant given as the mediant of the mediants of two sides of a subrhombus adjacent at an acute angle. If the vertices at the obtuse angles are labelled $(a,b,c)$ and $(b,a,c)$ (using the symmetry of coordinates about an altitude of $T_n$), and the vertex at the acute angle is labelled $(d,d,f)$, this gives between $(a+b,\ a+b,\ 2c)$ and $(d,d,f)$ the point which is the mediant of $(a+d,\ b+d,\ c+f)$ and $(b+d,\ a+d,\ c+f)$, which is $(a+b+2d,\ a+b+2d,\ 2c+2f)$. □

We remark that, since the short leg of the fundamental triangle is the extended hypotenuse of another triangle symmetric with the fundamental triangle, a similar pattern holds for it as well.

The next two results are based on the observation that once points occur in $T_n$ they remain as points in $T_m$ for every $m \geq n$. Rows that are adjacent in $T_n$ are separated by 1 row in $T_{n+1}$, 3 in $T_{n+2}$, and generally by $2^k - 1$ rows in $T_{n+k}$. Going backwards, we see that counting $2^k$ rows up from the base of $T_n$ gives a row which has points from the earlier triangle $T_{n-k}$. There are, of course, extra points inserted in this row when it is seen as a row of $T_n$, which arise as mediants are inserted in the $k$ steps from $T_{n-k}$ to $T_n$. The sequence of first components follows a pattern independent of $n$:

$$
\begin{array}{ccccccccccccccccc}
1 & & & & & & & & 1 & & & & & & & & 1 \quad \ldots \\
1 & & & & 2 & & & & 1 & & & & 2 & & & & 1 \quad \ldots \\
1 & & 3 & & 2 & & 3 & & 1 & & 3 & & 2 & & 3 & & 1 \quad \ldots \\
1 & 4 & 3 & 5 & 2 & 5 & 3 & 4 & 1 & 4 & 3 & 5 & 2 & 5 & 3 & 4 & 1 \quad \ldots \\
 & & & & & & & & \vdots & & & & & & & &
\end{array}
$$

**Theorem 9.** *Consider horizonatal row* $2^k$ *above the long leg in the fundamental triangle of* $T_n$. *The first components of row* $2^k$ *are given by that sequence formed by inserting mediants between the first components of row* $2^{k-1}$. *In the triangle with vertices (a,b,c) in row* $2^k$ *and (0,e,f) and (0,h,i) on the long leg of the fundamental triangle,* $b = e + h - a$ *and* $c = f + i - a$.

**Proof.** The key is the observation above that the rows indexed by powers of 2 were first rows in earlier triangles. In $T_1$ the first row up has endpoints (1,0,1) and (1,1,0). In $T_2$ this is the second row up when (2,1,1) is inserted, and the new first row, with endpoints (1,0,2) and (1,2,0), has all 1's for its first components. Continuing the subdivision accounts for the behavior of the first components. For the second part, note that $b = e + h - a$ and $c = f + i - a$ for $(0,e,f) = (0,0,1)$, $(0,h,i) = (0,1,0)$, and $(a,b,c) = (1,0,0)$, that the relationship is preserved in subtriangles when mediants are inserted to form $T_n$ from $T_{n-1}$, and further that the relationship is preserved as new, overlapping triangles are formed when corresponding mediants, inserted between the vertices of triangles on row 0 and row $2^k$, are connected. □

The observation about rows $2^k$ in this proof also holds for diagonals, and leads to the following result.

**Theorem 10.** *Consider diagonal number* $2^k$ *in* $T_n$ *connecting the bottom to the left hand side. The endpoints of this diagonal are* $(0,1,n-k)$ *and* $(1,0,n-k)$, *and the points on the diagonal are of the form* $(a,b,(n-k)(a+b))$ *where the pairs (a,b) arise as the entries (0,a,b) in the bottom row of* $T_k$.

## Open Questions

There are many avenues to explore in this generalization of Farey series to triples of points in triangles. Here are several questions that have not been answered.

(1) Given a point in the interior of $T_n$, what is the nature of the TFA triples near it? Points near the center of $T_n$ tend to have coordinates

nearly equal, and points near an edge to have at least one coordinate relatively small. Is there an immediate connection between TFA coordinates and barycentric coordinates, or does inserting mediants at midpoints of lines distort the Euclidean metric and make this relationship too complicated?

(2) How many triples in the fundamental triangle end in a given digit? The corresponding question for Farey series involves only Euler's $\phi$ function.

(3) What extensions are there to higher dimensions? 4–tuples in nested tetrahedrons is a natural first step to consider.

(4) It is an easy matter to insert $n$ geometric means, or $n$ arithmetic means, between two given real numbers. How should one insert $n$ mediants between two rational numbers, or between two triples in TFAs? This is relevant to understanding the patterns of triples in the interior of the fundamental triangle.

(5) This paper has examples of families of triples of fractions for which irreducibility in the central triangles holds. Find all of those initial triples of fractions for which irreducibility of central triangles must hold.

(6) Must every irreducible triple $(a,b,c)$ consisting of two odds and an even or two evens and an odd occur in $T_n$ for some $n$?

## References

[1] *L.E. Dickson*, History of the Theory of Numbers. Chelsea, New York, 1971.

[2] *H.W. Gould*, A ternary Farey process. Unpublished notes, January, 1986.

[3] *G.H. Hardy* and *E.M. Wright*, An Introduction to the Theory of Numbers. Oxford, 1960.

[4] *A. Hurwitz*, Über die angenäherte Darstellung der Zahlen durch rationale Brüche. Math. Annalen **44** (1894), 417–436.

[5] *A. Hurwitz*, Über die Reduktion der binären quadratischen Formen. Math. Annalen **45** (1894), 85–117.

---

Department of Mathematics, West Virginia University,
Morgantown, West Virginia 26506, U.S.A.

# Solution of the Class Number One Problem For Real Quadratic Fields of Extended Richaud–Degert Type (With One Possible Exception)

*R.A. Mollin* [1] *and H.C. Williams* [2]

## 1. Introduction

A conjecture of Gauss, the solution of which seems intractable to this day, says there are infinitely many real quadratic fields with class number one. On the other hand, Gauss conjectured that there are only finitely many complex quadratic fields with class number one. The latter conjecture has been affirmatively settled for some time. For an excellent survey of the problem and its solution, the reader is referred to Goldfeld's article [3]. The early part of this century saw the first significant progress toward a solution of Gauss' conjecture for complex quadratic fields. In particular, in 1918, Landau [6] published a result attributed to Hecke. This result showed that Gauss' conjecture for a complex quadratic field follows from the generalized Riemann hypothesis (GRH) for $L(s,\chi)$ where $\chi$ is a real non–principal character.

It is the purpose of this article to give an overview of the solution to the class number one problem for **real** quadratic fields of Richaud–Degert (R–D) type, under the GRH assumption, and to show that the GRH can now be removed yielding an unconditional result for the more general extended R–D types with the possibility of only one other value having class number one.

[1] This author's research is supported by NSERC Canada Grant #A8484
[2] This author's research is supported by NSERC Canada Grant #A7649

## 2. Real Quadratic Fields of Richaud–Degert Type

Throughout, $d$ will denote a positive square–free integer and $h(d)$ will denote the class number of $\mathbb{Q}(\sqrt{d})$. A real quadratic field $\mathbb{Q}(\sqrt{d})$, or simply $d$, is said to be of R–D type (see[2] and [18]) if $d = l^2 + r \neq 5$ where $-l < r \leq l$ and $r$ divides $4l$. If $|r| \in \{1,4\}$ then $d$ is said to be of **narrow** R–D type. In [1], S. Chowla conjectured that $h(p) > 1$ for primes $p$ of the form $l^2 + 1$, with $l > 26$.

In [9], Mollin provided three necessary and sufficient conditions for $h(d)$ to be 1 when $d = l^2 + 1$. Although these conditions provided strong evidence for the validity of the Chowla conjecture, no unconditional proof has yet been provided. However, in [12], Mollin and Williams were able to use the GRH to prove the Chowla conjecture. Moreover, it follows from this paper that $n - 1$ of the $n$ conjectures concerning R–D types of class number one (including the Chowla conjecture) are true, with the remainder failing for possibly only one value.

It follows from the celebrated Brauer–Siegel Theorem that there are only finitely many real quadratic fields of R–D type with class number one. The problem is that there is no known effective bound for $d$. This is where the GRH comes in. It provides a bound for $d$ beyond which $h(d) > 1$ when $d$ is of R–D type.

In [13], Mollin and Williams found what appeared to be general equivalent conditions for class number one as follows.

**Theorem 1.** *(Mollin–Williams) Let* $d \equiv 1 (\mathrm{mod}\, 4)$ *and let* $\alpha = \frac{\sqrt{d} - 1}{2}$. *Then the following are equivalent.*

I) $h(d) = 1$.

II) $f_d(x) = -x^2 + x + \frac{(d-1)}{4} \not\equiv 0 (\mathrm{mod}\, p)$ *for all integers x and primes p such that* $0 \leq x < p < \frac{\sqrt{d} - 1}{2}$.

III) $f_d(x)$ *is prime for all integers x with* $1 < x < \alpha$.

IV) *p is inert in* $\mathbb{Q}(\sqrt{d})$ for all primes $p < \alpha$.

Observe that the equivalence of (I) and (III) in Theorem 1 is strikingly similar to the well–known Rabinowitsch result, in [16] and [17], for imaginary quadratic fields which says that for $d \equiv 3 (\mathrm{mod}\, 4)$, $h(-d) = 1$ if and only if $x^2 - x + \frac{(d+1)}{4}$ is prime for all integers $x$ with $1 \le x \le \frac{(d-3)}{4}$.

Mollin and Williams discovered that Theorem 1 is not so general. In [13] they also proved the surprising:

**Theorem 2.** *(Mollin–Williams) If part* (III) *of Theorem 1 holds for* $d > 13$ *then* $d \equiv 1 (\mathrm{mod}\, 4)$ *is of narrow R–D type.*

The following table illustrates the above.

**Table 1: $h(d) = 1$**

| $d$ | prime values of $f_d(x)$ for $1 < x < \alpha$ | | | | | | | | | |
|---|---|---|---|---|---|---|---|---|---|---|
| 17 | — | | | | | | | | | |
| 21 | 3 | | | | | | | | | |
| 29 | 5 | | | | | | | | | |
| 37 | 7 | | | | | | | | | |
| 53 | 7 | 11 | | | | | | | | |
| 77 | 7 | 13 | 17 | | | | | | | |
| 101 | 13 | 19 | 23 | | | | | | | |
| 173 | 13 | 23 | 31 | 37 | 41 | | | | | |
| 197 | 19 | 29 | 37 | 43 | 47 | | | | | |
| 293 | 17 | 31 | 43 | 53 | 61 | 67 | 71 | | | |
| 437 | 19 | 37 | 53 | 67 | 79 | 83 | 97 | 103 | 107 | |
| 677 | 37 | 59 | 79 | 97 | 113 | 127 | 139 | 149 | 157 | 163 | 167 |

Mollin and Williams were able to invoke GRH in [13] to show that Table 1 effectively tells the whole story, together with the appropriate $d < 13$.

**Theorem 3.** *(Mollin and Williams) If GRH holds and* $d$ *is of narrow R–D type then* $h(d) = 1$ *if and only if* $d \in \{2, 3, 17, 21, 29, 37, 53, 77, 101, 173, 197, 293, 437, 677\}$.

**Remark 1.** In [5] Lachaud states without proof that, under the assumption of a suitable Riemann hypothesis, if $d = l^2 + 1 \equiv 1 (\mathrm{mod}\, 4)$ then $h(d) = 1$ if and only if $d \in \{5, 7, 37, 101, 197, 677\}$. This was independently obtained by Mollin and Williams in [12].

Thus Theorem 3 represents the solution (modulo GRH) of the class number one problem for real quadratic fields of narrow R–D type. Inspired by the intimate link between prime–producing quadratic polynomials and the class number one problem uncovered in Theorems 1 and 2, they turned their attention to the remaining R–D types in [14] and were able to establish several connections between prime–producing polynomials and the class number one problem. Among them is the following.

**Theorem 4.** *(Mollin–Williams)* (a) *If* $d = 4l^2 \pm 2 > 2$ *and* $f_d(x) = -2x^2 + \frac{d}{2}$ *is prime or* 1 *for all integers* $x$ *with* $0 \leq x < \frac{\sqrt{d}}{2}$, *then* $h(d) = 1$.

(b) *If* $d = (2l+1)^2 \pm 2$ *with* $l > 0$ *and* $f_d = -2x^2 + 2x + \frac{d-1}{2}$ *is prime or 1 for all integers* $x$ *with* $0 < x < \frac{\sqrt{d}+1}{2}$, *then* $h(d) = 1$.

The following tables illustrate Theorem 4.

**Table 2**

| $d$ | $f_d(x) = -2x^2 + \frac{d}{2}$ for $0 \leq x < \frac{\sqrt{d}}{2}$ | | | | | | | | | |
|---|---|---|---|---|---|---|---|---|---|---|
| 6 | 3 | | | | | | | | | |
| 14 | 7 | 5 | | | | | | | | |
| 38 | 19 | 17 | 11 | 1 | | | | | | |
| 62 | 31 | 29 | 23 | 13 | | | | | | |
| 398 | 199 | 197 | 191 | 181 | 167 | 149 | 127 | 101 | 71 | 37 |

**Table 3**

| $d$ | $f_d(x) = -2x^2 + 2x + \frac{d-1}{2}$ for $0 < x < \frac{\sqrt{d}+1}{2}$ | | | | | | | |
|---|---|---|---|---|---|---|---|---|
| 7 | 3 | | | | | | | |
| 11 | 5 | 1 | | | | | | |
| 23 | 11 | 7 | | | | | | |
| 47 | 23 | 19 | 11 | | | | | |
| 83 | 41 | 37 | 29 | 17 | 1 | | | |
| 167 | 83 | 79 | 71 | 59 | 43 | 23 | | |
| 227 | 113 | 109 | 101 | 89 | 73 | 53 | 29 | 1 |

In [10]—[11], Mollin had been able to show that if $d \not\equiv 1 (\bmod 4)$ and $d > 3$ is of non-narrow R–D type, then $h(d) = 1$ implies that $d = l^2 \pm 2$. Thus, with the aid of GRH, Mollin and Williams [14] were able to show that Tables 2 and 3 contain all the relevant values.

**Theorem 5.** *(Mollin–Williams) If GRH holds and* $d > 3$ *is of non–narrow R–D type with* $d \not\equiv 1 (\bmod 4)$ *then* $h(d) = 1$ *if and only if* $d \in \{6, 7, 11, 14, 23, 47, 62, 83, 167, 227, 398\}$.

This completes the solution of the class number one problem for all real quadratic fields $\mathbb{Q}(\sqrt{d})$ of narrow R–D type, and for $d \not\equiv 1 (\bmod 4)$ when $d$ is non–narrow. We now turn to non–narrow R–D types $d \equiv 1 (\bmod 4)$. We begin by considering R–D types $d = l^2 + r$ with $r$ dividing $2l$. In [10]—[11] Mollin showed that for such $d$, if $d \equiv 1 (\bmod 8)$, then $h(d) = 1$ if and only if $d = 33$. On the other hand, if $d \equiv 5 (\bmod 8)$ for such $d$, Mollin showed that $h(d) = 1$ implies that $d = t^2 - p$ where $p$ is an odd prime dividing $t$, whence $d = pq$ where $p \equiv q \equiv 3 (\bmod 4)$ are primes. The remaining R–D types are those of the form $d = l^2 \pm 4m \equiv 1 (\bmod 4)$ with $m > 1$, $m$ dividing $l$, $l$ odd and $m$ odd. Therefore, if $h(d) = 1$, then $p = m$ must be prime as must $q = \frac{d}{m}$ and $p \equiv q \equiv 3 (\bmod 4)$.

In [15] Mollin and Williams were able to use reduced ideal theory to prove the following result which proved two conjectures posed in [14].

**Theorem 6.** *(Mollin–Williams) Let* $d = pq$, $p < q$ *where* $p \equiv q \equiv 3 (\bmod 4)$ *are primes and* $d \equiv 5 (\bmod 8)$. *If* $\left| px^2 + px + \frac{(p-q)}{4} \right|$ *is prime or* 1 *for all integers* $x$ *with* $0 \leq x < \frac{\sqrt{d} - 1}{4} - \frac{1}{2}$ *then* $h(d) = 1$.

Although seemingly a general result the authors conjecture that if the hypothesis of Theorem 6 holds then $d$ is of R–D type. They proved this conjecture under the GRH assumption in [15].

Tables 4 and 5 following illustrate Theorem 6 and the aforementioned conjecture.

**Table 4**

| $d$ | $p$ | $q$ | $t$ | $\left\lvert px^2 + px - \frac{d - p^2}{4p} \right\rvert$ for $0 \le x < \frac{\sqrt{d} - 1}{4} - \frac{1}{2}$ | | | | | | | | | |
|---|---|---|---|---|---|---|---|---|---|---|---|---|---|
| 141 | 3 | 47 | 12 | 11 | 5 | 7 | | | | | | | |
| 573 | 3 | 191 | 24 | 47 | 41 | 29 | 11 | 13 | 43 | | | | |
| 1293 | 3 | 431 | 36 | 107 | 101 | 89 | 71 | 47 | 17 | 19 | 61 | 109 | |
| 1757 | 7 | 251 | 42 | 61 | 47 | 19 | 23 | 79 | 149 | 233 | 331 | 443 | 569 |

**Table 5**

| $d$ | $p$ | $q$ | $t$ | $\left\lvert px^2 + px - \frac{d - p^2}{4p} \right\rvert$ for $0 \le x < \sqrt{\frac{d - 1}{4}} - \frac{1}{2}$ | | | | | | | | | |
|---|---|---|---|---|---|---|---|---|---|---|---|---|---|
| 213 | 3 | 71 | 15 | 1 | 11 | 17 | 19 | 43 | 109 | 151 | | | |
| 237 | 3 | 79 | 15 | 1 | 13 | 17 | 19 | 41 | 71 | 107 | 149 | 197 | |
| 413 | 7 | 59 | 21 | 1 | 13 | 43 | 71 | 127 | 197 | 281 | 379 | 491 | 617 |
| | | | | 757 | | | | | | | | | |
| 453 | 3 | 151 | 21 | 1 | 19 | 23 | 31 | 37 | 53 | 89 | 131 | 173 | 233 |
| | | | | 293 | 359 | | | | | | | | |
| 717 | 3 | 239 | 27 | 23 | 31 | 37 | 41 | 53 | 59 | 67 | 109 | 157 | 233 |
| | | | | 271 | 337 | 409 | 487 | | | | | | |
| 1077 | 3 | 359 | 33 | 1 | 29 | 37 | 53 | 71 | 79 | 83 | 181 | 241 | 307 |
| | | | | 379 | 457 | 541 | 631 | 727 | 829 | 937 | | | |
| 1133 | 11 | 103 | 33 | 1 | 23 | 43 | 109 | 197 | 307 | 439 | 539 | 769 | 967 |
| | | | | 1187 | 1429 | 1693 | 1979 | 2287 | 2617 | 2969 | 3343 | | |
| 1253 | 7 | 179 | 39 | 13 | 29 | 41 | 43 | 97 | 167 | 251 | 349 | 461 | 587 |
| | | | | 727 | 881 | 1049 | 1231 | 1427 | 1637 | 1861 | 2079 | 2351 | |

Mollin and Williams invoked the GRH again in [14] to settle the class number one problem for real quadratic fields of R–D type; namely that Tables 4 and 5 complete the story.

**Theorem 7.** *(Mollin–Williams) Suppose that the GRH holds. If* $d \equiv 1 (\mathrm{mod}\, 4)$ *is of non–narrow R–D type then* $h(d) = 1$ *if and only if* $d \in \{33, 141, 213, 237, 413, 453, 573, 717, 1077, 1133, 1253, 1293, 1757\}$.

The following complete solution is the amalgamation of Theorems 2, 5, and 6.

**Theorem 8.** *(Mollin–Williams) Suppose that the GRH holds. and that d is of R–D type. Then* $h(d) = 1$ *if and only if* $d \in \{2, 3, 6, 7, 11, 14, 21, 23, 29, 33, 37, 38,$

47, 53, 62, 77, 83, 101, 141, 167, 173, 197, 213, 227, 237, 293, 398, 413, 437, 453, 573, 677, 717, 1077, 1133, 1253, 1293, 1757}.

Subsequent to the discovery of Theorem 7, it was learned by Mollin and Williams that S. Louboutin had obtained the result using the GRH (but missed the value $d = 573$), in his Ph.D. thesis [7].

**Remark 2.** 5 is not included in Theorem 7 since it does not fit the fundamental unit pattern for R–D types, and so is not generally considered to be an R–D type. Moreover, $13 = 3^2 + 4$, $69 = 9^2 - 12$ and $93 = 9^2 + 12$ are excluded because they are of the form $l^2 + r$ with $|r| > l$. We consider the condition $-l < r \leq l$ to be somewhat artificial. Therefore, we remove this condition, include $d = 5$, and call such $d$ of **extended R–D type.**

We now show that $h(d) > 1$ for all $d = e^{16}$, with one possible exception. (This follows the method of Kim et al [4].) We do this by using the following result of Tatuzawa [19]:

**Theorem 9.** *Let* $\frac{1}{2} > \epsilon > 0$ *and* $k \geq \max\left(e^{\frac{1}{\epsilon}}, e^{11.2}\right)$, *then with one possible exception*

$$L(1,\chi) > 0.655\frac{\epsilon}{k^{\epsilon}}$$

*where* $\chi$ *is a real, non-principal primitive character modulo* $k$.

Let $\chi$ be a modulo $\Delta$ character where $\Delta$ is the fundamental discriminant of $\mathbb{Q}(\sqrt{d})$ and let $R$ be the regulator of $\mathbb{Q}(\sqrt{d})$. It is easy to show that the regulator $R$ of $\mathbb{Q}(\sqrt{d})$ satisfies $R < \log 3d$. Thus if we put $\epsilon = \frac{1}{16}$ we can see by the analytic class number formula

$$2h(d)R = \sqrt{\Delta}\, L(1,\chi)$$

that with one possible exception

$$h(d) > \frac{0.655\Delta^{\frac{7}{16}}}{32\log 3d} \geq \frac{0.655 d^{\frac{7}{16}}}{32\log 3d} > 1$$

Thus $h(d) > 1$ for all $d > e^{16}$, $d$ of R–D type. This result is not conditional on the GRH. Since all such numbers less than $e^{16} < 9 \times 10^6$ have been checked on a computer (see [13]—[14] for details) we have now shown:

**Theorem 10.** *If $d$ is of extended R–D type then, with possibly only one more value, the following set contains all such $d$ with $h(d) = 1$:* {2, 3, 5, 6, 7, 11, 13, 14, 17, 21, 23, 29, 33, 37, 38, 47, 53, 62, 69, 77, 83, 93, 101, 141, 167, 173, 197, 213, 227, 237, 293, 398, 413, 437, 453, 573, 677, 717, 1077, 1133, 1253, 1293, 1757}.

In [15] the authors have made headway with the more general class number one problem for real quadratic fields, and have established criteria for certain non R–D types to have class number one in terms of certain prime–producing quadratic polynomials. This extends the work [9]—[14] and shows that, not surprisingly, the more general criteria are not as neat as those in Theorems 1 and 4, for example. The aim is to ultimately provide a **general** Rabinowitsch–like result for real quadratic fields, different and more specific than the straightforward Theorem 4 of [8] for example.

**Acknowledgement:** The authors would like to thank the referee for valuable comments.

## References

[1] *S. Chowla* and *J. Friedlander*, Class numbers and quadratic residues. Glasgow Math. J. **17** (1976), 47—52.

[2] *G. Degert*, Uber die bestimmung der grundeinheit gewisser reell–quadratischen zahlkorper. Abh. Math. Sem. Univ. Hamburg **22** (1958), 92—97.

[3] *D. Goldfeld*, Gauss' class number problem for imaginary quadratic fields. Bull. Amer. Math. Soc. **13** (1985), 23—37.

[4] *H.K. Kim*, *M.G. Lev* and *T. Ono*, On two conjectures on Real Quadratic Fields. Proc. Japan Acad. Sci. **63**, Ser. A (1987), 222—224.

[5] *G. Lachaud*, On real quadratic fields. Bull. Amer. Math. Soc. **17** (1987), 307—311.

[6] *E. Landau*, Uber die klassenzahl imaginar–quadratischen zahlkorper. Gottinger Nachr. (1918), 285—295.

[7] *S. Louboutin* , Arithmétique des corps quadratique reels et fractions continues. Université Paris VII, These de Doctorat, 22—06—87.

[8] *S. Louboutin,* Continued fractions and real quadratic fields. To appear: J. Number Theory.

[9] *R. Mollin* , Necessary and sufficient conditions for the class number of a real quadratic field to be one and a conjecture of S. Chowla. Proc. Amer. Math. Soc. **102** (1988), 17—21.

[10] *R. Mollin* , Class number one criteria for real quadratic fields I. Proc. Japan Acad. Sci. **16**, Ser. A, (1987), 121—125.

[11] *R. Mollin* , Class number one criteria for real quadratic fields II. Proc. Japan Acad. Sci. **16**, Ser. A, (1987), 162—164.

[12] *R. Mollin* and *H. Williams* , A conjecture of S. Chowla via the generalized Riemann hypothesis. Proc. Amer. Math. Soc. **102** (1988), 794—796.

[13] *R. Mollin* and *H. Williams* , On prime–valued polynomials and class numbers of real quadratic fields. Nagoya Math. J., (to appear).

[14] *R. Mollin* and *H. Williams* , Prime–producing quadratic polynomials and real quadratic fields of class number one. Proc. Internat. Number Theory Conf. (Québec City), (to appear).

[15] *R. Mollin* and *H. Williams* , Class number one for real quadratic fields, continued fractions and reduced ideals. Proc. NATO ASI on Number Theory and Applic., (Banff, 1988), (to appear).

[16] *G. Rabinowitsch* , Eindeutigkeit der zerlegung in primzahlfaktoren in quadratischen zahlkörpern. Proc. Fifth Internat. Congress Math. (Cambridge) Vol. I (1913), 418—421.

[17] *G. Rabinowitsch* , Eindeutigkeit der zerlegung in primzahlfaktoren in quadratischen zahlkörpern. J. rein. angew. Math. **142** (1913), 153—164.

[18] *C. Richaud* , Sur la résolution des équations $x^2 - Ay^2 = \pm 1$. . Atti. Acad. Pontif. Nuovi Lincei (1866), 177—182.

[19] *T. Tatuzawa* , On a theorem of Siegel. Japan J. Math. **21** (1951), 163—178.

---

Dept. of Mathematics, University of Calgary, Calgary, AB T2N 1N4, CANADA

Computer Science Dept., University of Manitoba, Winnipeg, MN R3T 2N2, CANADA

# On Simple Zeros of Certain $L$-Series

*M. Ram Murty* [1]

## 1. Introduction

Let $E$ be an elliptic curve defined over $\mathbb{Q}$ and denote by **III** the Tate–Shafarevic group of $E/\mathbb{Q}$. It is an outstanding unresolved conjecture that this group is finite. If the rank of the Mordell–Weil group of $E$ over $\mathbb{Q}$ is zero, and $E$ has complex multiplication, then K. Rubin [12] showed that **III** is finite. Recently, V. Kolyvagin [7] (see also Washington [15] for an exposition of Kolyvagin's work) showed that if $E$ is a modular elliptic curve over $\mathbb{Q}$ with Mordell–Weil rank equal to zero, then **III** is finite, *provided* there is a "quadratic twist" of the $L$-series of the given elliptic curve having a simple zero at $s = 1$. In any given case, it is not difficult to produce such a twist. To demonstrate that, in general, such a twist exists seems to be difficult.

The purpose of this paper is to show that on the generalized Riemann hypothesis (GRH), such a twist always exists. Our method has other applications. For instance, if we do not confine ourselves to quadratic twists and consider twisting by any character $\chi \pmod q$, then the density of such twists with a zero at $s = 1$ is at most 1/2, under the same hypothesis. The method applied to a classical context shows that if $L(s,\chi)$ is the classical Dirichlet $L$-series, then $L(1/2,\chi) \neq 0$ for at least $\phi(q)/2$ characters $\chi \pmod q$. (It is generally conjectured that $L(1/2,\chi) \neq 0$ for all characters $\chi \pmod q$, but this is, as yet, unproved. I am not sure of the source of this conjecture but I think it originates from S. Chowla.) To state the results more precisely, we need the following background.

1 Research partially supported by an NSERC grant.

Let $f$ be a cusp form of weight 2 for $\Gamma_0(N)$. Let

$$f(z) = \sum_{n=1}^{\infty} a_n e^{2\pi i n z}$$

be its Fourier expansion at the cusp $i\infty$. Let

$$L(f, s) = \sum_{n=1}^{\infty} \frac{a_n}{n^s}$$

and

$$L(f, D, s) = \sum_{n=1}^{\infty} \frac{a_n}{n^s}\left(\frac{D}{n}\right)$$

where $\left(\frac{D}{n}\right)$ denotes the Legendre symbol. It is well–known that [13] by the theory of modular forms, both $L(f, s)$ and $L(f, D, s)$ have an analytic continuation to the entire complex plane. Moreover, if $D$ is a fundamental discriminant (that is, $D$ is the discriminant of a quadratic field) and $f$ is an eigenfunction of the Hecke operators, then $L(f, s)$ and $L(f, D, s)$ can be written as Euler products and have functional equations. More precisely, if we let

$$\Lambda(s) = N^{s/2}(2\pi)^{-s}\Gamma(s)L(f, s)$$

then $\Lambda(s)$ is entire and satisfies

$$\Lambda(s) = w\,\Lambda(2 - s)$$

where $w = \pm 1$. If $w = -1$, then $L(f, s)$ has a zero of odd order at $s = 1$ and if $w = 1$, it has a zero of even order at $s = 1$. As $\Gamma(s)$ has simple poles at $s = 0$, $-1, \ldots,$ $L(f, s)$ has (trivial) zeros at these points. If we let

$$\Lambda(s, D) = D^s N^{s/2}(2\pi)^{-s}\Gamma(s)L(f, D, s)$$

then $\Lambda(s, D)$ is entire and satisfies

$$\Lambda(s, D) = w\left(\frac{D}{-N}\right)\Lambda(2 - s, D).$$

Again, $L(f, s)$ has trivial zeros at $s = 0, -1, -2, \ldots.$ Moreover, $L(f, s)$ has an Euler product of the form:

$$L(f,s)=\prod_{p|N}\left(1-\frac{a_p}{p^s}\right)^{-1}\prod_{p\nmid N}\left(1-\frac{\alpha_p}{p^s}\right)^{-1}\left(1-\frac{\bar{\alpha}_p}{p^s}\right)^{-1}.$$

$L(f, D, s)$ has a similar Euler product.

If $E$ is a modular elliptic curve of conductor $N$, then the $L$ series of the elliptic curve is given by $L(f, s)$ for some cusp eigenform $f$ on $\Gamma_0(N)$. In his proof of the finiteness of **III**, Kolyvagin needs to show that there is a fundamental discriminant $D$ of an imaginary quadratic field $K$ such that all of the prime divisors of $N$ split in $K$ and $L(f, D, s)$ has a simple zero at $s = 1$. Accordingly, we prove:

**Theorem 1.** *Let $N$ be a fixed natural number. Suppose that for each fundamental discriminant $D$, $L(f, D, s)$ and the classical Dirichlet $L$–series*

$$L(D,s)=\sum_{n=1}^{\infty}\left(\frac{D}{n}\right)\frac{1}{n^s}$$

*satisfy GRH. That is, all of the non-trivial zeros of $L(f, D, s)$ and $L(D, s)$ satisfy $\Re s = 1$ and $\Re s = 1/2$ respectively. Then, there is a fundamental discriminant $D$ of an imaginary quadratic field $K$ such that $L(f, D, s)$ has a simple zero at $s = 1$ and all the prime divisors of $N$ split in $K$.*

Let $\chi$ be a Dirichlet character mod $q$ and set

$$L(f,\chi,s)=\sum_{n=1}^{\infty}\frac{a_n\chi(n)}{n^s}.$$

Then

**Theorem 2.** *Suppose each $L(f, \chi, s)$ satisfies GRH. Then $L(f, \chi, 1) = 0$ for at most $\phi(q)/2$ characters $\chi$ (mod $q$).*

**Corollary.** *Under GRH, $L(f, \chi, 1) \neq 0$ for at least $\phi(q)/2$ characters $\chi$ (mod $q$).*

An analogous method in the classical situation leads to the interesting:

**Theorem 3.** *The number of* $\chi \pmod q$ *such that*

$$\sum_{n=1}^{\infty} \frac{\chi(n)}{\sqrt{n}} \neq 0$$

*is at least* $\phi(q)/2$ *on the GRH.*

## 2. Lemmas

Our proof is based on Weil's explicit formula. Goldfeld [5] had utilized a similar method to show that under GRH, the order of the zero of $L(f, D, s)$ at $s = 1$ is bounded by 3 for almost all $D$. Our proof is a modification of his method.

Throughout the rest of the paper, $f$ is a normalized cusp eigenform of weight 2 on $\Gamma_0(N)$. $L(f, s)$ and $L(f, D, s)$ are then as defined in section 1. Let us define

$$c_n = \begin{cases} \alpha_p^m + \bar{\alpha}_p^m & \text{if } n = p^m \text{ and } p \nmid N \\ a_p^m & n = p^m \text{ and } p \mid N \\ 0 & \text{otherwise.} \end{cases}$$

Then we can write

$$\frac{L'}{L}(f, s) = \sum_{n=1}^{\infty} \frac{c_n \Lambda(n)}{n^s}$$

Also,

$$\frac{L'}{L}(f, D, s) = \sum_{n=1}^{\infty} \frac{c_n \Lambda(n)}{n^s} \left(\frac{D}{n}\right).$$

**Lemma 1.** *(Weil's explicit formula.) Let* $F : \mathbb{R} \to \mathbb{R}$ *satisfy the following conditions:*

(A) *there is an* $\epsilon > 0$ *such that* $F(x)\exp\{(1+\epsilon)x\}$ *is integrable and of bounded variation,*

(B) *the function*

$$\frac{F(x) - F(0)}{x}$$

*is of bounded variation.*

*Define*

$$\phi(\gamma) = \int_{-\infty}^{\infty} F(x) e^{i\gamma x}\, dx .$$

*Then*

$$\sum_{\gamma} \phi(\gamma) = 2F(0)\log \frac{\sqrt{N}}{2\pi} - \frac{1}{\pi}\int_{-\infty}^{\infty} \frac{\Gamma'}{\Gamma}(1+it)\phi(t)\,dt - 2\sum_{n=1}^{\infty} \frac{c_n}{n}\Lambda(n) F(\log n) \tag{1}$$

*where the sum on the left hand side is over* $\gamma$ *such that* $L(f, 1+i\gamma) = 0$ *and* $1 \le \Re(1+i\gamma) \le 3/2$.

**Remark.** An analogous formula holds for $L(f, D, s)$. Namely,

$$\sum_{\gamma} \phi(\gamma) = 2F(0)\log \frac{\sqrt{N}\,D}{2\pi} - \frac{1}{\pi}\int_{-\infty}^{\infty} \frac{\Gamma'}{\Gamma}(1+it)\phi(t)\,dt - 2\sum_{n=1}^{\infty} \frac{c_n}{n}\left(\frac{D}{n}\right)\Lambda(n) F(\log n) \tag{2}$$

where the sum on the left hand side is over $\gamma$ satisfying $L(f, D, 1+i\gamma) = 0$ and $1 \le \Re(1+i\gamma) \le 3/2$.

**Proof.** See Mestre [8], p. 215.

**Lemma 2.** *Let* $T > 0$, *and define*

$$F(x) = \begin{cases} 2T - |x| & \text{if } |x| \le 2T \\ 0 & \text{otherwise.} \end{cases}$$

*Then, F satisfies the conditions of Lemma 1 and*

$$\phi(\gamma) = \left(\frac{2\sin(\gamma T)}{\gamma}\right)^2 .$$

**Proof.** It is clear that $F$ satisfies (A) and (B) of Lemma 1. The computation of $\phi$ follows from the straightforward integration.

**Lemma 3.** *Let $T > 1$. Then,*

$$\int_0^\infty \frac{\Gamma'}{\Gamma}(1+it)\left(\frac{\sin(Tt)}{t}\right)^2 dt \ll T.$$

**Proof.** We use the well–known estimate (see Davenport [4], p. 73)

$$\frac{\Gamma'}{\Gamma}(1+it) = O(\log(|t|+2)),$$

and decompose the integral

$$\int_0^\infty = \int_0^{1/T} + \int_{1/T}^\infty .$$

Since, $\sin x \leq x$ if $x > 0$, the first integral is $O(T)$ as the gamma function is bounded in this range. The second integral is

$$\ll \int_{1/T}^\infty \frac{\log(t+2)}{t^2} dt \leq T \log(2+\frac{1}{T}) + \int_{1/T}^\infty \frac{dt}{t(t+2)} \ll T,$$

on using the cited property of the gamma function.

**Lemma 4.** *Let $p$ be a prime and $N$ a natural number. Assuming GRH,*

$$\sum_{\substack{q \leq z \\ q \equiv 3 (\bmod 4)}} \left(\frac{-q}{p}\right) = O(z^{\frac{1}{2}} \log(pz)),$$

*where the summations over primes $q \equiv 3 (\mathrm{mod}\ 4)$ satisfying $(-q/r) = 1$ for all prime divisors $r$ of $N$.*

**Proof.** Let $\upsilon$ denote the number of prime factors of $N$. We can write the sum on the left hand side as

$$\frac{1}{2^{\upsilon+1}} \sum_{q \leq z} \left(\frac{-q}{p}\right)\left(1 - \left(\frac{-1}{q}\right)\right) \prod_{r|N} (1 + (-q/r))$$

and by the classical estimates, (see Davenport [4], p. 125) the result follows.

**Lemma 5.**

$$\sum_{p^{2m}\le x} \frac{\alpha_p^{2m}+\bar{\alpha}_p^{2m}}{p^{2m}}(\log p)\log\frac{x}{p^{2m}} \sim -\frac{1}{4}\log^2 x$$

*as* $x \to \infty$.

**Proof.** Set

$$L_2(f, s)=\prod_{p|N}(1-\frac{a_p^2}{p^s})^{-1}\prod_{p\nmid N}(1-\frac{\alpha_p^2}{p^s})^{-1}(1-\frac{\bar{\alpha}_p^2}{p^s})^{-1}.$$

It is well-known [14] that $\zeta(s-1)L_2(f, s)$ is entire. Therefore, $L_2(f, s+1)$ has a simple zero at $s=1$. Moreover, $L_2(f, s)$ does not vanish on the line $\Re s = 1$ (see Rankin [11]). Therefore, if we write $\alpha_p = \sqrt{p}\exp(i\,\theta_p)$, then by the classical Tauberian theorem,

$$\sum_{p^m\le x} 2(\cos 2m\theta_p)\log p \sim -x .$$

By partial summation, it follows that

$$\sum_{p^{2m}\le x}\frac{2\cos 2m\theta_p}{p^m}\log p \sim -\int_1^{\sqrt{x}}\frac{dt}{t} \sim -\frac{1}{2}\log x ,$$

as $x \to \infty$. Similarly,

$$\sum_{p^{2m}\le x}\frac{2\cos 2m\theta_p}{p^m}(\log p)(\log p^{2m}) \sim \frac{-\log^2 x}{4}.$$

Thus,

$$\sum_{p^{2m}\le x}\frac{2\cos 2m\theta_p}{p^m}\log p\,\log\frac{x}{p^{2m}} \sim \frac{-\log^2 x}{4}.$$

## 3. Proof of Theorem 1

We shall choose

$$F(u)=\begin{cases} 2T-|u| & \text{if } |u|\le 2T\\ 0 & \text{otherwise.}\end{cases}$$

Thus, by Lemma 2,

$$\phi(\gamma) = \left(\frac{2\sin(\gamma T)}{\gamma}\right)^2$$

so that if $\gamma \in \mathbb{R}$, then $\phi(\gamma) \geq 0$. We will choose $T = \log\sqrt{x}$. Let $q$ be a prime $\equiv 3 \pmod 4$. Then $-q$ is a fundamental discriminant. Let $r_q$ be the order of zero at $s = 1$ of $L(f, -q, s)$. With the above choice of $F$ and $\phi$, we apply the explicit formula (Lemma 1) to $L(f, -q, s)$. Assuming GRH, all of the $\gamma$ are real and so, we obtain the inequality

$$r_q (\log x)^2 \leq 2\left(\log \frac{\sqrt{N}\, q}{2\pi}\right)(\log x) + \frac{1}{\pi}\int_{-\infty}^{\infty} \left|\frac{\Gamma'}{\Gamma}(1+it)\right| \left(\frac{\sin^2(t \log\sqrt{x})}{t^2}\right) dt$$
$$- 2\sum_{n \leq x} \frac{c_n}{n}\left(\frac{-q}{n}\right)\Lambda(n)\log\frac{x}{n}.$$

The integral is, by Lemma 3,

$$\ll \log x .$$

Let $P(z)$ denote the number of primes $q \leq z$ satisfying $q \equiv 3 \pmod 4$, $(-q/r) = 1$ for all prime divisors $r$ of $N$. We obtain

$$\sum_{q \leq x}{}' \, r_q (\log x)^2 \leq 2(\log z)(\log x) P(z) + O(P(z)\log x)$$
$$- 2\sum_{n \leq x} \frac{c_n}{n}\Lambda(n)\log\frac{x}{n}\sum_{q \leq z}{}' \left(\frac{-q}{n}\right)$$

where the dash on the summation indicates we sum over primes $q \leq z$, $q \equiv 3 \pmod 4$, satisfying $(-q/r) = 1$ for all prime divisors $r$ of $N$. The last sum is by Lemma 4,

$$\sum_{q \leq x}{}' \left(\frac{-q}{n}\right) \ll z^{\frac{1}{2}} \log(nz)$$

if $n$ is prime. If $n = p^m$ and $m$ is odd, a similar estimate holds. If $m$ is even, the inner sum is equal to $P(z)$. Thus, if we split the sum over $n = p^m$ into two parts $S_1$ and $S_2$, where $S_1$ corresponds to the odd powers and $S_2$ corresponds to the even powers, then it is now clear that

$$S_1 \ll x^{\frac{1}{2}+\epsilon} z^{\frac{1}{2}}.$$

The remaining sum $S_2$ is equal to

$$P(z)\sum_{p^{2m}\le x}\frac{\alpha_p^{2m}+\bar{\alpha}_p^{2m}}{p^{2m}}(\log p)\log\frac{x}{p^{2m}}$$

which by Lemma 5 is

$$\approx -\frac{1}{4}(\log^2 x)P(z).$$

We therefore obtain the inequality

$$\sum_{q\le z}{}' r_q(\log x)^2 \le 2P(z)\log z\,\log x + \frac{P(z)(\log^2 x)}{2}$$

$$+\,O(P(z)\log x) + O(z^{\frac{1}{2}}x^{\frac{1}{2}+\epsilon}).$$

Hence,

$$\sum_{q\le z}{}' r_q \le \frac{2P(z)\log z}{\log x} + \frac{1}{2}P(z) + O\left(\frac{P(z)}{\log x}\right) + O(z^{\frac{1}{2}}x^{\frac{1}{2}+\epsilon}).$$

If there is no $q$ satisfying the hypothesis, then each $r_q \ge 3$ in the above sum. We will choose $z = x^{\alpha}$, where $\alpha$ satisfies

$$1 < \alpha < \frac{5}{4}.$$

Dividing the entire expression by $P(z)$ we get that

$$3 \le 2\alpha + .5 + O\left(\frac{1}{\log x}\right) + O(x^{-\frac{\alpha}{2}+\frac{1}{2}+\epsilon})$$

which is a contradiction since $\alpha < 5/4$. This proves the result.

## 4. Proof of Theorem 2

For the sake of notational simplicity, we shall consider the case when $q$ is prime. The general case is only slightly complicated in that we must deal carefully with imprimitive characters. The case when $q$ is prime, we have only one imprimitive character, namely the principal character. Let $r_\chi$ denote the order of the zero at $s = 1$

of $L(f, \chi, s)$. From Lemma 1, and the choice of $\phi$ as in Lemma 2, with $T = (\log x)/2$, it follows that

$$\sum_{\chi \bmod q} r_\chi (\log x)^2 \le \phi(q)(\log x)\log\frac{\sqrt{N}\,q}{2\pi}$$

$$+2\phi(q)\sum_{\substack{n\le x\\ n\equiv 1 (\bmod q)}}\frac{\Lambda(n)}{\sqrt{n}}\log\frac{x}{n}-\frac{(\log x)(\log q)}{\sqrt{q}},$$

where the last term is the correction term in the explicit formula for $L(f, \chi, s)$ when $\chi$ is the principal character. But,

$$\sum_{\substack{n\le x\\ n\equiv 1 (\bmod q)}}\frac{\Lambda(n)}{\sqrt{n}}\log\frac{x}{n} \ll \log^2 x \sum_{\substack{n\le x\\ n\equiv 1 (\bmod q)}}\frac{1}{\sqrt{n}}.$$

Since the number of primes $p \le x$, $p \equiv 1 \pmod q$ is

$$\ll \frac{x}{\phi(q)\log(x/q)}, \quad q < x,$$

by the Brun–Titchmarsh theorem, the above sum is seen to be

$$\ll \frac{\sqrt{x}\log^2 x}{\phi(q)\log(x/q)}.$$

Thus, choosing $x = \phi(q)^2$, we obtain

$$\sum_{\chi \bmod q} r_\chi \le \frac{\phi(q)}{2} + O\left(\frac{\phi(q)}{\log q}\right).$$

This proves the result.

The corollary is now immediate. To establish Theorem 3, there is only a slight variation on the above proof. The explicit formula is applied to the Dedekind zeta function of the cyclotomic field $\mathbb{Q}(\zeta_q)$. As the zeta function has a simple pole at $s = 1$, there is an additional contribution of $O(x^{1/2})$ arising from $\phi(0)$ and $\phi(1)$. This causes no problem upon division by $\log^2 x$. Moreover, the analogue of the sum dealt with above can be discarded as it appears with a negative sign. Since the proof is now completely analogous to the above, we leave the details to the reader.

## 5. Concluding Remarks

To eliminate the GRH from the above proofs seems to be difficult. In Theorem 1, it is used in two places, though, there is some possibility of removing it. More precisely, the use of GRH in Lemma 4 can be eliminated if instead of summing over prime discriminants, we sum over all fundamental discriminants satisfying the quadratic conditions of Kolyvagin. In such a case, we would require

$$\sum_{D \le z}{}' \left(\frac{D}{n}\right) \ll z^{\frac{1}{2}} n^{\frac{1}{8}+\epsilon}$$

when $n$ is prime. This is only slightly stronger than the Burgess estimate [3].

The second usage of the GRH is in establishing the positivity of $\phi(\gamma)$ as $1 + i\gamma$ runs over the zeros of $L(f, D, s)$. This can be circumvented by using a remark of Odlyzko (see Poitou [10], p. 148). We only need a function $\phi$ which satisfies $\Re\phi(\gamma) \ge 0$. Thus if $F(x)$ is a non-negative function satisfying the conditions of Lemma 1, then

$$\frac{F(x)}{\cosh x}$$

has the property that its Fourier transform has non–negative real part. This follows from the fact that the real part of the analytic function

$$\int_{-\infty}^{\infty} \frac{F(x)}{\cosh x} e^{(s-1)x} dx$$

is a harmonic function which is non–negative on $\Re s = 0$ and $\Re s = 2$ and hence non–negative throughout the critical strip. But this modification in the above proof eliminates one of the log factors and the resulting inequality is insufficient to deduce anything significant about the orders of zeros.

But in the case of Theorems 2 and 3, there is some possibility though this requires a great deal of technical prowess. Indeed Theorem 3 was recently established without

GRH by R. Balasubramanian and V. Kumar Murty [2] with a constant smaller than 1/2. It would be interesting to determine if by these methods one could prove that almost all of $L(s,\chi)$ with $\chi \pmod q$ do not vanish at $s = 1/2$.

## References


[1] *R. Ayoub,* An Introduction to the Analytic Theory of Numbers. Amer. Math. Soc. Math. Surveys **10** (1963), Providence.

[2] *R. Balasubramanian* and *V. Kumar Murty,* Zeros of the Dirichlet $L$-functions. (in preparation).

[3] *D.A. Burgess,* On character sums and $L$-series II. Proc. London Math. (3), **13** (1963), 524—536.

[4] *H. Davenport,* Multiplicative Number Theory. Springer–Verlag.

[5] *D. Goldfeld,* Conjectures on elliptic curves over quadratic fields. Proc. Southern Illinois Number Theory Conf., Springer Lecture Notes **751** (1979), Springer–Verlag.

[6] *D. Goldfeld* and *C. Viola,* Mean values of $L$-functions associated to elliptic, Fermat, and other curves at the centre of the critical strip. J. Number Theory **11** (1979), 305—320.

[7] *V.A. Kolyvagin,* On the Mordell–Weil and Shafarevic–Tate groups of elliptic Weil curves. Preprint (in Russian).

[8] *J.–F. Mestre,* Formules explicites et minorations de conducteurs de variétés algébriques. Comp. Math. **58** (1986), 209—232.

[9] *M.R. Murty, V.K. Murty* and *N. Saradha,* Modular forms and the Chebotarev density theorem. Amer. J. Math. **110** (1988), 253—281.

[10] *G. Poitou,* Minorations de discriminants. Séminaire Boubaki, Exposé 479, Springer Lecture Notes **567** (1976), Springer–Verlag.

[11] *R.A. Rankin,* Contributions to the theory of Ramanujan's function $\tau(n)$ and similar arithmetic functions. Proc. Camb. Phil. Soc. **35** (1939), 357–375.

[12] *K. Rubin,* Tate–Shafarevic groups and the $L$-functions of elliptic curves with complex multiplication. Inven. Math. **89** (1987), 527—560.

[13] *G. Shimura,* An introduction to the arithmetic theory of automorphic functions. Publications of Math. Soc. of Japan **11** (1971), Princeton University Press.

[14] *G. Shimura,* On the holomorphy of certain Dirichlet series. Proc. London Math. Soc. **31** (1975), 79—98.

[15] *L. Washington,* Number fields and elliptic curves. Proc. NATO ASI on Number Theory and Applications, Banff, Canada, Kluwer Academic Publishers; Editor: R.A. Mollin, 245—278.

---

Department of Mathematics, McGill University, Montréal, H3A 2K6 CANADA

# Asymptotic Distribution and Independence of Sequences of $g$–Adic Integers, II

*Kenji Nagasaka and Jau–Shyong Shiue*

**Dedicated to Professor L. Kuipers on the occasion of his 80th birthday**

## 1. Introduction

The notion of independence plays an important role in probability theory. One of the standard ways to introduce the notion of independence of two random variables $X$ and $Y$ is as follows: Consider an event $G$ defined by

$$G = \{\omega : \omega \in \Omega, \quad X(\omega) \leq x, \quad Y(\omega) \leq y\}, \tag{1}$$

where $X$ and $Y$ are real–valued random variables on a probability space $(\Omega, \mathfrak{F}, P)$. We also consider two events $E$ and $F$ defined by

$$E = \{\omega : \omega \in \Omega, \quad X(\omega) \leq x\}, \tag{2}$$

$$F = \{\omega : \omega \in \Omega, \quad Y(\omega) \leq y\}, \tag{3}$$

respectively.

We say that $X$ and $Y$ are independent if

$$P(G) = P(E) \bullet P(F) \tag{4}$$

holds for every pair of real numbers $(x, y)$; in other words, the joint distribution function is the marginal product of its distribution functions.

Now suppose that $X$ and $Y$ are two independent random variables, then we have:

$$E(XY) = E(X) \bullet E(Y), \tag{5}$$

where $E(Z)$ stands for the expectation of a random variable $Z$. Through a standard argument in probability theory, we also have, under the same assumption of independence,

$$E(e^{2\pi i h(X+Y)}) = E(e^{2\pi i hX}) \bullet E(e^{2\pi i hY}). \tag{6}$$

Instead of random variables $X$ and $Y$, Hugo Steinhaus [14] considered two real–valued functions $f$ and $g$ and introduced the notion of independence for $f$ and $g$.

Interesting connections between continuous uniform distribution mod 1 and the notion of independence were pointed out by Steinhaus [14] and Lauwerens Kuipers [3].

Indeed, Kuipers proved the continuous uniform distribution mod 1 for the function $f(t) = \alpha t + \beta \bullet \sin t$ with $\beta \neq 0$ whenever $\pi\alpha$ is irrational by deriving the independence of $\alpha t$ and $\beta \bullet \sin t$ then applying (6) through Weyl's criterion for continuous uniform distribution mod 1.

Inspired by this splendid idea, Kuipers and the second author of this note, Jau–Shyong Shiue [7], [8] introduced the notion of independence of $k$ sequences of rational integers and the mean value (expectation) of a sequence of rational integers, then obtained fundamental properties of the mean value. They established a criterion for independence of $k$ sequences of rational integers and gave some applications to the case of uniformly distributed sequences of rational integers mod $m$.

The uniform distribution mod $m$ of sequences of rational integers is a special case of the uniform distribution of $g$-adic integer sequences, introduced by H. Meijer [9] and [10]. Here $g$ stands for an arbitrary positive rational integer, which can be composite, therefore the theory of uniform distribution of $g$–adic integer sequences is on one hand a generalization of the uniform distribution of $p$–adic integer sequences for $p$ prime by M. Cugiani [2]; see also Kuipers and Harald Niederreiter [4].

Shiue [13] pointed out an important connection between the uniform distribution of $g$–adic integer sequences and the notion of independence. Meijer and Shiue deepened the results above [10] in their succeeding work.

Several years later, Kuipers and Niederreiter [5], [6] obtained some new results on asymptotic distributions mod $m$ of rational integers and independence. We, Kenji Nagasaka and Shiue [12] recently succeeded in proving three theorems on $g$–adic integer sequences which correspond to Theorems 1, 2 and 3 in [5]. This note is a direct continuation of our previous joint work [12] and gives further generalization of Example 1 and Theorems in [5] and a relationship among uniform distributions of a $g$-adic integer sequence.

Finally we give a slight improvement of conditions of uniform distribution of polynomial sequences with $g$–adic integer coefficients as a $g$–adic integer sequence.

## 2. Examples of Independent $g$–Adic Integer Sequences

Throughout this paper, notations and definitions are the same as in [12], unless otherwise stated.

Kuipers and Niederreiter [5] gave an example of two independent sequences of rational integers $a = \{a_n\}_{n=1}^{\infty}$ and $b = \{b_n\}_{n=1}^{\infty}$ that are uniformly distributed mod $m$, due to Kuipers and Shiue [7]. Here we shall prove the $g$–adic version of this example.

**Theorem 1.** *Let $\gamma = \{c_n\}_{n=1}^{\infty}$ be a sequence of g–adic integers and suppose that $\gamma$ is k–uniformly distributed in $\mathbb{Z}_{g^2}$ for a positive integer k. The g–adic representation of $c_n$ can be written as*

$$c_n = \sum_{i=0}^{\infty} c_{n_i} (g^2)^i , \tag{7}$$

*where $c_{n_i}$ is an element of $G_2 = \{0, 1, \ldots, g^2 - 1\}$. Let us write the i–th digit of $c_n$ as*

$$c_{n_i} = a_{n_i} + b_{n_i} \cdot g , \tag{8}$$

*where $a_{n_i}$ and $b_{n_i}$ are contained in $G_1 = \{0, 1, \ldots, g - 1\}$.*

*If we define two g–adic integers $a_n$ and $b_n$ by*

$$a_n = \sum_{i=0}^{\infty} a_{n_i} g^i \tag{9}$$

*and*

$$b_n = \sum_{i=0}^{\infty} b_{n_i} g^i, \tag{10}$$

*then two sequences of g–adic integer sequences* $\alpha = \{a_n\}_{n=1}^{\infty}$ *and* $\beta = \{b_n\}_{n=1}^{\infty}$ *are both k–uniformly distributed in* $\mathbb{Z}_g$ *and they are also k–independent in* $\mathbb{Z}_g$ .

**Proof.** We define a $k$–neighbourhood $U_k(j)$ of $j$ by

$$U_k(j) = \{x : x \in \mathbb{Z}_g ,|x - j|_g \le g^{-k}\}, \tag{11}$$

where $\mathbb{Z}_g$ is the ring of $g$–adic integers and $|\bullet|_g$ stands for the $g$–adic pseudo–valuation. For any $j \in G_k = \{0, 1, \ldots, g^k - 1\}$ and $N \in \mathbb{N}$, $A_N(U_k(j); \alpha)$ denotes the number of terms $a_n$ with $1 \le n \le N$ satisfying $a_n \in U_k(j)$, then we have

$$A_n(U_k(j); \alpha) = \sum_{\ell=0}^{g^k-1} A_N(U_k(j); \alpha, U_k(\ell); \beta), \tag{12}$$

since

$$\mathbb{Z}_g = \bigcup_{\ell=0}^{g^k-1} U_k(\ell). \tag{13}$$

Consider the ordinary expansion to the base $g$ of $j$ and of $\ell$ as

$$j = j_0 + j_1 g + \ldots + j_{k-1} g^{k-1} \tag{14}$$

and

$$\ell = \ell_0 + \ell_1 g + \ldots + \ell_{k-1} g^{k-1}, \tag{15}$$

with the $j_r$'s and $\ell_r$'s in $G_1$. Put

$$m_r = j_r + \ell_r \cdot g, \tag{16}$$

for $r = 0, 1, \ldots, k - 1$, then

$$m = \sum_{r=0}^{k-1} m_r (g^2)^r \tag{17}$$

is a positive integer developed to the base $g^2$ with $m_r$ 's in $G_2$.

Thus we get

$$A_N (U_k(j); \alpha, U_k(\ell); \beta) = A_N (U_k(m); \gamma). \tag{18}$$

Dividing (12) by $N$ and letting $N$ go to infinity, we have

$$\|A(\alpha \in U_k(j))\| = \sum_{\ell=0}^{g^k-1} \|A(\gamma \in U_k(m))\|. \tag{19}$$

Since we assume the $k$–uniform distribution of $\gamma$ in $\mathbb{Z}_{g^2}$,

$$\|A(\gamma \in U_k(m))\| = \frac{1}{g^{2k}}, \tag{20}$$

for all $m$ in $G_k$. Substituting (20) in (19), we get

$$\|A(\alpha \in U_k(j))\| = \sum_{\ell=0}^{g^k-1} \frac{1}{g^{2k}} = \frac{1}{g^k}, \tag{21}$$

which signifies the $k$–uniform distribution of $\alpha$ in $\mathbb{Z}_g$.

Similarly we deduce the $k$–uniform distribution of $\beta$ in $\mathbb{Z}_g$.

From (18) and (20), we obtain immediately

$$\left\|A(\alpha \in U_k(j), \beta \in U_k(\ell))\right\| = \|A(\gamma \in U_k(m))\| = \frac{1}{g^{2k}}. \tag{22}$$

On the other hand, the $k$–uniform distribution of $\alpha$ and $\beta$ in $\mathbb{Z}_g$ assures that, for each pair $(j, \ell) \in G_k \times G_k$,

$$\|A(\alpha \in U_k(j))\| = \left\|A(\beta \in U_k(\ell))\right\| = \frac{1}{g^{2k}}. \tag{23}$$

Combining (22) and (23), we get

$$\|A(\alpha \in U_k(j), \beta \in U_k(\ell))\| = \|A(\alpha \in U_k(j))\| \cdot \|A(\beta \in U_k(\ell))\|, \quad \frac{1}{g^k} \tag{24}$$

that means $k$-independence of $\alpha$ and $\beta$ in $\mathbb{Z}_g$.

■

The converse of this Theorem is also true.

**Theorem 2.** *Let $\alpha = \{a_n\}_{n=1}^{\infty}$ and $\beta = \{b_n\}_{n=1}^{\infty}$ be two k-independent g-adic integer sequences in $\mathbb{Z}_g$ and suppose further that both $\alpha$ and $\beta$ are k-uniformly distributed g-adic integer sequences in $\mathbb{Z}_g$. Let us define a g-adic integer $c_n$ by*

$$c_n = \sum_{i=0}^{\infty} c_{n_i} (g^2)^i, \tag{25}$$

*with*

$$c_{n_i} = a_{n_i} + b_{n_i} \cdot g, \tag{26}$$

*where $a_{n_i}$ and $b_{n_i}$ are the i-th digits of g-adic representations of $a_n$ and $b_n$, respectively.*

*Then a g-adic integer sequence $\gamma = \{c_n\}_{n=1}^{\infty}$ is k-uniformly distributed in $\mathbb{Z}_{g^2}$.*

**Proof.** From $k$-independence of $\alpha$ and $\beta$ in $\mathbb{Z}_g$, we have, for every pair $(j, \ell) \in G_k \times G_k$,

$$\|A(\alpha \in U_k(j), \beta \in U_k(\ell))\| = \|A(\alpha \in U_k(j))\| \cdot \|A(\beta \in U_k(\ell))\| = \frac{1}{g^k} \cdot \frac{1}{g^k} = \frac{1}{g^{2k}}, \tag{27}$$

since both $\alpha$ and $\beta$ are $k$-uniformly distributed in $\mathbb{Z}_g$.

From (26), we derive (18) again for these $\alpha$, $\beta$ and $\gamma$ and combining (27), we get the conclusion.

■

We shall give, in the final section, other necessary and sufficient conditions on $g$–adic integer sequences treated in this section.

## 3. Independence of Asymptotic Distribution Functions

In our preceding joint work [12], we gave three criteria concerning the independence of two sequences of $g$–adic integers in terms of their asymptotic distribution functions. Here we give another criterion of independence.

**Theorem 3.** *Let* $\alpha = \{a_n\}_{n=1}^{\infty}$ *be a* $g$*–adic integer sequence and suppose that* $\alpha$ *has* $\{\phi_k(j)\}$ *as its asymptotic* $k$*–distribution function.*

*Then* $\alpha$ *is* $(k_1, k_2)$*–independent in* $\mathbb{Z}_g$ *of any* $g$*–adic integer sequence* $\beta = \{b_n\}_{n=1}^{\infty}$ *having an asymptotic distribution function in* $\mathbb{Z}_g$ *if and only if, either*

(I) $\phi_{k_1}(j) = 1$, *for some* $j \in G_{k_1}$, *if* $k_1 \le k_2$,

*or*

(II) $\phi_{k_2}(j) = 1$, *for some* $j \in G_{k_2}$, *if* $k_2 \le k_1$,

*where* $\{\phi_{k_1}(j)\}$ *and* $\{\phi_{k_2}(j)\}$ *are asymptotic* $k_1$*–distribution functions of* $\alpha$ *and asymptotic* $k_2$*–distribution functions of* $\alpha$*, respectively.*

**Proof.** First we show the necessity. Consider the contrapositive of the necessity statement, that is, we assume that $0 \le \phi_{k_1}(j) < 1$ for all $j \in G_{k_1}$ if $k_1 \le k_2$, and $0 \le \phi_{k_2}(j) < 1$ for all $j \in G_{k_2}$ if $k_2 \le k_1$. For any case, $\alpha = \{a_n\}_{n=1}^{\infty}$ and $\alpha$ are not $(k_1, k_2)$–independent, nor $(k_2, k_1)$–independent in $\mathbb{Z}_g$, respectively by Theorem 3 [12]. This leads to a contradiction to the assumption of the theorem, thus the necessity is proved.

Now we proceed to the sufficiency part of Theorem 3 if $k_1 \le k_2$. So we suppose that

$$\phi_{k_1}(j) = 1, \tag{28}$$

for some $j \in G_{k_2}$ and $\beta$ has its asymptotic $k_2$–distribution function. Let us take two integers $\ell \in G_{k_1}$ and $m \in G_{k_2}$, then we get immediately

$$A_N\,(U_{k_1}(\ell);\, \alpha,\, U_{k_2}(m);\, \beta) \le A_N\,(U_{k_1}(\ell);\, \alpha), \tag{29}$$

for all positive integers $N$.

Consider first the case where $\ell = j$, then for all $N \ge 1$,

$$A_N\,(U_{k_2}(m);\, \beta) - \sum_{\substack{r=0 \\ r \ne \ell}}^{g^{k_1}-1} A_N\,(U_{k_1}(r);\, \alpha) \tag{30}$$

$$\le A_N\,(U_{k_1}(\ell);\, \alpha,\, U_{k_2}(m);\, \beta)$$

$$\le A_N\,(U_{k_2}(m);\, \beta).$$

Now for almost all $a_n$ in the sense of natural density,

$$a_n = j + \sum_{i=k_2}^{\infty} a_{n_i}\, g^i, \tag{31}$$

where $a_{n_i} = 0$ for all $i$ with $k_1 \le k_2 - 1$. Then we have

$$\lim_{N \to \infty} \sum_{\substack{r=0 \\ r \ne \ell}}^{g^{k_1}-1} \frac{A_N\,(U_{k_1}(r);\, \alpha)}{N} = 0. \tag{32}$$

Dividing (30) by $N$ and letting $N$ go to infinity, we obtain, from (32)

$$\|A\,(\alpha \in U_{k_1}(\ell), \beta \in U_{k_2}(m))\| \tag{33}$$

$$= \|A\,(\beta \in U_{k_2}(m))\|$$

$$= \|A\,(\alpha \in U_{k_1}(\ell))\| \cdot \|A\,(\beta \in U_{k_2}(m))\|,$$

since $\phi_{k_1}(j) = 1$ for some $j \in G_{k_1}$ implies $\phi_{k_1}(j) = \phi_{k_1}(\ell) = 1$. The equation (33) is indeed the conclusion of Theorem 3 for the sufficiency part.

Second consider the case where $\ell \ne j$ for all $N \ge 1$. Then, for each pair $(\ell, m) \in G_{k_1} \times G_{k_2}$,

$$0 = \|A(\alpha \in U_{k_1}(\ell), \beta \in U_{k_2}(m))\| \tag{34}$$
$$= \|A(\alpha \in U_{k_1}(\ell))\| \cdot \|A(\beta \in U_{k_2}(m))\|,$$

since $\phi_{k_1}(j) = 1$ for some $j \in G_{k_1}$ implies

$$A_N(U_{k_1}(\ell); \alpha, U_{k_2}(m); \beta) = o(N), \tag{35}$$

for all sufficiently large $N$, because $U_{k_2}(j) \subset U_{k_1}(j)$ and $U_{k_1}(\ell) \cap U_{k_1}(j) = \phi$.

Thus the sufficiency part is completed, if $k_1 \leq k_2$. For the case of $k_2 \leq k_1$, we trace arguments above by interchanging $k_1$ and $k_2$, which completes the proof of this Theorem.

■

## 4. Relationship Among Uniform Distributions in $\mathbb{Z}_g$

In this section, we establish a relationship between $k$-uniform distribution in $\mathbb{Z}_g$ and $2k$-uniform distribution in $\mathbb{Z}_g$.

**Theorem 4.** *Let $\gamma = \{c_n\}_{n=1}^{\infty}$ be a g-adic integer sequence. $\gamma$ is k-uniformly distributed in $\mathbb{Z}_{g^2}$ if and only if $\gamma$ is 2k-uniformly distributed in $\mathbb{Z}_g$.*

**Proof.** We prove first the necessity. Since $\gamma$ is $k$-uniformly distributed in $\mathbb{Z}_{g^2}$, we have, for every $j \in G_{2k}$,

$$\|A(\gamma \in U_k(j))\| = \frac{1}{g^{2k}}. \tag{36}$$

The $n$-th element of $\gamma$, $c_n$ is in $\mathbb{Z}_{g^2}$ and represented by

$$c_n = \sum_{i=0}^{\infty} c_{n_i} (g^2)^i, \tag{37}$$

with $c_{n_i} \in G_2$.

Similarly to the proof of Theorem 1, we write

$$c_{n_i} = a_{n_i} + b_{n_i} \cdot g\ , \tag{38}$$

where the $a_{n_i}$'s and $b_{n_i}$'s are contained in $G_1$ and

$$\psi_k(c_n) = \sum_{i=0}^{k-1} c_{n_i} (g^2)^i = j\ , \tag{39}$$

when $c_n \in U_k(j)$. Then we get

$$\begin{aligned} &c_{n_0} + c_{n_1} g^2 + \ldots + c_{n_{(k-1)}} g^{2(k-1)} \\ &= a_{n_0} + b_{n_0} g + a_{n_1} g^2 + b_{n_1} g^3 + \ldots + a_{n_{(k-1)}} g^{2k-2} + b_{n_{(k-1)}} g^{2k-1}. \end{aligned} \tag{40}$$

On the other hand, if we consider $c_n$ in $\mathbb{Z}_g$, then the representation of $c_n$ can be written as

$$c_n = \sum_{i=0}^{\infty} (a_{n_i} + b_{n_i} \cdot g)(g^2)^i = \sum_{i=0}^{\infty} (a_{n_i} \cdot g^{2i} + b_{n_i} \cdot g^{2i+1}). \tag{41}$$

From (40), we have

$$\psi_{2k}(c_n) = a_{n_0} + b_{n_0} g + \ldots + a_{n_{(k-1)}} g^{2k-2} + b_{n_{(k-1)}} g^{2k-1}, \tag{42}$$

which signifies that the number $A_N(U_k(j); \gamma)$ in $\mathbb{Z}_{g^2}$ is identical to $A_N(U_{2k}(j); \gamma)$ in $\mathbb{Z}_g$. Hence we deduce

$$\lim_{N \to \infty} \frac{A_N(U_{2k}(j); \gamma)}{N} = \lim_{N \to \infty} \frac{A_N(U_k(j); \gamma)}{N} = \frac{1}{g^{2k}}, \tag{43}$$

which proves $2k$ –uniform distribution of $\gamma$.

In order to prove the sufficiency, we only need to trace the above arguments backward. ∎

By modifying slightly the proof of Theorem 4, we obtain,

**Corollary 1.** *Let $k_1$ and $k_2$ be positive integers. A $g$–adic integer sequence $\alpha = \{a_n\}_{n=1}^{\infty}$ is $k_1$–uniformly distributed in $\mathbb{Z}_{g^{k_2}}$ if and only if $\alpha$ is $k_1 \cdot k_2$–uniformly distributed in $\mathbb{Z}_g$.*

**Corollary 2.** *A $g$–adic integer sequence $\alpha = \{a_n\}_{n=1}^{\infty}$ is uniformly distributed in $\mathbb{Z}_g$ if and only if $\alpha$ is 1–uniformly distributed in $\mathbb{Z}_{g^k}$ for every positive integer $k$.*

Since rational integers are $g$–adic integers, we may consider a sequence of rational integers as a $g$–adic integer sequence. Further, a 1–uniformly distributed sequence of rational integers considered as a $g$–adic integer sequence in $\mathbb{Z}_{g^k}$ is uniformly distributed mod $g^k$. Thus we get

**Theorem 5.** *A sequence of rational integers $a = \{a_n\}_{n=1}^{\infty}$ is uniformly distributed in $\mathbb{Z}$ if and only if $a$ is uniformly distributed mod $g^k$ for every positive integer $g$ greater than one and for all positive integers $k$.*

Now we give an example of Theorem 4.

**Example 1.** For every positive integer $n$, we develop $n$ to the base 9 as

$$n = n_0 + n_1 \cdot 9 + \ldots + n_\ell \cdot 9^\ell . \tag{44}$$

Let us define 9–adic integer $x_n$ by

$$x_n = n_0 n_1 n_2 \ldots n_\ell j\, 12\overline{12}, \tag{45}$$

where $n_i \in \{0,1,\ldots,8\}$ in (44) for $i = 0,1,\ldots,\ell$ and

$$\overline{12} = 121212\ldots, \tag{46}$$

in the sense of $p$–adic valuation; see Bachman [1].

Then $\chi = \{x_n\}_{n=1}^{\infty}$ is $k$–uniformly distributed in $\mathbb{Z}_9$ for $1 \le k \le j+1$ and also $2k$ –uniformly distributed in $\mathbb{Z}_3$.

Combining Theorems 1, 2 and 4 in this note and Theorem 3 in [13], we get

**Theorem 6.** *Let* $\gamma = \{c_n\}_{n=1}^{\infty}$ *be a g–adic integer sequence and let* $\alpha = \{a_n\}_{n=1}^{\infty}$ *and* $\beta = \{b_n\}_{n=1}^{\infty}$ *also be g–adic integer sequences determined as in Theorems* 1 *and* 2. *Then the following are equivalent:*

(I) $\gamma$ *is k–uniformly distributed in* $\mathbb{Z}_{g^2}$.

(II) $\gamma$ *is* 2*k–uniformly distributed in* $\mathbb{Z}_g$.

(III) $(\alpha,\beta)$ *is* $(k, k)$*–uniformly distributed in* $\mathbb{Z}_g \times \mathbb{Z}_g$.

Theorem 1 can be generalized by a slight modification of its proof.

**Corollary 1 of Theorem 1.** *Let* $\alpha = \{a_n\}_{n=1}^{\infty}$ *be a g–adic integer sequence. For two positive integers* $k_1$ *and* $k_2$, *we suppose* $\alpha$ *is* $k_1$*–uniformly distributed in* $\mathbb{Z}_{g^{k_2}}$. *The g–adic integers* $a_n$ *can be represented by*

$$a_n = \sum_{i=0}^{\infty} a_{n_i} (g^{k_2})^i \tag{47}$$

*with* $a_{n_i} \in G_{k_2}$. *Define* $a_{n_i}^{(1)}, a_{n_i}^{(2)}, \ldots, a_{n_i}^{(k_2)}$ *by*

$$a_{n_i} = a_{n_i}^{(1)} + a_{n_i}^{(2)} \cdot g + \ldots + a_{n_i}^{(k_2)} \cdot g^{k_2 - 1}, \tag{48}$$

*with* $a_{n_i}^{(j)} \in G_1$ *for* $j = 1,2,\ldots,k_2$. *Then determine* $a_n^{(j)}$ *by*

$$a_n^{(j)} = \sum_{i=0}^{\infty} a_{n_i}^{(j)} g^i, \tag{49}$$

*and consider* $k_2$ *g–adic integer sequences* $\alpha_j = \{a_n^{(j)}\}_{n=1}^{\infty}$ *for* $j = 1,2,\ldots,k_2$.

*Then* $\alpha_1, \alpha_2, \ldots, \alpha_{k_2}$ *are all* $k_1$*–uniformly distributed in* $\mathbb{Z}_g$ *and* $k_1$*–independent in* $\mathbb{Z}_g$.

Finally we remark that a sequence of $g$–adic integers $\alpha = \{f(n)\}_{n=1}^{\infty}$, where

$$f(n) = a_\ell n^\ell + a_{\ell-1} n^{\ell-1} + \ldots + a_1 n + a_0, \tag{50}$$

with $a_i$ $g$–adic integer for $i = 0,1,\ldots,\ell - 1$ is uniformly distributed in $\mathbb{Z}_g$ if and only if $|a_i|_g < 1$ for $i = 2,3,\ldots,\ell$ and $a_1$ is a unit of $\mathbb{Z}_g$. For the proof of this remark, we need only to follow the arguments in the proof of Theorem 4 in [11] with Corollary 1 of Theorem 1.

**Acknowledgment:** The authors are indebted to the referees for their many helpful suggestions.

## References

[1] *G. Bachman,* Introduction to $p$–Adic numbers and Valuation Theory. Academic Press, New York–London (1964).

[2] *M. Cugiani,* Successioni uniformente distribuite nei domini $p$–adici. Ist Lombardo Accad. Sci. Lett. Rend., A **96** (1962), 351—372.

[3] *L. Kuipers,* Continuous distribution mod 1 and independence of functions. Nieuw. Arch. voor Wisk., (3) **11** (1963), 1—3.

[4] *L. Kuipers* and *H. Niederreiter,* Uniform Distribution of Sequences. John Wiley and Sons, New York–London–Sydney–Toronto (1974).

[5] *L. Kuipers* and *H. Niederreiter,* Asymptotic distribution mod $m$ and independence of sequences of integers, I. Proc. Japan Acad., **50** (1974), 256—260.

[6] *L. Kuipers* and *H. Niederreiter,* Asymptotic distribution mod $m$ and independence of sequences of integers, II. Proc. Japan Acad., **50** (1974), 261—265.

[7] *L. Kuipers* and *J.–S. Shiue,* Asymptotic distribution modulo $m$ of sequences of integers and the notion of independence, I. Atti Accad. Naz. Lincei, Mem. (8) **11** (1972), 63—82.

[8] *L. Kuipers* and *J.–S. Shiue,* Asymptotic distribution modulo $m$ of sequences of integers and the notion of independence, II. Atti Accad. Naz. Lincei, Mem. (8) **11** (1972), 83—90.

[9] *H.G. Meijer,* Uniform distribution of $g$–adic numbers. Thesis, Universiteit van Amsterdam (1967).

[10] *H.G. Meijer,* Uniform distribution of $g$–adic numbers. Indag. Math., **29** (1967), 535—546.

[11] *H.G. Meijer* and *J.–S. Shiue,* Uniform distribution in $\mathbb{Z}_g$ and $\mathbb{Z}_{g_1} \times \ldots \times \mathbb{Z}_{g_t}$. Indag. Math., **38** (1976), 200—212.

[12] *K. Nagasaka* and *J.–S. Shiue,* Asymptotic distribution and independence of sequences of $g$–adic integers. Prospects of Math. Sci., T. Mitsui, K. Nagasaka *et al.* (ed.), World Sci. (1988), 157—171.

[13] *J.–S. Shiue,* On a theorem of uniform distribution of sequences of $g$–adic integers and a notion of independence. Rend. Accad. Naz. Lincei, **50** (1971), 90—93.

[14] *H. Steinhaus,* Fonctions indépendantes, IV. Studia Math., **9**(1940), 121—131.

---

University of the Air, 2–11, Wakaba, Chiba–shi, Chiba, 260, JAPAN.

Department of Mathematical Sciences, University of Nevada, Las Vegas, 4505 Maryland Parkway, Las Vegas, Nevada 89154, USA.

# On the Piltz Divisor Problem With Congruence Conditions

*Werner Georg Nowak*

## 1. Introduction

The classical Piltz' divisor problem concerns the arithmetic function $d_N(n)$, defined as the number of positive integer $N$ – tuples $(u_1, \dots, u_N)$ with $u_1, \dots, u_N = n$, in particular the order of the error term $\Delta_N(x)$ in the asymptotic formula

$$\sum_{n \le x} d_N(n) = \sum_{j=0}^{N-1} c_j \, x \, (\log x)^j + \Delta_N(x). \tag{1}$$

A classic result (*cf.* Titchmarsh [9], p. 266) asserts that, for $N \ge 4$,

$$\Delta_N(x) = O\left(x^{(N-1)/(N+2)+\varepsilon}\right), \tag{2}$$

and slight improvements can be found in Ivić [3], p. 355 f. Concerning lower bounds, already Szegö and Walfisz [7], [8] proved that

$$\Delta_N(x) = \Omega((x \log x)^{(N-1)/2N} (\log \log x)^{N-1}) \tag{3}$$

and this was sharpened only recently (see Hafner [2]) by a loglog-factor.

As in many other problems of analytic number theory (*cf.* e.g., the prime number theorem), it is a natural idea to look for appropriate generalizations of these results with respect to arithmetic progressions. One way to do this would be to restrict $n$ to a certain residue class $l$ modulw $m$ ($l, m \in \mathbb{N}$, fixed). Another possibility is to impose such a congruence condition to each (or some) of the factors $u_j$.

In both cases, it is straightforward to obtain the same $O$-results as in the classical Piltz divisor problem, since the theory of $\zeta(s)$ developed in the textbooks [3] and [9] may be readily generalized to $L$-series.

To derive an $\Omega$-estimate of about the same accuracy as (3) is , however, much less obvious, and even almost hopeless both in the first case ($n \equiv l \bmod m$) and in the case that on each of the factors $u_j$ a nontrivial congruence condition is imposed. (*Cf.* the details of the subsequent argument, in particular the "concluding remark" at the end. This could yield some explanation for the fact that results in this direction apparently never occurred in the literature, although the classical problem itself is a very old one.)

It turns out, however, that one can establish a result of the desired type if one imposes a congruence condition not to all but only to some of the factors $u_j$. This task is to be carried out in the present note, by an appropriate modification and combination of classic [7], [8] and modern [2] techniques.

## 2. Problem and Result.

Let $p, q \in \mathbb{N}$, $N = p + q$, $m_j, l_j \in \mathbb{N}$, $l_j < m_j$, $\lambda_j = \frac{l_j}{m_j}$ for $j = 1, \ldots, p$, and let $d^*(n)$ denote the number of positive integer $N$-tuples $(u_1, \ldots, u_N)$ satisfying $u_1 \ldots u_N = n$ and $u_j \equiv l_j \pmod{m_j}$ for $j = 1, \ldots, p$. We consider the error term $E(x)$ in the asymptotic formula

$$D^*(x) = \sum_{n \le x} d^*(n) = \sum_{s_0 = 0, 1} \operatorname*{Res}_{s = s_0} (F(s) x^s M^{-s} s^{-1}) + E(x), \tag{4}$$

where $M = m_1 \ldots m_p$ and $F(s)$ is the generating function

$$F(s) = M^s \sum_{n=1}^{\infty} d^*(n) n^{-s} = (\zeta(s))^q \prod_{j=1}^{p} \zeta(s, \lambda_j) \qquad (\operatorname{Re} s > 1).) \tag{5}$$

Here $\zeta(s)$ and $\zeta(s, .)$ denote the Riemann and Hurwitz zeta-function, respectively. Throughout the paper, all estimates and $O$-, $\ll$-symbols refer to $x \to \infty$, all other variables given above are considered fixed. (Obviously, the main term in (4) is of the same form as that in (1), up to an $O(1)$.)

**Theorem 1.** *Let* $P = \{1,2,\ldots,p\}$, $J' = P - J$ *for any* $J \subset P$ *and suppose that*

$$Z(1) := \sum_{J \subset P} \cos\left(\frac{\pi}{4}(N-3) - \frac{\pi}{2}(p - |J|)\right)$$
$$\times \prod_{j \in J}(-\log(2\sin(\pi\lambda_j))) \prod_{j' \in J'}\left(\frac{\pi}{2} - \pi\lambda_{j'}\right) \neq 0.$$

*Then*

$$E(x) = \Omega_*\left((x \log x)^{(N-1)/2N}(\log_2 x)^{q-1}(\log_3 x)^{-(N-1)/2N}\right)$$

*where* $\log_k$ *denotes the* $k$-fold *iterated logarithm and* $*$ *is the sign of* $Z(1)$.

**Remark 1.** The special case $N = 2$, $p = q = 1$ (the problem of divisors lying in a prescribed arethmetic progression) was treated earlier [6] by a different method, with a slightly sharper result (uniformity in the modulus).

**Remark 2.** The somewhat cumbersome condition $Z(1) \neq 0$ also appears in the case $N = 2$ ($p = q = 1$) where the "forbidden" values of $\lambda = \frac{l}{k}$ have been determined explicitly [6], as well as in the classic work [7] on the Piltz' divisor problem where it simply reduces to $N \not\equiv 1 \pmod 4$. It is likely that one could get rid of this condition (and replace $\Omega_*$ by $\Omega_\pm$ at the same time) for dimensions $N \geq 4$, but probably not for $N = 2$ or 3, applying the yet deeper method of [8]. This is to be postponed for a subsequent work.

## 3. Proof of the Theorem

Denoting by $G_r$ the Liouville–Riemann integral of order $r$ for any function $G$ integrable on every bounded subinterval of $\mathbb{R}_0^+$ ($r \in \mathbb{N}$), we have

$$D_r^*(x) = \Gamma(r)^{-1}\int_0^x (x-u)^{r-1} D^*(u)du = (r!)^{-1}\sum_{n \leq x}(x-n)^r d^*(n)$$
$$= (2\pi i)^{-1} x^r \int_{c-i\infty}^{c+i\infty} \frac{F(s)}{s \ldots (s+r)}\left(\frac{x}{M}\right)^s ds \tag{6}$$

for $c > 1$ and $r \in \mathbb{N}$ sufficiently large, by a general version of Perron's formula.

Let $H(x)$ denote the main term of (4), then obviously

$$H_r(x) = \sum_{s_0=0,1} \operatorname{Res}_{s=s_0}\left(\frac{F(s)}{s\ldots(s+r)}\frac{x^{s+r}}{M^s}\right)$$

and the residue theorem yields (in view of well-known growth properties of the Riemann and Hurwitz zeta-functions)

$$D_r^*(x) = H_r(x) + (2\pi i)^{-1} x^r \int_{-\omega-i\infty}^{-\omega+i\infty} \frac{F(s)}{s\ldots(s+r)}\left(\frac{x}{M}\right)^s ds$$

for any (small) $\omega \in \mathbb{R}^+$. Thus by (4) and a change of variable,

$$E_r(x) = (2\pi i)^{-1} x^r \int_{-\omega-i\infty}^{-\omega+i\infty} \frac{F(1-s)}{(1-s)\ldots(1-s+r)}\left(\frac{x}{M}\right)^{1-s} ds. \tag{7}$$

We recall the functional equations (see e.g., [1], p. 257 and 259)

$$\zeta(1-s,\lambda) = (2\pi)^{-s}\, 2\Gamma(s)\sum_{h=1}^{\infty} h^{-s}\cos(2\pi h\lambda - \tfrac{\pi}{2}s) \quad (\operatorname{Re} s > 1)$$

$$\zeta(1-s) = (2\pi)^{-s}\, 2\Gamma(s)\zeta(s)\cos\left(\frac{\pi}{2}s\right),$$

to conclude that, for $\operatorname{Re} s > 1$,

$$F(1-s) = 2^N (\Gamma(s))^N (2\pi)^{-Ns} \sum_{\substack{a+b=N\\ a\geq q}} \sum_{h=1}^{\infty} \beta(a,b;h) h^{-s} (\cos\tfrac{\pi}{2}s)^a (\sin\tfrac{\pi}{2}s)^b, \tag{8}$$

$$\beta(a,b;h) := \sum_{\substack{(h_1,\ldots,h_N)\in\mathbb{N}^N\\ h_1\ldots h_N = h}} \sum_{\substack{J\subset P\\ |J| = a-q}} \prod_{\substack{j\in J\\ j'\in J'}} \cos(2\pi h_j \lambda_j)\sin(2\pi h_{j'}\lambda_{j'}) \tag{8'}$$

(here $a, b \in \mathbb{N}_0$, $a \geq q$; $P, J, J'$ have the meaning as in the theorem). Thus, from (7),

$$E_r(x) = (2\pi i)^{-1} M^r \pi^{N/2-N} \sum_{h=1}^{\infty} h^{-r-1} \sum_{\substack{a+b=N\\ a\geq q}} \beta(a,b;h)$$

$$\times \int_{1+\omega-i\infty}^{1+\omega+i\infty} G_{a,b}(s)\left(\frac{x}{M}h\pi^N\right)^{r+1-s} \frac{ds}{(1-s)\ldots(1+r-s)}$$

where

$$G_{a,b}(s) := (2^{1-s}\,\Gamma(s))^{a+b}\,\pi^{-(a+b)/2}(\cos\frac{\pi}{2}s)^a(\sin\frac{\pi}{2}s)^b$$

$$= \Delta_{a,b}(s)\,(\Delta_{a,b}(1-s))^{-1},$$

$$\Delta_{a,b}(s) := (\Gamma(\frac{s}{2}))^a\,(\Gamma(\frac{s}{2}+\frac{1}{2}))^b,$$

by well known properties of the $\Gamma$–function (see e.g., Nielsen [5]). To simplify notation further, we put

$$J_r^{(a,b)}(y) := (2\pi i)^{-1}\int_{1+\omega-i\infty}^{1+\omega+i\infty} W_{a,b;r}(s)\,y^{r+1-s}\,ds,$$

$$W_{a,b;r}(s) := \frac{G_{a,b}(s)}{(1-s)\ldots(1+r-s)}$$

and obtain

$$E_r(x) = M^r\,\pi^{N/2-Nr}\sum_{h=1}^{\infty} h^{-r-1}\sum_{\substack{a+b=N\\ a\geq q}} G\,\beta(a,b;h)\,J_r^{(a,b)}(\pi^N h\frac{x}{M}). \tag{9}$$

By Stirling's formula, $W_{a,b;r}(\sigma+it) = O(|t|^{(\sigma-1/2)N-r-1})$ for $a+b=N$, $|t|\to\infty$, uniformly in any strip $\sigma_1\leq\sigma\leq\sigma_2$. Since $W_{a,b;r}(s)$ is regular at $s=1$ for every $r\in\mathbb{N}_0$ (because of $a\geq 1$), we may conclude that, for $r$ sufficiently large,

$$J_r^{(a,b)}(y) = I_r^{(a,b)}(y) := \frac{1}{2\pi i}\int_{\frac{1}{4}-i\infty}^{\frac{1}{4}+i\infty} W_{a,b;r}(s)\,y^{r+1-s}\,ds, \tag{10}$$

where the last integral converges absolutely for every $r\in\mathbb{N}_0$ and satisfies

$$\frac{d}{dy}(I_{r+1}^{(a,b)}(y)) = I_r^{(a,b)}(y).$$

We now put $X = X(t) = K_1(\log t)^{-1}\log_3 t$ and $k = k(t) = t^2 X^{-2/N}$ where $t$ is a large real parameter and $K_1$ a positive constant to be specified later ($X\to 0$ and $k\to\infty$ for $t\to\infty$). Following the classic examples of [7] and [2], we consider the Borel mean–value

$$B(t) := (\Gamma(k+1))^{-1} \int_0^\infty u^k e^{-u} E((2u)^\alpha X)\,du \tag{11}$$

$$= (\alpha\Gamma(k+1))^{-1} \int_0^\infty \phi(v) E(2^\alpha X v)\,dv$$

where $\alpha = \frac{N}{2}$ and $\phi(v) = v^{(k+1)/\alpha - 1}\exp(-v^{1/\alpha})$ for short. By an iterated integration by parts, we obtain for any $r \in \mathbb{N}$ (with $t$ and thus $k$ sufficiently large)

$$B(t) = (\alpha\Gamma(k+1))^{-1}(-2^{-\alpha}X^{-1})^r \int_0^\infty \phi^{(r)}(v) E_r(2^\alpha X v)\,dv.$$

We choose $r$ so large that (9) is valid (and the series involved is absolutely convergent), insert (9) and (10) and obtain

$$B(t) = M^r \pi^{N/2-Nr}(\alpha\Gamma(k+1))^{-1}(-2^{-\alpha}X^{-1})^r \sum_{h=1}^\infty h^{-r-1} \sum_{\substack{a+b=N \\ a\geq q}} \beta(a,b;h)$$

$$\times \int_0^\infty \phi^{(r)}(v) I_r^{(a,b)}\left(\pi^N h \frac{2^\alpha X}{M} v\right) dv$$

$$= \pi^{N/2}(\alpha\Gamma(k+1))^{-1} \sum_{h=1}^\infty h^{-1} \sum_{\substack{a+b=N \\ a\geq q}} \beta(a,b;h)$$

$$\times \int_0^\infty \phi(v) I_0^{(a,b)}\left(\pi^N h \frac{2^\alpha X}{M} v\right) dv$$

$$= \pi^{N/2}(\Gamma(k+1))^{-1} \sum_{h=1}^\infty h^{-1} \sum_{\substack{a+b=N \\ a\geq q}} \beta(a,b;h)$$

$$\times \int_0^\infty e^{-u} u^k I_0^{(a,b)}\left(\pi^N h \frac{X}{M}(2u)^\alpha\right) du, \tag{12}$$

inverting the repeated integration by parts and the change of variable. (The convergence of the last series will be clear in the sequel.)

We are now able to apply results due to Hafner [2], p. 50, observing that our definition of $I_0^{(a,b)}$ is consistent with his of $I_0$, in view of the regularity of the integrand in (10) in the half–plane $\mathrm{Re}\, s > 0$.

**Lemma 1.** *(Hafner) Let $\delta > 0$ be some small constant, $y \in \mathbb{R}$, $k$ a sufficiently large real, and define*

$$U_{(k)}(y) = U_{(k)}^{(a,b)}(y) := (\Gamma(k+1))^{-1} \int_0^\infty e^{-u} u^k I_0^{(a,b)}((2uy)^\alpha)du.$$

*then, for* $k^{-\delta} \le y \le k^{\delta}$,

(i) $$U_{(k)}(y) = c_1 e^{-y} (ky)^{\alpha\theta} \cos(c_2 (ky)^{1/2} + \gamma_b \pi) + O((ky)^{\alpha\theta} k^{-1/2+2\delta}) + O((ky)^{\alpha\theta - 1/2}),$$

*where* $\theta = \frac{1}{2}(1 - \frac{1}{N})$, $\gamma_b = \frac{1}{4}(N-3) - \frac{b}{2}$ *and* $c_i$ *denotes positive constants throughout the sequel. Furthermore, for* $y > k^{\delta}$,

(ii) $$U_{(k)}(y) = O(y^{-1/2\delta}).$$

We want to insert this into (12), with $y = \pi^2 (hX)^{2/N} M^{-2/N}$ and $k, X$ as defined earlier. To deal with the error terms arising, observe that $y > k^{\delta}$ implies $h \gg k^{\alpha\delta}$, and $\beta(a, b; h) = o(h^{\varepsilon})$ for any $\varepsilon > 0$, consequently the contribution of the terms in the sum (12), to which (ii) applies, is

$$\ll \sum_{h \gg k^{\alpha\delta}} h^{-1+\varepsilon-1/2\alpha\delta} k^{\varepsilon} \ll k^{-1/2+\varepsilon(1+\alpha\delta)} = k^{-1/2+\varepsilon'}$$

(where $\varepsilon' > 0$ can be make arbitrarily small). On the other hand, $y \le k^{\delta}$ implies $h \ll k^{\alpha\delta} X^{-1}$, thus the contribution of the order terms in (i) to the sum in (12) is

$$\ll \sum_{h \ll k^{\alpha\delta+\varepsilon}} h^{-1+\varepsilon} (h^{\theta} k^{\alpha\theta - 1/2 + 3\delta} + h^{\theta - 1/2\alpha} k^{\alpha\theta - 1/2 + \varepsilon})$$

$$\ll k^{\alpha\delta - 1/2 + 3\delta} \sum_{h \ll k^{\alpha\delta+\varepsilon}} h^{-1+\theta+\varepsilon} \ll k^{\alpha\theta - 1/2 + \delta(\alpha\theta + 3) + \varepsilon''}$$

$$\ll k^{\alpha\theta - 1/4}$$

if $\delta$ is chosen sufficiently small. Altogether we thus derive from (12) the asymptotic formula

$$B(t) = c_3 \sum_{h \le H} h^{-1} \sum_{\substack{a+b=N \\ a \ge q}} \beta(a,b;h) \exp(-\pi^2 M^{-1/\alpha} (hX)^{1/\alpha}) (hX)^{\theta} k^{\alpha\theta}$$

$$\times \cos(c_2 \pi M^{-1/N} (hX)^{1/N} k^{1/2} + \gamma_b \pi) + O(k^{\alpha\theta - 1/4}) \tag{13}$$

where $H = c' k^{\alpha\delta} X^{-1}$, $c'$ a suitable constant. In order to extend the range of summation to $1 \le h < \infty$, we observe that

$$\sum_{h > H} h^{\theta-1} \sum_{\substack{a+b=N \\ a \ge q}} \beta(a,b;h) \exp(-c_4 (hX)^{1/\alpha})$$

$$\ll \exp(-c_4 (HX)^{1/\alpha}) + \int_H^{\infty} \exp(-c_4 (Xu)^{1/\alpha})\, du$$

$$\ll \exp(-c_4 k^{\delta}) + \exp(-\tfrac{1}{2} c_4 (HX)^{1/\alpha}) \int_H^{\infty} (Xu)^{-2}\, du \ll k^{-C}$$

for every $C \in \mathbb{R}^+$, thus (13) simplifies to

$$B(t) = c_3 X^{\theta} k^{\alpha\theta} \sum_{h=1}^{\infty} h^{\theta-1} \sum_{\substack{a+b=N \\ a \le q}} \beta(a,b;h) \exp(-c_4 (hX)^{1/\alpha})$$

$$\times \cos(c_5 (hX)^{1/N} k^{1/2} + \gamma_b \pi) + O(k^{\alpha\theta - 1/4}). \tag{14}$$

We recall the definition (8') of $\beta(a,b;h)$, keep $(h_1, \dots, h_N) \in \mathbb{N}^N$ fixed for a moment, such that $h_1, \dots, h_N = h$, and compute (with $Z = c_5 (hX)^{1/N} k^{1/2} + \frac{\pi}{4}(N-3)$ for short)

$$\sum_{\substack{a+b=N \\ a \ge q}} \left\{ \sum_{\substack{J \subset P \\ |J| = a-q}} \prod_{\substack{j \in J \\ j' \in J'}} \cos(2\pi h_j \lambda_j) \sin(2\pi h_{j'} \lambda_{j'}) \right\} \cos(Z - \tfrac{\pi}{2} b)$$

$$= \sum_{J \subset P} (\cos Z \cos(\tfrac{\pi}{2}(p - |J|)) + \sin Z \sin \tfrac{\pi}{2}(p - |J|))$$

$$\times \prod_{\substack{j \in J \\ j' \in J'}} \cos(2\pi h_j \lambda_j) \sin(2\pi h_{j'} \lambda_{j'})$$

$$= \cos(Z - 2\pi \sum_{j=1}^{p} h_j \lambda_j),$$

by the general addition theorems

$$\cos\left(\sum_{j=1}^{p}\phi_j\right) = \sum_{J\subset P}\cos\left(\frac{\pi}{2}(p-|J|)\right)\prod_{\substack{j\in J\\ j'\in J'}}\cos\phi_j\ \sin\phi_{j'}\ ,$$

$$\sin\left(\sum_{j=1}^{p}\phi_j\right) = \sum_{J\subset P}\sin\left(\frac{\pi}{2}(p-|J|)\right)\prod_{\substack{j\in J\\ j'\in J'}}\cos\phi_j\ \sin\phi_{j'}\ .$$

Inserting this and the definitions of $k(t)$, $X(t)$ into (14), we obtain

$$\begin{aligned} B(t) = {} & c_3 t^{2\alpha\theta}\sum_{h=1}^{\infty} h^{\theta-1}\exp(-c_4(hX)^{1/\alpha}) \\ & \times \sum_{\substack{(h_1,\dots,h_N)\in\mathbb{N}^N:\\ h_1\dots h_N = h}} \cos(c_5 h^{1/N} + \frac{\pi}{4}(N-3) \\ & - 2\pi\sum_{j=1}^{p} h_j\lambda_j) + O(t^{2\alpha\theta-1/8}). \end{aligned} \tag{15}$$

The next step is the usual application of Dirichlet's approximation theorem, in a slightly modified way: Let $T$ be a given real number (arbitrarily large) and $A > \log T$ an integer, furthermore $Q = [c_0(\log A)^p]$, $c_0$ a suitable positive constant. Then we may choose a value of $t$ such that

$$T \le t \le Q^A T \tag{16}$$

and

$$\left\{\frac{1}{2\pi}c_5 h^{1/N}\, t\right\} < \frac{1}{Q}\quad \text{for } h = 1,\dots,A, \tag{17}$$

where $\{.\}$ denotes the distance from the nearest integer. It follows from (16) and the choice of $Q$ that

$$A \gg \frac{\log t}{\log Q} \gg \frac{\log t}{\log_2 A},\qquad \text{hence } A \gg \frac{\log t}{\log_3 t}.$$

Let us put $A' = c_6 \log t\,(\log_3 t)^{-1}$ with $c_6$ so small that $A' \le A$; certainly $A' \ge X^{-1}$ for $K_1$ sufficiently large. Define a sum $S$ like the main term of (15), with $\sum_{h=1}^{\infty}$ replaced by $\sum_{h\le A'}$ and put

$$S' := c_3 t^{2\alpha\theta} \sum_{h\le A'} h^{\theta-1} \exp(-c_4 (hX)^{1/\alpha}) \times \sum_{\substack{(h_1,\dots,h_N)\in\mathbb{N}^N:\\ h_1\dots h_N=h}} \cos\Big(\frac{\pi}{4}(N-3) - 2\pi\sum_{j=1}^{p} h_j\lambda_j\Big), \tag{18}$$

then (by (17) and the mean–value theorem),

$$S - S' \ll \frac{1}{Q} t^{2\alpha\theta} \sum_{h\le A'} h^{\theta-1} \exp(-c_4(hX)^{1/\alpha}) d_N(h)$$
$$= \frac{1}{Q} t^{2\alpha\theta} \int_0^{A'} \exp(-c_4(Xu)^{1/\alpha})\, df(u)$$

where

$$f(u) := \sum_{h\le u} h^{\theta-1} d_N(h) \sim c_7 u^{\theta} (\log u)^{N-1} \tag{19}$$

(*cf.* [7], p. 146). We infer from the above definitions that

$$|\log X| \sim \log A' \sim \log\log t. \tag{20}$$

Thus integration by parts yields

$$S - S' \ll \frac{1}{Q} t^{2\alpha\theta} \Big( \exp(-c_4 (XA')^{1/\alpha}) A'^{\theta} (\log A')^{N-1}$$
$$+ X^{1/\alpha} (\log A')^{N-1} \int_0^{A'} \exp(-c_4 (Xu)^{1/\alpha}) u^{\theta-1+1/\alpha} du \Big)$$
$$\ll \frac{1}{Q} t^{2\alpha\theta} X^{-\theta} |\log X|^{N-1} \Big(1 + \int_0^{A'X} \exp(-c_4 w^{1/\alpha}) w^{\theta-1+1/\alpha} dw\Big)$$
$$\ll c_0^{-1} t^{2\alpha\theta} X^{-\theta} |\log X|^{q-1}. \tag{21}$$

We are now going to establish a precise lower bound for $S'$. For $\nu \in \mathbb{N}$, we put

$$\eta(v) := \sum_{\substack{(h_1,\ldots,h_p)\in\mathbb{N}^p : \\ h_1\ldots h_p = v}} \cos\left(\frac{\pi}{4}(N-3) - 2\pi\sum_{j=1}^{p} h_j \lambda_j\right),$$

$$Z(s) := \sum_{v=1}^{\infty} \eta(v) v^{-s} \qquad (\mathrm{Re}\, s > 1)$$

$$\xi(h) := \sum_{uv=h} d_q(u)\eta(v) \qquad (h \in \mathbb{N})$$

thus

$$(\zeta(s))^q Z(s) = \sum_{h=1}^{\infty} \xi(h) h^{-s} \qquad (\mathrm{Re}\, s > 1)$$

From the trigonometric addition theorems stated earlier, we derive for Re $s > 1$ (with $y = \frac{\pi}{4}(N-3)$ for short) that

$$\begin{aligned} Z(s) = & \sum_{(h_1,\ldots,h_p)\in\mathbb{N}^p} \Big(\cos y \cos\Big(2\pi\sum_{j=1}^{p} h_j \lambda_j\Big) \\ & + \sin y \sin\Big(2\pi\sum_{j=1}^{p} h_j \lambda_j\Big)\Big)(h_1\ldots h_p)^{-s} \\ = & \sum_{J \subset P} \sum_{(h_1,\ldots,h_p)\in\mathbb{N}^p} \Big(\cos y \cos\frac{\pi}{2}(p-|J|) + \sin y \sin\frac{\pi}{2}(p-|J|)\Big) \\ & \times \prod_{\substack{j\in J \\ j'\in J'}} \cos(2\pi\lambda_j h_j)\sin(2\pi\lambda_{j'} h_{j'})\ (h_1\ldots h_p)^{-s} \\ = & \sum_{J\subset P} \cos\Big(\frac{\pi}{4}(N-3) - \frac{\pi}{2}(p-|J|)\Big) \\ & \times \prod_{\substack{j\in J \\ j'\in J'}} \sum_{h_j=1}^{\infty} \cos(2\pi\lambda_j h_j)\, h_j^{-s} \sum_{h_{j'}=1}^{\infty} \cos(2\pi\lambda_{j'} h_{j'})\, h_{j'}^{-s}. \end{aligned} \tag{22}$$

The series in this last representation converge for Re $s > 0$, hence $Z(s)$ possesses an analytic continuation in this half-plane. Appealing to [4], p.225, we obtain just the formula for $Z(1)$ in our theorem. Furthermore, we readily establish the following.

**Lemma 2.** *If* $Z(1) \neq 0$, *we have for* $w \to \infty$

$$V(w) := \sum_{h \le w} \xi(h) \sim Z(1) w (\log w)^{q-1}.$$

**Proof.** The generating function $g(s) = (\zeta(s))^q Z(s)$ has a pole of order $q$ at $s = 1$. Applying summation by parts to each of the series in the last representation for $Z(s)$, we see at once that $Z(\sigma + it) = O(|t|^P)$, as $|t| \to \infty$, uniformly in every half–plane $\sigma \ge \varepsilon'$, $\varepsilon' > 0$ arbitrary. By the Phragmén–Lindelöf principle (and absolute convergence in $\sigma > 1$),

$$g(\sigma + it) = O(|t|^{N(1-\sigma)+\varepsilon}) + O(1) \tag{23}$$

uniformly e.g., in $\frac{3}{4} \le \sigma \le 1 + \omega$, $\omega > 0$, $\varepsilon > 0$ arbitrarily small (*cf.* the corresponding result for the Riemann zeta–function in [9], p. 81). According to Perron's truncated formula, we get (for $w$ half an odd integer, w. l. o. g.)

$$V(w) = (2\pi i)^{-1} \int_{1+\omega-i\tau}^{1+\omega+i\tau} g(s) \frac{w^s}{s}\, ds + O(w^{1+\omega} \tau^{-1}), \tag{24}$$

where we choose $\tau = [w^{1/2N}]$, $\omega = \frac{1}{3N}$ (see [9],p. 53). We shift the line of integration to $\mathrm{Re}\, s = \frac{3}{4}$, deriving from (23) the estimates

$$\int_{\frac{3}{4} \pm i\tau}^{1+\omega \pm i\tau} g(s) \frac{w^s}{s}\, ds \ll \tau^{N-1+\varepsilon} \int_{\frac{3}{4}}^{1+\omega} \left( \frac{w}{\tau^N} \right)^{\sigma} d\sigma + \tau^{-1} w^{1+\omega} = o(w)$$

and

$$\int_{\frac{3}{4} - i\tau}^{\frac{3}{4} + i\tau} g(s) \frac{w^s}{s}\, ds \ll w^{3/4} \int_1^{\tau} t^{N/4 - 1 + \varepsilon}\, dt \ll w^{7/8 + \varepsilon/2N} = o(w).$$

Thus we infer from (24), by the residue theorem, that

$$V(w) = \operatorname*{Res}_{s=1} \left( g(s) \frac{w^s}{s} \right) + o(w) = Z(1) w (\log w)^{q-1} (1 + o(1)),$$

the assertion of Lemma 2.

Applying summation by parts, we readily derive the consequence

$$V_\theta(w) = \sum_{h \le w} h^{\theta-1} \xi(h) \sim c_8 Z(1) w^\theta (\log w)^{q-1}. \tag{25}$$

Recalling (18), we thus obtain a lower bound for $S'$: For $Z(1) \neq 0$,

$$
\begin{aligned}
Z(1)S' &= c_3 t^{2\alpha\theta} Z(1) \int_{1-}^{A'} \exp(-c_4 (wX)^{1/\alpha}) dV_\theta(w) \\
&= c_3 Z(1) t^{2\alpha\theta} \Big( \exp(-c_4 (c_6 K_1)^{1/\alpha}) V_\theta(A') \\
&\quad + c_4 X^{1/\alpha} \int_1^{A'} \exp(-c_4 (wX)^{1/\alpha}) V_\theta(w) dw \Big) \\
&\geq c_9 t^{2\alpha\theta} X^{1/\alpha} \int_{X^{-1/2}}^{X^{-1}} w^{1/\alpha - 1 + \theta} (\log w)^{q-1} dw \\
&\geq c_{10} t^{2\alpha\theta} X^{-\theta} |\log X|^{q-1}. 
\end{aligned} \tag{26}
$$

In a similar way, we conclude from (15) that

$$
\begin{aligned}
B(t) - S &= c_3 t^{2\alpha\theta} \int_{A'+}^{\infty} \exp(-c_4 (wX)^{1/\alpha}) dV_\theta(w) + O(t^{2\alpha\theta - 1/8}) \\
&\ll t^{2\alpha\theta} \Big( \exp(-c_4 (c_6 K_1)^{1/\alpha}) A'^{\theta} (\log A')^{q-1} \\
&\quad + X^{1/\alpha} \int_{A'}^{\infty} w^{1/\alpha - 1 + \theta} \exp(-c_4 (wX)^{1/\alpha}) (\log w)^{q-1} dw \Big) + t^{2\alpha\theta - 1/8} \\
&\ll t^{2\alpha\theta} X^{-\theta} |\log X|^{q-1} (c_6 K_1)^{\theta} \exp(-c_4 (c_6 K_1)^{1/\alpha}) \\
&\quad + t^{2\alpha\theta} X^{1/\alpha} (\log A')^{q-1} \exp(-\tfrac{1}{2} c_4 (c_6 K_1)^{1/\alpha}) \\
&\quad \times \int_{A'}^{A'^2} w^{1/\alpha - 1 + \theta} \exp(-\tfrac{1}{2} c_4 (wX)^{1/\alpha}) dw + t^{2\alpha\theta - 1/8} \\
&\ll \varepsilon_0(K_1) t^{2\alpha\theta} X^{-\theta} |\log X|^{q-1}
\end{aligned} \tag{27}
$$

where $\varepsilon_0(K_1)$ can be made arbitrarily small by making $K_1$ large (*cf.* also the calculation in [7],p. 155 f).

We now combine the results (21), (26) and (27), choose $c_0$ and $K_1$ ( in the definitions of $Q$ and $X$, respectively) sufficiently large, and obtain (for $Z(1) \neq 0$)

$$Z(1)B(t) \geq c_{11} t^{2\alpha\theta} X^{-\theta} |\log X|^{q-1}$$

$$\geq c_{12} t^{2\alpha\theta} (\log t)^{\theta} (\log_2 t)^{q-1} (\log_3 t)^{-\theta} \tag{28}$$

(for a certain sequence of reals $t$ tending to $\infty$).

The proof is now completed on classical lines: We suppose that, for some small positive constant $K_2$,

$$Z(1)E(x) \leq K_2 (x \log x)^{\theta} (\log_2 x)^{q-1} (\log_3 x)^{-\theta} \tag{29}$$

for all sufficiently large $x$ ( $\theta = \frac{1}{2}\left(1 - \frac{1}{N}\right)$ as before, $Z(1) \neq 0$ throughout). This would imply (recalling definition (11)) that, for every real $t$,

$$Z(1)B(t) \leq K_2 (\Gamma(k(t)+1))^{-1}$$
$$\times \int_0^{\infty} u^{k(t)+\alpha\theta} e^{-u} L(2^{\alpha} u^{\alpha} X(t)) du\, 2^{\alpha\theta} X(t)^{\theta} + O(1)$$

where $L(w) := (\log w)^{\theta} (\log_2 w)^{q-1} (\log_3 w)^{-\theta}$ for $w \geq 30$, $L(w) = 0$ else. Estimating this integral by Hafner's lemma 2.3.6 in[2], p.51, we get

$$Z(1)B(t) \leq c_{13} K_2 k(t)^{\alpha\theta} X(t)^{\theta} L(X(t) k(t)^{\alpha})$$
$$\leq c_{14} K_2 t^{2\alpha\theta} (\log t)^{\theta} (\log_2 t)^{q-1} (\log_3 t)^{-\theta}.$$

For $K_2$ sufficiently small, this contradicts (28), making the assumption (29) impossible. This completes the proof of our theorem.

## 4. Concluding Remarks

It is apparent from the proof that we really need $q \geq 1$, i.e., not all the factores in the decompostion $u_1 \ldots u_N = n$ may be subject to a congruence condition $u_j = l_j \pmod{m_j}$; otherwise the lemma is no longer true and the proof would fail.

On the other hand, the condition $Z(1) \neq 0$ might be replaced by " $Z(s)$ has a zero of order $z \leq q - 1$ at $s = 1$." Without altering the argument significantly, one would obtain the same result, with $(\log_2 x)^{q-1-z}$ instead of $(\log_2 x)^{q-1}$, and $\Omega_*$ where $*$ is the sign of $Z^{(z)}(1)$.

## References

[1] *T.M. Apostol* , Introduction to Analytic Number Theory. New York–Heidelberg–Berlin, Springer 1976.

[2] *J.L. Hafner* , On the average order of a class of arithmetic functions. J. Number Theory **15** (1982), 36—76.

[3] *A. Ivić* , The Riemann Zeta–Function. New York, J. Wiley & Sons, 1985.

[4] *E. Landau* , Elementary Number Theory, 2nd ed. New York, Chelsea Publ. Co., 1966.

[5] *N. Nielsen* , Die Gammafunktion. New York, Chelsea Publ. Co., 1965.

[6] *W.G. Nowak* , On a divisor problem in arithmetic progressions. To appear in J. Number Theory.

[7] *G. Szegö* and *A. Walfisz* , Über das Piltz'sche Teilerproblem in algebraischen Zahlkörpern (Erste Abhandlung). Math. Z. **26** (1927), 138—156.

[8] *G. Szegö* and *A. Walfisz* , Über das Piltz'sche Teilerproblem in algebraischen Zahlkörpern (Zweite Abhandlung). Math. Z. **26** (1927), 467—486.

[9] *E.C. Titchmarsh* , The Theory of the Riemann Zeta–Function. Oxford, Clarendon Press, 1951.

---

Institut für Mathematik, Universität für Bodenkultur, Gregor Mendel-Straße 33,
A—1180 Vienna, AUSTRIA.

# On the Pair Correlation of Zeros of Dirichlet $L$–Functions

*Ali E. Özlük*

In the course of investigating the vertical distribution of zeros of the Riemann Zeta Function $\zeta(s)$ on the critical line, Montgomery [5] introduced the form function

$$F(\alpha,\mathrm{T}) = \left(\frac{\mathrm{T}\log\mathrm{T}}{2\pi}\right)^{-1} \sum_{\substack{0<\gamma\le\mathrm{T}\\ 0<\gamma'\le\mathrm{T}}} \mathrm{T}^{i\alpha(\gamma-\gamma')} w(\gamma-\gamma'), \tag{1}$$

where $\alpha,\mathrm{T}\in\mathbb{R}$, $\mathrm{T}\ge 2$, $w(u) = \dfrac{4}{4+u^2}$ and $\zeta(\frac{1}{2}+i\gamma) = \zeta(\frac{1}{2}+i\gamma') = 0$.

Assuming the Riemann Hypothesis, his analysis has lead to the estimate (see [4]):

$$F(\alpha,\mathrm{T}) = \mathrm{T}^{-2|\alpha|}\log\mathrm{T} + |\alpha| + O\left(|\alpha|\mathrm{T}^{|\alpha|-1} + \mathrm{T}^{-\frac{3}{2}|\alpha|} + \frac{1}{\log\mathrm{T}}\right) \tag{2}$$

as $\mathrm{T}\to+\infty$, uniformly on each interval $|\alpha|\le 1-\varepsilon$. For the range $|\alpha|\ge 1$, Montgomery conjectured that

$$F(\alpha,\mathrm{T}) = 1 + o(1) \tag{3}$$

as $\mathrm{T}\to+\infty$ uniformly in $1\le a\le|\alpha|\le b<+\infty$, for any constants $a$, $b$. If (3) were true, then the following (essentially equivalent) statement would follow:

**Montgomery Pair Correlation Conjecture.** *For fixed* $0<\alpha<\beta<+\infty$,

$$\left(\frac{\mathrm{T}\log\mathrm{T}}{2\pi}\right)^{-1} \left|\left\{(\gamma,\gamma') : 0<\gamma,\gamma'\le\mathrm{T},\ \frac{2\pi\alpha}{\log\mathrm{T}}\le\gamma-\gamma'\le\frac{2\pi\beta}{\log\mathrm{T}}\right\}\right|$$

$$\sim\int_{\alpha}^{\beta}\left(1-\left(\frac{\sin\pi u}{\pi u}\right)^2\right)du \tag{4}$$

*as* $\mathrm{T}\to+\infty$.

(for recent work on the subject, see [2], [3], [4], [6] and the references therein.)

In this paper, we introduce the $q-$analogue of Montgomery's form function (1), and announce a result that supports this conjecture. In order to make the analogy more transparent we write $F(\alpha, T)$ as

$$F(\alpha, T) = \frac{4}{T \log T} \int_{-\infty}^{+\infty} \left| \sum_{0<\gamma\leq T} k(t, \gamma) T^{i\alpha\gamma} \right|^2 dt, \tag{5}$$

where

$$k(t, \gamma) = \frac{1}{1 + (t - \gamma)^2}. \tag{6}$$

Now let $K(s)$ be analytic in a strip $\mathcal{D}$ containing the critical strip, $K(\sigma + it)\log(|t| + 2) \in L^1(-\infty, +\infty)$ for all fixed $s \in \mathcal{D} \cap \mathbb{R}$ with further restrictions that the Mellin transform $a(u)$ of $K(s)$,

$$a(u) = \frac{1}{2\pi i} \int_{c-i\infty}^{c+i\infty} K(s) u^{-s} ds, \quad c \in \mathcal{D} \cap \mathbb{R}, \tag{7}$$

exists, has compact support $\operatorname{Supp} a(u) \subset [A, B] \subset \mathbb{R}^+$, and is of bounded total variation. The $q-$analogue of (1) we wish to consider, assuming the Generalized Riemann Hypothesis (GRH), is

$$F_K(\alpha, Q) = \frac{1}{N_K(Q)} \sum_{q \leq Q} \frac{1}{\phi(q)} \sum_{\chi (\mathrm{mod}\ q)} \left| \sum_{\gamma} K\left(\frac{1}{2} + i\gamma\right) Q^{i\alpha\gamma} \right|^2, \tag{8}$$

where the inner–most sum runs over the imaginary parts $\gamma$ of zeros of $L(s, \chi)$, and where

$$N_K(Q) = \frac{Q \log Q}{2\pi} \int_{-\infty}^{+\infty} \left| K\left(\frac{1}{2} + it\right) \right|^2 dt. \tag{9}$$

A special case of (8), for a particular choice of $K(s)$, had been introduced in [7]. We remark that $N_K(Q)$ is the asymptotic size of the diagonal terms in the triple sum in (8) and serves as a normalization factor. Stated more precisely,

$$\sum_{q\le Q}\frac{1}{\phi(q)}\sum_{\chi(\mathrm{mod}\ q)}\sum_{\gamma}\left|K\left(\tfrac{1}{2}+i\gamma\right)\right|^2 \sim N_K(Q) \text{ as } Q\to+\infty. \tag{10}$$

Our analysis of $F_K(\alpha, Q)$ depends on an explicit formula contained in the following:

**Lemma.** *Assume GRH. For $x \ge 1$, and any character $\chi$ (mod $q$), we have*

$$\sum_{\gamma} K\left(\tfrac{1}{2}+i\gamma\right)x^{i\gamma} = E(\chi)K(1)x^{\frac{1}{2}} - x^{-\frac{1}{2}}\sum_{n=1}^{\infty} a\left(\frac{n}{x}\right)\Lambda(n)\chi(n)$$
$$+\, x^{-\frac{1}{2}}a\left(\frac{1}{x}\right)\log\frac{q}{\pi} + O\left(\min\left(x^{\frac{1}{2}}, x^{-\frac{1}{2}}\log q\log x\right)\right). \tag{11}$$

*where $E(\chi) = 0$ or $1$ according as $\chi \ne \chi_0$ or $\chi = \chi_0$, and where $\gamma$ ranges over non-trivial zeros of $L(s,\chi)$. The implicit constant in (11) depends only on the kernel $K$, and should be interpreted as $O(1)$ when $x = 1$.*

**Proof.** We follow the proof of the classical explicit formula (as in [1]) connecting the zeta zeros to prime numbers by considering the evaluation of

$$I = \frac{1}{2\pi i}\int_{c-i\infty}^{c+i\infty} -\frac{L'}{L}(s,\chi)K(s)x^{s}\,ds, \tag{12}$$

where $c \in \mathcal{D}$, $c > 1$ and $\chi$ primitive. We move the integral to the contour $C$ consisting of

$C_1$: The line segment $[c - iT, d - iT]$

$C_2$: The line segment $[d - iT, d + iT]$

$C_3$: The line segment $[d + iT, c + iT]$,

where $d \in \mathcal{D}$, $d < 0$ and $T$ is chosen so that the horizontal sides of $C$ avoid zeros of $L(s,\chi)$.

We can assume that by varying $T$ by a bounded amount, we have $|\gamma - T| \gg (\log qT)^{-1}$ for all the zeros. The only poles of the integrand are at the non-trivial zeros $\rho$ of $L(s,\chi)$ and at $s = 1$ if $\chi = \chi_0$ with residues of $-K(\rho)X^{\rho}$ and $K(1)$

respectively. Integrals along $C_1$ and $C_3$ vanish as $T \to +\infty$ by virtue of the estimate $\frac{L'}{L}(s,\chi) \ll \log^2 qT$ and the conditions imposed on $K(s)$. On $C_2$, we have

$$\frac{L'}{L}(s,\chi) = -\log\frac{q}{\pi} + O(\log |t| + 2)$$

and

$$\frac{1}{2\pi i}\int_{d-iT}^{d+iT} (\log\frac{q}{\pi})K(s)x^s\, ds \to a(\frac{1}{x})\log\frac{q}{\pi} \tag{13}$$

as $T \to +\infty$. We remark that the possible zero of $L(s,\chi)$ at $s=0$ (when $\chi(-1)=1$) is absorbed into the error term in (11). Suppose now that $\chi$ is imprimitive $(\text{mod } q)$ and is induced by the primitive character $\chi_1 \pmod{q_1}$, $q = q_1 r$. The error thus caused in the second term on the right hand side of (11) is

$$\ll \sum_{(n,r)>1} a\left(\frac{n}{x}\right)\Lambda(n) \ll \sum_{n} a\left(\frac{n}{x}\right)\Lambda(n) \ll x.$$

On the other hand, it is also

$$\ll \sum_{\substack{p^k \le Bx \\ p \mid r}} \log p \ll (\log q)(\log x).$$

The assertion of the lemma follows from computing $I$ based on the observations above.

We now announce our main theorem on $F_K(\alpha, Q)$ and one of its corollaries, including a sketch of the proof of the former. The details well appear elsewhere.

**Theorem.** *If GRH holds, then*

$$F_K(\alpha, Q) = \begin{cases} \delta_Q(\alpha)\left(1+O\left(\frac{1}{\log Q}\right)\right)+\alpha+O\left(\frac{1}{\log Q}\right) & \text{if } |\alpha| \le 1 \\ 1 + O\left(\frac{1}{\log Q}\right) & \text{if } 1 \le |\alpha| \le 2 - \frac{14\log\log Q}{\log Q} \end{cases} \tag{14}$$

*uniformly as* $Q \to +\infty$, *where*

$$\delta_Q(\alpha) = \frac{Q^{1-\alpha} a^2(Q^{-\alpha})\log^2 Q}{N_K(Q)}. \tag{15}$$

**Corollary.** *Assuming the validity of GRH, the proportion of simple zeros of all Dirichlet L–functions is greater than or equal to $\frac{11}{12}$ in the sense of the inequality*

$$\frac{1}{N_K(Q)}\sum_{q\le Q}\frac{1}{\phi(q)}\sum_{c\,(\mathrm{mod}\ q)}\ \sum_{\substack{\gamma\\ \text{simple}}}\left|K\left(\tfrac{1}{2}+i\gamma\right)\right|^2\ge\frac{11}{12}+o(1) \tag{16}$$

*as* $Q\to+\infty$ (*See* (10).)

**Sketch of the proof of the Theorem.** We write (11) as $L(x,\chi)=R(x,\chi)$ and observe that $F_K(\alpha,Q)=\frac{1}{N_K(Q)}\sum_{q\le Q}\frac{1}{\phi(q)}\sum_{\chi(\mathrm{mod}\ q)}\left|L(Q^{\alpha},\chi)\right|^2$.. Hence, in order to estimate $F_K(\alpha,Q)$, it suffices to analyze $\sum_{q\le Q}\frac{1}{\phi(q)}\sum_{\chi(\mathrm{mod}\ q)}\left|R(Q^{\alpha},\chi)\right|^2$.

On expanding this particular expression using the orthogonality of characters, one finds that the crux of the problem lies in estimating

$$J(x,Q)=\sum_{Q<q\le x}\ \sum_{\substack{m<n\\ m\equiv n\,(\mathrm{mod}\ q)}}a\left(\frac{m}{x}\right)a\left(\frac{n}{x}\right)\Lambda(m)\Lambda(n) \tag{17}$$

We write this as

$$J(x,Q)=\int_0^1\left|S(x,\alpha)\right|^2W(x,\alpha)\,d\alpha, \tag{18}$$

with

$$S(x,\alpha)=\sum_m a\left(\frac{m}{x}\right)\Lambda(m)e(m\alpha) \tag{19}$$

and

$$W(x,\alpha)=\sum_{\substack{1\le k<Bx/q\\ Q<q\le x}}e(-qk\alpha)$$

We then evaluate the integral in (18) by the Hardy–Littlewood–Vinogradov circle method.

It should be remarked that $\delta_Q(\alpha)$ appearing in (14) behaves like a Dirac $\delta$–function. In fact, as $Q\to+\infty$,

$$\delta_Q(0) \approx \left(\frac{1}{2\pi}\int_{-\infty}^{+\infty} \left|K\left(\frac{1}{2} + it\right)\right|^2 dt\right)^{-1} a^2(1)\log Q,$$

$$\delta_Q(0) \sim \left(\frac{1}{2\pi}\int_{-\infty}^{+\infty} \left|K\left(\frac{1}{2} + it\right)\right|^2 dt\right)^{-1} a^2(1)\log Q, \tag{20}$$

and if $\alpha \neq 0$, $\delta_Q(\alpha) \to 0$. Furthermore

$$\int_{-\infty}^{+\infty} \delta_Q(\alpha)\, d\alpha = \frac{Q\log Q}{N_K(Q)}\int_0^{+\infty} a^2(u)du = 1 \quad \text{for all } Q, \tag{21}$$

the last step being achieved through the use of Plancherel's Theorem for the Mellin transform in the form

$$\frac{1}{2\pi}\int_{-\infty}^{+\infty} \left|K\left(\frac{1}{2} + it\right)\right|^2 dt = \int_0^{+\infty} a^2(u)du. \tag{22}$$

## References

[1] *H. Davenport,* Multiplicative Number Theory, Second edition, Berlin—Heidelberg—New York 1980.

[2] *P.X. Gallagher,* Pair correlation of zeros of the zeta function. J. reine angew. Math. **362** (1985), 72—86.

[3] *D.A. Goldston,* On the pair correlation conjecture for zeros of the Riemann zeta-function. J. reine angew. Math. **385** (1988), 24—40.

[4] *D.R. Heath-Brown,* Gaps between primes, and the pair correlation of zeros of the zeta-function. Acta Arithmetica **41** (1982), 85—99.

[5] *H.L. Montgomery,* The pair correlation of zeros of the zeta function. Proc. Sympos. Pure. Math. Vol. 24, Amer. Math. Soc., Providence, RI, 1973, 181—193.

[6] *A.M. Odlyzko,* On the distribution of spacings between zeros of the zeta function. Mathematics of Computation, Vol. 48, **117** (1987), 273—308.

[7] *A.E. Özlük,* Pair correlation of zeros of Dirichlet $L$-functions. Ph.D. Thesis, University of Michigan, 1982.

---

Department of Mathematics, University of Maine, 418 Neville Hall, Orono, ME 04469

# Computational Methods For the Resolution of Diophantine Equations

*Attila Pethö* [1]

## 1. Introduction

Recently, considerable progress was made in the practical resolution of large classes of diophantine equations. Several authors worked out methods based on combinations of results in the following fields.

1. Applications of lower bounds for linear forms in the logarithms of algebraic numbers to establish effective upper bounds for the solutions of large classes of diophantine equations.
2. New algorithms for the solution of diophantine approximation problems.
3. New algorithms in algebraic number theory.

The most important tool of the methods are the lower bounds for linear forms in the logarithms as well as $p-$adic logarithms of algebraic numbers. Gelfond [21] proved a bound in the complex case for two algebraic numbers. This was completely generalized by Baker [1]. He himself [3] and [4], Waldschmidt [41] and recently Blass, Glass, Manski, Meronk and Steiner [8], [9] gave improvements and refinements. From a computational point of view, the last three papers are the most important because the occurring absolute constants are not too large.

Similarly to the complex case lower bounds for linear forms of $p-$adic logarithms of algebraic numbers were found. Schinzel [35] proved such a result for two numbers,

[1]Research supported by Hungarian National Foundation for Scientific Research grant no. 273/86.

Kaufman [24] obtained the general case. Van der Poorten [34] and Yu [40] gave the best lower bounds, so far.

Using these results several authors found effective upper bounds for solutions of large classes of diophantine problems. For references we refer to the books of Baker [5], Győry [22] and Shorey and Tijdeman [36]. These types of results make it theoretically possible to find all solutions because one has to check only finitely many possibilities. But finitely many may be so many that a direct search is hopeless. As we shall see later a typical upper bound is $10^{30}$ even in the most modest cases.

Baker and Davenport [6], Ellison [14] and Ellison *et al.* [15] used continued fraction expansion of suitable real numbers to reduce Baker's upper bound to a much smaller one, and finally to solve some diophantine problems. Although Ellison [14] pointed out that his method is applicable in higher dimensions too, and a lot of interesting applications of Baker's method were found, only a little progress was made in the numerical resolution of diophantine problems.

The lattice basis reduction algorithm of Lenstra, Lenstra Jr. and Lovász [25] solves multidimensional diophantine approximation problems. Several mathematicians realized independently that this algorithm is applicable combined with Baker–type upper bounds for the complete resolution of diophantine equations.

Baker and Davenport [6] determined all common terms in two second order linear recurrence sequences and so solved a system of Pell's equations. Pethö [27], [28] computed all third and fifth powers in the Fibonacci sequence. These are the only cases when forward searches were used to exclude large solutions. Ellison [14], Ellison *et al* [15] and Steiner [37] solved third degree; Blass *et al* [7], Pethö and Schulenberg [32], Tzanakis and de Weger [39] and Zagier [46] solved fourth degree Thue equations. In many of the above papers the results were used to find all integer points on elliptic curves. Gaál [18] described a method to solve third degree inhomogeneous Thue equations. Gaál and Schulte [19] and Gaál *et al* [20] computed all power bases in several third and fourth degree number fields by solving completely the corresponding index form equations. Cherubini and Walliser [11] used the reduction method to find all imaginary quadratic fields with class number one.

$P$-adic linear form estimates and computer search were applied by Pethö [29], and Pethö and de Weger [33] and by de Weger [42] to find prime powers as well as products of prime powers in second order linear recurrences. De Weger [43] solved $S$-unit equations over $\mathbb{Z}$ and tested numerically the Oesterlé–Masser conjecture. Finally de Weger [45] used the combination of complex and $p$-adic arguments to solve third degree Thue–Mahler equations.

In this paper we describe the common ideas in the methods of the above papers. In section 2 we give the outline of the method. Sections 3 and 4 deal with the general components of the method; we cite the best known lower bound for linear forms in the logarithms of algebraic numbers, as well as the application of the lattice basis reduction algorithm of Lenstra, Lenstra Jr. and Lovász [25] to the reduction of a large upper bound for the solution of diophantine inequalities. The method of section 2 has two problem–specific components, we illustrate them in section 5 on Thue equations. Finally, in section 6, we report on a conjecture on the representation of one by cubic forms.

## 2. General Description of the Method

In the sequel we shall deal only with the classical complex case, for $p$-adic variants we refer to the thesis of de Weger [44].

Let $K$ be an algebraic number field of degree $k$ over $\mathbb{Q}$—the field of rational numbers—and let $G$ be the normal closure of $K$. Let $\mathbb{Z}_K$ denote the ring of integers of $K$ and $\alpha^{(1)},\dots,\alpha^{(k)}$ denote the conjugates of $\alpha \in K$. Finally, let $\varepsilon_1,\dots,\varepsilon_r$ be a system of independent units of $\mathbb{Z}_K$. With this notation the methods used in the papers mentioned above can be divided into four steps.

1. Transformation of the original problem to finitely many unit equations of type

$$\alpha_1\left(\frac{\varepsilon_1^{(i)}}{\varepsilon_1^{(q)}}\right)^{n_1}\cdots\left(\frac{\varepsilon_r^{(i)}}{\varepsilon_r^{(q)}}\right)^{n_r}+\alpha_2\left(\frac{\varepsilon_1^{(j)}}{\varepsilon_1^{(q)}}\right)^{m_1}\cdots\left(\frac{\varepsilon_r^{(j)}}{\varepsilon_r^{(q)}}\right)^{m_r}=1, \qquad (1)$$

where $1\le i, j, q\le k$, $n_h, m_h \in \mathbb{Z}$ $(h=1,\dots,r)$ and $\alpha_1,\alpha_2$ are fixed elements from $G$.

2. If $N_0 \leq N = \max\{|n_1|,\dots,|n_r|\} \leq M = \max\{|m_1|,\dots,|m_r|\}$, then taking the logarithm we get finitely many inequalities

$$\left| n_1 \log\left(\frac{\varepsilon_1^{(i)}}{\varepsilon_1^{(q)}}\right) + \dots + n_r \log\left(\frac{\varepsilon_r^{(i)}}{\varepsilon_r^{(q)}}\right) + \log \alpha_1 \right| < c_1 \exp(-c_2 N), \tag{2}$$

where $c_1, c_2$ are constants. If $N \geq M$, then we have to exchange the role of $N$ and $M$.

Using a suitable effective lower bound for linear forms in the logarithms of algebraic numbers compute an upper bound $N_1$ for $N$ from (2).

3. Reduce $N_1$ iteratively until either the new bound will be smaller than $N_0$ or the iteration does not give a better bound. For the reduction, one can use numerical diophantine approximation techniques.

4. Search for the actual solutions either solving (2) in the remaining range or using specific properties of the original problem.

We remark that steps 1 and 4 depend strongly on the original problem, while the other two steps can be done automatically.

## 3. A Lower Bound For Linear Forms in the Logarithms of Algebraic Numbers

The first general, non–trivial component in the method is the lower bound for linear forms in the logarithms of algebraic numbers. In this section we cite the best known general bound due to Blass *et al* [8]. We shall mention that until today most of the applications used a weaker theorem of Waldschmidt [41].

Let $\alpha_1,\dots,\alpha_n$, $\beta_0, \dots, \beta_n$ be algebraic numbers with $\alpha_1,\dots,\alpha_n$ non–zero. Let

$$\Lambda = \beta_0 + \beta_1 \log \alpha_1 + \dots + \beta_n \log \alpha_n$$

and $D = [K : \mathbb{Q}]$, where $K = \mathbb{Q}(\alpha_1,\dots,\alpha_n, \beta_0,\dots,\beta_n)$. Let $h(\alpha)$ be the absolute logarithmic height of the algebraic number $\alpha$. Define

$$V_1 = \max\left\{ h(\alpha_1), \frac{1}{D}, \frac{|\log \alpha_1|}{D} \right\}$$

and

$$V_{j+1} = \max\left\{ h(\alpha_{j+1}), V_j, \frac{|\log \alpha_{j+1}|}{D} \right\} \quad (1 \le j \le n-1).$$

Let $a_j = \dfrac{DV_j}{|\log \alpha_j|}$ $(1 \le j \le n)$ and $\dfrac{1}{a} = \dfrac{1}{n}\sum_{j=1}^{n} \dfrac{1}{a_j}$. Let $V_j' = jV_j$, $V_0^+ = \overline{V}_0 = 1$, $V_j^+ = \max\{V_j, 1\}$ and $\overline{V}_j = \max\{V_j', 1\}$. Let $W = h(\beta_j)$ $(0 \le j \le n)$ and $q$ be a prime number. Let $\overline{E}_2 = \min\{e^{qDV_1}, 2qa\}$ and $\overline{M} = 2(2^6 q^2 nD\overline{V}_{n-1}\overline{E}_2)^n$.

Finally, let

$$x_n^* = \begin{cases} \log(2^{13}\overline{V}_1), & \text{if } n = D = 1 \\ n^2(n+1)\log\left(\dfrac{6n}{\log D}\right) + n(n+1)\log(n!) + \log n, & \text{if } D \ge 2 \\ n^2(n+1)\log(9n) + n(n+1)\log(n!) + \log n, & \text{if } D = 1 < n. \end{cases}$$

**Theorem 1.** *(Blass et al* [8]) *If* $\Lambda \ne 0$, *then*

$$|\Lambda| > \exp\left\{ -C_1(n) D^{n+2} \frac{V_1 \cdots V_n}{(\log \overline{E}_2)^{n+1}} (\log \overline{M})(W + C_2(n)) \right\}$$

*where*

$$C_1(n) = \begin{cases} n^{2n+1}(24e^2)^n 2^{20}, & \text{if } n \ge 3 \\ n^{2n+1}(24e^2)^n 2^{21}, & \text{if } n < 3 \end{cases}$$

*and*

$$C_2(n) = n(n+1)\log(D^3 \overline{V}_n) + \frac{x_n^*}{n}.$$

## 4. Reduction of the Large Upper Bound

Now let us turn our attention to step 3 of the algorithm of section 2. In the sequel we assume that the left hand side of (2) does not vanish, $\alpha_1 \ne 1$ and $r \ge 2$. Dividing (2) by $\log(\varepsilon_r^{(i)} / \varepsilon_r^{(q)})$ we get

$$0 \neq \left| n_1\delta_1 + \ldots + n_{r-1}\delta_{r-1} + n_r + \delta_{r+1} \right| < c_3 \exp(-c_2 N), \tag{3}$$

where $\delta_h = \dfrac{\log\left(\dfrac{\varepsilon_h^{(i)}}{\varepsilon_h^{(q)}}\right)}{\log\left(\dfrac{\varepsilon_r^{(i)}}{\varepsilon_r^{(q)}}\right)}$, $h = 1,\ldots,r-1$ and $\delta_{r+1} = \dfrac{\log \alpha_1}{\log\left(\dfrac{\varepsilon_r^{(i)}}{\varepsilon_r^{(q)}}\right)}$.

The following lemma is the generalization of a lemma of Baker and Davenport [6]. For the proof see Pethö and Schulenberg [32].

**Lemma 1.** *Let $Q_1, Q_2$ and $Q_3$ be real numbers such that $Q_2 \geq 1$, $Q_1 > 2^{r-1}((r-1)Q_2 + 1)$. If there exists an integer $q$ with*

$$1 \leq q \leq Q_1 Q_3 \tag{4}$$

$$\left\| q\delta_i \right\| \leq Q_2 (Q_1 Q_3)^{-1/(r-1)}, \quad i = 1,\ldots,r-1 \tag{5}$$

*and*

$$\left\| q\delta_{r+1} \right\| \geq ((r-1)Q_2 + 1) Q_1^{-1/(r-1)} \tag{6}$$

*then* (3) *has no solutions $n_1,\ldots,n_r \in \mathbb{Z}$ with*

$$\frac{\log\left(Q_1^{r/(r-1)} Q_3 c_3\right)}{\log c_2} < N \leq Q_3^{\frac{1}{r-1}}, \tag{7}$$

*where $N = \max\left\{|n_1|,\ldots,|n_r|\right\}$ and $\|x\|$ denotes the distance of the real number $x$ to the nearest integer.*

If $r = 2$ then the $t$-th denominator $q_t$ with $q_t \leq Q_1 Q_3 < q_{t+1}$ of the continued fraction expansion of $\delta_1$ solves (4) and (5) with $Q_2 = 1$. In the general case one can use the LLL lattice basis reduction algorithm of Lenstra, Lenstra Jr. and Lovász [25] or its modified version by de Weger [43]. The following theorem is a reformulation of Proposition (1.39) of Lenstra, Lenstra Jr. and Lovász [25].

**Theorem 2.** *Let* $b_1 = p_1 e_1 + \ldots + p_{r-1} e_{r-1} + q d$, $b_2, \ldots, b_r$ *be an LLL–reduced basis of the lattice spanned by the column vectors* $e_i$, $i = 1, \ldots, r-1$ *whose i–th coordinate is* 1 *all others* 0 *and by*

$$d = (-\delta_1, \ldots, -\delta_{r-1}, 2^{r/4} (Q_1 Q_3)^{-r/(r-1)})^T .$$

*Then* $q$ *solves* (4) *and* (5) *with* $Q_2 = 2^{r/4}$.

The reduction procedure works in practice in the following way. Assume that we want to solve (1) with $N \le M$ and $N_0 \le N \le N_1$, where $N_1$ is much larger than $N_0$.

(i) Compute $\delta_1, \ldots, \delta_{r-1}, \delta_{r+1}$ with the required (high) accuracy, see Pethö and Schulenberg [32] Lemma 3. Put $Q_2 = 2^{r/4}$ and $Q_1 = (10r\, Q_2^r)^{r-1}$.

(ii) Put $Q_3 = N_1^{r-1}$ and solve the diophantine approximation problem (4), (5) using the LLL–reduction.

(iii) If (6) holds, then let $S$ be the smallest value of $Q_1$ with (4) and (5) and put $N_2 = \dfrac{\log (S^{r/(r-1)} Q_3 c_3)}{\log c_2}$, otherwise let $N_0 = N_1$ and terminate.

If the algorithm terminates at (iii) then, as Baker and Davenport [6] pointed out, the solutions of (3) can be found by solving a linear diophantine equation. The occurrence of this case was never reported in the literature.

De Weger [43] discussed the cases, when $r = 1$ or $\alpha_1 = 1$ or when $\delta_1, \ldots, \delta_{r-1}, \delta_{r+1}$ are linearly dependent over $\mathbb{Q}$. Tzanakis and de Weger [39] used another reduction technique, which was also based on the LLL basis reduction algorithm. The key idea of their reduction technique is, that small values of the linear form

$$n_1 \delta_1 + \ldots + n_r \delta_r + \delta_{r+1}$$

correspond to short vectors of an appropriately defined lattice.

## 5. Thue Equations

So far we focussed our attention on the general elements of the method from section 2. In this section we shall describe, using the example of Thue equations, how to transform them into finitely many unit equations, and how to find their "small" solutions. Here, small means the magnitude of $10^{100}$, because the reduction procedure of section 4 cannot give a better upper bound in this case.

Let $F(x,y) = a_0x^k + a_1x^{k-1}y + \ldots + a_ky^k \in \mathbb{Z}[x,y]$ be irreducible over $\mathbb{Q}[x,y]$, $k \geq 3$ and $0 \neq m \in \mathbb{Z}$. The diophantine equation

$$F(x,y) = m \tag{8}$$

is called a Thue equation. Thue [38] proved that (8) has finitely many solutions $x, y \in \mathbb{Z}$. Baker [2] has given an effectively computable upper bound for $\max\{|x|,|y|\}$. In the transformation of (8) into finitely many unit equations we use Baker's method, which was refined by Győry and Papp [23].

### 5.1 Transformation of (8) to finitely many unit equations

Let $\beta$ be a root of $F(x,1)$ and $K = \mathbb{Q}(\beta)$, then $[K:\mathbb{Q}] = k$. To avoid technical difficulties, we assume in the sequel that $K$ is totally real. Let $r = k-1$, and $\xi$ be the group generated by the multiplicatively independent units $\varepsilon_1, \ldots, \varepsilon_r$ of norm 1 of $\mathbb{Z}_K$. Let $\left|\overline{\tau}\right| = \max\left\{\left|\tau^{(i)}\right|,\ 1 \leq i \leq k\right\}$. Take

$$c_4 = \max\left\{\log\left|\overline{\varepsilon_i}\right|,\ 1 \leq i \leq r\right\},$$

$$c_5 = \prod_{j=1}^{r} \max\left\{\log\left|\overline{\varepsilon_j}\right|,\ 1\right\},$$

$$M = \frac{m}{a_0}.$$

The following lemma is easy to prove using the geometrical representation of $K$ (see Győry and Papp [23]).

**Lemma 2.** *There exists a finite set* $\mathcal{A} \subset K$ *with the following properties*

(i) *If* $x, y \in \mathbb{Z}$ *is a solution of* (8) *then there exist* $\gamma \in \mathcal{A}$ *and* $b_1, \ldots, b_r \in \mathbb{Z}$ *with*

$$x - By = \gamma \varepsilon_1^{b_1} \ldots \varepsilon_r^{b_r}, \tag{9}$$

(ii) *For all* $\gamma \in \mathcal{A}$ $\mathrm{Norm}_{K/\mathbb{Q}}(\gamma) = M$ *and* $O \log \left| \overline{M^{-1/k_\gamma}} \right| \leq \frac{r c_4}{2}$.

Let $I = \{1, \ldots, k\}$ and fix a $u \in I$. Let $T_{u,j} = \left| \beta^{(u)} - \beta^{(j)} \right|$ for all $j \in I$, $0 < t_u < \min\{T_{u,j}, j \in I, j \neq u\}$ and $T_u = T_{u,q} = \max\{T_{u,j}, j \in I\}$. With this notation it is easy to prove the following lemma.

**Lemma 3.** *Let* $x, y \in \mathbb{Z}$, $y \neq 0$ *be a solution of* (8) *with* $\left| x - \beta^{(u)} y \right| < \left| x - \beta^{(j)} y \right|$ *for all* $j \in I \setminus \{u\}$ *and with* $\left| x - \beta^{(u)} y \right| \leq T_u |y|$. *Then*

$$\left| x - \beta^{(u)} y \right| \leq \left( |M| \prod_{\substack{j=1 \\ j \neq u}}^{k} \frac{1}{T_{u,j} - t_{u,j}} \right) \frac{1}{|y|^{k-1}} = c_6 |y|^{-k+1}. \tag{10}$$

*Further, if* $|y| > Y_0 \geq \max\left\{ \left( \frac{2c_6}{T_u} \right)^{1/(k-1)}, \left( \frac{8c_6}{t_u} \right)^{1/k}, T_u \right\}$ *also holds, then*

$$\left| x - \beta^{(u)} y \right| \leq (1 + T_{u,j}) |y|, \text{ for all } j \in I \tag{11}$$

*and*

$$\left| x - \beta^{(q)} y \right| > \frac{T_u |y|}{2}. \tag{12}$$

In the sequel denote by $R$ the regulator of $\xi$, (9) and (11) imply

$$\left| \gamma^{(j)} \varepsilon_1^{(j)^{b_1}} \ldots \varepsilon_r^{(j)^{b_r}} y \right| \leq (1 + T_u) |y|$$

for all $j \in I$. Taking the logarithm we get

$$\log |y| \geq \frac{R}{r! c_5} B - \log\left( \left| \gamma^{-1} \right| (1 + T_u) \right), \tag{13}$$

where $B = \max\left\{ |b_1|, \ldots, |b_r| \right\}$. Now fix a $j \in I \setminus \{u, q\}$, then

$$(\beta^{(q)} - \beta^{(j)})(x - \beta^{(u)} y) + (\beta^{(u)} - \beta^{(q)})(x - \beta^{(j)} y) = (\beta^{(u)} - \beta^{(j)})(x - \beta^{(q)} y)$$

holds. Using the conjugates of (9) and dividing the last equation by $(\beta^{(u)} - \beta^{(j)}) \times (x - \beta^{(q)})$ we get the required unit equation. This implies, from (10) and (12), that

$$\left| \frac{\beta^{(u)} - \beta^{(q)}}{\beta^{(u)} - \beta^{(j)}} \frac{\gamma^{(j)}}{\gamma^{(q)}} \left( \frac{\varepsilon_1^{(j)}}{\varepsilon_1^{(q)}} \right)^{b_1} \cdots \left( \frac{\varepsilon_r^{(j)}}{\varepsilon_r^{(q)}} \right)^{b_r} - 1 \right| \leq \frac{4c_6}{t_u} |y|^{-k}.$$

Now let

$$\xi_\nu (u, q, j, \gamma) = \xi_\nu = \begin{cases} \dfrac{\varepsilon_\nu^{(j)}}{\varepsilon_\nu^{(q)}}, & \text{if } 1 \leq \nu \leq r \\ \dfrac{\beta^{(u)} - \beta^{(q)}}{\beta^{(u)} - \beta^{(j)}} \dfrac{\gamma^{(j)}}{\gamma^{(q)}}, & \text{if } \nu = r + 1 \end{cases}$$

$$c_7 = \log\left( \frac{8c_6}{t_u} \right) + k \log(|\overline{\gamma^{-1}}|(1 + T_u))$$

$$c_8 = \frac{kR}{r!\, c_5}$$

$$B_0 = \frac{r!\, c_5}{R} (\log Y_0 + \log(|\overline{\gamma^{-1}}|(1 + T_u))),$$

Then we have the following:

**Theorem 2.** *Let* $x, y \in \mathbb{Z}$ *be a solution of* (8) *and let* $\gamma \in \mathcal{A}$, $b_1, \ldots, b_r \in \mathbb{Z}$ *defined by* (9). *If* $B = \max\{|b_1|, \ldots, |b_r|\} > B_0$, *then there exist pairwise distinct indices* $1 \leq u,\ q,\ j \leq k$ *such that*

$$0 < \left| b_1 \log|\delta_1| + \ldots + b_r \log|\delta_r| + \log|\delta_{r+1}| \right| < \exp(c_8 - c_9 B). \tag{14}$$

The example of Thue equations shows that the method of section 2 is applicable for polynomial diophantine equations, too. Furthermore, this is the only known general method for the complete resolution of Thue equations. Although we must remark that the transformation is very redundant. Namely, Bombieri and Schmidt [10] proved that the number of solutions of (8) is $O(k)$. Further, Everste and Győry [17] proved that if $K$ and $m$ are fixed then there exist only finitely many inequivalent forms $F(x,y) \in \mathbb{Z}$ with splitting field $K$ such that the number of solutions of (9) is larger than 2.

In spite of this, the above described transformation yields $O(k^2)$ essentially different unit equations because we have to choose two parameters $u$ and $q$ independently. Hence most of the solutions of the linear form inequalities (14) correspond, by (9), either to the same solutions or do not give solutions of (8).

To find the solutions of (8) with some hundred or thousand decimal digits there is a much more economical method. Using it we have to solve only $O(k)$ diophantine approximation problems. We shall describe it in the next section.

## 5.2 Continued fraction method for the computation of small solutions of (8)

It is clear form the preceding section that (8) implies (14) only if $\max\{|x|,|y|\}$, and consequently $B$, is large enough. Furthermore, the reduced upper bound for $B$ implies by (9) an upper bound for $\max\{|x|,|y|\}$ which is of magnitude from $10^{100}$ to $10^{1000}$. To find the solutions of (8) up to such an upper bound one can use another reduction technique based on the continued fraction expansion of the real roots of $F(x,1)$.

Let $u \in I$ be fixed such that $\beta^{(u)}$ is real and for $h > 0$ define the polynomial

$$H_h(t) = \prod_{\substack{j=1 \\ j \neq u}}^{k} (T_{u,j} - t) - t^{k-2} \left| \frac{m}{a_0} \right|^{2/k} \frac{1}{h}.$$

$\beta^{(u)} = [b_0; b_1, \ldots]$ will denote the simple continued fraction expansion of the irrational number $\beta$, while $\frac{p_n}{q_n}$ the $n$-th convergent to $\beta$. With this notation we have:

**Theorem 3** [30]. *Let $y_0$ be a given real number, and $(x,y) \in \mathbb{Z}^2$ a solution of the inequality*

$$|F(x,y)| \leq m$$

*with* $\left|x - \beta^{(u)} y\right| \le \left|x - \beta^{(i)} y\right|$ $(i = 1,\ldots,k)$, *such that* $y \neq 0$, $(x, y) = 1$ *and* $|y| \le y_0$. *Let* $\beta^{(u)} = [b_0; b_1, \ldots, b_\nu, \ldots]$, *where* $\nu$ *is chosen so that* $q_{\nu-1} > y_0$. *Let* $w \ge 1$, $B = \max_{w \le j \le \nu} b_j$, *and let* $T = T_{\alpha^{(u)}}$ *be the smallest positive root of* $H_{1/2}(t)$. *Then either*

$$|y| \le \min\left\{ y_0, \frac{1}{T}\left|\frac{m}{a_0}\right|^{1/k} \right\}$$

*or* $\frac{x}{y}$ *is a convergent to* $\alpha$ *with*

$$|y| \le \max\left\{ q_{w-1}, \frac{1}{T}\left|\frac{m}{a_0}\right|^{1/k} \left(\frac{B+2}{2}\right)^{1/(k-2)} \right\}.$$

We remark that in the reduction based on Theorem 3 we do not compute the exact value of the denominators of the convergents, only the partial quotients. This observation speeds up the method essentially.

### 5.3 Results

Using variants of the method described in the preceding sections Ellison *et al* [15], Steiner [37] and Pethö and Schulenberg [32] solved several third degree Thue equations. Gaál and Schulte [19] computed all power bases in totally real cubic fields with discriminant at most 3137 solving also third degree Thue equations.

Pethö and Schulenberg [32] and Tzanakis and de Weger [39] solved the following fourth degree Thue equations:

| $F(x, y)$ | m | solutions $(x, y)$ |
|---|---|---|
| $x^4 + 5x^3y + 4x^2y^2 - 5xy^3 - y^4$ | 1 | (±1,0); (±2,+1) |
| | -1 | (0,+1) |
| $x^4 - 4x^3y + 8xy^3 - y^4$ | 1 | (±1,0) |
| | -1 | (±2,±1); (0,+1) |
| $x^4 + x^3y - 3x^2y^2 - xy^3 + y^4$ | 1 | (±1,0); (0,±1) |
| | -1 | (±2,+1); (±1,±2); (±1,±1); (±1,+1) |
| $x^4 - 4x^3y - 12x^2y^2 + 4y^4$ | 1 | (±1,0) |
| $x^4 - 12x^3y - 8xy^3 + 4y^4$ | 1 | (±1,0); (±1,+1); (±1,±3); (±3,+1) |

## 6. Representation of One by Cubic Forms

Numerical methods are useful not only to solve completely diophantine equations but also for finding solutions up to a prescribed large upper bound. Using the continued fraction reduction of section 5.2 we computed (*cf.* Pethő [31]) the solutions with $|y| < 10^{41}$ of approximately 3000 equations of type

$$f(x,y) = 1,$$

where the discriminant $D_f$ of $f(x,y) = ax^3 + bx^2y + cxy^2 + dy^3 \in \mathbb{Z}[x,y]$ is positive. A form with $a = d = 1$ will be called reversible.

Two cubic forms $f_1(x,y)$, $f_2(x,y) \in \mathbb{Z}[x,y]$ are called equivalent if there exist integers $a_1, a_2, a_3, a_4$ with $|a_1a_4 - a_2a_3| = 1$ such that

$$f_2(x,y) = f_1(a_1x + a_2y,\ a_3x + a_4y).$$

Summarizing the observations we conjecture the following connection between cubic forms $f(x,y)$ with $D_f > 0$ and the number of solutions $N_f$ of (1)

$$N_f = \begin{cases} 0,1,2 \text{ or } 3, & \text{if } f \text{ is not equivalent to a reversible form} \\ 2,3,4 \text{ or } 5, & \text{if } f \text{ is equivalent to a reversible form} \\ 6, & \text{if } D_f = 81, 229, 257, 361,\ ? \\ 7, & \text{none} \\ 8, & \text{none} \\ 9, & \text{if } D_f = 49. \end{cases}$$

Analogous results for cubic forms with negative discriminant were proved by Delone [12] and Nagell [26], see also Delone and Faddeev [13].

## References


[1] *A. Baker,* Linear forms in the logarithms of algebraic numbers. Mathematica **13** (1966), 204—216.

[2] *A. Baker,* Contribution to the theory of Diophantine equations I. On the representation of integers by binary forms. Philos. Trans. Roy. Soc. London Ser. A **263** (1968), 173—191.

[3] *A. Baker,* A sharpening of the bounds for linear forms in logarithms I. Acta Arith. **21** (1972), 117—129.

[4] *A. Baker*, The theory of linear forms in logarithms. In: Transcendence Theory: Advances and Applications, Academic Press, London, 1977, 1—27.

[5] *A. Baker*, Transcendental Number Theory. Cambridge Univ. Press, Cambridge, 1975.

[6] *A. Baker* and *H. Davenport*, The equations $3x^2 - 2 = y^2$ and $8x^2 - 7 = z^2$. Quart. J Math. Oxford **20** (1969), 129—137.

[7] *J. Blass, A.M.W. Glass, D.B. Meronk and R.P. Steiner*, Practical solutions to Thue equations over the rational integers. To appear.

[8] *J. Blass, A.M.W. Glass, D. Manski, D.B. Meronk* and *R.P. Steiner*, Constants for lower bounds for linear forms in the logarithms of algebraic numbers I: The general case. Acta Arith., to appear.

[9] *J. Blass, A.M.W. Glass, D. Manski, D.B. Meronk and R.P. Steiner*, Constants for lower bounds for linear forms in the logarithms of algebraic numbers II: The rational case. Acta Arith., to appear.

[10] *E. Bombieri* and *W. M. Schmidt*, On Thue's equation. Invent. Math. **88** (1987), 69—82.

[11] *J.M. Cherubini* and *R.V. Wallisser*, On the computation of all imaginary quadratic fields of class number one. Math. Comp. **49** (1987), 295—299.

[12] *B.N. Delone (Delaunay)*, Über die Darstellung der Zahlen durch die binäre kubischen Formen von negativer Diskriminante. Math. Zeitschr. **31** (1930), 1—26.

[13] *B.N. Delone* and *D.K. Faddeev*, The Theory of Irrationalities of the Third Degree. Amer. Math. Soc. Transl. of Math. Monographs 10. Providence, 1964.

[14] *W.J. Ellison*, Recipes for solving diophantine problems by Baker's method. Sem. Th. Nombr., 1970–1971. Exp. No. 11. Talence: Lab. Theorie Nombres, C.N.R.S.

[15] *W.J. Ellison, F. Ellison, J. Pesek, C.E. Stahl* and *D.S. Stall*, The diophantine equation $y^2 + k = x^3$. J. Number Theory **4** (1972), 107—117.

[16] *J.H. Evertse*, On the representation of integers by binary cubic forms of positive discriminant. Invent Math. **73** (1983), 117—138.

[17] *J.H. Evertse* and *K. Györy*, Thue–Mahler equations with a small number of solutions. To appear.

[18] *I. Gaál*, On the resolution of inhomogeneous norm form equations in two dominating variables. Math. Comp., **51** (1988), 359—373.

[19] *I. Gaál* and *N. Schulte*, Computing all power integral bases to cubic fields. Math. Comp., to appear.

[20] *I. Gaál, A. Pethö* and *M. Pohst,* On the resolution of index form equations corresponding to biquadratic number fields I, and II. To appear.

[21] *A.O. Gelfond,* On the approximation of transcendental numbers by algebraic numbers. Dokl. Akad. Nauk. SSSR. **2** (1935), 177—182. (Russian)

[22] *K. Györy,* Resultats Effectives Sur la Representation des Entiers Par des Formes Decomposables. Queen's Papers in Pure and Applied Math., No. 56, Kingston, Canada, 1980.

[23] *K. Györy* and *Z.Z. Papp,* Norm form equations and explicit lower bounds for linear forms with algebraic coefficients. In: Studies in Pure Mathematics (To the Memory of Paul Turán), Akadémiai Kiadó, Budapest (1983), 245—267.

[24] *R.M. Kaufman,* A bound for linear forms in logarithms of algebraic numbers in $P$-adic metric. Vest. Mosk. Univ. Ser. Mat. Meh. **2** (1971), 3—10. (Russian)

[25] *A.K. Lenstra, H.W. Lenstra Jr.* and *L. Lovász,* Factoring polynomials with rational coefficients. Math. Ann., **261** (1982), 515—534.

[26] *T. Nagell,* Darstellung ganzer Zahlen durch binäre kubische Formen mit negativer Diskriminante. Math. Zeitschr. **28** (1928), 10—29.

[27] *A. Pethö,* Perfect powers in second order recurrences. In: Topics in Classical Number Theory, Colloq. Math. Soc. János Bolyai, Vol. 34, Budapest, 1981, 1217—1227.

[28] *A. Pethö,* Full cubes in the Fibonacci sequences. Publ. Math. Debrecen, **30** (1983), 117—127.

[29] *A. Pethö,* On the solution of the diophantine equation $G_n = p^z$. In: Proceedings EUROCAL '85, Vol. 2, Lecture Notes in Comput. Sci. Vol. 204, Springer–Verlag, 1985, 503—512.

[30] *A. Pethö,* On the resolution of Thue inequalities. J. Symbolic Computation **4** (1987), 103—109.

[31] *A. Pethö,* On the representation of 1 by binary cubic forms with positive discriminant. In: Proceedings Journees Arithmetiques, Ulm, to appear.

[32] *A. Pethö* and *R. Schulenberg,* Effektives Lösen von Thue Gleichungen. Publ. Math. Debrecen **34** (1987), 189—196.

[33] *A. Pethö* and *B.M.M. de Weger,* Products of prime powers in binary recurrence sequences I: The hyperbolic case, with an application to the generalized Ramanujan–Nagell equation. Math. Comp. **47** (1986), 713—727.

[34] *A.J. van der Poorten,* Linear forms in logarithms in the $p$-adic case. In: Transcendence Theory: Advances and Applications, Academic Press, London, 1977, 29—57.

[35] *A. Schinzel,* On two theorems of Gelfond and some of their applications. Acta Arith. **13** (1967), 177—236.

[36] *T.N. Shorey* and *R. Tijdeman,* Exponential Diophantine Equations. Cambridge Univ. Press, Cambridge, 1986.

[37] *R.P. Steiner,* On Mordell's equation $y^2 - k = x^3$: A problem of Stolarsky. Math. Comp. **46** (1986), 703—714.

[38] *A. Thue,* Annäherungswarte algebraischer Zahlen. J. reine angew. Math. **135** (1909), 284—305.

[39] *N. Tzanakis* and *B.M.M. de Weger,* On the practical solution of the Thue equation. Memorandum **668**, Faculty of Applied Mathematics, University of Twente (1987).

[40] *K. Yu,* Linear forms in the $p$-adic logarithms. MPI/87-20.

[41] *M. Waldschmidt,* A lower bound for linear forms in logarithms. Acta Arith. **37** (1980), 257—283.

[42] *B.M.M. de Weger,* Products of prime powers in binary recurrence sequences II: The elliptic case, with an application to a mixed exponential equation. Math. Comp. **47** (1986), 729—739.

[43] *B.M.M. de Weger,* Solving exponential diophantine equations using lattice basis reduction algorithms, J. Number Theory **26** (1987), 325—367.

[44] *B.M.M. de Weger,* Algorithms for diophantine equations, Ph.D. Thesis, Amsterdam, 1987.

[45] *B.M.M. de Weger,* On the practical solution of Thue–Mahler equations, an outline. Memorandum **649**, Faculty of Applied Mathematics, University of Twente (1987).

[46] *D. Zagier,* Large integral points on elliptic curves. Math Comp. **48** (1987), 425—436.

---

Mathematical Institute, Kossuth Lajos University, 4010 Debrecen, P.O. Box 12, HUNGARY.

# Some Confirming Instances of the Birch–Swinnerton–Dyer Conjecture Over Biquadratic Fields

*Michael I. Rosen*

## Introduction

Let $E$ be an elliptic curve defined over a number field $K$, and let $L(E/K,s)$ be the corresponding $L$–function. Assuming the analytic continuation is possible, at least past $s = 1$, the weak version of the conjecture of Birch–Swinnerton–Dyer asserts that the rank of the finitely generated group $E(K)$ is equal to the order of the zero at $s = 1$ of $L(E/K,s)$. The strong version gives a conjectural formula for the leading coefficient of the power series expansion about $s = 1$. We will only be concerned with the weak version, and will denote it by $B - SwD$.

Suppose that $E$ is an elliptic curve defined over $\mathbb{Q}$ with complex multiplication (CM) by $\sqrt{-d}$, $d$ positive and square–free. Since the class number of $\mathbb{Q}(\sqrt{-d})$ must be 1 in these cases, $d$ can only assume the values 1,2,3,7,11,19,43,67, and 163. In what follows $-D$ will always represent a fundamental discriminant of an imaginary quadratic number field. For such a $D$, let $K_D$ be the biquadratic field $\mathbb{Q}(\sqrt{-D},\sqrt{-d},\sqrt{dD})$. Then we have

**Main Theorem.** *With the above assumptions and notations, assume in addition that* $L(E/\mathbb{Q},1) \neq 0$. *Then there are infinitely many fundamental discriminants* $-D < 0$ *with the property that rank* $E(K_D) = \operatorname*{ord}_{s=1} L(E/K,s) = 2$.

The proof of this result is not difficult, but it makes use of a number of deep theorems relating to the conjectures of Birch and Swinnerton–Dyer due to J. Coates and A. Wiles [2], B. Gross and D. Zagier [3], and K. Rubin [6]. We will also make use of a highly non-trivial analytic result (which we will describe later) due independently to D. Bump, S. Friedberg, and J. Hoffstein [1], and M.R. Murty and V.K. Murty [5].

The hypothesis of the theorem is satisfied, for example, by the curve $y^2 = x^3 - x$. This curve has CM by $\sqrt{-1}$. The fact that $L(E/\mathbb{Q},1) \neq 0$ is proved in [4] on page 96. The hypothesis is in fact satisfied by infinitely many $E/\mathbb{Q}$ which are non-isomorphic over $\mathbb{Q}$. This can be seen as follows. If $E/\mathbb{Q}$ has $CM$, it follows from a theorem of G. Shimura [8] that $E$ is modular, i.e. $L(E,s) = L(f,s)$ where $f$ is a newform of weight 2 on $\Gamma_0(N)$ (recall that $N$ is the conductor of $E$). It follows from a theorem of J.-L. Waldspurger [9] that there are infinitely many fundamental discriminants $c$ such that $L(E_c/\mathbb{Q},1) \neq 0$. Here $E_c$ is the quadratic twist of $E$ by $c$; if $E$ is defined by the Weierstrass equation $y^2 = x^3 + Ax + B$, then $E_c$ is defined by $cy^2 = x^3 + Ax + B$, or, via the substitution $(x,y) \longrightarrow (x/c,\ y/c^2)$, by $y^2 = x^3 + c^2Ax + c^3B$. Thus, if $E$ doesn't satisfy the hypothesis of the theorem, just replace it by an appropriate twist.

Our main theorem was discovered in the course of joint work with E. Kani on "idempotent relations among mathematical objects". Although no trace of that work appears in the present exposition, it did provide the original motivation and may indeed point the way to generalizations. In addition to Kani, thanks are due to J. Hoffstein, K. Kramer, J. Silverman, and the referee for helpful discussions and suggestions.

## 2. Proof of the Main Theorem

We will need some preliminary lemmas on twists. For the basic definitions and properties of twists, see J. Silverman's book [7], Chapter 10, Sections 2 and 5.

Let $K$ be a field of characteristic different from 2, and $c$ an element of $K^*$ which is not a square. If $E$ is an elliptic curve over $K$ defined by

$y^2 = f(x)$, let $E_c$ be defined by $cy^2 = f(x)$. $E_c$ is the quadratic twist of $E$ by $c$. Let $L = K(\sqrt{c})$ and $G = \langle\sigma\rangle$ the Galois group of $L/K$.

**Lemma 1.** *There is an exact sequence*

$$(0) \rightarrow E(K) \rightarrow E(L) \rightarrow E_c(K) \rightarrow H^1(G,E(L)) \rightarrow (0).$$

**Proof.** Define $E(L)^- = \{P \in E(L) \mid P^\sigma = -P\}$. It is easy to check that $P = (x,y) \rightarrow P^* = (x,y\sqrt{c})$ is a bijection of $E_c(K)$ with $E(L)^-$. If $P$, $Q$, and $R$ are on a straight line, then $P^*$, $Q^*$, and $R^*$ are also on a straight line. Thus $P \rightarrow P^*$ is a group isomorphism of $E_c(K)$ with $E(L)^-$. Map $E(L)$ to $E(L)^-$ by $P \rightarrow P - P^\sigma$. The kernel is $E(K)$ and the cokernel is $H^{-1}(G,E(L)) \cong H^1(G,E(L))$.

**Lemma 2.** *Suppose $K$ is a number field. Then*

1) *rank $E(K(\sqrt{c}))$ = rank $E(K)$ + rank $E_c(K)$.*
2) *$L(E/K(\sqrt{c}),s) \approx L(E/K,s)L(E_c/K,s)$ where $\approx$ means equality up to finitely many Euler factors which vanish at $s = 1$.*

**Proof.** Both parts of the lemma are well known, so we only sketch the proof.

Part 1 is immediate from Lemma 1. Just tensor the exact sequence with $\mathbb{Q}$ and count dimensions. The cohomology group is finite and so doesn't contribute.

In Part 2 it suffices to check matters locally. Consider primes $\mathcal{P}$ of $K$ at which $E$ has good reduction and $c$ is a unit at $\mathcal{P}$. Assume also that $\mathcal{P}$ does not divide 2. One knows that $\mathcal{P}$ splits in $K(\sqrt{c})$ if and only if $c$ is a square modulo $\mathcal{P}$. Thus, if $\mathcal{P}$ splits the equality of the Euler factors above $\mathcal{P}$ on both sides of 2, it is obvious. If $\mathcal{P}$ doesn't split, the equality comes down to showing $|\tilde{E}(\tilde{L})| = |\tilde{E}(\tilde{K})|\ |\tilde{E}_c(\tilde{K})|$, where $\tilde{E}$ is the reduction of $E$ at $\mathcal{P}$, $\tilde{K}$ is the residue class field of $K$ at $\mathcal{P}$, and $\tilde{L}$ is the residue class field of $L$ at the unique prime above $\mathcal{P}$. This equality follows from

Lemma 1 and a well known result of *S.* Lang which asserts that $H^1(G, \tilde{E}(\tilde{L}))$ is trivial when $\tilde{L}/\tilde{K}$ is an extension of finite fields with Galois group $G$. (Lang's result applies to algebraic groups in general, not just elliptic curves).

**Lemma 3.** *Let* $E/\mathbb{Q}$ *be an elliptic curve with CM by* $\sqrt{-d}$, $d > 1$, *and* $d$ *not a square. Then,* $E$ *and* $E_{-d}$ *are isogenous, but not isomorphic, over* $\mathbb{Q}$. *If* $E$ *has CM by* $\sqrt{-1}$, *then* $E$ *and* $E_{-1}$ *are isomorphic over* $\mathbb{Q}$.

Proof. Considering $\sqrt{-d}$ as an element of $\mathrm{End}(E)$, we have the exact sequence

$$(0) \longrightarrow C \longrightarrow E(\overline{\mathbb{Q}}) \xrightarrow{\sqrt{-d}} E(\overline{\mathbb{Q}}) \longrightarrow (0).$$

$C$ is a subgroup of $E(\overline{\mathbb{Q}})$ of order $d$ and stable under the action of $\mathrm{Gal}(\overline{\mathbb{Q}}/\mathbb{Q})$. It follows that there is an elliptic curve $E^*$ defined over $\mathbb{Q}$ and a $\mathbb{Q}$–isogeny $\varphi : E \longrightarrow E^*$ such that $\ker \varphi = C$.

Over $\mathbb{Q}(\sqrt{-d})$ we have $E/C \approx E$, and so $E \approx E^*$ over $\mathbb{Q}(\sqrt{-d})$. Thus $E^*$ is isomorphic to either $E$ or $E_{-d}$ over $\mathbb{Q}$. If $d > 1$ we will show that $E$ cannot be isomorphic to $E^*$ over $\mathbb{Q}$. Suppose $\lambda : E^* \longrightarrow E$ is a $\mathbb{Q}$ isomorphism. Then, $\lambda \circ \varphi \in \mathrm{End}_{\mathbb{Q}}(E)$. However, $\mathrm{End}_{\mathbb{Q}}(E) = \mathbb{Z}$ and $\deg(\lambda \circ \varphi) = d$ which is not a square. This is a contradiction.

If $E$ has CM by $\sqrt{-1}$ then $E$ has a Weierstrass model of the form $y^2 = x^3 - Ax$. Since $E_{-1}$ is given by $-y^2 = x^3 - Ax$ it is clear that $(x,y) \longrightarrow (-x,y)$ is a $\mathbb{Q}$–isomorphism of $E$ with $E_{-1}$. The last thing we need to do before beginning the proof of the main theorem is to state the analytic result alluded to in the introduction. Let $-D < 0$ be a fundamental discriminant, and let $\chi_{-D}$ be the Dirichlet character associated to the extension $\mathbb{Q}(\sqrt{-D})/\mathbb{Q}$. The following theorem follows immediately from the much more general result proved by Bump, Friedberg, and Hoffstein in [1] and Murty and Murty in [5].

**Theorem.** *Let $E/\mathbb{Q}$ be a modular elliptic curve and suppose that $L(E/\mathbb{Q},1) \neq 0$. Let $N$ be the conductor of $E$. Then, there are infinitely many fundamental discriminants $-D < 0$ such that* 1) $L(E_{-D},1) \neq 0$ *and* 2) $\chi_{-D}(N) = 1$.

We are now in a position to begin the proof of the main theorem. Recall that $E$ is an elliptic curve defined over $\mathbb{Q}$ with CM by $\sqrt{-d}$, where $d$ is square free, and $K_D = \mathbb{Q}(\sqrt{-D}, \sqrt{dD})$. We set $r$, $r_{-D}$, $r_{-d}$, and $r_{dD}$ equal to the ranks of $E(\mathbb{Q})$, $E_{-D}(\mathbb{Q})$, $E_{-d}(\mathbb{Q})$, and $E_{dD}(\mathbb{Q})$ respectively. Similarly, we set $\rho$, $\rho_{-D}$, $\rho_{-d}$, $\rho_{dD}$ equal to the order at $s = 1$ of $L(E/\mathbb{Q},s)$, $L(E_{-D}/\mathbb{Q},s)$, $L(E_{-d}/\mathbb{Q},s)$, and $L(E_{dD}/\mathbb{Q},s)$ respectively. We will assume that $L(E/\mathbb{Q},1) \neq 0$, $L(E_{-D},1) \neq 0$, and that $\chi_{-D}(N) = 1$. Having assumed $L(E/\mathbb{Q},1) \neq 0$, the above theorem shows we have infinitely many "$D$"s at our disposal.

**Step 1.** $r = p = 0$.

**Proof.** $\rho = 0$ by hypothesis. It follows that r $= 0$ by a theorem of Coates–Wiles [2].

**Step 2.** $r_{-D} = \rho_{-D} = 1$.

**Proof.** We recall the functional equations of $L(E/\mathbb{Q},s)$ and $L(E_{-D}/\mathbb{Q},s)$. Let $\Lambda(E/\mathbb{Q},s) = (2\pi)^{-s}\Gamma(s)\, N^{s/2}L(E/\mathbb{Q},s)$. Since $E$ is modular, $\Lambda(E/\mathbb{Q},s)$ can be analytically continued to the whole complex plane and one has $\Lambda(E/\mathbb{Q},s) = \epsilon\Lambda(E/\mathbb{Q},2-s)$ where $\epsilon$ is 1 or $-1$. The same remarks apply to $L(E_{-D},s)$ except we must replace $N$ by $D^2N$ and $\epsilon$ by $-\chi_{-D}(N)\epsilon$ (assuming $(D,N) = 1$).

Since by hypothesis $L(E/\mathbb{Q},1) \neq 0$ it follows that $\epsilon = 1$. The assumption $\chi_{-D}(N) = 1$ then implies that the sign of the functional equation for $L(E_{-D},s)$ is $-1$, and so $L(E_{-D},s)$ must have a zero at $s = 1$. Finally, the assumption that $L'(E_{-D},1) \neq 0$ implies that $L(E_{-D},s)$ has a simple zero at $s = 1$, i.e. $\rho_{-D} = 1$. Combining a theorem of Gross–Zagier with a theorem of Rubin it follows that $r_{-D} = 1$ (see [6], Corollary C).

**Step 3.** $r_{-d} = \rho_{-d} = 0$.

**Proof.** By Lemma 3 we know that $E$ and $E_{-d}$ are isogenous over $\mathbb{Q}$. Thus, $r_{-d} = r = 0$ and $\rho_{-d} = \rho = 0$.

**Step 4.** $r_{dD} = \rho_{dD} = 1$.

**Proof.** $E_{dD} = (E_{-D})_{-d}$, so by Lemma 3 we have $E_{dD}$ and $E_{-D}$ are isogenous over $\mathbb{Q}$. Thus, $\rho_{dD} = \rho_{-D} = 1$ and $r_{dD} = r_{-D} = 1$.

Finally, applying Lemma 2 we find

$$\begin{aligned} \operatorname{rank} E(K_D) &= \operatorname{rank} E(\mathbb{Q}(\sqrt{-d}) + \operatorname{rank} E_{-D}(\mathbb{Q}\sqrt{-d}) \\ &= r + r_{-d} + r_{-D} + r_{dD} = 2. \end{aligned}$$

Similarly,

$$\operatorname*{ord}_{s=1} L(E/K_D,s) = \operatorname*{ord}_{s=1} L(E/\mathbb{Q}(\sqrt{-d}),s) + \operatorname*{ord}_{s=1} L(E_{-D}/\mathbb{Q}(\sqrt{-d}),s)$$

$$= \rho + \rho_{-d} + \rho_{-D} + \rho_{dD} = 2.$$

This completes the proof.

## References

[1] *D. Bump, S. Friedberg* and *J. Hoffstein,* Non–vanishing theorems for *L*–functions of modular forms and their derivatives. To appear.

[2] *J. Coates* and *A. Wiles,* On the Conjecture of Birch and Swinnerton–Dyer. Invent. Math. **39** (1977), 223–251.

[3] *B. Gross* and *D. Zagier,* Heegner Points and Derivatives of *L*–Series. Invent. Math. **84** (1986), 225–320.

[4] *N. Koblitz*, Introduction to Elliptic Curves and Modular Forms. Grad. Texts in Math. Vol 97, Springer–Verlag, 1986.

[5] *M.R. Murty* and *K.M. Murty*, Mean values of derivatives of modular forms. To appear.

[6] *K. Rubin*, Tate–Shaferevich groups and *L*-functions of elliptic curves with complex multiplication. Invent. Math. **89** (1987), 527–560.

[7] *J. Silverman*, The Arithmetic of Elliptic Curves. Grad. Texts in Math. Vol 106, Springer–Verlag, 1986.

[8] *G. Shimura*, On elliptic curves with complex multiplication as factors of the Jacobians of modular function fields. Nagoya Math. J. **43** (1971), 199–208.

[9] *J.–L. Waldspurger*, Correspondences de Shimura. J. Math. Pure et Appl. **59** (1980), 1–132.

---

Mathematics Department, Box 1917, Brown University
Providence, Rhode Island 02912, USA

# Criteria for the Class Number of Real Quadratic Fields to be One

*Ryuji Sasaki*

## 1. Introduction

As a criterion for complex quadratic fields to have a class number one, G. Rabinowitsch [8] proved the following:

**Theorem.** *The class number of the complex number field* $\mathbb{Q}(\sqrt{1-4m})$ $(m > 0)$ *is one if and only if the quadratic polynomial* $x^2 + x + m$ *takes only prime values for integers* $x = 0, 1, \dots m-2$.

In [9] we generalized this result to the complex quadratic fields with class number two (*cf.* Theorem 1). For real quadratic fields, we obtained in [11] a criterion which is similar to this theorem. We call this criterion Rabinowitsch's theorem for real quadratic fields. In this paper, we shall give criteria for real quadratic fields of several types, which contain fields of Richaud–Degert type, to have class number one.

Mollin and Williams [6] have found all real quadratic fields of Richaud–Degert type with class number one (with one possible exception). This completes their work in [5] where the Generalized Riemann Hypothesis was used to get the result. Thus they showed that the real quadratic field $\mathbb{Q}(\sqrt{d})$ of Richaud–Degert type has class number one if and only if $d$ is one of {2, 3, 6, 7, 11, 14, 17, 21, 23, 29, 33, 37, 38, 47, 53, 67, 77, 83, 101, 141, 167, 173, 197, 213, 227, 237, 293, 398, 413, 437, 453, 573, 677, 717, 1077, 1133, 1253, 1293, 1757}, with possibly one more value.

## 2. Criteria for Class Number One

Throughout this paper we fix the following notation. Let $d$ be a square–free integer. We set

$$\omega = \begin{cases} \dfrac{(1+\sqrt{d}}{2} & \text{for } d \equiv 1 \pmod 4 \\ \sqrt{d} & \text{for } d \equiv 2,3 \pmod 4 \end{cases}$$

and

$$P(x) = x^2 + \mathrm{Tr}(\omega)x + \mathrm{Nm}(\omega)$$

We denote by $h(d)$ and $\Delta = \Delta(d)$ the class number and the discriminant of the quadratic field $\mathbb{Q}(\sqrt{d})$.

When $d$ is negative, T. Ono introduced the following invariant:

$$q(d) = \max_{0 \leq a \leq [|\Delta|/4-1]} \deg P(a)$$

where $[\alpha]$ is the largest integer not exceeding a real number $\alpha$, $\deg N$ is the number of prime divisors of a positive integer $N > 1$ and $\deg 1 = 1$. The following properties of $q(d)$ are proved in [9]. Theorem 1 (2) is a slight generalization of the theorem of Rabinowitsch stated above.

**Theorem 1.** (1) $q(d) \leq h(d)$.

(2) $q(d) = 1 \Leftrightarrow h(d) = 1$.

(3) $q(d) = 2 \Leftrightarrow h(d) = 2$.

Now we shall explain Rabinowitsch's theorem for real quadratic fields in [11]. From now on $d$ is a square–free positive integer. The positive quadratic irrational $\omega$ can be expanded into the continued fraction:

$$\omega = a_0 + \cfrac{1}{a_1 + \cfrac{1}{a_2 + \cfrac{1}{a_3 + \cdots}}} .$$

As is well known, there is a positive integer $k = k(d)$, which we call the period of $\omega$ or $d$, satisfying $a_i = a_{i+k}$ for $i \leq 1$. We shall define integers $A_i$ and $B_i$ by

$$A_0 = 1 \qquad B_0 = \mathrm{Tr}\,(a_0 - \omega)$$

$$2\omega_{i+1} A_{i+1} = B_i + \sqrt{\Delta} \quad (>0) \qquad B_i + B_{i+1} = 2a_{i+1} \quad (i = 0, 1, \dots).$$

By the periodicity of $\omega$, we have $A_i = A_{i+k}$ and $B_i = B_{i+k}$. Define the exceptional set for $d$ by

$$\mathcal{E}(d) = \{A_0 = 1, A_1, \dots, A_{k-1}\}.$$

Making use of this, we shall generalize the notion of the degree: We set $\deg_{\mathcal{E}(d)} 1 = 1$ and for a positive integer $N \geq 2$

$$\deg_{\mathcal{E}(d)} N = \begin{cases} \text{the largest length } l \text{ of a sequence } \{N_1, N_2, \dots, N_l\} \\ \text{of divisors of } N \text{ satisfying} \\ (1)\ \ N_i > 1. \\ (2)\ \ N_i \text{ divides } N_{i+1} \text{ for } 1 \leq i \leq l-1. \\ (3)\ \ \min\{N_j/N_i,\ NN_i/N_j\} \notin \mathcal{E}(d) \text{ for } 1 \leq i \leq j \leq l. \end{cases}$$

In the case $d > 0$ we shall define two invariants by

$$p(d) = \max_{0 \leq a \leq [(\sqrt{2\Delta} - \mathrm{Tr}(\omega))/2]} \deg_{\mathcal{E}(d)} |P(a)|$$

and

$$p'(d) = \max_{0 \leq a \leq [\sqrt{\Delta}/2 - 1]/2} \deg_{\mathcal{E}(d)} |P(a)|$$

Following is the main theorem in [11]. The proof of the first part in (2) is not given there. Combining the proofs of Theorem 1 (3) and the second part of Theorem 2 (2), we get the proof.

**Theorem 2.** (1) $p'(d) \leq p(d) \leq h(d)$.

(2) $p'(d) = 1 \Leftrightarrow p(d) = 1 \Leftrightarrow h(d) = 1$.

**Corollary 1.** *Assume* $d \equiv 2, 3 \pmod 4$ *and* $2 \in \mathcal{E}(d)$. *Let*

$$p_+(d)\ (\textit{resp. } p_-(d)) = \max_{\substack{0 \leq a \leq [\sqrt{\Delta}/2], \\ a \text{ is even (resp. odd)}}} \deg_{\mathcal{E}(d)} |P(a)|$$

*Then* $p_+(d) = 1 \Leftrightarrow p_-(d) = 1 \Leftrightarrow h(d) = 1$.

## 3. Applications

In this section, we shall give simple criteria, which are easily obtained by Theorem 2 and Corollary 1, for the class number of real quadratic fields of certain types to be one.

We start with the following which is a further developed version of the corollary to Theorem 2 in [1].

**Theorem 3.** *Assume* $d \equiv 1 \pmod 4$. *If* $-\left(x^2 + x + \frac{(1-d)}{4}\right)$ *is prime for* $x = 0, 1, \ldots, [[\sqrt{d}/2 - 1]/2]$, *then* $h(d) = 1$.

**Proof.** If $P(x) = -\left(x^2 + x + \frac{(1-d)}{4}\right)$ is prime, then $\deg_{\mathcal{E}(d)} P(x) = 1$; hence $p'(d) = 1$. By Theorem 2, we get $h(d) = 1$.

Q.E.D.

The next is well known (*cf.* e.g., Theorem 2 in [10]) and it comes from Theorem 2 immediately.

**Theorem 4.** *Assume* $d = m^2 + 1$ ($m$: *odd*). *Then* $h(d) = 1$ *if and only if* $d = 2$.

The following two theorems generalize some results in [2], [4], [10], [11], and [12].

**Theorem 5.** *Assume* $d = m^2 + 4$ ($m \geq 1$: *odd*) (*resp.* $d = m^2 - 4$ ($m \geq 5$: *odd*)). *Let* $P(x) = x^2 + x + (1-d)/4$. *Then the following are equivalent:*

(1) $h(d) = 1$.

(2) $|P(x)|$ *is prime or* 1 *for* $0 \leq x \leq [\sqrt{d}/2 - 1]$.

(3) $|P(x)|$ *is prime or* 1 *for* $0 \leq x \leq [\sqrt{d}/2 - 1]$ (*resp. and* $m - 2$ *is prime*).

**Proof.** When $d = m^2 + 4$, $k(d) = 1$ and $\mathcal{E}(d) = 1$; hence $\deg N = \deg_{\mathcal{E}(d)} N$ for all $N$. By Theorem 2, we get the theorem. If $d = m^2 - 4$, then $k(d) = 3$ and $\mathcal{E}(d) = \{1, m-2\}$. If $m - 2$ is prime and divides $P(x)$ with $0 \leq x \leq [\sqrt{d}/2 - 1]$, then $x = \frac{m-3}{2}$ and $P\left(\frac{m-3}{2}\right) = -(m-2)$. This establishes the result.

Q.E.D.

**Examples.** $d = m^2 + 4 = 5, 13, 29, 53, 173, 293.$
$d = m^2 - 4 = 21, 77, 437.$

**Theorem 6.** *Assume* $d = 4m^2 + 1$ $(m > 1)$. *Let* $P(x) = x^2 + x - m^2$. *Then the following are equivalent:*

(1) $h(d) = 1$.
(2) $|P(x)|$ *is prime for* $1 \leq x \leq [\sqrt{2d}/2 - 1]$.
(3) $|P(x)|$ *is prime for* $1 \leq x \leq [(m - 1))/2]$ *and* $m$ *is prime.*

**Proof.** In this case we have $k(d) = 3$ and $\mathcal{E}(d) = \{1, m\}$. Since $P(0) = -m^2$, it follows that $\deg_{\mathcal{E}(d)} |P(0)| = 1$ if and only if $m$ is prime. If $m$ is prime and divides $P(x)$ with $1 \leq x [\sqrt{2d}/2 - 1]$, then $x$ is $m$ - 1 or $m$, and $P(m) = m = -P(m - 1)$. Therefore the result follows from Theorem 2.

Q.E.D.

**Examples.** $d = 17, 37, 101, 197, 677.$

**Remark 1.** R. Mollin [3] gives versions of Theorems 3–6 and shows that if $d$ is of Richaud–Degert type; i.e., $d = m^2 + r$ with $r | 4m$, $-m < r \leq m$, $|r| \notin \{1, 4\}$; then $h(d) = 1$ implies $|r| = 2$. This then gives the reason for studying such types in the following Theorem 7 and 8.

**Theorem 7.** *Assume* $d = m^2 \pm 2$ $(m \geq 3\text{: odd})$. *Let* $P(x) = x^2 - d$. *Then the following are equivalent:*

(1) $h(d) = 1$.
(2) $|P(x)|/2^{\varepsilon(x)}$ *is prime or* 1 *for* $0 \leq x \leq [\sqrt{2d}]$, *where* $\varepsilon(x) = 1$ *or* 0 *if* $x$ *is even or odd.*
(3) $|P(x)|$ *is prime for* $x = 0, 2, 4, \ldots, m - 1$.
(4) $|P(x)|/2$ *is prime for* $x = 1, 3, 5, \ldots, m - 2$.

**Proof.** If $d = m^2 + 2$, then the period $k(d) = 1$ and $\varepsilon(d) = \{1, 2\}$. If $x$ is odd, then $P(x) \equiv 2 \pmod 4$. Moreover $P(x)$ is odd for an even $x$. These and Theorem 2 yield (1) $\Leftrightarrow$ (2). (1) $\Leftrightarrow$ (3) $\Leftrightarrow$ (4) follows from Corollary 1.

If $d = m^2 - 2$, then $k(d) = 4$ and $\mathcal{E}(d) = \{1, 2, 2m - 3\}$. $P(m - 1) = -2m + 3$ and $P(m - 2) = -2(2m - 3)$. If $2m - 3$ is prime and divides $P(x)$ with $0 \leq x \leq$

$[\sqrt{2d}]$, then $x = m - 1$ or $m - 2$. On the other hand, we have $P(x) \equiv 2 \pmod 4$ for odd $x$. Thus we get the theorem by the same reason as above.

Q.E.D.

**Examples.** $d = m^2 + 2 = 11, 83, 227.$
$d = m^2 - 2 = 7, 23, 47, 167.$

The following theorem is proved the same way as above so we shall omit the proof.

**Theorem 8.** *Assume* $d = m^2 + 2$ $(m \geq 2$: *even*$)$ (*resp.* $d = m^2 - 2$ $(m \geq 4$: *even*$)$). *Let* $P(x) = x^2 - d$. *Then the following are equivalent:*

(1) $h(d) = 1$.
(2) $|P(x)|/2^{1-\varepsilon(x)}$ *is prime or* 1 *for* $0 \leq x \leq [\sqrt{2d}]$.
(3) $|P(x)|$ *is prime for* $x = 1, 3, \ldots, m - 1$.
(4) $|P(x)|/2$ *is prime for* $x = 0, 2, \ldots, m - 2$.

**Examples.** $d = m^2 + 2 = 6, 38.$
$d = m^2 - 2 = 14, 62, 398.$

**Remark 2.** In the above two theorems, (4)⇒(1) is proved by Mollin and Williams [5].

**Theorem 9.** *Assume* $d = (mn)^2 \pm 4m$ $(m, n$: *odd*$)$. *Let* $P(x) = x^2 + x + (1 - d)/4$. *Then the following are equivalent:*

(1) $h(d) = 1$.
(2) $m$ *is prime and* $|P(x)|/m^{\mu(x)}$ *is prime or* 1 *for* $0 \leq x \leq [(\sqrt{2d} - 1)/2]$, *where* $0 \leq x \leq \mu(x) = 1$ *if* $2x + 1 \equiv 0 \pmod m$ *and* $\mu(x) = 0$ *otherwise.*
(3) $m$ *is prime and* $|P(x)|/m^{\mu(x)}$ *is prime for* $0 \leq x \leq [(mn - 3)/4]$.

**Proof.** When $d = (mn)^2 + 4m$, then period $k(d) = 2$ and $\mathcal{E}(d) = \{1, m\}$. Assume $m$ is prime. Then $P(x) \equiv 0 \pmod m$ if and only if $2x + 1 \equiv 0 \pmod m$. On the other hand $P(x) \not\equiv 0 \pmod{m^2}$. Therefore we get the theorem by Theorem 2.

If $d = (mn)^2 = 4m$, then $k(d) = 4$ and $\mathcal{E}(d) = \{1, m, mn - m - 1\}$. The rest is the same as above.

Q.E.D.

**Examples.** $d = (mn)^2 + 4m = 93, 237, 1133, 1253.$
$d = (mn)^2 - 4m = 69, 213, 413, 717, 1077.$

**Theorem 10.** *Assume* $d = (mn)^2 - m$ $(n \geq 2,\ m \equiv 0 \pmod 4)$. *Let* $P(x) = x^2 + x + (1 - d)/4$. *Then the following are equivalent:*

(1) $h(d) = 1$.

(2) $m$, $mn - (m+1)/4$ *and* $mn + (m+1)/4$ *are prime, and* $|P(x)|/m^{\mu(x)}$ *is prime for* $0 \leq x \leq [(\sqrt{2d} - 1)/2]$, *and* $x \neq (4m - 3)/4$.

(3) $m$, $mn - (m+1)/4$ and $mn + (M + 1)/4$ *are prime, and* $|P(x)|/m^{\mu(x)}$ *is prime for* $0 \leq x \leq [(mn - 2)/2]$ and $x \neq (m - 3)/4$.

**Proof.** In this case, $k(d) = 4$ and $e(d) = \{1, m, mn - (m+1)/4\}$. Since $P(mn - (m+1)/2) = -m(mn - (m+1)/4)$ and $m$ and $mn - (m+1)/4$ are relatively prime, it follows that $\deg_{\mathcal{E}(d)} P(mn - (m+1)/2) = 1$ if and only if $m$ and $(m+1)/4$ are prime. By the assumption $n \geq 2$, $m$ must be an odd prime. In this case $P(x) \equiv 0 \pmod m$ if and only if $2x + 1 \equiv 0 \pmod m$. When $p = mn - (m+1)/4$ is prime, $P(x) \equiv 0 \pmod p$ if and only if $4(2x+1)^2 \equiv (m-1)^2 \pmod p$. If $0 \leq x \leq [(\sqrt{2d} - 1)/2]$, then $x = (m-3)/4$, $mn - (m+1)/4$ or $mn - 1$. Thus we get the theorem by Theorem 2.

Q.E.D.

**Examples.** $d = 141, 573, 1293, 1757.$

**Remark 3.** Mollin and Williams [7] give generalizations of parts of Theorems 9 and 10, and also give criteria for *non*–Richaud–Degert types of prime–producing polynomials similar to results in this paper.

## References

[1] *M. Kutsuna*, On a criterion for the class number of a quadratic number field to be one. Nagoya Math. J. **79** (1980), 123—129.

[2] *S. Louboutin*, Critères de principalité et minoration des nombres de classes d'idéaux des corps quadratiques réels a l'aide de la théorie des fractions continues. (Preprint).

[3] *R.A. Mollin*, Class number one criteria for real quadratic fields, I. Proc. Japan Acad. **63** (1983), 121—125; II, *ibid*, 162—164.

[4] *R.A. Mollin* and *H.C. Williams*, On prime valued polynomials and class numbers of real quadratic fields. Nagoya Math. J. (to appear).

[5] *R.A. Mollin* and *H.C. Williams*, Prime producing quadratic polynomials and real quadratic fields of class number one. Proceedings of the International Conference on Number Theory at Quebec 1987, (to appear).

[6] *R.A. Mollin* and *H.C. Williams*, Class number one for real quadratic fields, continued fractions and reduced ideals. Proceedings of the NATO ASI on Number Theory and Applications at Banff Canada, (to appear).

[7] *G. Rabinowitsch*, Eindeutigkeit der Zerlegung in Primzahlfactoren in quadratischen Zahlkörpern. J. reine angew. Math. **142** (1913), 153—164.

[8] *R. Sasaki*, On a lower bound for the class number of an imaginary quadratic field. Proc. Japan Acad. **62** A (1986), 37–39.

[9] *R. Sasaki*, A characterization of certain real quadratic fields. Proc. Japan Acad. **62** A (1986), 97—100.

[10] *R. Sasaki*, Generalized Ono invariant and Rabinowitsch's theorem for real quadratic fields. Nagoya Math. J. **109** (1988), 117—124.

[11] *H. Yokoi*, Class number one problem for certain kinds of real quadratic fields. Proc. International Conference on Class numbers and fundamental units of algebraic number fields, Katata, Japan (1986), 125—137.

---

Department of Mathematics, College of Science and Technology, Nihon University, Kanda, Tokyo 101, Japan.

# An Analog of Hilbert's Irreducibility Theorem

*A. Schinzel*

Nearly a hundred years ago Hilbert proved his irreducibility theorem. In a slightly refined, but not the most general form the theorem runs as follows.

**Hilbert (1892).** Let $F(x_1,\dots,x_r,t_1,\dots,t_s)$ be irreducible over $\mathbb{Q}$ as a polynomial in $r+s$ variables. For almost all (in the sense of density) integer vectors $[t_1^*,\dots,t_s^*]$ the polynomial $F(x_1,\dots,x_r,t_1^*,\dots,t_s^*)$ is irreducible over $\mathbb{Q}$ as a polynomial in $r$ variables.

The idea behind this theorem is to obtain from an irreducible polynomial in many variables by a substitution an irreducible polynomial in a smaller number of variables. Ten years before Hilbert's discovery the same idea was used by Kronecker. To study irreducibility of polynomials in several variables he applied the substitution $(t_1,\dots,t_s)\to(t,t^d,\dots,t^{d^{s-1}})$ and indicated the connection between the factorization of $F(t_1,\dots,t_s)$ and that of $F(t,t^d,\dots,t^{d^{s-1}})$. I am going to prove a theorem in this direction, but first I need two definitions.

**Definition 1.** For a rational function $\phi\in K(t_1,\dots,t_s)$ such that $\phi=f\prod_{i=1}^{s}t_i^{\alpha_i}$, $f\in K[t_1,\dots,t_s]$, $(f,t_1t_2\dots t_s)=1$, $\alpha_i\in\mathbb{Z}$, we set $J_{\bar t}\,\phi=f$.

**Definition 2.** A polynomial $F\in K[t_1,\dots,t_s]$ is reciprocal with respect to $t_1,\dots,t_s$ if

$$J_{\bar t}\,F(t_1^{-1},\dots,t_s^{-1})=\pm F(t_1,\dots,t_s).$$

**Theorem 1.** *Let $F(x_1,\dots x_r,t_1^d,\dots,t_s^d)$ be irreducible for all positive integers $d$ and let $F(x_1,\dots x_r,t_1,\dots,t_s)$ be nonreciprocal with respect to $t_1,\dots,t_s$. If $F(x_1,\dots x_r,t_1^{n_1},\dots,t_s^{n_s})$ is reducible over $\mathbb{Q}(t)$ then the exponents $n_1,\dots,n_s$ (positive integers) satisfy a relation*

$$\gamma_1 n_1 + \ldots\gamma + \gamma_s n_s = 0, \tag{1}$$

*where* $\gamma_i$ *are integers and* $0 < \max_{1 \le i \le s} |\gamma_i| < C(F)$.

**Definition 3.** For a vector $\bar{\gamma} = [\gamma_1, \ldots, \gamma_s]$ we put $h(\bar{\gamma}) = \max_{1 \le i \le s} |\gamma_i|$, $t^{\bar{\gamma}} = [t^{\gamma_1}, \ldots, t^{\gamma_s}]$.

**Corollary 1.** *The number of vectors* $\bar{n} = [n_1, \ldots, n_s] \in \mathbb{N}^s$ *such that* $h(\bar{n}) \le N$ *and* $F(\bar{x}, t^{\bar{n}})$ *is reducible over* $\mathbb{Q}(t)$ *is* $O(N^{s-1})$.

**Corollary 2.** *If* $q > C(F)$ *then* $F(x_1, \ldots, x_r, t, t^q, \ldots, t^{q^{s-1}})$ *is irreducible over* $\mathbb{Q}(t)$.

**Lemma 1.** *Let* $k_i$ $(0 \le i \le l)$ *be an increasing sequence of integers. Let* $k_{j_p} - k_{i_p}$ $(1 \le p \le p_0)$ *be all the numbers that appear only once in the double sequence* $k_j - k_i$ $(0 \le i \le j \le l)$. *Suppose that* $[k_{j_1} - k_{i_1}, \ldots, k_{j_{p_0}} - k_{i_{p_0}}] = \bar{n}C$, *where* $C \in \mathfrak{M}_{s,p_0}(\mathbb{Z})$ *and* $h(C) \le c$. *Then either there exist matrices* $K = [\kappa_{qi}] \in \mathfrak{M}_{s,l}(\mathbb{Z})$ *and a vector* $\bar{v} \in \mathbb{Z}^5$ *such that*

$$\begin{gathered}[k_1 - k_0, \ldots, k_l - k_0] = \bar{v}K, \quad \bar{n} = \bar{v}\Lambda; \\ h(K) \le c_1(s, l, c), \\ |\Lambda| > 0, \quad h(\bar{v}) \le 2^{\ell},\end{gathered} \tag{2}$$

*or there is a vector* $\bar{\gamma} \in \mathbb{Z}^s$ *such that* $\bar{\gamma}\bar{n} = 0$ *and* $0 < h(\bar{\gamma}) < c_2(s, l, c)$.

Lemma 1 is proved in [1] (Lemma 7) with corrections in [2] (pp. 263—264) and [3] (p. 291).

**Lemma 2.** *If* $b, b_0, \ldots, b_l \in \mathbb{Z}[\bar{x}]$, $b_i \ne 0$, $\sum_{i=0}^{l} b_i^2 = b$ *then* $l < c_3(b)$, *and* $b_i \in S(b)$, *where* $S(b)$ *is a finite set.*

**Proof.** We take integrals over the unit cube $C = [0,1]^r$ and we obtain

$$\sum_{i=0}^{l} \int_C b_i^2 \, dx_1 \ldots dx_r = \int_C b \, dx_1 \ldots dx_r.$$

Since

$$\int_C b_i^2 \, dx_1 \dots dx_r \geq \frac{1}{\prod_{\rho=1}^{r} (\deg_{x_\rho} b_i + 1)!}; \quad \deg_{x_\rho} b_i \leq \frac{1}{2} \deg_{x_\rho} b$$

we obtain

$$l + 1 \leq \int_C b \, dx_1 \dots dx_r \prod_{\rho=1}^{r} \left(2\left[\frac{\deg_{x_\rho} b}{2}\right] + 1\right)!.$$

For $S(b)$ we take the set

$$\left\{a \in \mathbb{Z}[\bar{x}] : 2 \deg a \leq \deg b, a(x)^2 \leq b(x) \quad \text{for all } x \in \mathbb{Z}^r\right\}.$$

**Lemma 3.** *Let $P, Q \in \mathbb{Q}[\bar{x}, \bar{t}]$, $\bar{n} \in \mathbb{Z}^s$. If $(P, Q) = 1$, but $(P(\bar{x}, t^{\bar{n}}), Q(\bar{x}, t^{\bar{n}})) \notin \mathbb{Q}[t]$ then there exists a vector $\bar{\beta} \in \mathbb{Z}^s$ such that*

$$\bar{\beta}\bar{n} = 0 \quad \text{and} \quad 0 < h(\bar{\beta}) \leq c_4(P, Q). \tag{3}$$

**Proof.** If $D = J_t P(\bar{x}, t^{\bar{n}}), J_t Q(\bar{x}, t^{\bar{n}})) \notin \mathbb{Q}[t]$ we may assume that it is of positive degree in $x_r$ and take an irreducible factor $D_0$ of $D$ positive degree in $x_r$. If $D_0 \in \mathbb{Q}[x]$ then either (3) holds or $D_0 | (P, Q)$, contrary to the assumption. Thus let $\deg_t D_0 > 0$. We consider $D_0$ as an irreducible polynomial in $x_r, t$ over the field $\mathbb{Q}(x_1, \dots, x_{r-1}) = k_0$ defining an algebraic function field $k_0(x_r, t)$. Let $v$ be a valuation of this field trivial on $k_0$ with $v(t) \neq 0$ and let $v_0$ be the corresponding normalized valuation on $k_0(x_r)$ with the ramification index $v$ over $v_0$ equal to $e$. Since $D_0 | P(\bar{x}, t^{\bar{n}})$ there are at least two terms in $P(\bar{x}, t^{\bar{n}})$ with the same valuation.

Hence, denoting the terms in question by $A_i(\bar{x}) t^{\bar{\alpha}_i}$ $(i = 1,2)$ $\bar{n}(\bar{\alpha}_1 - \bar{\alpha}_2) v(t) = e(v_0(A_0) - v_0(A_1))!$, where $h(\bar{\alpha}_1 - \bar{\alpha}_2) \leq \deg_{\bar{t}} P$, $|v_0(A_2) - v_0(A_1)| \leq \deg_{x_r} P$.

We may assume without loss of generality that $\bar{\alpha}_1 - \bar{\alpha}_2$ has the $s$ th coordinate different from 0. We take the resultant $R$ of $P$ and $Q$ with respect to $t_s$ and find that $D_0(\bar{x}, t) | R(\bar{x}, t^{n_1}, \dots, t^{n_{s-1}})$.

By the same argument as before it follows that

$$\bar{n}(\bar{\beta}_1 - \bar{\beta}_2) v(t) = e(v_0(B_2) - v_0(B_1)),$$

where

$$h(\bar\beta_1 - \bar\beta_2) \le \deg_{\bar t} R\,, \quad |v_0(B_2) - v_0(B_1)| \le \deg_{x_r} R\,.$$

($B_i(\bar x)t^{\beta_i}$ $(i = 1,2)$ are two terms with equal valuation.)

Eliminating $v(t)$ and $e$ we obtain the desired relation (3) with

$$c_4(P,Q) = 2\deg_{\bar x} P \deg_{\bar t} P\,(\deg_{\bar x} Q + \deg_{\bar t} Q\,).$$

if $v_0(B_2) - v_0(B_1) \ne 0$, otherwise we take $\bar\gamma = \bar\beta_1 - \bar\beta_2$

**Proof of the theorem.** Let

$$F(\bar x, \bar t\,) = \sum_{i=0}^{L} a_i(\bar x)t_1^{\alpha_{i1}} \ldots t_s^{\alpha_{i2}} \in \mathbb{Z}[\bar x, \bar t\,],$$

where $a_i$ are polynomials $\ne 0$ and the vectors $\bar\alpha_i$ are all distinct. Let further $F(\bar x, t^{\bar n}) = f(\bar x, t)g(\bar x, t)$, where $f, g \in \mathbb{Z}[\bar x, t]$ are of positive degree in $\bar x$.

We consider two cases:

1) $J_t f$ is reciprocal with respect to $t$,
2) $J_t f$ is not reciprocal with respect to $t$.

In the first case we take Lemma 3: $P = F(\bar x, t)$, $Q = J_t F(\bar x, t^{-1})$ and obtain (1) provided $C(F) \ge c_4(P,Q)$.

In the second case we set

$$f(\bar x, t^{-1})g(\bar x, t) = \sum_{i=0}^{l} b_i(\bar x)t^{k_i} \quad (b_i \ne 0,\ k_0 < k_1 < \ldots < k_l).$$

and consider two expressions for $F(\bar x, t^{\bar n})F(\bar x, t^{-\bar n})$

$$F(\bar x, t^{\bar n})F(\bar x, t^{-\bar n}) = \sum_{i=0}^{L} a_i(\bar x)^2 + \sum_{\substack{i,j=0,\\ i\ne j}}^{L} a_i a_j t^{\bar n\bar\alpha_j - \bar n\bar\alpha_i}$$

$$(f(\bar x, t^{-1})g(\bar x, t))(f(\bar x, t)g(\bar x, t^{-1})) = \sum_{i=0}^{l} b_i^2 + \sum_{\substack{i,j=0,\\ i\ne j}}^{\ell} b_i b_j t^{k_j - k_i}\,.$$

If for any pair $\langle i, j \rangle$

$$i \neq j \quad \text{and} \quad \bar{n}\,\bar{\alpha}_i - \bar{n}\,\bar{\alpha}_j = 0 \tag{4}$$

we have (1) with $h(\bar{\gamma}) \leq \deg_{\bar{t}} F$. If no pair $i, j$ satisfies (4) it follows that

$$\sum_{i=0}^{l} b_i^2 = \sum_{i=0}^{L} a_i^2$$

and by Lemma 2: $l \leq c_3\left(\sum a_i^2\right) = c_5(F)$.

Moreover, each number $k_j - k_i$ which appears only once in the double sequence $k_j - k_i$ $(0 \leq i \leq j \leq l)$ has the value $\sum n_q d_q$ with $|d_q| \leq \deg_{\bar{t}} F$. Applying Lemma 1 with $c = \deg_{\bar{t}} F$ we find matrices $K = [\kappa_{qi}]$, $\Lambda = [\lambda_{q\sigma}]$ and a vector $\bar{v}$ satisfying (2). Let us set

$$P(\bar{x}, z_1, \dots, z_s) = J_{\bar{z}} \sum_{i=0}^{L} a_i(\bar{x}) \prod_{q=1}^{s} z_q^{\lambda_q \bar{\alpha}_q}$$

$$Q(\bar{x}, z_1, \dots, z_s) = J_{\bar{z}} \sum_{i=0}^{l} b_i(\bar{x}) \prod_{q=1}^{s} z_q^{\kappa_{qi}}.$$

The polynomial $P$ is irreducible. Indeed making the substitution

$$z_\sigma = \prod_{\rho=1}^{s} t_\rho^{\mu_{\sigma\rho}} \quad \text{where } [\mu_{\sigma\rho}] = |\Lambda| \bullet \Lambda^{-1}$$

we obtain from

$$\sum_{i=0}^{l} a_i(\bar{x}) \prod_{q=1}^{s} z_q^{\lambda_q \bar{\alpha}_i}$$

the polynomial $F(\bar{x}, t_1^d, \dots, t_s^d)$ which is irreducible by assumption. Therefore we have two possibilities:

1) $P | Q$,
2) $(P, Q) = 1$.

Since $J_t P(\bar{x}, t^{\bar{v}}) = J_t F(\bar{x}, t^{\bar{n}}) = J_t(f(\bar{x}, t) g(\bar{x}, t))$, $J_t Q(\bar{x}, t^{\bar{v}}) = J_t(f(\bar{x}, t^{-1}) g(\bar{x}, t))$, the first possibility gives $J_t f(\bar{x}, t) | J_t f(\bar{x}, t^{-1})$ contrary to the assumption that $J_t f$ is not reciprocal. The second possibility gives by virtue of Lemma 3 and Lemma 2 $\bar{\beta}\bar{v} = 0$ with $h(\bar{\beta}) \le c_4(P, Q) \le c_6(F)$.

Since by (2) $\bar{v} = \bar{n}\Lambda^{-1}$ the formula $\bar{v}\bar{\beta}^T = 0$ implies $\bar{n}\Lambda^{-1}\bar{\beta}^{-T} = 0$, hence $\bar{n}\bar{\gamma}^T = 0$, where $\bar{\gamma} = \bar{\beta}\Lambda^A$. The estimate for $h(\bar{\gamma})$ follows from (2) and the estimate for $h(\bar{\beta})$.

**Remark 1.** In a much more complicated way one can show an analogous theorem concerning irreducibility over $\mathbb{Q}$. In this case one has to exclude from the polynomial $F(x, t^{\bar{n}})$ all cyclotomic factors and the result is weaker: the exceptional vectors $\bar{n}$ form a set of density 0, but not necessarily contained in a union of hyperplanes. The proof in the case $r = 0$ will appear in [4].

**Remark 2.** Irreducibility of $F(\bar{x}, t_1^d, \ldots, t_r^d)$ for all $d > 0$ cannot be replaced as the assumption by irreducibility of $F$ itself, even if one wants merely the existence of $\bar{n}$ such that $F(\bar{x}, t^{\bar{n}})$ is irreducible over $Q(t)$. Here is a counterexample (see [4])

$$F(x, t_1, t_2) = t_1^2 + t_2^2 - 2t_1t_2 - 2x^2t_1 - 2x^2t_2 + x^4.$$

**Remark 3.** The field $\mathbb{Q}$ can be replaced in the theorem by any totally real algebraic number field.

**Remark 4.** The assumption that $F(\bar{x}, \bar{t})$ is non-reciprocal with respect to $\bar{t}$ may be unnecessary, but I cannot settle even the case

$$F(x, \bar{t}) = a(x) + b(x)t_1 + b(x)t_2 + a(x)t_1t_2.$$

## References

[1] *A. Schinzel*, Reducibility of lacunary polynomials, I. Acta Arith. **16** (1970), 123—159.

[2] *A. Schinzel*, Reducibility of lacunary polynomials, III. Acta Arith. **34** (1978), 227—266.

[3] *A. Schinzel*, Reducibility of lacunary polynomials, VI. Acta Arith. **47** (1986), 277—293.

[4] *A. Schinzel*, Reducibility of lacunary polynomials, X. Acta Arith. **53** (to appear).

---

Mathematical Institute PAN, P.O. Box 137, 00–950 Warszawa, Poland.

# On Partitions of the Positive Integers With No $x, y, z$ Belonging to Distinct Classes Satisfying $x + y = z$

*J. Schönheim* [1]

## 1. Introduction

Let $\mathcal{P}^s = \mathcal{A}_1 \cup \ldots \cup \mathcal{A}_s$ be a partition of the positive integers into $s$ non-empty classes. Let $n$ be a positive integer and denote by $\mathcal{P}_n^s$ a partition, as above, of the integers 1, 2, ..., $n$.

The question whether for any given $\mathcal{P}_n^3$ the equation

$$x + y = z \tag{1}$$

has a solution with $x, y, z$ belonging to the same class has been answered in the affirmative for sufficiently large $n$, by Schur [2].

To consider the opposite point of view, let us call a partition as above admissible, if the equation (1) has no solution with $x, y, z$ belonging to three distinct classes. Then, imposing some conditions on $\mathcal{P}_n^3$ a solution, as above, of (1) can be forced.

For instance, C.J. Smyth [3] raised a question and V.E. Alekseev and S. Savchev [1] recently proposed it, as a problem in Kvant's contest, to prove:

**Proposition 1.** *There is no admissible* $\mathcal{P}_{3n}^3$ *with* $|\mathcal{A}_1| = |\mathcal{A}_2| = |\mathcal{A}_3|$.

E. and G. Szekeres [4] proved the same assertion from a weaker assumption, namely:

---

[1]This research was completed while the author was a visiting professor at The University of Calgary.

**Proposition 2.** *There is no admissible* $\mathcal{P}_n^3$ *with* $\min(|\mathcal{A}_1|,|\mathcal{A}_2|,|\mathcal{A}_3|) > \frac{1}{4}n$.

In Section 3 of this paper, we shall study the admissible partitions $\mathcal{P}^3$. Our result, Theorem 2, implies that one of the classes $\mathcal{A}$, $\mathcal{B}$, $\mathcal{C}$ is a subset of the multiples of a certain integer greater than 1. As an application, Proposition 2 appears to be a corollary of that fact.

It turns out that the properties of an admissible partition $\mathcal{P}^3$ depend on a very particular structure, a nontrivial subset of the integers 1, 2, ..., $n$ - 1 may have. This structure will be defined and investigated in Section 2.

Our main results concern the general case $\mathcal{P}^s$ and $\mathcal{P}_n^s$, and appear in Sections 4 and 5.

Theorem 2 admits a straightforward generalization while Proposition 2 becomes: There is no admissible $\mathcal{P}_n^s$ with $\min_i|\mathcal{A}_i| > 2^{1-s}n$.

As a so called Anti–Ramsey result, it is shown (Theorem 3) that if $\mathcal{P}_n^s$ is admissible, then $s \leq \log_2 n + 1$ and equality can occur.

## 2. On $a^{(m)}$–Coverings of Subsets of $[1,m-1]$

Small letters shall denote positive integers, $(x\ ,y\ )$ is the g.c.d. of $x$ and $y$. Capitals shall denote sets of positive integers. The set $\{1,2,\ldots,n\}$ will be denoted $[1,n]$.

### 2.1 Definitions, notation and main properties

We shall be interested in properties of a subset $S$ of $[1,m-1]$, when the following three structures are superposed.

(i) Central symmetry in $[1,m-1]$:

$$x \in S \Rightarrow m - x \in S \tag{2}$$

(ii) Periodicity modulo $a$:

$$a \in S \tag{3}$$

$$x \in S \Rightarrow x + a \in S \text{ , provided } x + a < m \tag{4}$$

$$x - a \in S \text{ ,provided } x - a > 0$$

(iii) Central symmetry in $[1,a-1]$:

$$x \in S \Rightarrow a - x \in S \text{ , provided } a - x > 0 \tag{5}$$

**Definition 1.** *A subset S of* $[1, m-1]$ *is called* $a^{(m)}$*–covered if* (2), (3), (4), *and* (5) *hold.*

**Example 1.** Let $x, y, a, m$ be integers $0 < a < m$. Define:

$$T = \{xa + ym;\, 0 < xa + ym < m\}, \tag{6}$$

then $T$ is $a^{(m)}$–covered.

An important role in obtaining the results of Section 3 is played by the following Theorem 1 and its corollaries.

**Theorem 1.** *Let S be a subset of* $[1, m-1]$, $a \in S$ *and* $(a, m) = d$. *Then if S is* $a^{(m)}$*–covered it is also* $d^{(m)}$*–covered.*

**Corollary 1.** *If* $a$ *is the minimal element for which* $S$ *is* $a^{(m)}$*–covered and if* $S$ *is also* $c^{(m)}$*–covered then* $c$ *is a multiple of* $a$.

**Proof of Corollary 1.** Suppose $(a, b) = c < a$. Set $T = S \cap [1, b-1]$ then $T$ is $a^{(b)}$–covered, therefore also $c^{(b)}$–covered, it follows by periodicity modulo $b$ that $S$ is $c^{(m)}$–covered. A contradiction.

**Corollary 2.** *If* $S \subset [1, m-1]$ *is* $a^{(m)}$*–covered then in* $\bar{S}$, *the complement of* $S$ *in* $[1, m-1]$, *there is no* $b$ *for which* $\bar{S}$ *is* $b^{(m)}$*–covered.*

**Proof of Corollary 2.** Suppose w.l.o.g. $a > b$, then $a - b$ must be in $\bar{S}$ otherwise $b \in S$. But $(a - b) \in \bar{S}$ implies $a \in \bar{S}$ a contradiction.

## 2.2 Proof of Theorem 1

One can suppose that $m$ is not a multiple of $a$. Consider $m = at + r$, with $0 \le r < a$, then for some $s, u, v$ $a = sd$, $m = ud$ and $r = ud$, moreover $(v, s) = 1$.

A first step in the proof is to show that $S$ contains all positive multiples of $d$ which are smaller than $m$. The next and final step is to show that if $x$ is positive and smaller than $d$ and for some $l$ the integer $ld + x$ is a member of $S$, then $id \pm x$ is also a member of $S$ for every $i$ such that $id \pm x$ is positive and smaller than $m$. In fact, because of the periodicity modulo $a$, it is enough to prove the above statements replacing $m$ by $a$.

Define now the integers $a_k, b_k, c_k$ by:

$$a_k = \left(s\left\lceil \frac{kv}{s} \right\rceil - kv\right)d \quad \text{for } k = 1,2,\ldots,s-1 \quad a_0 = 0,\ a_s = a \tag{7}$$

$$b_k = \left(kv - \left\lfloor \frac{kv}{s} \right\rfloor s\right)d \quad \text{for } k = 1,2,\ldots,s-1 \quad b_0 = a,\ b_s = 0 \tag{8}$$

$$c_k = a_k + \left(t + \left\lfloor \frac{v(k+1)}{s} \right\rfloor - \left\lceil \frac{kv}{s} \right\rceil\right)a \quad \text{for } k = 0,1,\ldots,s-2. \tag{9}$$

A direct verification shows that

$$a_k + b_k = a, \quad m - a < c_k < m \tag{10}$$

$$b_{k+1} + c_k = m \tag{11}$$

$$b_k = a_{s-k} \tag{12}$$

and for $k = 1,2,\ldots,s-1$, $0 < a_k < a$, $0 < b_k < a$, moreover, using $(v, s) = 1$ and (12) it is easy to see that

$$\{a_k\}_{k=1}^{s} = \{b_k\}_{k=0}^{s-1} = \{id\}_{i=1}^{s}. \tag{13}$$

The first step in the proof of Theorem 1 is done by proving the following lemma and observing that $a_s$ and $b_0$ are members of $S$.

**Lemma 1.** *The sets* $\{a_k\}_{k=1}^{s-1}$, $\{b_k\}_{k=1}^{s-1}$ *and* $\{c_k\}_{k=0}^{s-2}$ *are subsets of* $S$.

**Proof of Lemma 1.** The integer $c_0$ is an element of $S$, since it is a multiple of $a$, consequently by (11) and (2) $b_1$ is an element of $S$ and therefore, by (10) and (5) also $a_1 \in S$. Apply induction and suppose $c_j \in S$, then as above $b_{j+1}$ and $a_{j+1}$ are elements of $S$, thus also $c_{j+1}$ is an element of $S$, by (9) and (4).

Notice that the above first step in the proof of Theorem 1 shows that if $a$ and $m$ are given, then all positive multiples of $d$ can be obtained from $a$ by central symmetry in

$[1, m-1]$ and $[1, a-1]$ and periodicity modulo $a$. This is the converse of the statement in Example 1.

For the second step, let $x$ be positive and smaller than $d$. Define $a_k(x) = a_k + x$, $b_k(x) = b_k + x$ and $c_k(x) = c_k + x$. With that notation, relation (13) is improved to

$$\{a_k(x)\}_{k=0}^{s-1} = \{id + x\}_{i=1}^{s-1}$$
$$\{b_k(x)\}_{k=1}^{s} = \{id + x\}_{i=1}^{s}.$$

The proof of the theorem is then achieved by proving the following lemma by arguments similar to those used in the proof of Lemma 1.

**Lemma 2.** *If for some* $l$ *and* $x$ *as above the integer* $ld + x$ *or* $ld - x$ *is an element of* $S$ *then* $\{a_k(x)\}_{k=0}^{s-1}$ *and* $\{b_k(x)\}_{k=1}^{s}$ *are subsets of* $S$.

**Definition 2.** A subset $S$ of $[1, m-1]$ is called *bad* if $S$ is $a^{(m)}$–covered for some $a$. The multiples of $a$, which are members of $S$ are called bad elements. The subsets of $S$ containing all members of $S$ which are not bad will be denoted $S'$.

We shall use the above defined concepts in the next section and in Section 5.

## 3. On Admissible Tripartitions of the Positive Integers

Let $\mathcal{P}^3 = \mathcal{A} \cup \mathcal{B} \cup \mathcal{C}$ be an admissible partition of the positive integers. Then each of the classes contains a smallest element; let us denote the largest of them by $m$. We shall choose the notation so that $\mathcal{A} \supset 1$ and $\mathcal{C} \supset m$. Notice that $m \geq 4$.

Define $A = \{a;\ \ a \in \mathcal{A} \text{ and } a < m\}$ and similarly $B = \{b;\ \ b \in \mathcal{B} \text{ and } b < m\}$.

**Proposition 3.** (i) *A and B are complementary in* $[1, m-1]$.

(ii) *A and B both have the central symmetry property in* $[1, m-1]$.

(iii) *A and B can not be both bad.*

**Proof.** For (i), $A \cup B$ contains all positive integers, smaller than $m$. For (ii), the contrary contradicts the admissibility of $\mathcal{P}$. For (iii), the statement is implied by Corollary 2 and (i).

Further structural properties of $A$ and $B$ are established in the following proposition.

**Proposition 4.** *If $c$ is the minimal bad element, then*

(i) $c \geq 2$

(ii) *for $c = 2$ it follows that $m = 2w$ for some $w$, $B$ is bad and* $B = \{2i\}_{i=1}^{w-1}$

(iii) *for $c = 3$ it follows that $m = 3v$ for some $v$, $B$ is bad and* $B = \{3i\}_{i=1}^{v-1}$.

**Proof.** For (i), otherwise $B = \phi$.

For (ii), $A$ contains 1; if $2 \in A$ and 2 is bad then $A = [1, m-1]$ and $B = \phi$ which is impossible.

For (iii), A contains 1; if $3 \in A$ and 3 is bad then $2 \in A$ and again $B = \phi$.

The main result of this section, the following theorem, can be now formulated.

The long assumption in the theorem is only to be self–contained. It can be replaced by: "using the above defined notation."

**Theorem 2.** *Let $\mathcal{P}^3 = \mathcal{A} \cup \mathcal{B} \cup \mathcal{C}$ be an admissible partition of the positive integers and let $m$, the smallest element of $\mathcal{C}$, be larger than the smallest element of $\mathcal{A}$ and of $\mathcal{B}$. Let, moreover, $A$ and $B$ be the subsets of $\mathcal{A}$ and $\mathcal{B}$ respectively containing the elements smaller than $m$. Furthermore, denote by $A'$, respectively $B'$, the subsets of $A$, respectively $B$, containing all the elements which are not bad, then*

$$\mathcal{A} \supset A \cup \{A' + im\}_{i=1}^{\infty} \tag{13a}$$

*and*

$$\mathcal{B} \supset B \cup \{B' + im\}_{i=1}^{\infty}. \tag{13b}$$

**Proof.**

Case 1. $A = A'$ i.e., $A$ is not bad.

a) First we shall prove that

$$a \in A \Rightarrow a + m \in \mathcal{A}$$

Since $A$ is not bad $a$ is not a bad element, hence one of the following two must hold

(i) $a = a' + b$, $a' \in A$, $b \in B$

(ii) $\exists i$ , $\exists a'$ such that $a' + ja \in A$ for $j = 0, 1, \ldots . i - 1$, but $a' + ia \in B$.

If (i) holds then $a' = a - b$; and we claim that $a' + m \in \mathcal{A}$. Indeed,

$$a' + m = a + (m - b),$$

the left side is not in $\mathcal{B}$, by the admissibility of $\mathcal{P}$, while the right side is not in $\mathcal{C}$, by the same reason, noticing that $m - b$ is in $B$ by central symmetry. Since $\mathcal{P}$ is a partition $a' + m$ must be in one of the classes so it is in $\mathcal{A}$.

Now consider $a + m$:

$$a + m = (a' + m) + b.$$

As above, the left side can not be in $\mathcal{B}$, while the right side can not be in $\mathcal{C}$, since $a' + m$ is in $\mathcal{A}$, so $a + m \in \mathcal{A}$

If (ii) holds then $a + m$ can be written as

$$a + m = [m - a' - (i - 1)a] + [a' + ia),$$

the left side shows that $a + m$ is not in $\mathcal{B}$; while the right side shows that it is not in $\mathcal{C}$, since the first brackets are in $\mathcal{A}$ and the second in $\mathcal{B}$. So $a + m$ is in $\mathcal{A}$.

(b) Now we shall prove

$$a \in A \Rightarrow a + jm \in \mathcal{A} \text{ for } j = 1, 2, \ldots \qquad (14)$$

The proof is by induction. For $j = 1$ (14) is true, suppose true for $j - 1$. It follows that $jm - a \in \mathcal{A}$ since $jm - a = (m - a) + (j - 1)m$. Now, if $a$ is as in (i), then

$$a' + (j - 1)m + m = a' + jm = [a + (j - 1)m] + (m - b)$$

therefore, $a' + jm$ is in $\mathcal{A}$; then by

$$a + (j - 1)m + m = a + jm = (a' + jm) + b$$

it follows that $a + jm$ is also in $\mathcal{A}$.

If $a$ is as in (ii), then

$$a+(j-1)m+m=a+jm=[jm-(a'+(i-1)a)]+(a'+ai),$$

hence $a+jm$ is in $\mathcal{A}$.

Case 2. $A \neq A'$, i.e., $A$ is bad.

The arguments used in this case are similar to those used in Case 1, but one has to avoid bad elements in $A$, on the other hand the arguments are even simpler since one can use the fact that $B$ is not bad, and Case 1 applies to $B$.

a) Again we first prove

$$a \in A' \Rightarrow a+m \in \mathcal{A}. \tag{15}$$

Now $a$ is not bad by assumption and (i) or (ii) holds.

If (i) holds, $a=a'+b$ as before and $a+m=a'+(b+m)$ showing that (15) holds.

For (ii), the proof is exactly as in Case 1.

b) We shall now prove the analogue of (14):

$$a \in A' \Rightarrow a+jm \in \mathcal{A}. \quad j=1,2,\ldots$$

This is true for $j=1$, suppose by induction true for $j-1$, then for (i),

$$[a+(j-1)m]+m=a+jm=a'+(b+jm)$$

where the left–hand bracket is in $\mathcal{A}$ by induction while the right–hand bracket is in $\mathcal{B}$ by Case 1. Therefore, $a+jm \in \mathcal{A}$

For (ii),

$$[a+(j-1)m]+m=a+jm=[m-(a'+(i-1)a)]+[(a'+ia)+(j-1)m]$$

proving that $a+jm \in \mathcal{A}$ Notice the brackets are not set in the same way as in Case 1.

This proves (13a). The proof of (13b) is similar.

We shall formulate below three corollaries exhibiting the effect on $\mathcal{C}$, the obtained results have.

**Corollary 3.** *If* $c$ *is the minimal bad element, then* $\mathcal{C} \subset \{ic\}_{i=1}^{\infty}$.

**Corollary 4.** *If neither of* A, B *is bad then* $\mathcal{C} \subset \{im\}_{i=1}^{\infty}$.

**Corollary 5.** *If in the admissible partition* $\mathcal{P}_n^3$ *neither one of* A *and* B *is bad, then*

$$|\mathcal{C}| \leq \frac{1}{4} \cdot n \tag{16}$$

**Proof.** $m$ is at least 4.

Proposition 2 is a consequence of the above results.

If none of $A, B$ is bad, the proposition holds by Corollary 5. If one of $A, B$ is bad but $c \geq 4$, then it follows from Corollary 3.

The remaining cases are $c = 2$ or 3. For $c = 2$, by Proposition 4, $m = 2w$, $B = \{2i\}_{i=1}^{w-1}$. Therefore, $|\mathcal{A}| \geq \frac{n}{2}$ and $\mathcal{B}$ and $\mathcal{C}$ cannot both satisfy the requirement. For $c = 3$, by Proposition 4, $m = 3v$, $\mathcal{B} = \{3i\}_{i=1}^{v-1} |\mathcal{A}| \geq \frac{2}{3} n$ and again $\mathcal{B}$ and $\mathcal{C}$ cannot both satisfy the requirement.

## 4. On Admissible Partitions $\mathcal{P}^s$ and $\mathcal{P}_n^s$

### 4.1 An anti-Ramsey theorem

If $n$ is given, admissible partitions $\mathcal{P}_n^s$ cannot exist for too large $s$, for instance, if $s = n$ each class contains a single integer and if $x + y = z$ then $x, y, z$ are from different classes. Hence it is a natural question to ask for the smallest $s$ which enforces a solution in distinct classes. The answer will be formulated in Theorem 3.

Let $\mathcal{P}_n^s = \mathcal{A}_1 \cup \mathcal{A}_2 \cup \ldots \cup \mathcal{A}_s$ be an admissible partition of $[1, n]$. Denote the smallest element of $\mathcal{A}_i$ by $m_i$ and choose the notation so that $m_1 < m_2 < \ldots < m_s$.

Then $m_1 = 1$ and for convenience put $m_s = m$.

**Proposition 5.** $m_{i+1} \geq 2m_i$, $i = 1, 2, \ldots, s-1$

**Proof.** Suppose, to the contrary, that $m_{i+1} < 2m_i$ , then

$$0 < m_{i+1} - m_i < m_i ,$$
$$m_{i+1} - m_i \in \mathcal{A}_j , \quad j < i$$

and
$$(m_{i+1} - m_i) + m_i = m_{i+1}$$

contradicts the admissibility of $\mathcal{P}_n^s$.

**Corollary 6.**
$$m \geq 2^{s-1} . \tag{17}$$

**Theorem 3.** (i) *If* $\mathcal{P}_n^s$ *is admissible, then*

$$s \leq 1 + \log_2 n$$

(ii) *there is an admissible* $\mathcal{P}_n^s$ *with* $s = 1 + \lfloor \log_2 n \rfloor$.

**Proof.** For (i), from (17) $s - 1 \leq \log_2 m \leq \log_2 n$.

For (ii), choose $m_i = 2^i$ , $s = 1 + \lfloor \log_2 n \rfloor$ and

$$\mathcal{A}_i = \{ a \mid a \in [1, n], a \equiv m_i \pmod{m_{i+1}}, \ i \in [1, s-1] \}$$
$$\mathcal{A}_s = \{ a \mid a \in [1, n], a \equiv 0 \pmod{m_s} \}.$$

### 4.2 A generalization

We shall prove the following generalization of Proposition 2.

**Theorem 4.** *No admissible* $\mathcal{P}_n^s$ *exists with* $\min_i |\mathcal{A}_i| > 2^{1-s} n$.

We are adopting a definition and using a notation as in [4]:

Let the set $\mathcal{A}_s$ be $\mathcal{A}_s \{z_1, z_2, \ldots, z_\zeta\}$ arranged in increasing order so $z_1 = m_s = m$, and define $r$ to be the smallest difference between (successive) members of $\mathcal{A}_s$ and let $k$ be the smallest suffix for which $z_{k+1} - z_k = r$.

The following lemma is an immediate generalization of a lemma used in [4]:

**Lemma 3.** $r \geq m_{s-1}$.

**Proof.** The proof is also a straightforward generalization of that in [4]; we shall give the proof only for completeness, [4] being not too accessible.

Suppose $r < m_{s-1}$. Then $r \in \mathcal{A}_j$ with $j < s-1$ and $m_{s-1} - r \in \mathcal{A}_l$, also with $l < s-1$. Observe that by admissibility

$$t < s-1, \ \mathcal{A}_t \not\ni z_k + r - m_{s-1} = z_k - (m_{s-1} - r) \notin \mathcal{A}_{s-1}$$

therefore $z_k + r - m_{s-1} \in \mathcal{A}_s$ and consequently

$$t < s-1, \ \mathcal{A}_t \not\ni z_k - m_{s-1} = (z_k + r - m_{s-1}) \notin \mathcal{A}_{s-1}$$

therefore $z_k - m_{s-1} \in \mathcal{A}_s$

The difference between these above defined members of $\mathcal{A}_s$ being $r$ and both being smaller than $z_k$ leads to a contradiction.

**Observation.** Defining analogously the smallest difference $r_i$ in set $\mathcal{A}_i$ one can prove in the same way that $r_i \geq m_{i-1}$.

**Proof of Theorem 4.**

Case 1. $r \geq s^{s-1}$.

In this case

$$n \geq z_\zeta \geq m + (\zeta - 1)2^{s-1} \geq \zeta 2^{s-1}$$

therefore $\min_i \mathcal{A}_i \leq \zeta \leq 2^{1-s} n$.

Notice that in the remaining cases $m_{s-1} \leq r < 2^{s-1}$, and since, as shown below, Case 2 cannot happen, one has $r = m_{s-1}$.

Case 2. $2^{s-1} > r = m_{s-1} + \beta$, $0 < \beta < m_{s-1}$.

Let us denote $\bigcup_{i=1}^{s-z} \mathcal{A}_i$ by $U$. Then the integers $[z_k+1, z_k+r-1]$ are not in $\mathcal{A}_s$, since $z_{k+1} = z_k + r$.

Moreover, the integers $[z_k+1, z_k+m_{s-1}-1]$ are also not in $\mathcal{A}_{s-1}$, since the integers $[1, m_{s-1}-1]$ being smaller than $m_{s-1}$ are in $U$. In particular, $\beta$ is in $U$ and $z_k + \beta$

is not in $\mathcal{A}_s$ and not in $\mathcal{A}_{s-1}$. So it must be in $U$ but this is impossible since $(z_k + \beta) + m_{s-1} = z_{k+1}$ contradicts the admissibility.

Case 3. $r = m_{s-1}$.

In this case we shall prove first that all non-multiples of $r$ are in $U$.

Observe that $[z_k+1,\ z_k+r-1] \cap \mathcal{A}_s = \phi$ since $z_{k+1} = z_k + r$. Furthermore,

$$[1, r-1] \subset U \tag{18}$$

since the first integer not in $U$ is $m_{s-1} = r$. Therefore $[z_k+1,\ z_k+r-1] \cap \mathcal{A}_{s-1} = \phi$ and $[z_k+1,\ z_k+r-1] \subset U$. Moreover $[z_k-1,\ z_k-m+1] \subset U$ since none of the elements is in $\mathcal{A}_s$ and none is in $\mathcal{A}_{s-1}$.

We shall prove our claim using induction.

By (18) one can start induction. Let $0 < \mu < r$ and suppose $qr + \mu \in U$ for $q = 1,2,\ldots,t-1$. Consider $tr + \mu$. If

$$z_k - [(t-1)r + \mu] > 0 \tag{19}$$

then
$$z_k - [(t-1)r + \mu] = (z_k - r) - [(t-2)r + \mu]. \tag{20}$$

The left-hand side shows that this integer cannot be in $\mathcal{A}_{s-1}$ while the right-hand side shows that it cannot be in $\mathcal{A}_s$ either, since the first term must be in $\mathcal{A}_{s-1}$. So it is in $U$. But this implies that

$$t_{r+\mu} = (z_k + r) - (z_k - (t-1)r - \mu) = [(t-1)r + \mu + r]$$

must be in $U$. Indeed middle terms show that it cannot be in $\mathcal{A}_{s-1}$ and the right-hand side shows that it cannot be in $\mathcal{A}_s$.

If the expression in (19) is negative, then

$$(t-1)r + \mu - z_k$$

turns out to be $U$ by simply reversing the order of the terms in the differences in (20) and finally

$$tr + \mu = (z_k + r) + [(t-1)r + \mu - z_k] = [(t-1)r + \mu] + r$$

shows that all non-multiples of $r$ are in $U$.

This proves the theorem also in Case 3, hence it follows that $|U| \geq \frac{r-1}{r} n$ but then

$$|\mathcal{A}_{s-1} \cup \mathcal{A}_s| \leq \frac{1}{r} n = \frac{1}{m_{s-1}} n \leq 2^{z-s} n$$

and therefore not both $|\mathcal{A}_{s-1}|$ and $|\mathcal{A}_s|$ can be larger than $2^{1-s} n$.

## 5. Some Structures in $\mathcal{P}^S$

The structural properties of $\mathcal{P}^3$ based on concepts introduced in Section 2 and given in Section 3 generalized straightforward for $\mathcal{P}^s$ for general $s$. Here the formulation of the necessary definitions and results.

Define $A_i = \mathcal{A}_i \cap [1, m-1]$ for $i \in [1, s-1]$ while $A'_i$ is the set of non–bad elements of $\mathcal{A}_i$.

Proposition 3 becomes:

**Proposition 6.** *If $\mathcal{P}^s$ is an admissible partition of the integers, then*

(i) $\bigcup_{i=1}^{s-1} A_i = [1, m-1]$

(ii) $A_i$ *has the central symmetry property in* $[1, m-1]$ *for each* $i \in [1, s-1]$

(iii) $A_i$ *can be bad for at most a single value of* $i \in [1, s-1]$

The analogue to Theorem 2 is:

**Theorem 5.** *If $\mathcal{P}^s$ is an admissible partition of the integers then for each $i$, $i \in [1, s-1]$*

$$\mathcal{A}_i \supset A_i + \{A'_i + jm\}_{j=1}^{\infty}.$$

The proofs are omitted.

**Acknowledgement.** I am grateful to Professor E.C. Milner for drawing my attention to the mentioned problem of G. Szekeres, and for the contribution he made in discussing with me the difficulties which arose while obtaining the above results.

I am also grateful to Dr. G.B. Sands for showing me the Alekseev–Savchev problem and to Professor G. Szekeres for his comments and communications.

Finally, thanks are due to the referee, for completing the list of references.

## References

[1] *V.E. Alekseev* and *S. Savchev,* Problem M. 1040. Kvant **4** (1987), 23.

[2] *I. Schur,* Uber die Kongruenz, $x^m + y^m \equiv z^m \pmod p$. Jber. Deutsch Math. Verein. **25**, 114—116.

[3] *G.J. Smyth,* On additive and multiplicative relations connecting algebraic numbers. J. Number Theory **23** (1986), 243—254.

[4] *E.* and *G. Szekeres,* Adding numbers. James Cook, Math. Notes **4**, no. 35, (1984), 4073—4075.

School of Mathematical Sciences, Tel–Aviv University, Tel–Aviv 69978, Israel.

# Algebraic surfaces derived from unit groups of quaternion algebras

*Kisao Takeuchi*

**Dedicated to Professor Michio Kuga on his 60th birthday.**

## Introduction

Let $k$ be a totally real algebraic number field of degree $n \geqq 2$. Let $d(k)$ be its discriminant. We have $n$ distinct isomorphisms $v_j$ $(1 \leqq j \leqq n)$ of $k$ into the real number field $\mathbb{R}$. Let $v$ be a place of $k$ and $k_v$ be the completion of $k$ at $v$. We may identify the infinite (resp. finite) places with $\{v_1,\ldots,v_n\}$ (resp. the prime ideals of $k$). Let $A$ be a quaternion algebra over $k$ (i.e. a central simple algebra over $k$ of dimension 4). Let $A_v = A \otimes_k k_v$. If $A_v$ is division algebra, we say that $A$ ramifies at $v$ and otherwise we say that $A$ splits at $v$. We have the following

**Theorem 0.1.** (Hasse). *Let $A$ be a quaternion algebra over an algebraic number field $k$ of finite degree. Then the set of all places of $k$ where $A$ ramifies is of finite even cardinality. Conversely, for a given finite set $P$ of the places of $k$ with even cardinality there exists a quaternion algebra $A$ over $k$ such that $A$ ramifies at each place belonging to $P$ and splits at any other place. $A$ is uniquely determined up to $k$–isomorphism.*

Let $D(A)$ be the product of all finite prime ideals where $A$ ramifies. If there exists no such ideals, we put $D(A) = (1)$. $D(A)$ is called the discriminant of $A$.

From now on, we assume that $A$ splits at $v_1$, $v_2$ and ramifies at any other infinite place of $k$. Then we have

$$A \otimes_{\mathbb{Q}} \mathbb{R} = M_2(\mathbb{R}) \oplus M_2(\mathbb{R}) \oplus \mathbb{H}^{n-2},$$

where $\mathbb{H}$ is the Hamilton quaternion algebra over $\mathbb{R}$.

Denote by $\rho_j$ $(1 \leqq j \leqq n)$ the canonical embedding of $A$ into $A_{v_j}$. Then we have $\rho_j|_k = v_j$. Let $\rho$ be the isomorphism of $A$ into $M_2(\mathbb{R}) \oplus M_2(\mathbb{R})$ defined by

$$\rho(\alpha) = (\rho_1(\alpha), \rho_2(\alpha)).$$

Let $O$ be a maximal order of $A$ and let $U(O) = \{e \in O \mid n_{A/k}(e) = 1\}$, where $n_{A/k}(\ )$ is the canonical norm of $A$ over $k$. $U(O)$ is called the unit group of $O$ of norm 1. Let $\Gamma(O) = \rho(U(O))$ be the image of $U(O)$ under $\rho$. Then it is known that $\Gamma(O)$ is a discrete subgroup of $SL_2(\mathbb{R}) \times SL_2(\mathbb{R})$. The group $SL_2(\mathbb{R}) \times SL_2(\mathbb{R})$ operates on the product $H \times H$ of the upper half plane $H = \{z \in \mathbb{C} \mid Im(z) > 0\}$. $\Gamma(O)$ operates on $H \times H$ as a properly discontinuous group. Let $X(\Gamma(O)) = \Gamma(O)\backslash H \times H$ be the quotient space of $H \times H$ by $\Gamma(O)$. In the case $A = M_2(k)$, where $k$ is a quadratic field, $X(\Gamma(O))$ is known as "the Hilbert modular surface". In the case $A$ is a division quaternion algebra $X(\Gamma(O))$ is called "the Shimura surface".

From now on, we assume that $A$ is a division quaternion algebra. Then it is known that $X(\Gamma(O))$ is compact.

M. Kuga [4] raised the following problem:

> Make a complete list of groups $\Gamma(O)$ such that $X(\Gamma(O))$ is a smooth algebraic surface with geometric genus $p_g = 0$.

The case where $k$ is quadratic is treated by Shaval [11] and by Otsubo [6]. The aim of this paper is to answer the above problem.

Let $L$ be an algebraic number field and $K$ be an extension of $L$ of degree $n$. Let $P$ be a prime ideal of $K$. We denote by $f_P$, $e_P$, $n_{K/L}(P)$ the relative degree, the ramification index and norm of $P$ in $K/L$.

For a prime number $p$ we denote by $P_p$ the prime ideal of $L$ above $(p)$.

The author is grateful to Professor Michio Kuga for many valuable suggestions and to the referee for supplying references for a table of $\zeta_k(2)$ which makes our arguments short.

## 1. Main Theorem

1.0. As an answer to the above Kuga's problem we shall prove the following

**Theorem 1.1.** *Let the notations be as above. The complete list of all $\Gamma(O)$ such that $X(\Gamma(O))$ is a smooth algebraic surface with geometric genus $p_g = 0$ is given as follows:*

| $n$ | $f(x)$ | $\alpha$ | $d(k)$ | $D(A)$ |
|---|---|---|---|---|
| 2 | $x^2-2$ | $\sqrt{2}$ | 8 | $P_2P_5$ |
| 2 | $x^2-3$ | $\sqrt{3}$ | 12 | $P_2P_{13}$ |
| 3 | $x^3-3x^2+1$ | $2\cos(2\pi/9)$ | 81 | $P_{37}$ |
| 3 | $x^3-x^2-3x+1$ | | 148 | $P_{13}$ |
| 3 | $x^3+x^2-4x+1$ | $2\cos(2\pi/13) + 2\cos(10\pi/13)$ | 169 | $P_{13}$ |
| 4 | $x^4-5x^2+5$ | $2\cos(2\pi/20)$ | 2000 | $P_2P_5$ |
| 4 | $x^4-4x^2+1$ | $2\cos(2\pi/24)$ | 2304 | $P_2P_3$ |

*where $\alpha$ is a generator of $k$ over $\mathbb{Q}$ and $f(x)$ is the irreducible polynomial of $\alpha$ over $\mathbb{Q}$.*

The cases $n = 2$ are given in Shaval [11] and the case $n = 3$ with $d(k) = 169$ is given in Kuga [4].

We shall give the proof of our main theorem in §2–§8.

## 2. Several known results

**2.0.** Let $z = (z_1, z_2)$ be the canonical variable on $H \times H$, where $z_j = x_j + iy_j \in H$ $(1 \leqq j \leqq 2)$. We see that $y_1^{-2} y_2^{-2} dx_1 dy_1 dx_2 dy_2$ is a $SL_2(\mathbb{R}) \times SL_2(\mathbb{R})$ – invariant volume element of $H \times H$. Denote by $\mathrm{vol}(\Gamma(O))$ the volume of the fundamental domain of $\Gamma(O)$ with respect to the above volume element. As to $\mathrm{vol}(\Gamma(O))$ we have the following.

**Theorem 2.1.** (Shimizu [12]). *Let the notations be as above. The volume of the fundamental domain of* $\Gamma(O)$ *is given by the formula:*

$$vol(\Gamma(O)) = 2^{5-2n} \pi^{2-2n} d(k)^{3/2} \zeta_k(2) \prod_{P \mid D(A)} (n_k(P) - 1), \tag{1}$$

*where* $\zeta_k(2)$ *is the value of the Dedekind zeta function* $\zeta_k(s)$ *of* $k$ *at* $s = 2$ *and* $n_k(\ )$ *is the norm of* $k$ *over* $\mathbb{Q}$.

The following proposition is essential for our computations.

**Proposition 2.2.** (Shimura [13]). *Let* $A$ *be a quaternion algebra over a totally real algebraic number field* $k$ *of finite degree. Let* $O$ *be a maximal order of* $A$. *Let* $K$ *be a totally imaginary quadratic extension of* $k$. *Let* $O_k$ *be the ring of integer of* $k$. *Then the following conditions are equivalent.*

(*i*) *There exists a* $k$*–embedding* $f$ *of* $K$ *into* $A$.

(*ii*) *Each prime ideal* $P$ *dividing* $D(A)$ *does not split in* $K/k$. (*i.e.* $P$ *remains prime or ramifies in* $K/k$.)

*Moreover, if the above condition holds, the* $k$*–embedding* $f$ *can be taken so that* $f(O_K) \subset O$.

The space $X(\Gamma(O))$ is smooth if and only if $\Gamma(O)$ has no torsion element other than $\pm 1$. It is known that if $X(\Gamma(O))$ is smooth, then $X(\Gamma(O))$ is an algebraic surface. Denote by $p_g$, $p_a$ and $C_2$ the geometric genus, arithmetic genus, and the Chern number of $X(\Gamma(O))$ respectively. Then by Shaval [11] if $p_g = 0$, then $p_a = 1$ and $C_2 = 4$. Consequently, we have the following formula:

$$2(2\pi)^{-2n} d(k)^{3/2}\ \zeta_k(2)\ N(A) = 1, \tag{2}$$

where we denote

$$N(A) = \prod_{P \mid D(A)} (n_k(P)-1). \tag{3}$$

Now we shall prove the following

**Proposition 2.3.** *Let the notations be as above. If $\Gamma(O)$ is a smooth algebraic surface with geometric genus $p_g = 0$. Then the degree $n$ of $k$ is smaller than* 9.

**Proof.** We need the following

**Theorem 2.4.** (Odlyzko [5]). *Let $k$ be a totally real algebraic number field of degree $n$ with discriminant $d(k)$. Then the following inequality holds:*

$$(29.099)^n\ e^{-8.3185} < d(k). \tag{4}$$

Since $1 \leqq N(A)$ and $1 < \zeta_k(2)$, we have

$$n < (3*8.315 - 2*\log_e 2)/(3*\log_e 29.099 - 4*\log_e(2\pi)) = 8.5377\ldots .$$

Hence we have $n \leqq 8$. Q.E.D.

Assume that $N(A) = 1$. If $D(A) = (1)$, then by Proposition 2.2 we see easily that $\Gamma(O)$ contains an element of order 3. Hence we see that

$$D(A) = P_1 \ldots P_r \qquad n_k(P_i) = 2 \quad (1 \leqq i \leqq r).$$

Let $L = \mathbb{Q}(\sqrt{-3})$ and $K = Lk$. Since $f_2 = 2$ in $L/\mathbb{Q}$ and $f_{p_i} = 1$ in $k/\mathbb{Q}$, we see that $f_{p_i} = 2$ in $K/k$. Hence by Proposition 2.2 we see that $\Gamma(O)$ contains an element of order 3. Therefore, from now on we may see assume that

$$2 \leqq N(A). \tag{5}$$

Now we shall treat each $2 \leqq n \leqq 8$ case by case. Since the case $n = 2$ is treated by Shaval [10], we may assume that $3 \leqq n \leqq 8$.

## 3. The case $n = 3$

**3.0.** In this section we shall treat the case $n = 3$. Since $n = 3$ is odd, by Theorem 0.1 we see that the number of the prime ideals dividing $D(A)$ is odd. In particular, we have $D(A) \neq (1)$. By (2) and (5) we have

$$d(k) < 2^{-4/3}(2\pi)^4 = 618.509\ldots .$$

Hence we have

$$d(k) \leqq 618. \tag{6}$$

The list of all such fields $k$ is given by Delone–Faddeev [2]. The values $\zeta_k(2)$ for such $k$ are given by Cartier–Roy [1] and by Pohst [7]. Hence we have the following table:

| $d(k)$ | $f(x)$ | $\pi^{-6}d(k)^{3/2}\zeta_k(2)$ | $N(A)$ |
|---|---|---|---|
| 49 | $x^3+x^2-2x-1$ | $8/21$ | 84 |
| 81 | $x^3-3x+1$ | $8/9$ | 36 |
| 148 | $x^3-x^2-3x+1$ | $8/3$ | 12 |
| 169 | $x^3+x^2-4x+1$ | $8/3$ | 12 |
| 229 | $x^3-4x-1$ | $16/3$ | 6 |
| 257 | $x^3-x^2-4x+3$ | $16/3$ | 6 |
| 316 | $x^3-x^2-4x+2$ | $32/3$ | 3 |
| 321 | $x^3-x^2-4x+1$ | 8 | 4 |
| 361 | $x^3-x^2-6x+7$ | 8 | 4 |
| 404 | $x^3-x^2-5x-1$ | $40/3$ | $12/5$ |
| 469 | $x^3-x^2-5x+4$ | 16 | 2 |
| 473 | $x^3-5x-1$ | $40/3$ | $12/5$ |
| 564 | $x^3-x^2-5x+3$ | 24 | $4/3$ |
| 568 | $x^3-x^2-6x-2$ | $80/3$ | $6/5$ |

**Table 1**

where $f(x)$ is a defining equation of $k$ over $\mathbb{Q}$.

It is known that the field $k$ with $d(k)$ listed above is determined uniquely up to $\mathbb{Q}$–isomorphism and that $O = \mathbb{Z}[\alpha]$ for an element $\alpha$ of $k$ such that $f(\alpha) = 0$.

First of all, in view of Table 1 we can exclude

$$d(k) = 404,\ 473,\ 564,\ 568$$

because $N(A)$ are not integers for these cases.

**3.1.** The cases $d(k) = 49, 316, 321, 361, 469$. First consider the case $d(k) = 49$. In this case we see that $k = \mathbb{Q}(2\cos(2\pi/7))$ and we have

$$f_p = 3 \quad \text{for} \quad p = 2,3,5,11,17,19,$$

$$f_p = 1 \quad \text{for} \quad p = 7,13,29,43.$$

Since $N(A) = 84$, for $P$ dividing $D(A)$ we have

$$n_k(P) = 7,8,13,29,43.$$

Since the number of the prime ideals $P$ dividing $D(A)$ is odd, we have

$$N(A) \geqq 6*6*6 = 216.$$

This is a contradiction. By the same argument we can exclude the cases $d(k) = 316, 321, 361, 469$.

**3.2.** The cases $d(k) = 81, 148, 169$. First consider the case $d(k) = 81$. In this case we have $k = \mathbb{Q}(2\cos(2\pi/9))$.

Since $N(A) = 36$, for $P$ dividing $D(A)$ we have

$$n_k(P) = 2,3,4,5,7,13,19,37.$$

On the other hand, we have

$$f_p = 3 \quad \text{for} \quad p = 2,5,7,11,13,$$

$$f_p = 1 \quad \text{for} \quad p = 3,19,37.$$

Hence we have $P = P_3, P_{19}, P_{37}$. Since the number of the prime ideals dividing $D(A)$ is odd, we have

$$D(A) = P_{37}.$$

We shall show that this case is a solution of our problem. It suffices to show that in this case $\Gamma(O)$ has no torsion elements other than $\pm 1$.

Assume that $\Gamma(O)$ contains an element $w$ of order $m \geqq 3$. Let $K = k(w)$ and $L = \mathbb{Q}(w)$. Then we have $[K:\mathbb{Q}] = 6$ and $[L:\mathbb{Q}] | 6$. It implies that

$$m = 3,4,6,7,9,14,18.$$

Since $k$ is the maximal real subfield of $K$, we see that $\mathbb{Q}(2\cos(2\pi/m)) \subset k$. Hence we see that

$$L = \mathbb{Q}(\sqrt{-1})\ (m=4),\quad \mathbb{Q}(\sqrt{-3})\ \ (m=3,6)\ \text{ or }\ \mathbb{Q}(e^{2\pi i/9})\ \ (m=9,18).$$

It suffices to show that we get a contradiction for $m = 3$, 4. Assume that $m = 4$. Then we see that $L = \mathbb{Q}(\sqrt{-1})$. We have $d(L) = -4$. We need the following well-known

**Theorem 3.1.** *Let $K$ and $L$ be finite algebraic number fields of degree $m$ and $n$ respectively. Let $O_K$ and $O_L$ be their rings of integers respectively. Let $M$ be the composite of $K$ and $L$. Assume that $(d(K),d(L)) = 1$. Then the following identities hold:*

*(i)* $O_M = O_K O_L$

*(ii)* $d(M) = d(K)^n d(L)^m$, $d(L)^m = n_K(d(M/K))$,

*where $d(M/K)$ is the relative discriminant of the extension $M/K$.*

Since $(d(k),d(L)) = 1$, we can apply the above theorem to $\{k,L\}$. We have $O_K = O_k O_L$ and $n_k(d(K/k)) = d(L)^3 = -4^3$. Hence $P_{37}$ is unramified in $K/k$. Since $f_{37} = 1$ in $L/\mathbb{Q}$ and $f_{P_{37}} = 1$ in $k/\mathbb{Q}$, we see that $f_{P_{37}} = 1$ in $K/k$. Hence we see that $P_{37}$ splits in $K/k$. By Proposition 2.2 we get a contradiction. Assume that $m = 3$. Then we have $L = \mathbb{Q}(\sqrt{-3})$ and

$d(L) = -3$. We see easily that $K = \mathbb{Q}(e^{2\pi i/9})$. Hence we see that $P_{37}$ is unramified and $f_{p_{37}} = 1$ in $K/k$. By Proposition 2.2 we get a contradiction. This proves that this case is a solution of our problem.

Next consider the case $d(k) = 148$. We have

$$N(A) = 12.$$

We have the following table:

$$\begin{array}{lll}
(2) = P_2^{\ 3} & f_2 = 1 & e_2 = 3, \\
(3) = P_3 & f_3 = 3, & \\
(5) = P_{5,1}P_{5,2} & f_{5,1} = 1 & f_{5,2} = 2, \\
(7) = P_7 & f_7 = 3, & \\
(11) = P_{11} & f_{11} = 3, & \\
(13) = P_{13,1}P_{13,2} & f_{13,1} = 1 & f_{13,2} = 2.
\end{array}$$

Therefore, if $P|D(A)$ then $P = P_{2,1}, P_{5,1}$ or $P_{13,1}$. Since the number of the prime ideals dividing $D(A)$ is odd, we have

$$D(A) = P_{13,1} \cdot$$

We show that this is a solution of our problem. Assume that there exists an element $w$ of finite order $m \geqq 3$ in $\Gamma(O)$. By the same argument as in the case $d(k) = 81$ we see that

$$m = 3,\ 4 \text{ or } 6.$$

It suffices to show that we get a contradiction for $m = 3, 4$. Let $L = \mathbb{Q}(e^{2\pi i/m})$ and $K = Lk$. Assume that $m = 3$. Then we have $L = \mathbb{Q}(\sqrt{-3})$ and $d(L) = -3$. Since $(d(k), d(L)) = 1$, we can apply Theorem 3.1. Hence we have $O_K = O_L O_k$ and $n_k(d(K/k)) = -3^3$. Hence we see that $P_{13,1}$ is unramified in $K/k$. Since $f_{13} = 1$ in $L/\mathbb{Q}$ and $f_{p_{13,1}} = 1$ in $k/\mathbb{Q}$, we see that $P_{13,1}$ splits in $K/k$. By Proposition 2.2 we get a contradiction. Next assume that $m = 4$. In this case we have $L = \mathbb{Q}(\sqrt{-1})$ and $d(L) = -4$. Let $D(K/k)$ be the different of the extension $K/k$. Since $K = k(\sqrt{-1})$ and $f(x) = x^2 + 1$ is the irreducible polynomial of $\sqrt{-1}$ over $k$, we see that $D(K/k) \mid f'(\sqrt{-1}))$. Let

$$(f'(\sqrt{-1})) = D(K/k)F.$$

Then we see that $(F,P_{13}) = 1$. By the following congruence equation:

$$f(x) \equiv (x-5)(x+5) \pmod{13},$$

we see that $P_{13}$ splits in $K/k$. By Proposition 2.2 we get a contradiction. This shows that this case is a solution of our problem.

By the same argument as above we can prove that for $d(k) = 169$ the case $D(A) = P_{13}$ is a solution of our problem.

**3.3.** The cases $d(k) = 229, 257$. First consider the case $d(k) = 229$. In this case we have

$$N(A) = 6.$$

We have the following table:

$$\begin{array}{lll} (2) = P_{2,1}P_{2,2} & f_{2,1} = 1 & f_{2,2} = 2, \\ (3) = P_3 & f_3 = 3, & \\ (5) = P_5 & f_5 = 3, & \\ (7) = P_{7,1}P_{7,2} & f_{7,1} = 1 & f_{7,2} = 2. \end{array}$$

If $P|D(A)$ then we have $n_k(P) = 2,3,4$ or $7$. Since the number of the prime ideals dividing $D(A)$ is odd, we see that

$$D(A) = P_{7,1}\,.$$

Let $L = \mathbf{Q}(\sqrt{-1})$ and $K = Lk$. Since $f_7 = 2$ in $L/\mathbf{Q}$ and $f_{7,1} = 1$ in $k/\mathbf{Q}$, we see that $f_{P_{7,1}} = 2$ in $K/k$. By Proposition 2.2 we see that $\Gamma(O)$ contains an element of order 4. This is a contradiction.

By the same argument we can exclude the case $d(k) = 257$. This completes the case $n = 3$.

## 4. The case $n = 4$

**4.0.** Let us treat the case $n = 4$. In this case the number of the prime ideals dividing $D(A)$ is even. By (2) we have

$$N(A) = 2^7\pi^8 d(k)^{-3/2}\zeta_k(2)^{-1}. \tag{7}$$

Hence by (5) we have

$$d(k) \leqq 7171. \tag{8}$$

The list of the fields $k$ satisfying (8) is as follows: (Delone–Faddeev [2]) The value $\zeta_k(2)$ for some $k$ is calculated by Cartier–Roy [1] and by [15].

| $d(k)$ | $f(x)$ | subfield | $\pi^{-8}d(k)^{3/2}\zeta_k(2)$ | $N(A)$ |
|---|---|---|---|---|
| 725 | $x^4-x^3-3x^2+x+1$ | $\mathbb{Q}(\sqrt{5})$ | $2^5/15$ | 60 |
| 1125 | $x^4-x^3-4x^2+x+1$ | $k=\mathbb{Q}(2\cos(2\pi/15))$ | $2^6/15$ | 30 |
| 1600 | $x^4-6x^2+4$ | $k=\mathbb{Q}(\sqrt{2},\sqrt{5})$ | $2^4 7/15$ | 120/7 |
| 1957 | $x^4-4x^2+x+1$ | | $2^5/3$ | 12 |
| 2000 | $x^4-5x^2+5$ | $k=\mathbb{Q}(2\cos(2\pi/20))$ | $2^5/3$ | 12 |
| 2048 | $x^4-4x^2+2$ | $k=\mathbb{Q}(2\cos(2\pi/16))$ | $2^3 5/3$ | 48/5 |
| 2225 | $x^4-x^3-5x^2+2x+4$ | $\mathbb{Q}(\sqrt{5})$ | | |
| 2304 | $x^4-4x^2+1$ | $k=\mathbb{Q}(2\cos(2\pi/24))$ | $2^4$ | 8 |
| 2525 | $x^4-2x^3-4x^2+5x+5$ | $\mathbb{Q}(\sqrt{5})$ | | |
| 2624 | $x^4-2x^3-3x^2+2x+1$ | $\mathbb{Q}(\sqrt{2})$ | | |
| 2777 | $x^4-x^3-4x^2+x+2$ | | | |
| 3600 | $x^4-2x^3-7x^2+8x+1$ | $k=\mathbb{Q}(\sqrt{3},\sqrt{5})$ | | |
| 3981 | $x^4-x^3-4x^2+2x+1$ | | | |
| 4205 | $x^4-x^3+5x^2-x+1$ | $\mathbb{Q}(\sqrt{29})$ | | |
| 4225 | $x^4-9x^2+4$ | $k=\mathbb{Q}(\sqrt{5},\sqrt{13})$ | | |
| 4352 | $x^4-6x^2+4x+2$ | $\mathbb{Q}(\sqrt{2})$ | | |
| 4400 | $x^4-7x^2+11$ | $\mathbb{Q}(\sqrt{5})$ | | |
| 4525 | $x^4-x^3-7x^2+3x+9$ | $\mathbb{Q}(\sqrt{5})$ | | |
| 4752 | $x^4-2x^3-3x^2+4x+1$ | $\mathbb{Q}(\sqrt{3})$ | | |
| 4913 | $x^4-x^3-6x^2+x+1$ | $k=\mathbb{Q}(\sqrt{17+\sqrt{17}})$ | | |

| | | |
|---|---|---|
| 5125 | $x^4-2x^3-6x^2+7x+11$ | $\mathbb{Q}(\sqrt{5})$ |
| 5225 | $x^4-x^3-8x^2+x+11$ | $\mathbb{Q}(\sqrt{5})$ |
| 5725 | $x^4-x^3-8x^2+6x+11$ | $k=\mathbb{Q}(\sqrt{27+10\sqrt{5}})$ |
| 5744 | $x^4-5x^2+2x+1$ | |
| 6125 | $x^4-x^3-5x^2+2x+1$ | $\mathbb{Q}(\sqrt{5})$ |
| 6224 | $x^4-2x^3-4x^2+2x+2$ | |
| 6809 | $x^4-5x^2+x+1$ | |
| 7053 | $x^4-2x^3-4x^2+3x+3$ | |
| 7056 | $x^4-5x^2+1$ | $k=\mathbb{Q}(\sqrt{3},\sqrt{7})$ |
| 7168 | $x^4-6x^2+7$ | $\mathbb{Q}(\sqrt{2})$ |

**Table 2**

where $f(x)$ is a defining equation for $k$.

It is known that the field $k$ listed above is uniquely determined by $d(k)$ and for each $k$ the integral basis of $O_k$ is given. Hence we can make a table of decomposition of $(p)$ for $p = 2,3,\dots$ .

First of all, in view of Table 2 we can exclude the cases $d(k) = 1600$, $2048$ because $N(A)$ is not an integer for these cases.

**4.1.** The cases $d(k) = 725$, $1125$. First consider the case $d(k) = 725$. We have

$$N(A) = 60.$$

We have the following table of decomposition:

$$(2) = P_2 \qquad f_2 = 4,$$

$$(3) = P_3 \qquad f_3 = 4,$$

$$(5) = P_5^2 \qquad f_5 = 2,$$

$$(11) = P_{11,1}P_{11,2}P_{11,3} \qquad f_{11,1} = f_{11,2} = 1 \quad f_{11,3} = 2$$

Since $k \supset \mathbb{Q}(\sqrt{5})$ we see that

$$f_p \geqq 2 \quad \text{for} \quad p = 7,13,17.$$

If $P | D(A)$ then we have

$$n_k(P) = 16,31 \quad \text{or} \quad 61.$$

Hence $P = P_2, P_{31}, P_{61}$.

Since the number of the prime ideals dividing $D(A)$ is even, we get a contradiction for each combination of the above ideals.

By the same argument we can exclude the case $d(k) = 1125$.

**4.2.** The case $d(k) = 1957$. In this case by Table 2 we have

$$N(A) = 12.$$

We have the following table of decomposition:

$$\begin{array}{lll} (2) = P_2 & f_2 = 4, & \\ (3) = P_{3,1}P_{3,2} & f_{3,1} = 1 & f_{3,2} = 3, \\ (5) = P_5 & f_5 = 4, & \\ (7) = P_{7,1}P_{7,2} & f_{7,1} = 1 & f_{7,2} = 3. \end{array}$$

If $P | D(A)$ then $P = P_{3,1}, P_{7,1}$ or $P_{13}$.

Since the number of the prime ideals dividing $D(A)$ is even, we have

$$D(A) = P_{3,1}P_{7,1}.$$

Let $L = \mathbb{Q}(\sqrt{-1})$ and $K = Lk$. Since $f_3 = 2$, $f_7 = 2$ in $L/\mathbb{Q}$, we see that both of $P_{3,1}, P_{7,1}$ remain prime in $K/k$. By Proposition 2.2 we see that $\Gamma(O)$ contains an element of order 4.

**4.3.** The cases $d(k) = 2000, 2304$. First consider the case $d(k) = 2000$. By Table 2 we have

$$N(A) = 12.$$

We have the following table of decomposition:

$$\begin{array}{lll}
(2) = P_2^2 & f_2 = 2 & e_2 = 2, \\
(3) = P_3 & f_3 = 4, & \\
(5) = P_5^4 & f_5 = 1 & e_5 = 4, \\
(7) = P_7 & f_7 = 4, & \\
(11) = P_{11,1}P_{11,2} & f_{11,1} = f_{11,2} = 2, & \\
(13) = P_{13} & f_{13} = 4. &
\end{array}$$

If $P|D(A)$ then we have $n_k(P) = 4$ or 5. Hence we have

$$D(A) = P_2 P_5 .$$

We show that this case is a solution of our problem. Assume that $\Gamma(O)$ contains an element $w$ or order $m \geqq 3$. Let $L = \mathbb{Q}(e^{2\pi i/m})$ and $K = Lk$. Since $[K:\mathbb{Q}] = 8$, we have $\phi(m)|8$. Hence we have

$$m = 3,4,5,6,8,10,12 \quad \text{or} \quad 20.$$

It suffices to show that we get a contradiction for $m = 3,4$, or 5. Assume that $m = 3$. Then we have $L = \mathbb{Q}(\sqrt{-3})$. Since $d(L) = -3$, we see that $(d(k),\ d(L)) = 1$. By Theorem 3.1 we have $n_k(d(K/k)) = 3^4$ and $O_K = O_L O_k$. Hence $P_2$ is unramified in $K/k$. Moreover, since $f_2 = 2$ in $L/\mathbb{Q}$ and $f_{p_2} = 2$ in $k/\mathbb{Q}$ and $O_K = O_L O_k$, we see that $f_{p_2} = 2$ in $K/\mathbb{Q}$. Hence $f_{p_2} = 1$ in $K/k$. By Proposition 2.2 this is a contradiction. Assume that $m = 4$, 5. Then we see easily that $K = \mathbb{Q}(e^{2\pi i/20})$. Since $d(K) = 2^8 * 5^6 = d(k)^2$, we see that $K/k$ is unramified. Since $f_5 = 1$ in $K/\mathbb{Q}$, we see that $f_{p_5} = 1$ in $K/k$. Hence $P_5$ splits in $K/k$. By Proposition 2.2 we get a contradiction. This shows that this case is a solution of our problem.

For the case $d(k) = 2304$ by the same argument as above we can show that $D(A) = P_2P_3$ is a solution of our problem.

**4.4.** The cases $d(k) = 2225, 2525, 2624, 2777$. First consider the case $d(k) = 2225$. We have

$$N(A) < (2\pi)^8 d(k)^{-3/2}/2 = 11.572\ldots .$$

On the other hand, we have

$$(2) = P_{2,1}P_{2,2} \qquad f_{2,1} = f_{2,2} = 2.$$

Since $k \supset \mathbb{Q}(\sqrt{5})$, we have

$$f_3 \geqq 2.$$

Hence we have

$$N(A) \geqq (2^2-1)(2^2-1) = 9.$$

Thus we have

$$9 \leqq N(A) \leqq 11.$$

If $P | D(A)$ then $n_k(P) = 4, 11$. Hence $P = P_{2,1}, P_{2,2}, P_{11}$. Therefore, we have

$$D(A) = P_{2,1}P_{2,2}\,.$$

Let $L = \mathbb{Q}(\sqrt{-1})$ and $K = Lk$. Since $(2)$ is ramified in $L/\mathbb{Q}$ and both of $P_{2,1}, P_{2,2}$ are unramified in $k/\mathbb{Q}$, both of $P_{2,1}, P_{2,2}$ are ramified in $K/k$. By Proposition 2.2 we see that $\Gamma(O)$ contains an element of order 4.

By the same argument we can exclude the cases $d(k) = 2525, 2624, 2777$.

**4.5.** The cases $d(k) > 2777$. In these cases by Table 2 we have

$$d(k) \geqq 3600.$$

Hence we have

$$N(A) \leqq 5.$$

If $P|D(A)$, then we have

$$n_k(P) = 2,3,4,5.$$

Consider first the case where $f_{p_2} \geqq 2$ for all prime ideals $P_2$ in $k$ such that $P_2|(2)$. Hence we have

$$D(A) = P_{3,1}P_{3,2} \qquad f_{3,1} = f_{3,2} = 1.$$

Let $L = \mathbf{Q}(\sqrt{-1})$ and $K = Lk$. Since $f_3 = 2$ in $L/\mathbf{Q}$ and $f_{3,1} = f_{3,2} = 1$ in $k/\mathbf{Q}$, we see that both of $P_{3,1}, P_{3,2}$ remain prime in $K/k$. By Proposition 2.2 we see that $\Gamma(O)$ contains an element of order 4. Therefore, we may assume that there exists an ideal $P_2$ of $k$ such that $P_2|(2)$ and $f_{p_2} = 1$. Then by the above list we have the following cases:

$$\begin{aligned} d(k) &= 4352 & N(A) &\leqq 4, \\ d(k) &= 6809 & N(A) &\leqq 2, \\ d(k) &= 7168 & N(A) &\leqq 2. \end{aligned}$$

For each $k$ we calculate the decomposition of $(p)$ in $k$ for $p = 2,3,5$. It implies that

$$\begin{aligned} N(A) &\geqq 6 \quad \text{for} \quad d(k) = 4352, \\ N(A) &\geqq 4 \quad \text{for} \quad d(k) = 6809,\ 7168. \end{aligned}$$

This is a contradiction. We finish the case $n = 4$.

## 5. The case $n = 5$

5.0. Let us treat the case $n = 5$. Since $n = 5$ is odd, the number of the prime ideals dividing $D(A)$ is odd. We have

$$N(A) = 2^9\pi^{10}d(k)^{-3/2}\zeta_k(2)^{-1}. \tag{9}$$

It implies that

$$d(k) < 131981. \tag{10}$$

The list of such fields $k$ is given in Pohst–Weiler–Zassenhaus [10]. However, we need only a part of the list, the first five fields. This is as follows:

| $d(k)$ | $f(x)$ |
|---|---|
| 14641 | $x^5 + x^4 - 4x^3 - 3x^2 + 3x + 1$ |
| 24217 | $x^5 - 5x^3 - x^2 + 5x + 1$ |
| 36497 | $x^5 - 6x^3 - x^2 + 4x - 1$ |
| 38569 | $x^5 - 5x^3 + 4x - 1$ |
| 65657 | $x^5 - 7x^3 - 4x^2 + 8x + 5$ |

**Table 3**

where $f(x)$ is a defining equation for $k$ over $\mathbb{Q}$. The fields listed above are uniquely determined by $d(k)$. For each $k$ we have $d(k) = d(f)$ (= the discriminant of $f(x)$). Hence for an element $\alpha$ of $k$ such that $f(\alpha) = 0$ we see that $O_k = \mathbb{Z}[\alpha]$.

**5.1.** The case $d(k) = 14641 = 11^5$. We see that $k = \mathbb{Q}(2\cos(2\pi/11))$. It is known that (cf. [15])

$$\zeta_k(2) = 2^7 5 \pi^{10} d(k)^{-3/2}/33.$$

Hence we have

$$N(A) = 132/5.$$

This is not an integer, which is a contradiction.

**5.2.** The case $d(k) = 24217 = 61*397$. By Cartier–Roy [1] we have

$$\xi_k(2) = (2\pi)^{10} d(k)^{-3/2}/24.$$

Hence we have

$$N(A) = 12.$$

Since we have

$$f_2 = 5,$$
$$f_p \geqq 2 \quad \text{for} \quad p = 3,7,13,$$

we have

$$D(A) = P_5 \qquad n_k(P_5) = 5.$$

It implies that

$$N(A) = 4.$$

This is a contradiction.

**5.3.** The case $d(k) = 36497$. In this case we have

$$N(A) \leqq 6.$$

Since we see that

$$f_2 = 5 \quad \text{and} \quad f_p \geqq 2 \quad \text{for} \quad p = 5,7,$$

we have

$$D(A) = P_3 \qquad n_k(P_3) = 3.$$

Let $L = \mathbb{Q}(\sqrt{-1})$ and $K = Lk$. Since $f_3 = 2$ in $L/\mathbb{Q}$ and $f_{P_3} = 1$ in $k/\mathbb{Q}$, we see that $P_3$ remains prime in $K/k$. Hence we see that $\Gamma(O)$ contains an element of order 4. This is a contradiction.

**5.4.** The case $d(k) = 38569$. In this case we have

$$N(A) \leqq 6.$$

Since we see that

$$f_2 = 5 \quad \text{and} \quad f_3 \geqq 2,$$

we have

$$D(A) = P_5 \quad \text{or} \quad P_7 \quad (f_5 = 1,\ f_7 = 1).$$

If $D(A) = P_5$ (resp. $P_7$), put $L = \mathbb{Q}(\sqrt{-3})$ (resp. $\mathbb{Q}(\sqrt{-1})$). Then we see that $f_5$ (resp. $f_7$) $= 2$ in $L/\mathbb{Q}$. Hence we see that $\Gamma(O)$ contains an element of order 3 (resp. 4).

**5.5.** The cases $d(k) > 38569$. In these cases by the above table we see that

$$d(k) \geqq 65657 \tag{11}$$

Hence by (9) we have

$$N(A) = 2.$$

It implies that

$$D(A) = P_{2,1} \ldots P_{2,r} P_3 \quad (r: \text{even}), \; n_k(P_{2,i}) = 2 \quad (1 \leqq i \leqq r) \quad n_k(P_3) = 3.$$

If $r \geqq 2$, then we have

$$\zeta_k(2) > (4/3)(4/3)(9/8) = 2.$$

It follows that

$$d(k) < (2\pi)^{20/3}/4 = 52376.82\ldots$$

This contradicts to (11). Hence we have

$$D(A) = P_3 \qquad n_k(P_3) = 3.$$

Let $L = \mathbb{Q}(\sqrt{-1})$ and $K = Lk$. Since $f_3 = 2$ in $L/\mathbb{Q}$ and $f_{p_3} = 1$ in $k/\mathbb{Q}$, we see that $P_3$ remains prime in $K/k$. Hence we see that $\Gamma(O)$ contains an element of order 4. This completes the case $n = 5$. This shows that there exist no solutions in the case $n = 5$.

## 6. The case $n = 6$

**6.0.** Let us treat the case $n = 6$. In this case we see that the number of the prime ideals dividing $D(A)$ is even. By Pohst [8] it is known that

$$d(k) \geqq 300125. \tag{12}$$

Hence we have

$$N(A) \leqq 11.$$

**6.1.** The case $N(A) = 2$. We have

$$D(A) = P_{2,1} \cdots P_{2,r} P_3 \quad (r:\text{odd}) \quad n_k(P_{2,i}) = 2 \quad (1 \leqq i \leqq r) \quad n_k(P_3) = 3.$$

Since

$$\zeta_k(2) > (4/3)*(9/8) = 3/2,$$

we have

$$d(k) \leqq 735650. \tag{13}$$

The list of all such fields is as follows: (Pohst–Weiler–Zassenhaus [10])

| $d(k)$ | $f(x)$ | subfield |
|---|---|---|
| 300125 | $x^6+2x^5-7x^4-2x^3+7x^2+x-1$ | $k = \mathbb{Q}(\cos(2\pi/7),\sqrt{5})$ |
| 371293 | $x^6+x^5-5x^4-4x^3+6x^2+3x-1$ | $k = \mathbb{Q}(\cos(2\pi/13))$ |
| 434581 | $x^6+x^5-8x^4+2x^3+9x^2+3x-1$ | $\mathbb{Q}(\cos(2\pi/7))$ |
| 453789 | $x^6+x^5-6x^4-6x^3+8x^2+8x+1$ | $k = \mathbb{Q}(\cos(2\pi/21))$ |
| 485125 | $x^6-7x^4-x^3+6x^2+x-1$ | $\mathbb{Q}(\sqrt{5})$ |
| 592661 | $x^6+x^5-5x^4-4x^3+5x^2+2x-1$ | |
| 703493 | $x^6+x^5-7x^4-2x^3+14x^2-5x-1$ | $\mathbb{Q}(\cos(2\pi/7))$ |
| 722000 | $x^6+x^5-6x^4-7x^3+4x^2+5x+1$ | $\mathbb{Q}(\sqrt{5})$ |

**Table 4**

It is known that the field $k$ listed above is uniquely determined by $d(k)$. We have $d(k) = d(f)$ (= the discriminant of $f(x)$). Hence let $\alpha$ be an element of $k$ such that $f(\alpha) = 0$. Then we have $O_k = \mathbb{Z}[\alpha]$.

**Remark:** Using the defining equation $f(x)$ for $k$ we see that

$$(2) = P_2, \quad f_2 = 6 \quad \text{for all } k \text{ except for } k \text{ with } d(k) = 722000.$$

$$(2) = P_2^3 \quad f_2 = 2 \quad e_2 = 3 \quad \text{for } k \text{ with } d(k) = 722000.$$

It follows that $n_k(P) > 2$ for any prime ideal $P$. This is a contradiction.

**6.2.** The case $N(A) = 3$. We have

$$D(A) = P_{2,0}P_{2,1}\dots P_{2,r}\ (r: \text{odd})\ n_k(P_{2,0}) = 2^2\quad n_k(P_{2,i}) = 2\quad (1 \leqq i \leqq r).$$

Hence we have

$$\zeta_k(2) > 4/3.$$

Therefore, we see that

$$d(k) < (2\pi)^8/4 = 607265.984\dots .$$

Hence $k$ is one of the fields in Table 4. By the Remark we get a contradiction.

**6.3.** The cases $4 \leqq N(A) \leqq 11$. We have

$$d(k) \leqq 607265.$$

Hence $k$ is one of the fields in Table 4. By the Remark we have

$$(2) = P_2 \qquad f_2 = 6 \quad \text{in } k/\mathbb{Q}.$$

Using the fact that the number of the prime ideals dividing $D(A)$ is even, we have the following cases:

$$D(A) = P_{3,1}P_{3,2} \qquad n_k(P_{3,i}) = 3 \quad (1 \leqq i \leqq 2),$$

$$D(A) = P_3P_5 \qquad n_k(P_3) = 3 \quad n_k(P_5) = 5 \quad \text{for} \quad d(k) = 300125,\ 371293.$$

In the former case, let $L = \mathbb{Q}(\sqrt{-1})$ and $K = Lk$. Since $f_3 = 2$ in $L/\mathbb{Q}$ and $f_{P_{3,i}} = 1$ $(1 \leqq i \leqq 2)$ in $k/\mathbb{Q}$, we see that $P_{3,i}$ $(1 \leqq i \leqq 2)$ remains prime in $K/k$. Hence we see that $\Gamma(O)$ contains an element of order 4.

In the latter case we see easily that

$$f_3 \geqq 3 \quad \text{in} \quad k/\mathbb{Q} \quad \text{for} \quad d(k) = 300125,\ 371293.$$

This is a contradiction. This completes the case $n = 6$. We have no solutions in the case $n = 6$.

## 7. The case $n = 7$

7.0. Let us treat the case $n = 7$. We need the following

**Theorem 7.1** (Pohst [9]). *The minimum of the discriminant $d(k)$ of the totally real algebraic number fields $k$ of degree 7 is equal to* 20134393.

It follows from (2) and Theorem 7.1 that

$$\zeta_k(2)N(A) \leqq 2^{-1}(2\pi)^{14}(30134393)^{-3/2} = 0.82714... < 1.$$

This is a contradiction. We have no solutions in the case $n = 7$.

## 8. The case $n = 8$

8.0. Finally, we treat the case $n = 8$. since $n = 8$ is even, the number of the prime ideals $P$ dividing $D(A)$ is even. By (2), (3) we have

$$\zeta_k(2)\ N(A) < 2^{-1}(2\pi)^{16}\ e^{12.47775}(29.099)^{-12} = 2.100531... \tag{14}$$

Hence we have

$$N(A) = 2.$$

It implies that

$$D(A) = P_1P_2...P_r\ (r: \text{even})\quad n_k(P_1) = 3\quad n_k(P_i) = 2 \qquad (2 \leqq i \leqq r).$$

It follows that

$$\zeta_k(2) > (4/3)^*(9/8) = 3/2.$$

Hence we have

$$\zeta_k(2)N(A) > 3.$$

This contradicts to (14). This finishes the case $n = 8$. This completes the proof of our Theorem 1.1.

## 9. Some comments

**9.0.** In the Hilbert Modular cases there are many surfaces for which the defining equations are given explicitly (cf. Hirzebruch [3]).

In our cases Kuga [4] made some considerations how to obtain the defining equations and gave a conjecture. However, we do not know any examples with explicit defining equations. This is still an open problem.

## References

[1] *P. Cartier–Y. Roy*, Certains calculs numériques relatifs à l'interpolation *p*–adique des séries de Dirichlet, Modular functions of one variable III. Proceedings International Summer School, University of Antwerp, RUCA (1972), 269–349, L.N.M. **350**, Springer.

[2] *B.N. Delone* and *D.K. Faddeev*, The theory of irrationalities of the third degree. Trans. Math. Mono. **10**, Amer. Math. Soc. 1964.

[3] *F. Hirzebruch*, Hilbert modular surfaces. L'Ens. Math. **71** (1973), 183–291.

[4] *M. Kuga*, Sugaku, **32** (1980), 182–184, Math. Soc. of Japan (in Japanese).

[5] *A. Odlyzko*, Unconditional bounds for discriminants (A table of numerical lower bounds for the discriminants of totally real algebraic number fields), preprint (1976).

[6] *T. Otsubo*, Algebraic surfaces derived from quaternion algebras over real quadratic fields. Saitama Math. J. **3** (1985), 1–10.

[7] *M. Pohst*, Mehrklassige Geschlechter von Einheitsformen in total reellen algebraischen Zahlkörpern. J. fur die reine und angewandte Mathematik **262/263** (1973), 420–435.

[8] *M. Pohst*, Berechnung kleiner Diskriminanten total reeler algebraischer zahrkörper. J. fur die reine und angewandte Mathematik **278/279** (1975), 278–300.

[9] *M. Pohst*, The minimum discriminant of seventh degree totally real algebraic number field. Number theory and algebra collected papers dedicated to H.B. Manin (1977), 235–240.

[10] *M. Pohst*, *P. Weiler* and *H. Zassenhaus*, On effective computation of fundamental units II. Math. Comp. **38** (1982), 293–329.

[11] *I. Shaval*, A class of algebraic surfaces of general type constructed from quaternion algebras. Pacific J. of Math. **76** (1978), 221–245.

[12] *H. Shimizu*, On zeta functions of quaternion algebras. Ann. of Math. **81** (1965), 166–193.

[13] *G. Shimura*, Construction of class fields and zeta functions of algebraic curves. Ann. of Math. **85** (1967), 58–159.

[14] *C.L. Siegel*, Berechnung von Zetafunktionen an ganzzahligen Stellen. Nachr. Akad. Wiss. Göttingen, Math.–Phys. K1. (1969), 87–102.

[15] *K. Takeuchi*, Commensurability classes of arithmetic triangle groups. J. Fac. Sci. Univ. Tokyo Sect. I, **24** (1977), 201–212.

[16] *K. Takeuchi*, Totally real algebraic number fields of degree 5 and 6 with small discriminant. Saitama Math. J. **2** (1984), 21–32.

---

Department of Mathematics, Faculty of Science,
Saitama University, Urawa, Saitama, Japan 338

# Explicit evaluation of certain Eisenstein sums

*Kenneth S. Williams*[1], *Kenneth Hardy*[2] *and Blair K. Spearman*
*with the assistance of Nicholas Buck and Iain deMille*

## 0. Notation

The following notation will be used throughout this paper: $p$ is an odd prime, $m$ is an integer $\geq 2$ coprime with $p$, and $f$ is a positive integer such that $p^f - 1$ is divisible by $m$.

## 1. Introduction

The finite field with $q = p^f$ elements is denoted by $F_q$. The prime subfield of $F_q$ is $F_p = \{0,1,2,\ldots,p-1\}$. The trace of an element $\alpha \in F_q$ is defined by

$$tr(\alpha) = \alpha + \alpha^p + \alpha^{p^2} + \ldots + \alpha^{p^{f-1}} \in F_p . \tag{1.1}$$

The trace function has the following properties:

$$tr(\alpha+\beta) = tr(\alpha) + tr(\beta), \quad \forall\, \alpha, \beta \in F_q , \tag{1.2}$$

$$tr(k\alpha) = k\,tr(\alpha), \quad \forall\, \alpha \in F_q,\ k = 0,1,2,\ldots, \tag{1.3}$$

$$tr(\alpha^p) = tr(\alpha), \quad \forall\, \alpha \in F_q , \tag{1.4}$$

$$tr(k) = kf, \quad k = 0,1,2,\ldots . \tag{1.5}$$

---

[1]Research supported by Natural Sciences and Engineering Research Council of Canada Grant A–7233.

[2]Research supported by Natural Sciences and Engineering Research Council of Canada Grant A–7823.

We also set $F_q^* = F_q - \{0\}$. With respect to multiplication, $F_q^*$ is a cyclic group of order $q-1$. We fix once and for all a generator $\gamma$ of $F_q^*$. For $\alpha \in F_q^*$. For $\alpha \in F_q^*$ the unique integer $r$ such that $\alpha = \gamma^r$, where $0 \leq r \leq q-2$, is called the *index* of $\alpha$ with respect to $\gamma$ and is denoted by $\mathrm{ind}_\gamma(\alpha)$. We have $\mathrm{ind}_\gamma(-1) = (q-1)/2$. Since $\gamma$ generates $F_q^*$,

$$g = \gamma^{(q-1)/(p-1)} \in F_p \tag{1.6}$$

generates $F_q^*$. For $k \in F_q^*$ the unique integer $s$ such that $k = g^s$, where $0 \leq s \leq p-2$, is called the *index* of $k$ with respect to $g$ and is denoted by $\mathrm{ind}_g(k)$. The relationship between $\mathrm{ind}_\gamma(k)$ and $\mathrm{ind}_g(k)$ is given by

$$\mathrm{ind}_\gamma(k) \equiv \frac{(q-1)}{(p-1)}\,\mathrm{ind}_g(k) \pmod{q-1}. \tag{1.7}$$

For $m$ a positive integer ($\geq 2$) dividing $q-1$ and any integer $n$, the Eisenstein sum $E_q(w_m^n)$ is defined by

$$E_q(w_m^n) = \sum_{\substack{\alpha \in F_q^* \\ tr(\alpha)=1}} w_m^{n\,\mathrm{ind}_\gamma(\alpha)}, \tag{1.8}$$

where $w_m = \exp(2\pi i/m)$. Clearly $E_q(w_m^n)$ is an integer of the cyclotomic field $Q(w_m)$. We remark that if $n \equiv n' \pmod m$ then

$$E_q(w_m^n) = E_q(w_m^{n'}). \tag{1.9}$$

In particular, if $n \equiv 0 \pmod m$, we have

$$E_q(w_m^n) = E_q(1) = \sum_{\substack{\alpha \in F_q^* \\ tr(\alpha)=1}} 1 = p^{f-1}.$$

We also note that if GCD $(m,n) = d > 1$, say $m = m_1 d$, $n = n_1 d$, where GCD $(m_1,n_1) = 1$, then $m_1 | q-1$ and

$$E_q(w_m^n) = E_q(w_{m_1}^{n_1}). \tag{1.10}$$

Thus it suffices to consider only those $E_q(w_m^n)$ for which $1 \leq n < m$, GCD $(n,m) = 1$. Further, if $\sigma_n$ is the automorphism of $Q(w_m)$ such that $\sigma_n(w_m) = w_m^n$ (GCD $(n,m) = 1$), then $\sigma_n(E_q(w_m)) = E_q(w_m^n)$, and we can further restrict our attention to $E_q(w_m)$.

It is the purpose of this paper to evaluate explicitly the Eisenstein sums $E_q(w_m)$ for $m = 2,3,...,8$. The evaluation of $E_q(w_m)$ for $m = 2,3,4,5,6,7,8$ is given in Theorem 1, 2, 3, 4, 5, 6, 7, respectively. Examples are given in Tables 1–29 (§11). It is planned to treat additional values of $m$ in another paper, as well as to apply the results of this paper to the determination of cyclotomic numbers over $F_q$ and the determination of binomial coefficients modulo $p$.

These evaluations are accomplished by using the basic facts about Eisenstein sums established by Stickelberger [19] together with the theory of Gauss sums, including the important results on Gauss sums established by Davenport and Hasse in [4]. Our results include and extend those of Berndt and Evans in [2] in the case $f = 2$.

We make use of the Gauss sums $G_q(w_m^n)$, $g_q(w_m^n)$, and $g_p(w_m^n)$ defined for any integer $n$ by

$$G_q(w_m^n) = \sum_{\alpha \in F_q^*} w_m^{n\,\mathrm{ind}_\gamma(\alpha)} \exp(2\pi i\, tr(\alpha)/p), \tag{1.11}$$

$$g_q(w_m^n) = \sum_{k \in F_p^*} w_m^{n\,\mathrm{ind}_\gamma(k)} \exp(2\pi i k/p), \tag{1.12}$$

$$g_p(w_m^n) = \sum_{k \in F_p^*} w_m^{n\,\mathrm{ind}_g(k)} \exp(2\pi i k/p), \quad \text{provided } p \equiv 1 \pmod m, \tag{1.13}$$

as well as the Jacobi sum $J_p(w_m^r, w_m^s)$ defined for any integers $r$ and $s$ by

$$J_p(w_m^r, w_m^s) = \sum_{k=2}^{p-1} w_m^{r\,\mathrm{ind}_g(k) + s\,\mathrm{ind}_g(1-k)}, \quad \text{provided } p \equiv 1 \pmod m. \tag{1.14}$$

It is well-known (see for example [17: Chapter 5]) that

$$G_q(w_m^n)G_q(w_m^{-n}) = w_m^{n(q-1)/2}q, \quad \text{if } m \nmid n, \tag{1.15}$$

$$g_q(w_m^n)g_q(w_m^{-n}) = w_m^{n(q-1)/2}p, \quad \text{if } m \nmid n\left[\frac{q-1}{p-1}\right], \tag{1.16}$$

and if $p \equiv 1 \pmod{m}$

$$g_p(w_m^n)g_p(w_m^{-n}) = w_m^{n(p-1)/2}p, \quad \text{if } m \nmid n, \tag{1.17}$$

$$J_p(w_m^r, w_m^s) = \frac{g_p(w_m^r)g_p(w_m^s)}{g_p(w_m^{r+s})}, \quad m \nmid r,\ m \nmid s,\ m \nmid r+s. \tag{1.18}$$

We close this section by emphasizing that the Eisenstein sum $E_q(w_m^n)$ depends upon the generator $\gamma$ as well as upon $m$, $n$ and $q$. On the few occasions when we wish to indicate this dependence, we write $E_q(w_m^n, \gamma)$ for $E_q(w_m^n)$. If $\gamma'$ is another generator of $F_q^*$ we have

$$E_q(w_m^n, \gamma) = E_q\left(w_m^{n \operatorname{ind}_\gamma(\gamma')}, \gamma'\right), \tag{1.19}$$

as $\operatorname{ind}_\gamma(\alpha) \equiv \operatorname{ind}_g(\gamma') \operatorname{ind}_{\gamma'}(\alpha) \pmod{q-1}$, and so

$$E_q(w_m^n, \gamma) = E_q(w_m^n, \gamma'), \quad \text{if } \operatorname{ind}_\gamma(\gamma') \equiv 1 \pmod{m/\operatorname{GCD}(m,n)}. \tag{1.20}$$

## 2. Eisenstein sums

The following basic results concerning Eisenstein sums are implicit in the work of Stickelberger [19].

**Theorem A.** (Stickelberger)

(a) [19: p. 338] $E_q(w_m^n) = E_q(w_m^{np})$.

(b) [19: p. 339] $E_q(w_m^n) = \begin{cases} G_q(w_m^n)/g_q(w_m^n), & \textit{if} \quad m \nmid n\left[\frac{q-1}{p-1}\right], \\ -G_q(w_m^n)/p, & \textit{if } m \mid n\left[\frac{q-1}{p-1}\right],\ m \nmid n. \end{cases}$

(c) [19: p. 339] $$E_q(w_m^n)E_q(w_m^{-n}) = \begin{cases} p^{f-1}, & \text{if } m \nmid n\left[\frac{q-1}{p-1}\right], \\ (w_m^{n(q-1)/2}p^{f-2}, & \text{if } m \mid n\left[\frac{q-1}{p-1}\right],\ m \nmid n. \end{cases}$$

(d) [19: p. 361] *For* $i = 0,1,2,\dots, f-1$ *set*

$$a_i = \text{least positive residue of } p^i \ (\text{mod } m). \tag{2.1}$$

*Let*

$$A_0 = \text{least nonnegative residue of } \frac{q-1}{p-1} \ (\text{mod } m). \tag{2.2}$$

*Define the integer* $B_0$ *by*

$$B_0 = \left(\sum_{i=0}^{f-1} a_i - A_0\right) / m. \tag{2.3}$$

*Then, for some prime ideal* $\mathcal{P}$ *of* $Q(w_m)$ *dividing* $p$, *we have*

$$E_q(w_m) \equiv (-1)^{B_0} p^{f-1-B_0} \frac{\left[\frac{pa_0}{m}\right]! \dots \left[\frac{pa_{f-1}}{m}\right]!}{\left[\frac{pA_0}{m}\right]} \ (\text{mod } \mathcal{P}^{f-B_0}). \tag{2.4}$$

Next we relate $E_q(w_m^n)$ to $E_{p^\ell}(w_m^n)$, where $\ell$ is the least positive integer such that $m$ divides $p^\ell - 1$, so that $\ell$ is a divisor of $f$. The sum $E_{p^\ell}(w_m^n)$ is taken with respect to the generator $\gamma' = \gamma^{(q-1)/(p^\ell-1)}$. We prove

**Theorem B.** *Let* $\ell$ *denote the least positive integer such that* $p^\ell - 1$ *is divisible by* $m$, *so that* $\ell$ *is a divisor of* $f$. *Then*

$$E_q(w_m^n) = \begin{cases} (-1)^{\frac{f}{\ell}-1} \dfrac{g_{p^\ell}(w_m^n)^{f/\ell}}{g_{p^\ell}\left[w_m^{n\left[\frac{p^f-1}{p^\ell-1}\right]}\right]} (E_{p^\ell}(w_m^n))^{f/\ell}, & \text{if } m \nmid n\left[\frac{p^f-1}{p-1}\right], \\ (-1)^{f/\ell} \dfrac{(g_{p^\ell}(w_m^n))^{f/\ell}}{p} (E_{p^\ell}(w_m^n))^{f/\ell}, & \text{if } m \nmid n\left[\frac{p^\ell-1}{p-1}\right], m \mid n\left[\frac{p^f-1}{p-1}\right], \\ p^{\frac{f}{\ell}-1} (E_{p^\ell}(w_m^n))^{f/\ell}, & \text{if } m \mid n\left[\frac{p^\ell-1}{p-1}\right],\ m \nmid n. \end{cases}$$

**Proof.** From the Davenport–Hasse theorem [4: p. 153] (see also [17: p. 197]), we have

$$G_q(w_m^n) = (-1)^{\frac{f}{\ell}-1} (G_{p^\ell}(w_m^n))^{f/\ell}. \tag{2.5}$$

Also, as

$$\mathrm{ind}_\gamma(k) \equiv \left(\frac{q-1}{p^\ell-1}\right) \mathrm{ind}_{\gamma'}(k) \pmod{q-1},$$

where $\gamma' = \gamma^{(q-1)/(p^\ell-1)}$, we have appealing to (1.12)

$$g_q(w_m^n) = g_{p^\ell}(w_m^{n\left[\frac{p^f-1}{p^\ell-1}\right]}). \tag{2.6}$$

(a): $m \nmid n \left[\frac{p^f-1}{p-1}\right]$. We have

$$E_q(w_m^n) = \frac{G_q(w_m^n)}{g_q(w_m^n)} \quad \text{(by Theorem A(b))}$$

$$= \frac{(-1)^{f/\ell-1}(G_{p^\ell}(w_m^n))^{f/\ell}}{g_{p^\ell}\left(w_m^{n(q-1)/(p^\ell-1)}\right)} \quad \text{(by (2.5) and (2.6))}$$

$$= (-1)^{f/\ell-1} \frac{g_{p^\ell}(w_m^n)^{f/\ell}}{g_{p^\ell}\left(w_m^{n(q-1)/(p^\ell-1)}\right)} (E_{p^\ell}(w_m^n))^{f/\ell},$$

by Theorem A(b).

(b): $m \mid n\left(\frac{q-1}{p-1}\right)$, $m \nmid n\left(\frac{p^\ell-1}{p-1}\right)$. We have

$$E_q(w_m^n) = -\frac{G_q(w_m^n)}{p} \quad \text{(by Theorem A(b))}$$

$$= (-1)^{f/\ell}\left(G_{p^\ell}(w_m^n)\right)^{f/\ell}/p \quad \text{(by (2.5))}$$

$$= (-1)^{f/\ell} \frac{\left(g_{p^\ell}(w_m^n)\right)^{f/\ell}}{p} \left(E_{p^\ell}(w_m^n)\right)^{f/\ell},$$

by Theorem A(b).

(c): $m \mid n\left(\frac{p^\ell-1}{p-1}\right)$, $m \nmid n$ (so that $m \mid n\left(\frac{q-1}{p-1}\right)$). We have

$$\begin{aligned} E_q(w_m^n) &= -\frac{G_q(w_m^n)}{p} \quad \text{(by Theorem A(b))} \\ &= \frac{(-1)^{f/\ell}\left(G_{p^\ell}(w_m^n)\right)^{f/\ell}}{p} \quad \text{(by (2.5))} \\ &= \frac{(-1)^{f/\ell}\left(-pE_{p^\ell}(w_m^n)\right)^{f/\ell}}{p} \quad \text{(by Theorem A(b))} \\ &= p^{f/\ell-1}\left(E_{p^\ell}(w_m^n)\right)^{f/\ell}. \end{aligned}$$

The special case of Theorem B when $p \equiv 1 \pmod{m}$, so that $\ell = 1$, gives the following corollary.

**Corollary 1.** *If $p \equiv 1 \pmod{m}$ then*

$$E_q(w_m) = \begin{cases} (-1)^{f-1} \dfrac{g_p(w_m)^f}{g_p(w_m^f)}, & \text{if } m \nmid f, \\ (-1)^f \dfrac{(g_p(w_m))^f}{p}, & \text{if } m \mid f. \end{cases}$$

**Proof.** As $p \equiv 1 \pmod{m}$ we have $\ell = 1$. By Theorem B with $n = 1$ we obtain

$$E_q(w_m) = \begin{cases} (-1)^{f-1} \dfrac{g_p(w_m)^f}{g_p\left(w_m^{\left(\frac{p^f-1}{p-1}\right)}\right)} E_p(w_m))^f, & \text{if } m \nmid \dfrac{p^f-1}{p-1}, \\ (-1)^f \dfrac{(g_p(w_m))^f}{p} (E_p(w_m))^f, & \text{if } m \mid \dfrac{p^f-1}{p-1}. \end{cases}$$

The required result now follows as

$$E_q(w_m) = 1, \quad \frac{p^f-1}{p-1} = p^{f-1} + \dots + p+1 \equiv f \pmod m,$$

and $\gamma^{\frac{p^f-1}{p^\ell-1}} = \gamma^{\frac{p^f-1}{p-1}} = g.$

The next theorem gives the value of $E_q(w_m)$ when there is an integer $r$ such that $p^r \equiv -1 \pmod m$.

**Theorem C.** *Let $p$ be a prime for which there is an integer $r$ such that $p^r \equiv -1 \pmod m$. Let $\ell$ be the least positive integer such that*

$$p^\ell \equiv -1 \pmod m.$$

*Then, for $f \equiv 0 \pmod{2\ell}$, we have*

$$E_q(w_m) = (-1)^{\frac{f}{2\ell}\left(\frac{p^\ell-(m-1)}{m}\right)} p^{f/2-1}.$$

Proof. By Theorem A(a) we have

$$E_q(w_m) = E_q(w_m^p) = E_q(w_m^{p^2}) = \dots = E_q(w_m^{p^\ell}),$$

that is

$$E_q(w_m) = E_q(w_m^{m-1}), \tag{2.7}$$

showing that $E_q(w_m)$ is real. Next set

$$\frac{f}{2\ell} = 2^r s, \quad r \geq 0, \quad s \text{ odd}, \tag{2.8}$$

so that

$$p^{f/2^{r+1}} = p^{\ell s} = \left(p^{\ell}\right)^{s} \equiv (-1)^{s} \equiv -1 \pmod{m}. \tag{2.9}$$

Then we have

$$\frac{q-1}{p-1} = \left(\frac{p^{f/2^{r+1}} - 1}{p-1}\right) \left(p^{f/2^{r+1}} + 1\right) \left(p^{f/2^{r}} + 1\right) \dots \left(p^{f/2} + 1\right) \equiv 0 \pmod{m},$$

and

$$(q-1)/2 = \frac{\left(p^{f/2^{r+1}} - 1\right)}{2} \left(p^{f/2^{r+1}} + 1\right)\left(p^{f/2^{r}} + 1\right) \dots \left(p^{f/2} + 1\right) \equiv 0 \pmod{m},$$

so, by Theorem A(c), we have

$$E_q(w_m)E_q(w_m^{m-1}) = p^{f-2}. \tag{2.10}$$

From (2.7) and (2.10), we deduce that

$$E_q(w_m) = \theta p^{f/2-1}, \tag{2.11}$$

where $\theta = \pm 1$.

Next, since $p^{\ell} \equiv -1 \pmod{m}$, for $k = 0,1,2,\dots$, we have (with the notation of (2.1))

$$a_k = \begin{cases} a_{k-\ell[k/\ell]}, & \text{if } [k/\ell] \equiv 0 \pmod{2}, \\ m - a_{k-\ell[k/\ell]}, & \text{if } [k/\ell] \equiv 1 \pmod{2}. \end{cases} \tag{2.12}$$

Thus we have

$$\left[\frac{pa_0}{m}\right]!\left[\frac{pa_1}{m}\right]!\dots\left[\frac{pa_{f-1}}{m}\right]!$$

$$= \prod_{i=1}^{2^{r+1}s} \prod_{j=0}^{\ell-1} \left[\frac{pa_{(i-1)\ell+j}}{m}\right]!$$

$$= \prod_{\substack{i=1 \\ i \text{ odd}}}^{2^{r+1}s} \prod_{j=0}^{\ell-1} \left[\frac{pa_j}{m}\right]! \prod_{\substack{i=1 \\ i \text{ even}}}^{2^{r+1}s} \prod_{j=0}^{\ell-1} \left[\frac{p(m-a_j)}{m}\right]!$$

$$= \left(\prod_{j=0}^{\ell-1} \left[\frac{pa_j}{m}\right]!\right)^{2^r s} \left(\prod_{j=0}^{\ell-1} \left[p - \frac{pa_j}{m}\right]!\right)^{2^r s}$$

$$= \left(\prod_{j=0}^{\ell-1} \left[\frac{pa_j}{m}\right]!\left[p - \frac{pa_j}{m}\right]!\right)^{f/2\ell}$$

$$= \left(\prod_{j=0}^{\ell-1} \left[\frac{pa_j}{m}\right]!\left(p - 1 - \left[\frac{pa_j}{m}\right]\right)!\right)^{f/2\ell}$$

$$\equiv \left(\prod_{j=0}^{\ell-1} (-1)^{\left[\frac{pa_j}{m}\right]+1}\right)^{f/2\ell} \pmod p$$

$$\equiv (-1)^{f/2\ell\left(\sum_{j=0}^{\ell-1} \left[\frac{pa_j}{m}\right]+\ell\right)} \pmod p,$$

that is

$$\prod_{k=0}^{f-1} \left[\frac{pa_k}{m}\right]! \equiv (-1)^{f/2\ell \sum_{j=0}^{\ell-1}\left[\frac{pa_j}{m}\right]+f/2} \pmod p. \tag{2.13}$$

Clearly we have

$$a_k = p^k - m[p^k/m] \quad (k = 0,1,2,\dots)$$

so that

$$\left[\frac{pa_k}{m}\right] = \left[\frac{p^{k+1}}{m} - p[p^k/m]\right]$$

$$= \left[\frac{p^{k+1}}{m}\right] - p\left[\frac{p^k}{m}\right]$$

$$\equiv \left[\frac{p^{k+1}}{m}\right] - \left[\frac{p^{k}}{m}\right] \pmod 2,$$

and thus

$$\sum_{k=0}^{\ell-1} \left[\frac{p a_k}{m}\right] \equiv \left[\frac{p^{\ell}}{m}\right] \pmod 2,$$

that is

$$\sum_{k=0}^{\ell-1} \left[\frac{p a_k}{m}\right] \equiv \frac{p^{\ell} - (m-1)}{m} \pmod 2. \tag{2.14}$$

From (2.13) and (2.14), we obtain

$$\prod_{k=0}^{f-1} \left[\frac{p a_k}{m}\right]! \equiv (-1)^{(f/2\ell)\left(\frac{p^{\ell}-(m-1)}{m}\right)+f/2} \pmod p. \tag{2.15}$$

Next, with the notation of (2.3), we have $A_0 = 0$ and

$$\begin{aligned} mB_0 &= \sum_{k=0}^{f-1} a_k \\ &= \sum_{i=1}^{2^{r+1}s} \sum_{j=0}^{\ell-1} a_{(i-1)\ell+j} \\ &= \sum_{\substack{i=1\\ i \text{ odd}}}^{2^{r+1}s} \sum_{j=0}^{\ell-1} a_j + \sum_{\substack{i=1\\ i \text{ even}}}^{2^{r+1}s} \sum_{j=0}^{\ell-1} (m-a_j) \\ &= 2^r s \sum_{j=0}^{\ell-1} a_j + 2^r s \sum_{j=0}^{\ell-1} (m-a_j) \\ &= 2^r sm\ell, \end{aligned}$$

so that

$$B_0 = 2^r s\ell = f/2. \tag{2.16}$$

Hence, by Stickelberger's congruence (2.4), we have

$$E_q(w_m) \equiv (-1)^{(f/2\ell)\left(\frac{p^\ell-(m-1)}{m}\right)} p^{f/2-1} \pmod{p^{f/2}}. \tag{2.17}$$

From (2.11) and (2.17), we deduce that

$$\theta = (-1)^{(f/2\ell)\left(\frac{p^\ell-(m-1)}{m}\right)},$$

so that

$$E_q(w_m) = (-1)^{(f/2\ell)\left(\frac{p^\ell-(m-1)}{m}\right)} p^{f/2-1}.$$

This completes the proof of Theorem C.

We conclude this section with two lemmas which we will need later.

**Lemma 1.** *For* $k = 0,1,2,\ldots,m-1$ *we have*

$$\sum_{\substack{\alpha \in F_q \\ tr(\alpha)=1}} \alpha^{k(q-1)/m} = 0, \quad \text{if } m \,\Big|\, \frac{q-1}{p-1}.$$

**Proof.** Let

$$F_q^o = \{\beta \in F_q \mid tr(\beta) = 0\}.$$

It is easily checked that $F_q^o$ is a $(f-1)$-dimensional subspace of the vector space $F_q$ over $F_p$. We let $\beta_1,\ldots,\beta_{f-1}$ be a basis for $F_q^o$ over $F_p$, and let $\alpha_1$ be a fixed element of $F_q$ with $tr(\alpha_1) = 1$. Let $\alpha$ be any element of $F_q$ with $tr(\alpha) = 1$. Then we have $tr(\alpha-\alpha_1) = 0$, and so $\alpha - \alpha_1 \in F_q^o$, and thus $\alpha = \alpha_1 + \beta$, for some $\beta \in F_q^o$. Hence every element of $F_q$ having trace 1 is given uniquely by

$$\alpha = \alpha_1 + b_1\beta_1 + \ldots + b_{f-1}\beta_{f-1},$$

where $b_1,b_2,\ldots,b_{f-1} \in F_p$. Then we have, for $k = 0,1,2,\ldots,m-1$, by the Multinomial theorem,

$$\sum_{\substack{\alpha \in F_q \\ tr(\alpha)=1}} \alpha^{k(q-1)/m}$$

$$= \sum_{b_1,\dots,b_{f-1} \in F_p} (\alpha_1 + b_1\beta_1 + \dots + b_{f-1}\beta_{f-1})^{k(q-1)/m}$$

$$= \sum_{b_1,\dots,b_{f-1} \in F_p} \sum_{n_0+\dots+n_{f-1}=k(q-1)/m} \binom{k(q-1)/m}{n_0, n_1, \dots, n_{f-1}} \alpha_1^{n_0} (b_1\beta_1)^{n_1} \cdots (b_{f-1}\beta_{f-1})^{n_{f-1}}$$

$$= \sum_{n_0+\dots+n_{f-1}=k(q-1)/m} \binom{k(q-1)/m}{n_0, n_1, \dots, n_{f-1}} \alpha_1^{n_0} \beta_1^{n_1} \cdots \beta_{f-1}^{n_{f-1}} \prod_{i=1}^{f-1} \left( \sum_{b_i \in F_p} b_i^{n_i} \right)$$

$$= \sum_{\substack{n_0+\dots+n_{f-1}=k(q-1)/m \\ p-1|n_1, \dots, p-1|n_{f-1}}} \binom{k(q-1)/m}{n_0, n_1, \dots, n_{f-1}} \alpha_1^{n_0} \beta_1^{n_1} \cdots \beta_{f-1}^{n_{f-1}}.$$

As $m \left| \frac{q-1}{p-1} \right.$ we have $p-1 \left| \frac{q-1}{m} \right.$, and so $p-1|n_0$. Now Genocchi [9] has shown that

$$\binom{a_1 + a_2 + \dots + a_r}{a_1, a_2, \dots, a_r} \equiv 0 \pmod{p},$$

provided the nonnegative integers $a_1, a_2, \dots, a_r$ satisfy

$$p-1|a_1,\dots,p-1|a_r, \quad a_1 + a_2 + \dots + a_r < p^r - 1.$$

Thus we have

$$\binom{k(q-1)/m}{n_0, n_1, \dots, n_{f-1}} \equiv 0 \pmod{p}$$

for

$$p-1|n_0,\dots,p-1|n_{f-1}, \quad n_0 + n_1 + \dots + n_{f-1} = k(q-1)/m.$$

This completes the proof of Lemma 1.

**Lemma 2.** *Let h and k be integers such that*

$$h + k = \frac{1}{2}(p-1), \quad h \geq 0,\ k \geq 0.$$

*Then we have*

$$h! \equiv \frac{(-1)^k 2^{2k} k! \left(\frac{p-1}{2}\right)!}{(2k)!} \quad (mod\ p).$$

**Proof.** Modulo $p$ we have

$$\begin{aligned}
\left(\frac{p-1}{2}\right)! &\equiv 2^{p-1} \left(\frac{p-1}{2}\right)! \\
&\equiv \left(\frac{2}{p}\right) 2\cdot 4\cdot 6 \ldots (p-1) \\
&\equiv \left(\frac{2}{p}\right) 2\cdot 4 \ldots (2h)(2h+2)\ldots(p-1) \\
&\equiv \left(\frac{2}{p}\right) 2^h h!\,(2h+2)\ldots(p-1) \\
&\equiv \left(\frac{2}{p}\right) 2^h h!\,(-1)^{\frac{p-1}{2}-h} (p-(2h+2))\ldots(p-(p-1)) \\
&\equiv \left(\frac{2}{p}\right) 2^h h!\,(-1)^k (2k-1)(2k-3)\ldots 1 \\
&\equiv \left(\frac{2}{p}\right) 2^h h!\,(-1)^k \frac{(2k)!}{2^k k!} \\
&\equiv \left(\frac{2}{p}\right) 2^{\frac{p-1}{2}} h!\,(-1)^k \frac{(2k)!}{2^{2k} k!} \\
&\equiv (-1)^k \frac{h!\,(2k)!}{2^{2k} k!},
\end{aligned}$$

completing the proof of Lemma 2.

## 3. Evaluation of Eisenstein sums: $m = 2$.

We prove the following theorem.

**Theorem 1.**

$$E_q(w_2) = \begin{cases} (-1)^{\frac{p-1}{2}\cdot\frac{f}{2}} p^{\frac{f}{2}-1}, & \textit{if } f \equiv 0 \pmod 2, \\ (-1)^{\frac{p-1}{2}\cdot\frac{f-1}{2}} p^{\frac{f-1}{2}}, & \textit{if } f \equiv 1 \pmod 2. \end{cases}$$

**Proof.** The theorem follows immediately from Corollary 1 (with $m = 2$) and the classical result

$$g_p(w_2) = i^{\left(\frac{p-1}{2}\right)^2} p^{1/2}, \tag{3.1}$$

see for example [17: p. 199]. We remark that for $f \equiv 0 \pmod 2$ the result also follows from Theorem C.

## 4. Evaluation of Eisenstein sums: $m = 3$

In this case the condition $m | p^f - 1$ holds if and only if

$$\begin{cases} \text{(a) } p \equiv 1 \pmod 3, \text{ or} \\ \text{(b) } p \equiv 2 \pmod 3, \ f \equiv 0 \pmod 2. \end{cases} \tag{4.1}$$

**Case (a):** $p \equiv 1 \pmod 3$. By Corollary 1 with $m = 3$ we have

$$E_q(w_3) = \begin{cases} (-1)^{f-1} \dfrac{g_p(w_3)^f}{g_p(w_3^f)}, & \text{if } f \not\equiv 0 \pmod 3, \\ (-1)^f \dfrac{g_p(w_3)^f}{p}, & \text{if } f \equiv 0 \pmod 3. \end{cases} \tag{4.2}$$

As $p \equiv 1 \pmod 3$ there are integers $L$ and $M$ such that

$$4p = L^2 + 27M^2. \tag{4.3}$$

The positive integers $|L|$ and $|M|$ are determined uniquely by (4.3). We specify $L$ uniquely by choosing between $L$ and $-L$ so that

$$L \equiv -1 \pmod 3). \tag{4.4}$$

The two non–trivial cube roots of unity modulo $p$ are $\frac{L+9M}{L-9M}$ and $\frac{L-9M}{L+9M}$.

As $g^{\frac{p-1}{3}}$ is a non–trivial cube root of unity (mod $p$), we can distinguish between $M$ and $-M$ by choosing $M$ so that

$$g^{\frac{p-1}{3}} \equiv \frac{L+9M}{L-9M} \pmod p. \tag{4.5}$$

The integers $L$ and $M$ are uniquely determined by (4.3), (4.4) and (4.5). It is a classical result (see for example [10: pp. 443–444]) that

$$\begin{cases} J_p(w_3, w_3) = -\frac{1}{2}(L + 3M\sqrt{-3}), \\ g_p(w_3)^3 = -\frac{1}{2}(L+3M\sqrt{-3})\,p, \\ g_p(w_3^2)^3 = -\frac{1}{2}(L-3M\sqrt{-3})\,p, \\ g_p(w_3)\,g_p(w_3^2) = p. \end{cases} \tag{4.6}$$

**Subcase (i):** $f \equiv 0 \pmod 3$. From (4.2) and (4.6), we have

$$E_q(w_3) = (-1)^f g_p(w_3)^f/p = p^{f/3-1}\left(\frac{L+3M\sqrt{-3}}{2}\right)^{f/3}.$$

**Subcase (ii):** $f \equiv 1 \pmod 3$. From (4.2) and (4.6), we have

$$E_q(w_3) = (-1)^{f-1} g_p(w_3)^{f-1} = p^{\frac{f-1}{3}}\left(\frac{L+3M\sqrt{-3}}{2}\right)^{\frac{f-1}{3}}.$$

**Subcase (iii):** $f \equiv 2 \pmod 3$. From (4.2) and (4.6), we have

$$E_q(w_3) = (-1)^{f-1}\frac{g_p(w_3)^f}{g_p(w_3^2)}$$

$$= (-1)^{f-1}\frac{g_p(w_3)^{f+1}}{p}$$

$$= p^{\frac{f-2}{3}} \left(\frac{L + 3M\sqrt{-3}}{2}\right)^{\frac{f+1}{3}}.$$

**Case (b):** $p \equiv 2 \pmod 3$, $f \equiv 0 \pmod 2$. Taking $m = 3$ and $\ell = 1$ in Theorem C, we obtain

$$E_q(w_3) = (-1)^{\frac{f}{2}\cdot\frac{p-2}{3}} p^{f/2-1} = (-1)^{\frac{f}{2}} p^{\frac{f}{2}-1}.$$

This completes the proof of the following theorem.

**Theorem 2.** *(a) If $p \equiv 1 \pmod 3$ let $(L,M)$ be the unique solution of*

$$4p = L^2 + 27M^2, \quad L \equiv -1 \pmod 3,$$

$$M \equiv \left(\frac{g^{\frac{p-1}{3}} - 1}{g^{\frac{p-1}{3}} + 1}\right) \frac{L}{9} \pmod p.$$

*Then we have*

$$E_q(w_3) = p^{\alpha}\left(\frac{1}{2}(L + 3M\sqrt{-3})\right)^{\beta},$$

*where*

$$\alpha = \begin{cases} f/3 - 1, & \text{if } f \equiv 0 \pmod 3, \\ (f-1)/3, & \text{if } f \equiv 1 \pmod 3, \\ (f-2)/3, & \text{if } f \equiv 2 \pmod 3, \end{cases}$$

$$\beta = \begin{cases} f/3, & \text{if } f \equiv 0 \pmod 3, \\ (f-1)/3, & \text{if } f \equiv 1 \pmod 3, \\ (f+1)/3, & \text{if } f \equiv 2 \pmod 3. \end{cases}$$

*(b) If $p \equiv 2 \pmod 3$ then we have*

$$E_q(w_3) = (-1)^{f/2} p^{f/2-1}.$$

For some numerical examples illustrating Theorem 2(a) see Tables 1–5 at the end of the paper.

**5. Evaluation of Eisenstein sums:** m = 4.

In this case the condition $m|p^f - 1$ holds if and only if

$$\begin{cases} \text{(a)} & p \equiv 1 \pmod 4 \text{ or} \\ \text{(b)} & p \equiv 3 \pmod 4, \quad f \equiv 0 \pmod 2. \end{cases} \tag{5.1}$$

**Case (a):** $p \equiv 1 \pmod 4$. As $p \equiv 1 \pmod 4$ there are integers $A$ and $B$ such that

$$p = A^2 + B^2. \tag{5.2}$$

If $A$ is chosen to be odd and $B$ even, the relation (5.2) determines $|A|$ and $|B|$ uniquely. Replacing $A$ by $-A$, if necessary, we may specify $A$ uniquely by requiring

$$A \equiv 1 \pmod 4. \tag{5.3}$$

As $(\pm B/A)^4 \equiv 1 \pmod p$, $(\pm B/A)^2 \equiv -1$, we may choose between $B$ and $-B$ by requiring

$$B/A \equiv g^{\frac{p-1}{4}} \pmod p. \tag{5.4}$$

Thus $A$ and $B$ are determined uniquely by (5.2), (5.3) and (5.4). With this normalization we have [10: p. 443]

$$\begin{cases} g_p(w_4)^2 = -(A+Bi)p^{1/2} \\ g_p(w_4^3)^2 = -(A-Bi)p^{1/2} \\ g_p(w_4)g_p(w_4^3) = (-1)^{(p-1)/4}\, p. \end{cases} \tag{5.5}$$

Next, by Corollary 1 (with $m = 4$), we have

$$E_q(w_4) = \begin{cases} (-1)^{f-1} \dfrac{g_p(w_4)^f}{g_p(w_4^f)}, & \text{if } f \not\equiv 0 \pmod 4, \\ \dfrac{g_p(w_4)^f}{p}, & \text{if } f \equiv 0 \pmod 4. \end{cases} \tag{5.6}$$

**Subcase (i):** $f \equiv 0 \pmod 4$. From (5.5) and (5.6), we have

$$E_q(w_4) = (-(A + Bi)p^{1/2})^{f/2}/p$$
$$= p^{\frac{f}{4}-1}(A + Bi)^{f/2}.$$

**Subcase (ii):** $f \equiv 1 \pmod 4$. From (5.5) and (5.6), we have

$$E_q(w_4) = g_p(w_4)^{f-1}$$
$$= (-(A + Bi)p^{1/2})^{\frac{f-1}{2}}$$
$$= p^{\frac{f-1}{4}}(A + Bi)^{\frac{f-1}{2}}.$$

**Subcase (iii):** $f \equiv 2 \pmod 4$. From (5.5), (5.6) and (3.1), we have

$$E_q(w_4) = (-1)\,\frac{g_p(w_4)^f}{g_p(w_2)}$$
$$= (-1)(-(A + Bi)p^{1/2})^{f/2}/p^{1/2}$$
$$= p^{\frac{f-2}{4}}(A + Bi)^{f/2}.$$

**Subcase (iv):** $f \equiv 3 \pmod 4$. From (5.5) and (5.6), we have

$$E_q(w_4) = \frac{g_p(w_4)^f}{g_p(w_4^3)}$$
$$= \frac{g_p(w_4)^{f+1}}{(-1)^{\frac{p-1}{4}}p}$$
$$= (-(A + Bi)p^{1/2})^{\frac{f+1}{2}}(-1)^{\frac{p-1}{4}}/p$$
$$= (-1)^{\frac{p-1}{4}}p^{\frac{f-3}{4}}(A + Bi)^{\frac{f+1}{2}}.$$

**Case (b):** $p \equiv 3 \pmod 4$, $f \equiv 0 \pmod 2$. Taking $m = 4$ and $\ell = 1$ in Theorem C, we obtain

$$E_q(w_4) = (-1)^{\frac{f}{2}\cdot\frac{p-3}{4}}\, p^{\frac{f}{2}-1}\,.$$

This completes the proof of the following theorem.

**Theorem 3.** *(a) If $p \equiv 1\ (mod\ 4)$ let $(A,B)$ be the unique solution of*

$$\begin{cases} p = A^2 + B^2, & A \equiv 1\ (mod\ 4), \\ B \equiv g^{\frac{p-1}{4}} A\ (mod\ p). \end{cases}$$

*Then we have*

$$E_q(w_4) = \epsilon p^{\alpha}(A + Bi)^{\beta},$$

*where*

$$\alpha = \begin{cases} f/4 - 1, & \text{if } f \equiv 0\ (mod\ 4), \\ [f/4], & \text{if } f \not\equiv 0\ (mod\ 4), \end{cases}$$

$$\beta = \begin{cases} f/2, & \text{if } f \equiv 0\ (mod\ 2), \\ (f - (-1)^{\frac{f-1}{2}})/2, & \text{if } f \equiv 1\ (mod\ 2), \end{cases}$$

$$\epsilon = \begin{cases} 1, & \text{if } f \not\equiv 3\ (mod\ 4), \\ (-1)^{(p-1)/4}, & \text{if } f \equiv 3\ (mod\ 4). \end{cases}$$

*(b) If $p \equiv 3\ (mod\ 4)$ and $f \equiv 0\ (mod\ 2)$ then we have*

$$E_q(w_4) = (-1)^{\frac{f}{2}\cdot\frac{p-3}{4}}\, p^{\frac{f}{2}-1}\,.$$

Some numerical examples illustrating Theorem 3(a) are given in Tables 6–10.

## 6. Evaluation of Eisenstein sums: $m = 5$

The condition $m | p^f - 1$ in this case holds if and only if

$$\begin{cases} \text{(a) } p \equiv 1 \pmod 5, & \text{or} \\ \text{(b) } p \equiv 2,3 \pmod 5, & f \equiv 0 \pmod 4, \text{ or} \\ \text{(c) } p \equiv 4 \pmod 5, & f \equiv 0 \pmod 2. \end{cases} \tag{6.1}$$

**Case (a):** $p \equiv 1 \pmod 5$. As $p \equiv 1 \pmod 5$ there are integers $x,u,v,w$, (see [5]) such that

$$\begin{cases} 16p = x^2 + 50u^2 + 50v^2 + 125w^2, \\ xw = v^2 - 4uv - u^2. \end{cases} \tag{6.2}$$

If $(x,u,v,w)$ is a solution of (6.2), all solutions are given by

$$\pm(x,u,v,w),\ \pm(x,-v,u,-w),\ \pm(x,-u,-v,w),\ \pm(x,v,-u,-w). \tag{6.3}$$

Thus the diophantine equation system (6.2) determines $|x|$ uniquely. We distinguish between $x$ and $-x$ by choosing

$$x \equiv 1 \pmod 5. \tag{6.4}$$

Set

$$\begin{cases} R = R(x,w) = x^2 - 125w^2, \\ S = S(x,u,v,w) = 2xu - xv - 25vw, \\ e(x,u,v,w) \equiv \dfrac{R-10S}{R+10S} \pmod p. \end{cases} \tag{6.5}$$

Then (see for example [14: p. 72]) $e(x,u,v,w)$, $e(x,-v,u,-w) \equiv e(x,u,v,w)^2 \pmod p$, $e(x,v,-u,-w) \equiv e(x,u,v,w)^3 \pmod p$, $e(x,-u,-v,w) \equiv e(x,u,v,w)^4 \pmod p$ are the four primitive fifth roots of unity modulo $p$. Of the four solutions of (6.2) and (6.4), we choose $(x,u,v,w)$ to be the one such that

$$g^{\frac{p-1}{5}} \equiv e(x,u,v,w) \pmod p. \tag{6.6}$$

Then (6.2), (6.4) and (6.6) determine $x,u,v,w$ uniquely. For this solution we set

$$\tau(x,u,v,w) = \frac{1}{4}\left(x + (u+2v)\, i\sqrt{10+2\sqrt{5}} + (2u-v)\, i\sqrt{10-2\sqrt{5}} + 5w\sqrt{5}\right). \tag{6.7}$$

Then we have [14: Theorem 1]

$$\begin{cases} J_p(w_5,w_5) = J_p(w_5,w_5^3) = \tau(x,u,v,w), \\ J_p(w_5^2,w_5^2) = J_p(w_5,w_5^2) = \tau(x,v,-u,-w), \\ J_p(w_5^3,w_5^3) = J_p(w_5^3,w_5^4) = \tau(x,-v,u,-w), \\ J_p(w_5^4,w_5^4) = J_p(w_5^2,w_5^4) = \tau(x,-u,-v,w). \end{cases} \tag{6.8}$$

Now [12: Prop. 8.3.3]

$$g_p(w_5)^5 = pJ_p(w_5,w_5)J_p(w_5,w_5^2)J_p(w_5,w_5^3), \tag{6.9}$$

so that

$$\begin{cases} g_p(w_5)^5 = p\tau(x,u,v,w)^2\tau(x,v,-u,-w), \\ g_p(w_5^2)^5 = p\tau(x,v,-u,-w)^2\tau(x,-u,-v,w), \\ g_p(w_5^3)^5 = p\tau(x,-v,u,-w)^2\tau(x,u,v,w), \\ g_p(w_5^4)^5 = p\tau(x,-u,-v,w)^2\tau(x,-v,u,-w). \end{cases} \tag{6.10}$$

Next, from (1.17), we have

$$g_p(w_5)g_p(w_5^4) = g_p(w_5^2)g_p(w_5^3) = p. \tag{6.11}$$

Appealing to Corollary 1, we obtain

$$E_q(w_5) = \begin{cases} (-1)^{f-1}\dfrac{g_p(w_5)^f}{g_p(w_5^f)}, & \text{if } f \not\equiv 0 \pmod 5, \\ (-1)^f\dfrac{g_p(w_5)^f}{p}, & \text{if } f \equiv 0 \pmod 5. \end{cases} \tag{6.12}$$

**Subcase (i):** $f \equiv 0 \pmod 5$. From (6.10) and (6.12), we obtain

$$\begin{aligned} E_q(w_5) &= (-1)^f\left(p\tau(x,u,v,w)^2\tau(x,v,-u,-w)\right)^{f/5}/p \\ &= (-1)^f p^{f/5-1}\tau(x,u,v,w)^{2f/5}\tau(x,v,-u,-w)^{f/5}. \end{aligned}$$

**Subcase (ii):** $f \equiv 1 \pmod 5$. From (6.10) and (6.12), we obtain

$$\begin{aligned} E_q(w_5) &= (-1)^{f-1} g_p(w_5)^{f-1} \\ &= (-1)^{f-1}\Big(p\tau(x,u,v,w)^2\tau(x,v,-u,-w)\Big)^{\frac{f-1}{5}} \\ &= (-1)^{f-1}\, p^{\frac{f-1}{5}}\, \tau(x,u,v,w)^{\frac{2(f-1)}{5}}\tau(x,v,-u,-w)^{\frac{(f-1)}{5}}. \end{aligned}$$

**Subcase (iii):** $f \equiv 2 \pmod 5$. From (6.8), (6.10), (6.12) and (1.18), we obtain

$$\begin{aligned} E_q(w_5) &= (-1)^{f-1}\,\frac{g_p(w_5)^f}{g_p(w_5^2)} \\ &= (-1)^{f-1} g_p(w_5)^{f-2}\,\frac{g_p(w_5)^2}{g_p(w_5^2)} \\ &= (-1)^{f-1}\Big(p\tau(x,u,v,w)^2\tau(x,v,-u,-w)\Big)^{\frac{f-2}{5}} J_p(w_5,w_5) \\ &= (-1)^{f-1}\, p^{\frac{f-2}{5}}\, \tau(x,u,v,w)^{\frac{2f+1}{5}}\tau(x,v,-u,-w)^{\frac{f-2}{5}}. \end{aligned}$$

**Subcase (iv):** $f \equiv 3 \pmod 5$. From (1.18), (6.8), (6.10) and (6.12), we obtain

$$\begin{aligned} E_q(w_5) &= (-1)^{f-1}\,\frac{g_p(w_5)^f}{g_p(w_5^3)} \\ &= (-1)^{f-1} g_p(w_5)^{f-3} \cdot \frac{g_p(w_5)^2}{g_p(w_5^2)} \cdot \frac{g_p(w_5)\,g_p(w_5^2)}{g_p(w_5^3)} \\ &= (-1)^{f-1}\Big(g_p(w_5)^5\Big)^{\frac{f-3}{5}} J_p(w_5,w_5)J_p(w_5,w_5^2) \\ &= (-1)^{f-1}\Big(p\tau(x,u,v,w)^2\tau(x,v,-u,-w)\Big)^{\frac{f-3}{5}} \tau(x,u,v,w)\tau(x,v,-u,-w) \\ &= (-1)^{f-1}\, p^{\frac{f-3}{5}}\tau(x,u,v,w)^{\frac{2f-1}{5}}\, \tau(x,v,-u,-w)^{\frac{f+2}{5}}. \end{aligned}$$

**Subcase (v):** $f \equiv 4 \pmod 5$. From (6.10) and (6.12), we obtain

$$E_q(w_5) = (-1)^{f-1} \frac{g_p(w_5)^f}{g_p(w_5^4)}$$

$$= (-1)^{f-1} \frac{g_p(w_5)^{f+1}}{p}$$

$$= (-1)^{f-1} \Big(p\tau(x,u,v,w)^2 \tau(x,v,-u,-w)\Big)^{\frac{f+1}{5}} / p$$

$$= (-1)^{f-1}\, p^{\frac{f-4}{5}} (\tau(x,u,v,w))^{\frac{2f+2}{5}}\; \tau(x,v,-u,-w)^{\frac{f+1}{5}}.$$

**Case (b):** $p \equiv 2$ or $3 \pmod 5$, $f \equiv 0 \pmod 4$. Taking $m = 5$ and $\ell = 2$ in Theorem C, we obtain

$$E_q(w_5) = (-1)^{f/4(\frac{p^2-4}{5})} p^{f/2-1} = (-1)^{f/4}\, p^{f/2-1}.$$

**Case (c):** $p \equiv 4 \pmod 5$, $f \equiv 0 \pmod 2$. Taking $m = 5$ and $\ell = 1$ in Theorem C, we obtain

$$E_q(w_5) = (-1)^{f/2(\frac{p-4}{5})} p^{f/2-1} = (-1)^{f/2}\, p^{f/2-1}.$$

This completes the proof of the following theorem.

**Theorem 4.** *(a) If $p \equiv 1 \pmod 5$, let $(x,u,v,w)$ be the unique solution of*

$$\begin{cases} 16p = x^2 + 50u^2 + 50v^2 + 125w^2, \\ xw = v^2 - 4uv - u^2, \\ x \equiv 1 \pmod 5, \\ \dfrac{(x^2-125w^2) - 10(2xu-xv-25vw)}{(x^2-125w^2) + 10(2xu-xv-25vw)} \equiv g^{\frac{p-1}{5}} \pmod p. \end{cases}$$

*Set*

$$\tau(x,u,v,w) = \frac{1}{4}\Big(x + (u+2v)\, i\sqrt{10+2\sqrt{5}} + (2u-v)\, i\sqrt{10-2\sqrt{5}} + 5w\sqrt{5}\Big).$$

*Then we have*

$$E_q(w_5) = \epsilon p^{\alpha} \tau(x,u,v,w)^{\beta} \tau(x,v,-u,-w)^{\delta},$$

*where*

$$\alpha = \begin{cases} f/5-1, & \text{if } f \equiv 0 \pmod 5, \\ [f/5], & \text{if } f \not\equiv 0 \pmod 5, \end{cases}$$

$$\beta = \begin{cases} [2f/5], & \text{if } f \equiv 0,1,3 \pmod 5, \\ [2f/5] + 1, & \text{if } f \equiv 2,4 \pmod 5, \end{cases}$$

$$\delta = \begin{cases} [f/5], & \text{if } f \equiv 0,1,2 \pmod 5, \\ [f/5] + 1, & \text{if } f \equiv 3,4 \pmod 5, \end{cases}$$

$$\epsilon = \begin{cases} (-1)^{f}, & \text{if } f \equiv 0 \pmod 5, \\ (-1)^{f-1}, & \text{if } f \not\equiv 0 \pmod 5. \end{cases}$$

*(b) If $p \equiv 2$ or $3 \pmod 5$ and $f \equiv 0 \pmod 4$ then we have*

$$E_q(w_5) = (-1)^{f/4} \, p^{f/2-1}.$$

*(c) If $p \equiv 4 \pmod 5$ and $f \equiv 0 \pmod 2$ then we have*

$$E_q(w_5) = (-1)^{f/2} p^{f/2-1}.$$

For some numerical examples illustrating Theorem 4(a) see Tables 11–15.

## 7. Evaluation of Eisenstein sums: $m = 6$.

The condition $m | p^f - 1$ in this case holds if and only if

$$\begin{cases} \text{(a) } p \equiv 1 \pmod 6, & \text{or} \\ \text{(b) } p \equiv 5 \pmod 6, & f \equiv 0 \pmod 2. \end{cases} \tag{7.1}$$

Case (a): $p \equiv 1 \pmod 6$. As $p \equiv 1 \pmod 6$ we may determine $L$ and $M$ uniquely (as in §4) by (4.3), (4.4) and (4.5). By Jacobi's theorem [13: p. 167] (see also [5: p. 407]) we have

$$\begin{cases} w_3^{\mathrm{ind}_g(2)} g_p(w_6) g_p(w_3^2) = g_p(w_3) g_3(w_2), \\ w_3^{2\,\mathrm{ind}_g(2)} g_p(w_6^5) g_p(w_3) = g_p(w_3^2) g_p(w_2), \end{cases} \tag{7.2}$$

so that (by 3.1))

$$\begin{cases} g_{\mathrm{p}}(w_6) = w_3^{2\,\mathrm{ind}_{\mathrm{g}}(2)}\, i^{(\frac{p-1}{2})^2} p^{1/2} g_{\mathrm{p}}(w_3)/g_{\mathrm{p}}(w_3^2), \\ g_{\mathrm{p}}(w_6^5) = w_3^{\mathrm{ind}_{\mathrm{g}}(2)}\, i^{(\frac{p-1}{2})^2} p^{1/2} g_{\mathrm{p}}(w_3^2)/g_{\mathrm{p}}(w_3). \end{cases} \tag{7.3}$$

By (1.17) or (7.3) we have

$$g_{\mathrm{p}}(w_6)g_{\mathrm{p}}(w_6^5) = (-1)^{\frac{p-1}{2}}\, p. \tag{7.4}$$

Thus we have

$$\begin{aligned} J_{\mathrm{p}}(w_6,w_6) &= (g_{\mathrm{p}}(w_6)^2/g_{\mathrm{p}}(w_3) \quad \text{(by (1.18))} \\ &= w_3^{\mathrm{ind}_{\mathrm{g}}(2)}(-1)^{\frac{p-1}{2}}\, p\, g_{\mathrm{p}}(w_3)/(g_{\mathrm{p}}(w_3^2))^2 \quad \text{(by (7.3))} \\ &= w_3^{\mathrm{ind}_{\mathrm{g}}(2)}(-1)^{\frac{p-1}{2}} g_{\mathrm{p}}(w_3)^2/g_{\mathrm{p}}(w_3^2) \quad \text{(by (4.6))} \\ &= (-1)^{\frac{p-1}{2}}\, w_3^{\mathrm{ind}_{\mathrm{g}}(2)} J_{\mathrm{p}}(w_3,w_3) \quad \text{(by (1.18))}, \end{aligned}$$

that is (by (4.6)),

$$J_{\mathrm{p}}(w_6,w_6) = (-1)^{\frac{p+1}{2}}\, w_3^{\mathrm{ind}_{\mathrm{g}}(2)} \left(\frac{L+3M\sqrt{-3}}{2}\right). \tag{7.5}$$

Cubing the first equation in (7.3), and appealing to (4.6), we obtain

$$g_{\mathrm{p}}(w_6)^3 = i^{3(\frac{p-1}{2})^2}\, p^{3/2}\, g_{\mathrm{p}}(w_3)^6/p^3,$$

that is (by (4.6)),

$$g_{\mathrm{p}}(w_6)^3 = (-1)^{\frac{p-1}{2}}\, i^{(\frac{p-1}{2})^2} \left(\frac{1}{2}(L+3M\sqrt{-3})\right)^2 p^{1/2}. \tag{7.6}$$

Squaring (7.7) we deduce that

$$g_{\mathrm{p}}(w_6)^6 = (-1)^{\frac{p-1}{2}} \left(\frac{1}{2}(L+3M\sqrt{-3})\right)^4 p. \tag{7.7}$$

Also, by Corollary 1, we have

$$E_q(w_6) = \begin{cases} (-1)^{f-1} g_p(w_6)^f / g_p(w_6^f), & \text{if } f \not\equiv 0 \pmod 6, \\ g_p(w_6)^f / p, & \text{if } f \equiv 0 \pmod 6. \end{cases} \tag{7.8}$$

**Subcase (i):** $f \equiv 0 \pmod 6$. From (7.7) and (7.8), we obtain

$$\begin{aligned} E_q(w_6) &= (g_p(w_6)^6)^{f/6}/p \\ &= (-1)^{\frac{p-1}{2}\cdot\frac{f}{6}} \left(\frac{L + 3M\sqrt{-3}}{2}\right)^{2f/3} p^{f/6-1}. \end{aligned}$$

**Subcase (ii):** $f \equiv 1 \pmod 6$. From (7.7) and (7.8), we obtain

$$\begin{aligned} E_q(w_6) &= g_p(w_6)^{f-1} \\ &= \left((-1)^{\frac{p-1}{2}} \left(\frac{L + 3M\sqrt{-3}}{2}\right)^4 p\right)^{\frac{f-1}{6}} \\ &= (-1)^{\frac{p-1}{2}\cdot\frac{f-1}{6}} p^{\frac{f-1}{6}} \left(\frac{L + 3M\sqrt{-3}}{2}\right)^{\frac{2f-2}{3}}. \end{aligned}$$

**Subcase (iii):** $f \equiv 2 \pmod 6$. From (7.5), (7.7) and (7.8), we obtain

$$\begin{aligned} E_q(w_6) &= -\frac{g_p(w_g)^f}{g_p(w_6^2)} \\ &= -g_p(w_6)^{f-2} \frac{g_p(w_6)^2}{g_p(w_6^2)} \\ &= -\left\{(-1)^{\frac{p-1}{2}} \left(\frac{L + 3M\sqrt{-3}}{2}\right)^4 p\right\}^{\frac{f-2}{6}} J_p(w_6, w_6) \\ &= (-1)^{\frac{p-1}{2}\cdot\frac{f+4}{6}} w_3^{\operatorname{ind}_g(2)} p^{\frac{f-1}{6}} \left(\frac{L + 3M\sqrt{-3}}{2}\right)^{\frac{2f-1}{3}}. \end{aligned}$$

**Subcase (iv):** $f \equiv 3 \pmod 6$. By (3.1), (7.6) and (7.8), we have

$$E_q(w_6) = \frac{g_p(w_6)^f}{g_p(w_2)}$$

$$= \left\{(-1)^{\frac{p-1}{2}} i^{(\frac{p-1}{2})^2} \left(\frac{L+3M\sqrt{-3}}{2}\right)^2 p^{1/2}\right\}^{f/3} / \, i^{(\frac{p-1}{2})^2} p^{1/2}$$

$$= (-1)^{\frac{p-1}{2}\cdot\frac{f+3}{6}} p^{\frac{f-3}{6}} \left(\frac{L+3M\sqrt{-3}}{2}\right)^{2f/3}.$$

**Subcase (v):** $f \equiv 4 \pmod 6$. Appealing to (1.18), (4.6), (7.5), (7.7), and (7.8), we obtain

$$\mathrm{E}_q(w_6) = -\frac{g_p(w_6)^f}{g_p(w_6^4)}$$

$$= -g_p(w_6)^{f-4} \left(\frac{g_p(w_6)^2}{g_p(w_6^2)}\right)^2 \frac{g_p(w_6^2)^2}{g_p(w_6^4)}$$

$$= -g_p(w_6)^{f-4} \, (J_p(w_6,w_6))^2 \, J_p(w_6^2,w_6^2)$$

$$= -g_p(w_6)^{f-4} (J_p(w_6,w_6))^2 \, J_p(w_3,w_3)$$

$$= (-1)^{\frac{p-1}{2}\cdot\frac{f-4}{6}} w_3^{2\mathrm{ind}_g(2)} p^{\frac{f-4}{6}} \left(\frac{L+3M\sqrt{-3}}{2}\right)^{\frac{2f+1}{3}}.$$

**Subcase (vi):** $f \equiv 5 \pmod 6$. By (7.4), (7.7) and (7.8), we have

$$E_q(w_6) = \frac{g_p(w_6)^f}{g_p(w_6^5)}$$

$$= g_p(w_6)^{f+1} / (-1)^{\frac{p-1}{2}} p$$

$$= (-1)^{\frac{p-1}{2}\,\frac{f-5}{6}} p^{\frac{f-5}{6}} \left(\frac{L+3M\sqrt{-3}}{2}\right)^{\frac{2f+2}{3}}.$$

**Case (b):** $p \equiv 5 \pmod 6$, $f \equiv 0 \pmod 2$. Taking $m = 6$ and $\ell = 1$ in Theorem C, we obtain

$$E_q(w_6) = (-1)^{\frac{f}{2}\cdot\frac{p-5}{6}} p^{f/2-1}.$$

This completes the proof of the following theorem.

**Theorem 5.** *(a) If $p \equiv 1 \pmod 6$ let $(L,M)$ be the unique solution of*

$$\begin{cases} 4p = L^2 + 27M^2, \quad L \equiv -1 \pmod 3, \\ M \equiv \left[\dfrac{g^{\frac{p-1}{3}} - 1}{g^{\frac{p-1}{3}} + 1}\right] \dfrac{L}{9} \pmod p. \end{cases}$$

*Then we have*

$$E_q(w_6) = \epsilon p^{\alpha}\left(\frac{1}{2}(L + 3M\sqrt{-3})\right)^{\beta},$$

*where*

$$\alpha = \begin{cases} f/6 - 1, & \text{if } f \equiv 0 \pmod 6) \\ [f/6], & \text{if } f \not\equiv 0 \pmod 6), \end{cases}$$

$$\beta = \begin{cases} 2f/3, & \text{if } f \equiv 0 \pmod 3), \\ (2f-2)/3, & \text{if } f \equiv 1 \pmod 6), \\ (2f-1)/3, & \text{if } f \equiv 2 \pmod 6), \\ (2f+1)/3, & \text{if } f \equiv 4 \pmod 6), \\ (2f+2)/3, & \text{if } f \equiv 5 \pmod 6), \end{cases}$$

$$\epsilon = \begin{cases} (-1)^{\left(\frac{p-1}{2}\right)[f/6]}, & \text{if } f \equiv 0,1,5 \pmod 6) \\ (-1)^{\left(\frac{p-1}{2}\right)\left(\frac{f+3}{6}\right)}, & \text{if } f \equiv 3 \pmod 6), \\ (-1)^{\left(\frac{p-1}{2}\right)\left(\frac{f+4}{6}\right)} w_3^{\mathrm{ind}_g(2)}, & \text{if } f \equiv 2 \pmod 6), \\ (-1)^{\left(\frac{p-1}{2}\right)\left(\frac{f-4}{6}\right)} w_3^{2\,\mathrm{ind}_g(2)}, & \text{if } f \equiv 4 \pmod 6). \end{cases}$$

*(b) If $p \equiv 5 \pmod 6$ and $f \equiv 0 \pmod 2$ then we have*

$$E_q(w_6) = (-1)^{\frac{f}{2}\cdot\frac{p-5}{6}} p^{\frac{f}{2}-1}.$$

For some numerical examples illustrating Theorem 5(a) see Tables 16–19.

## 8. Evaluation of Eisenstein sums: $m = 7$

The condition $m|p^f - 1$ in this case holds if and only if

$$\begin{array}{lll} \text{(a)} & p \equiv 1 \pmod 7, & \text{or} \\ \text{(b)} & p \equiv 2,4 \pmod 7, & f \equiv 0 \pmod 3, \text{ or} \\ \text{(c)} & p \equiv 3,5 \pmod 7, & f \equiv 0 \pmod 6, \text{ or} \\ \text{(d)} & p \equiv 6 \pmod 7, & f \equiv 0 \pmod 2. \end{array} \tag{8.1}$$

**Case (a):** p ≡ 1 (mod 7). This case can be handled similarly to Case (a) of §6. The appropriate diophantine system and Jacobi sums are given in [6]. The details are complicated and will be included in a sequel to this paper.

**Case (b):** $p \equiv 2,4 \pmod 7$, $f \equiv 0 \pmod 3$. As $p \equiv 2,4 \pmod 7$ there are integers $G,H$ such that

$$p = G^2 + 7H^2. \tag{8.2}$$

If $(G,H)$ is a solution of (8.2), all four solutions are given by $(\pm G, \pm H)$. We distinguish between $G$ and $-G$ by requiring

$$G \equiv \begin{cases} 4 \pmod 7, & \text{if } p \equiv 2 \pmod 7, \\ 2 \pmod 7, & \text{if } p \equiv 4 \pmod 7. \end{cases} \tag{8.3}$$

Next we determine a unique solution $S \pmod p$ of

$$S^2 \equiv -7 \pmod p \tag{8.4}$$

by means of

$$S \equiv \gamma^{(\frac{q-1}{7})} + \gamma^{2(\frac{q-1}{7})} - \gamma^{3(\frac{q-1}{7})} + \gamma^{4(\frac{q-1}{7})} - \gamma^{5(\frac{q-1}{7})} - \gamma^{6(\frac{q-1}{7})} \pmod p. \tag{8.5}$$

Replacing $S$ by $S + p$, if necessary, we can suppose that $S$ is odd. As

$$H^2 \equiv -G^2/7 \equiv (SG/7)^2 \pmod p, \tag{8.6}$$

we can distinguish between $H$ and $-H$ by choosing

$$H \equiv SG/7 \pmod p. \tag{8.7}$$

The pair of integers $(G,H)$ is now uniquely determined by (8.2), (8.3) and (8.7).

As $p \equiv 2,4 \pmod 7$ the least positive integer $\ell$ such that $p^{\ell} \equiv 1 \pmod 7$ is $\ell = 3$. We first determine $E_{p^3}(w_7) = E_{p^3}(w_7, \gamma^{(q-1)/(p^3-1)})$. Appealing to Theorem A(a), we have

$$\begin{cases} E_{p^3}(w_7) = E_{p^3}(w_7^2) = E_{p^3}(w_7^4), \\ E_{p^3}(w_7^3) = E_{p^3}(w_7^5) = E_{p^3}(w_7^6). \end{cases} \tag{8.8}$$

Thus $E_{p^3}(w_7)$ is fixed under the automorphism $\sigma_2 : w_7 \to w_7^2$. Hence $E_{p^3}(w_7)$ belongs to the field $Q(w_7 + w_7^2 + w_7^4) = Q(\sqrt{-7})$. As $E_{p^3}(w_7)$ is an algebraic integer, we must have $E_{p^3}(w_7) \in Z + Z\left(\frac{-1+\sqrt{-7}}{2}\right)$ (the ring of integers of $Q(\sqrt{-7})$, which is a unique factorization domain). By (2.4), for some prime $\pi \in Z + Z\left(\frac{-1+\sqrt{-7}}{2}\right)$ dividing $p$, we have

$$E_{p^3}(w_7) \equiv -p\left(\left[\frac{p}{7}\right]!\left[\frac{2p}{7}\right]!\left[\frac{4p}{7}\right]!\right) \pmod{\pi^2}. \tag{8.9}$$

As $p = \pi\bar{\pi}$, from (8.2) we see that $\pi = \theta(G \pm H\sqrt{-7})$ for some unit $\theta$ of $Z + Z\left(\frac{-1+\sqrt{-7}}{2}\right)$, that is $\theta = \pm 1$. Replacing $\pi$ by $-\pi$, if necessary, we may suppose that

$$\pi = G + H_1\sqrt{-7}, \quad \text{where } H_1 = \pm H. \tag{8.10}$$

Next, for $p \equiv 2 \pmod 7$, we have

$$\begin{aligned} &\left[\frac{p}{7}\right]!\left[\frac{2p}{7}\right]!\left[\frac{4p}{7}\right]! \\ &\equiv \left(\frac{p-2}{7}\right)!\left(\frac{2p-4}{7}\right)!\left(\frac{4p-1}{7}\right)! \pmod p \\ &\equiv \frac{\left(\frac{p-2}{7}\right)!\left(\frac{2p-4}{7}\right)!}{\left(\frac{3p-6}{7}\right)!} \pmod p \quad \text{(by Wilson's theorem)} \end{aligned}$$

$$\equiv \begin{bmatrix} \frac{3p-6}{7} \\ \frac{p-2}{7} \end{bmatrix}^{-1} \pmod{p}$$

$$\equiv -1/2G \pmod{p} \qquad ([7: \text{p. } 126])$$

and, for $p \equiv 4 \pmod{7}$, we have

$$\left[\frac{p}{7}\right]!\left[\frac{2p}{7}\right]!\left[\frac{4p}{7}\right]!$$

$$\equiv \left(\frac{p-4}{7}\right)!\left(\frac{2p-1}{7}\right)!\left(\frac{4p-2}{7}\right)! \pmod{p}$$

$$\equiv -\frac{\left(\frac{p-4}{7}\right)!\left(\frac{2p-1}{7}\right)!}{\left(\frac{3p-5}{7}\right)!} \pmod{p} \quad \text{(by Wilson's theorem)}$$

$$\equiv -\begin{bmatrix} \frac{3p-5}{7} \\ \frac{p-4}{7} \end{bmatrix}^{-1} \pmod{p}$$

$$\equiv -1/2G \pmod{p} \qquad ([7: \text{p. } 126]).$$

Hence we have

$$E_{p^3}(w_7) \equiv \pi\bar{\pi}/(2G) \pmod{\pi^2}. \tag{8.11}$$

From (8.11) we see that

$$E_{p^3}(w_7) = \lambda\pi, \tag{8.12}$$

where $\lambda(\epsilon Z + Z\left(\frac{-1+\sqrt{-7}}{2}\right))$ is not divisible by $\pi$. Mapping $w_7 \longrightarrow w_7^6$ in (8.12), we obtain

$$E_{p^3}(w_7^6) = \bar{\lambda}\bar{\pi}. \tag{8.13}$$

Now, by Theorem A(c), we have

$$E_{p^3}(w_7)E_{p^3}(w_7^6) = p, \tag{8.14}$$

so that

$$\lambda\bar{\lambda} = 1. \tag{8.15}$$

This shows that $\lambda$ is a unit of $Z + Z\left(\frac{-1+\sqrt{-7}}{2}\right)$, thus

$$\lambda = \pm 1. \tag{8.16}$$

From (8.12) and (8.16) we obtain

$$E_{p^3}(w_7) = \pm\pi. \tag{8.17}$$

Appealing to (8.11) and (8.17), we deduce that the sign $\pm$ in (8.17) satisfies

$$\pm 1 \equiv \bar{\pi}/(2G) \equiv (\pi + \bar{\pi})/(2G) \equiv 2G/2G \equiv 1 \pmod{\pi}, \tag{8.18}$$

proving that the + sign holds in (8.17), that is

$$E_{p^3}(w_7) = \pi = G + H_1\sqrt{-7}. \tag{8.19}$$

The next step is to show that $H_1 = H$. As $p \equiv 2$ or $4 \pmod 7$, the cyclotomic polynomial

$$\phi_7(x) = \frac{x^7 - 1}{x - 1} = x^6 + x^5 + x^4 + x^3 + x^2 + x + 1$$

is congruent to the product of two distinct irreducible cubic polynomials modulo $p$, namely,

$$\phi_7(x) \equiv \phi_7^-(x)\phi_7^+(x), \tag{8.20}$$

where

$$\phi_7^-(x) = x^3 + \left(\frac{1-S}{2}\right)x^2 + \left(\frac{-1-S}{2}\right)x - 1, \tag{8.21}$$

$$\phi_7^+(x) = x^3 + \left(\frac{1+S}{2}\right)x^2 + \left(\frac{-1+S}{2}\right)x - 1. \tag{8.22}$$

Hence, by Kummer's theorem, the principal ideal $\langle p\rangle$ of the ring of integers $D$ of $Q(w_7)$ factors into the product of two prime ideals, namely,

$$\langle p\rangle = P_1P_2, \tag{8.23}$$

where

$$\begin{cases} P_1 = \langle p, \phi_7^-(w_7)\rangle, \quad P_2 = \langle p, \phi_7^+(w_7)\rangle, \\ N(P_1) = N(P_2) = p^3. \end{cases} \tag{8.24}$$

Thus $D/P_1$ is a finite field with $p^3$ elements. Hence there exists an isomorphism $\theta : D/P_1 \longrightarrow F_{p^3}$. Let $\lambda : D/P_1$ be the canonical homomorphism defined by

$$\lambda(\alpha) = \alpha + P_1 \quad (\alpha \in D). \tag{8.25}$$

Set $\tau = \theta \circ \lambda$ so that $\tau$ is a homomorphism such that

$$\tau : D \xrightarrow{\text{onto}} F_{p^3} . \tag{8.26}$$

Clearly we have $\tau(w_7) \neq 0$, otherwise $\tau(\alpha) = 0$ for all $\alpha \in D = Zw_7 + Zw_7^2 + \ldots + Zw_7^6$. Similarly $\tau(w_7) \neq 1$, otherwise $\tau(D) \subseteq F_p$. Hence $\tau(w_7) \in F_{p^3}^*\backslash\{1\}$ and so, as $F_{p^3}^* = \langle\gamma_1\rangle$, where $\gamma_1 = \gamma^{(q-1)/(p^3-1)}$, we have $\tau(w_7) = \gamma_1^{k_1}$ for some integer $k_1$ satisfying $1 \leq k_1 \leq p^3 - 2$. Then we have

$$\gamma_1^{7k_1} = \tau(w_7)^7 = \tau(1) = \theta(\lambda(1)) = \theta(1 + P_1) = 1,$$

and so $(p^3 - 1) | 7k_1$, that is $\frac{p^3-1}{7} | k_1$, say $k_1 = \left(\frac{p^3-1}{7}\right)k$, where $1 \leq k \leq 6$, showing that

$$\tau(w_7) = \gamma_1^{k\left(\frac{p^3-1}{7}\right)}, \quad 1 \leq k \leq 6. \tag{8.27}$$

Next, as

$$\begin{aligned} &2(w_7 + w_7^2 + w_7^4) + 1 - S \\ &= \left(\frac{(S^2+7)}{2p}(w_7 + w_7^2)\right)p + (2w_7 - 1 + S)\, \phi_7^-(w_7) \\ &\in \langle p, \phi_7^-(w_7)\rangle = P_1 , \end{aligned}$$

we have

$$\lambda(\sqrt{-7}) = \lambda(2(w_7 + w_7^2 + w_7^4) + 1) = (2(w_7 + w_7^2 + w_7^4) + 1) + P_1 = S + P_1,$$

and so

$$\tau(\sqrt{-7}) = \theta(S + P_1) = S. \tag{8.28}$$

But, from (8.27), we have

$$\tau(\sqrt{-7}) = \tau(2(w_7 + w_7^2 + w_7^4) + 1)$$

$$= 2\left(\gamma_1^{k\left(\frac{p^3-1}{7}\right)} + \gamma_1^{2k\left(\frac{p^3-1}{7}\right)} + \gamma_1^{4k\left(\frac{p^3-1}{7}\right)}\right) + 1$$

$$= 2\left(\gamma^{k\left(\frac{q-1}{7}\right)} + \gamma^{2k\left(\frac{q-1}{7}\right)} + \gamma^{4k\left(\frac{q-1}{7}\right)}\right) + 1$$

$$= \begin{cases} S, & \text{if } k = 1,2,4 \\ -S, & \text{if } k = 3,5,6 \end{cases} \qquad \text{(by (8.5))}.$$

Hence we must have

$$\tau(w_7) = \gamma_1^{k\left(\frac{p^3-1}{7}\right)}, \quad k = 1,2,4. \tag{8.29}$$

Applying the homomorphism $\tau$ to (8.19), we obtain

$$\sum_{\substack{\alpha \in F^*_{p^3} \\ tr(\alpha)=1}} \gamma_1^{k\left(\frac{p^3-1}{7}\right)\mathrm{ind}_{\gamma_1}(\alpha)} \equiv G + H_1 S \pmod p,$$

that is

$$\sum_{\substack{\alpha \in F^*_{p^3} \\ tr(\alpha)=1}} \alpha^{k\left(\frac{p^3-1}{7}\right)} \equiv G + H_1 S \pmod p. \tag{8.30}$$

But, by Lemma 1, the left hand side of (8.30) is $\equiv 0 \pmod p$, and, by (8.7), the right hand side is $\equiv G + H_1 7H/G \pmod p$. Thus we have

$$0 \equiv G^2 + 7HH_1 \pmod p,$$

that is (as $G^2 \equiv -7H^2 \pmod p$) $H_1 \equiv H \pmod p$ and so $H_1 = H$ as asserted. Thus (8.19) becomes

$$E_{p^3}(w_7) = G + H\sqrt{-7}. \tag{8.31}$$

Finally, by Theorem B, we obtain

$$E_q(w_7) = p^{f/3-1}(E_{p^3}(w_7))^{f/3} = p^{f/3-1}(G + H\sqrt{-7})^{f/3}.$$

**Case (c):** $p \equiv 3,5 \pmod 7$, $f \equiv 0 \pmod 6$. In this case $\ell = 3$ is the least positive integer such that $p^\ell \equiv -1 \pmod 7$. Hence, by Theorem C, we have, as $f \equiv 0 \pmod 6$,

$$E_q(w_7) = (-1)^{f/6(\frac{p^3-6}{7})} p^{f/2-1} = (-1)^{f/2}\, p^{f/2-1}.$$

**Case (d):** $p \equiv 6 \pmod 7$, $f \equiv 0 \pmod 2$. In this case $\ell = 1$ is the least positive integer such that $p^\ell \equiv -1 \pmod 7$. Hence, by Theorem C, we have as $f \equiv 0 \pmod 2$,

$$E_q(w_7) = (-1)^{f/2(\frac{p-6}{7})} p^{f/2-1} = (-1)^{f/2}\, p^{f/2-1}.$$

This completes the proof of the following theorem.

**Theorem 6.** *(a) If $p \equiv 2,4 \pmod 7$ and $f \equiv 0 \pmod 3$ let $(G,H)$ be the unique solution of*

$$\begin{cases} p = G^2 + 7H^2, \\ G \equiv \begin{cases} 4 \pmod 7, & \text{if } p \equiv 2 \pmod 7, \\ 2 \pmod 7, & \text{if } p \equiv 4 \pmod 7, \end{cases} \\ H \equiv \left( \sum_{k=1}^{6} \left(\frac{k}{7}\right) \gamma^{k(\frac{q-1}{7})} \right) G/7 \pmod p. \end{cases}$$

*Then we have*

$$E_q(w_7) = p^{f/3-1}(G + H\sqrt{-7})^{f/3}.$$

*(b) If $p \equiv 3,5 \pmod 7$ and $f \equiv 0 \pmod 6$, or $p \equiv 6 \pmod 7$ and $f \equiv 0 \pmod 2$, then we have*

$$E_q(w_7) = (-1)^{f/2}\, p^{f/2-1}.$$

Some numerical examples illustrating Theorem 6(a) are given in Tables 20–21.

## 9. Evaluation of Eisenstein sums: $m = 8$

The condition $m | p^f - 1$ in this case holds if and only if

$$\begin{cases} (a)\ p \equiv 1 \pmod 8, & \text{or} \\ (b)\ p \equiv 3 \pmod 8, & f \equiv 0 \pmod 2, \text{ or} \\ (c)\ p \equiv 5 \pmod 8, & f \equiv 0 \pmod 2, \text{ or} \\ (d)\ p \equiv 7 \pmod 8, & f \equiv 0 \pmod 2. \end{cases} \tag{9.1}$$

**Case (a):** $p \equiv 1 \pmod 8$. As $p \equiv 1 \pmod 8$ we can define integers $A, B, C, D$ uniquely as follows:

$$\begin{cases} p = A^2 + B^2, \\ A \equiv 1 \pmod 4, \quad B \equiv g^{\frac{p-1}{4}} A \pmod p, \end{cases} \tag{9.2}$$

and

$$\begin{cases} p = C^2 + 2D^2, \\ C \equiv 1 \pmod 4, \quad D \equiv \left(g^{\frac{p-1}{8}} + g^{\frac{3(p-1)}{8}}\right) C/2 \pmod p. \end{cases} \tag{9.3}$$

With this normalization, we show that

$$J_p(w_8, w_8^2) = (-1)^{\frac{p+7}{8}} i^{3\mathrm{ind}_g(2)}(A + Bi) \tag{9.4}$$

and

$$J_p(w_8, w_8) = -i^{3\mathrm{ind}_g(2)}(C + D\sqrt{-2}). \tag{9.5}$$

We first prove (9.4). We have

$$J_p(w_8, w_8^2) = \frac{g_p(w_8)\, g_p(w_8^2)}{g_p(w_8^3)} \quad \text{(by (1.18))}$$

$$= (-1)^{\frac{p-1}{8}} \frac{g_p(w_8)\, g_p(w_8^5)}{g_p(w_8^6)} \quad \text{(by (1.17))}$$

$$= (-1)^{\frac{p-1}{8}} w_8^{-2\mathrm{ind}_g(2)} \frac{g_p(w_8^2)\, g_p(w_8^4)}{g_p(w_8^6)} \quad \text{(by Jacobi's theorem [(13: p. 167)], [5: p. 407])}$$

$$= (-1)^{\frac{p-1}{8}} i^{3\mathrm{ind}_g(2)} \frac{g_p(w_4)g_p(w_2)}{g_p(w_4^3)}$$

$$= (-1)^{\frac{p-1}{8}} i^{3\mathrm{ind}_g(2)} \frac{g_p(w_4)^2 g_p(w_2)}{p} \quad \text{(by (1.17))}$$

$$= (-1)^{\frac{p-1}{8}} i^{3\mathrm{ind}_g(2)} \frac{(-1)(A+Bi)p^{1/2}\cdot p^{1/2}}{p} \quad \text{(by (3.1) and (5.5))}$$

$$= (-1)^{\frac{p+7}{8}} i^{3\mathrm{ind}_g(2)}(A+Bi),$$

completing the proof of (9.4).

Next we prove (9.5). We have

$$J_p(w_8,w_8) = \frac{g_p(w_8)^2}{g_p(w_4)} \quad \text{(by (1.18))}$$

$$= i^{3\mathrm{ind}_g(2)} \frac{g_p(w_8)g_p(w_8^4)}{g_p(w_8^5)} \quad \text{(by Jacobi's theorem [5] [13])}$$

$$= (-1)^{\frac{p-1}{8}} i^{3\mathrm{ind}_g(2)} \frac{g_p(w_8)g_p(w_8^3)}{g_p(w_8^4)} \quad \text{(by (1.17))},$$

that is

$$J_p(w_8,w_8) = (-1)^{\frac{p-1}{8}} i^{3\mathrm{ind}_g(2)} J_p(w_8,w_8^3) \quad \text{(by (1.18))}. \tag{9.6}$$

As

$$\sigma_3(J_p(w_8,w_8^3)) = J_p(w_8^3,w_8^9) = J_p(w_8,w_8^3),$$

we see that $J_p(w_8,w_8^3) \in Q(w_8 + w_8^3) = Q(\sqrt{-2})$. Hence, as $J_p(w_8,w_8^3)$ is an algebraic integer, it must be an integer of $Q(\sqrt{-2})$, that is

$$J_p(w_8,w_8^3) \in Z + Z\sqrt{-2}.$$

The domain $Z + Z\sqrt{-2}$ is a unique factorization domain. Thus, in view of

$$J_p(w_8,w_8^3)J_p(w_8^7,w_8^5) = p = (C+D\sqrt{-2})\,(C-D\sqrt{-2}),$$

we must have

$$J_p(w_8,w_8^3) = \theta(C+D_1\sqrt{-2}), \tag{9.7}$$

where $\theta$ is a unit of $Z + Z\sqrt{-2}$, that is $\theta = \pm 1$, and $D_1 = \pm D$. Putting (9.6) and (9.7) together, we obtain

$$J_p(w_8,w_8) = (-1)^{\frac{p-1}{8}} i^{3\mathrm{ind}_g(2)}\ \theta(C+D_1\sqrt{-2}). \tag{9.8}$$

We next show that $\theta = (-1)^{\frac{p+7}{8}}$. From (1.17), (1.18), and (9.7), we have

$$J_p(w_8,w_8^4) = (-1)^{\frac{p-1}{8}} J_p(w_8,w_8^3) = (-1)^{\frac{p-1}{8}} \theta(C+D_1\sqrt{-2}). \tag{9.9}$$

Further, in the ring $R$ of integers of $Q(w_8)$ ($R$ is a unique factorization domain), we have

$$\sum_{k=2}^{p-1}\left(w_8^{\mathrm{ind}_g(k)}-1\right)\left(w_8^{4\mathrm{ind}_g(1-k)}-1\right) \equiv 0 \pmod{2(w_8-1)}, \tag{9.10}$$

where $w_8-1$ is a prime such that

$$(w_8-1)(w_8^3-1) = -\sqrt{-2},\quad (w_8-1)(w_8^5-1) = 1-i,$$
$$(w_8-1)(w_8^7-1) = (\sqrt{2}-1)\sqrt{2},\quad (w_8-1)(w_8^3-1)(w_8^5-1)(w_8^7-1) = 2.$$

Expanding and summing the left hand side of (9.10), we obtain

$$J_p(w_8,w_8^4) \equiv -1 \pmod{2(w_8-1)}. \tag{9.11}$$

From (9.9) and (9.11), we obtain (as $C \equiv 1 \pmod 4$ and $D_1 \equiv 0 \pmod 2$)

$$(-1)^{\frac{p-1}{8}}\ \theta = -1 \pmod{2(w_8-1)},$$

proving that $\theta = (-1)^{\frac{p+7}{8}}$ as asserted. Hence (9.8) becomes

$$J_p(w_8,w_8) = -i^{3\mathrm{ind}_g(2)}(C+D_1\sqrt{-2}). \tag{9.12}$$

The next step is to show that $D_1 = D$. As $p \equiv 1 \pmod 8$, the cyclotomic polynomial $\phi_8(x) = \frac{x^8-1}{x^4-1} = x^4+1$ is congruent to the product of four distinct linear polynomials modulo $p$, namely,

$$\phi_8(x) \equiv (x-g^{\frac{p-1}{8}})(x-g^{3(\frac{p-1}{8})})(x-g^{5(\frac{p-1}{8})})(x-g^{7(\frac{p-1}{8})}) \pmod p. \tag{9.13}$$

Hence, by Kummer's theorem, the principal ideal $\langle p \rangle$ of $R$ is the product of four prime ideals, namely

$$\langle p \rangle = P_1P_3P_5P_7 , \tag{9.14}$$

where

$$P_k = \langle p, w_8 - g^{k(\frac{p-1}{8})} \rangle, \; N(P_k) = p, \quad k = 1,3,5,7. \tag{9.15}$$

Thus $R/P_1$ is a finite field with $p$ elements. Hence there exists an isomorphism $\theta : R/P_1 \to F_p$. Let $\lambda : R \to R/P_1$ be the canonical homomorphism defined by $\lambda(\alpha) = \alpha + P_1 (\alpha \in R)$. Set $\tau = \theta \circ \lambda$ so that $\tau$ is a homomorphism such that $\tau : R \xrightarrow{\text{onto}} F_p$. Clearly $\tau(w_8) \neq 0$ otherwise $\tau(R) = \{0\}$. Hence $\tau(w_8) \in F_p^* = \langle g \rangle$ and so there exists an integer $k_1$ $(0 \leq k_1 \leq p-1)$ such that $\tau(w_8) = g^{k}_1$. Hence we have

$$g^{8k_1} = (\tau(w_8))^8 = \tau(1) = \theta(\lambda(1)) = \theta(1 + P_1) \equiv 1 \pmod p,$$

so that $(p-1) | 8k$, that is, $k_1 = \left(\frac{p-1}{8}\right)k$, for some integer $k (0 \leq k \leq 7)$. Thus

$$\tau(w_8) \equiv g^{(\frac{p-1}{8})k} \pmod p \quad (0 \leq k \leq 7).$$

Next we have

$$\begin{aligned} &\sqrt{-2} - \left(g^{\frac{p-1}{8}} + g^{3(\frac{p-1}{8})}\right) \\ &= (w_8 + w_8^3) - \left(g^{\frac{p-1}{8}} + g^{3(\frac{p-1}{8})}\right) \\ &= \left(w_8 - g^{\frac{p-1}{8}}\right)\left(1 + g^{\frac{p-1}{4}} + g^{\frac{p-1}{8}} w_8 + w_8^2\right) \\ &\in P_1 , \end{aligned}$$

and so

$$\lambda(\sqrt{-2}) = \sqrt{-2} + P_1 = \left(g^{\frac{p-1}{8}} + g^{3(\frac{p-1}{8})}\right) + P_1,$$

giving

$$\tau(\sqrt{-2}) = \theta\left(g^{\frac{p-1}{8}} + g^{3\left(\frac{p-1}{8}\right)} + P_1\right) \equiv g^{\frac{p-1}{8}} + g^{3\left(\frac{p-1}{8}\right)} \pmod{p}.$$

But

$$\tau(\sqrt{-2}) = \tau(w_8 + w_8^3) \equiv g^{\left(\frac{p-1}{8}\right)k} + g^{3\left(\frac{p-1}{8}\right)k} \pmod{p},$$

proving that $k = 1$ or $3$. Hence we have

$$\tau(w_8) \equiv g^{\left(\frac{p-1}{8}\right)k} \pmod{p}, \quad k = 1 \text{ or } 3.$$

Applying the homomorphism $\tau$ to (9.12), we obtain

$$\begin{aligned}\sum_{s=0}^{p-1} s^{k\left(\frac{p-1}{8}\right)} (1-s)^{k\left(\frac{p-1}{8}\right)} &\equiv -2^{3\left(\frac{p-1}{4}\right)} \left(C + D_1\left(g^{\left(\frac{p-1}{8}\right)k} + g^{3\left(\frac{p-1}{8}\right)k}\right)\right) \pmod{p} \\ &\equiv -2^{3\left(\frac{p-1}{4}\right)k}(C + D_1(2D/C)) \pmod{p} \qquad \text{(by 9.3))}.\end{aligned}$$

By the Binomial theorem we have

$$\sum_{s=0}^{p-1} s^{k\left(\frac{p-1}{8}\right)}(1-s)^{k\left(\frac{p-1}{8}\right)} \equiv 0 \pmod{p},$$

as

$$\sum_{s=0}^{p-1} s^m \equiv \begin{cases} 0 \pmod{p}, & m = 1,2,\dots,p-2, \\ -1 \pmod{p}, & m = p-1, \end{cases}$$

Hence we have $C^2 + 2DD_1 \equiv 0 \pmod{p}$, that is (as $C^2 \equiv -2D^2 \pmod{p}$) $D \equiv D_1 \pmod{p}$, so $D_1 = D$, as asserted. The result (9.5) now follows from (9.12).

From (1.18), (5.5) and (9.12) we obtain

$$\frac{g_p(w_8)^2}{g_p(w_4)} = -i^{3\operatorname{ind}_g(2)} (C + D\sqrt{-2}), \tag{9.16}$$

$$g_p(w_8)^4 = -(A + Bi)(C + D\sqrt{-2})^2 p^{1/2}. \tag{9.17}$$

Next, by Corollary 1, we have

$$E_q(w_8) = \begin{cases} (-1)^{f-1} \dfrac{g_p(w_8)^f}{g_p(w_8^f)}, & \text{if } f \not\equiv 0 \pmod 8, \\ \dfrac{g_p(w_8)^f}{p}, & \text{if } f \equiv 0 \pmod 8. \end{cases} \tag{9.18}$$

**Subcase (i):** $f \equiv 0 \pmod 8$. We have

$$\begin{aligned} E_q(w_8) &= (g_p(w_8)^4)^{f/4}/p \quad \text{(by (9.18))} \\ &= p^{f/8-1}(A+Bi)^{f/4}(C+D\sqrt{-2})^{f/2} \quad \text{(by (9.17))}. \end{aligned}$$

**Subcase (ii):** $f \equiv 1 \pmod 8$. We have

$$\begin{aligned} E_q(w_8) &= g_p(w_8)^{f-1} \quad \text{(by (9.18))} \\ &= \left(-(A+Bi)(C+D\sqrt{-2})^2 p^{1/2}\right)^{\frac{f-1}{4}} \quad \text{(by (9.17))} \\ &= p^{\frac{f-1}{8}}(A+Bi)^{\frac{f-1}{4}}(C+D\sqrt{-2})^{\frac{f-1}{2}}. \end{aligned}$$

**Subcase (iii):** $f \equiv 2 \pmod 8$. We have

$$\begin{aligned} E_q(w_8) &= -\frac{g_p(w_8)^f}{g_p(w_4)} \quad \text{(by (9.18))} \\ &= -g_p(w_8)^{f-2} J_p(w_8, w_8) \quad \text{(by (1.18))} \\ &= i^{3\mathrm{ind}_g(2)} p^{\frac{f-2}{8}}(A+Bi)^{\frac{f-2}{4}}(C+D\sqrt{-2})^{f/2}, \end{aligned}$$

by (9.5) and (9.17).

**Subcase (iv):** $f \equiv 3 \pmod 8$. We have

$$\begin{aligned} E_q(w_8) &= \frac{g_p(w_8)^f}{g_p(w_8^3)} \quad \text{(by (9.18))} \\ &= g_p(w_8)^{f-3} J_p(w_8, w_8) J_p(w_8, w_8^2) \quad \text{(by (1.18))} \end{aligned}$$

$$= (-1)^{\frac{p-1}{8}} p^{\frac{f-3}{8}} (A+Bi)^{\frac{f+1}{4}} (C+D\sqrt{-2})^{\frac{f-1}{2}},$$

by (9.4), (9.5) and (9.17), as $\mathrm{ind}_g(2) \equiv 0 \pmod 2$.

**Subcase (v):** $f \equiv 4 \pmod 8$. We have

$$E_q(w_8) = (-1)\frac{g_p(w_8)^f}{g_p(w_2)} \quad \text{(by (9.18))}$$

$$= (-1)\frac{(g_p(w_8)^4)^{f/4}}{p^{1/2}} \quad \text{(by (3.1))}$$

$$= (-1)\frac{((-1)(A+Bi)(C+D\sqrt{-2})^2 p^{1/2})^{f/4}}{p^{1/2}} \quad \text{(by (9.17))}$$

$$= p^{\frac{f-4}{8}} (A+Bi)^{f/4}(C+D\sqrt{-2})^{f/2}.$$

**Subcase (vi):** $f \equiv 5 \pmod 8$. We have

$$E_q(w_8) = \frac{g_p(w_8)^f}{g_p(w_8^5)} \quad \text{(by (9.18))}$$

$$= (-1)^{\frac{p-1}{8}} g_p(w_8)^{f-5} \cdot \frac{g_p(w_8)^4}{g_p(w_4)^2} \cdot \frac{g_p(w_8)\, g_p(w_4)}{g_p(w_8^3)} \cdot \frac{g_p(w_8^3)^2}{g_p(w_8^6)} \quad \text{(by (1.17))}$$

$$= (-1)^{\frac{p-1}{8}} g_p(w_8)^{f-5}(J_p(w_8,w_8))^2 J_p(w_8,w_8^2) J_p(w_8^3,w_8^3) \quad \text{(by (1.18))}$$

$$= p^{\frac{f-5}{8}} (A+Bi)^{\frac{f-1}{4}} (C+D\sqrt{-2})^{\frac{f+1}{2}},$$

by (9.4), (9.5) and (9.17), as $\sigma_3\,(J_p(w_8,w_8)) = J_p(w_8^3,w_8^3)$.

**Subcase (vii):** $f \equiv 6 \pmod 8$. We have

$$E_q(w_8) = (-1)\frac{g_p(w_8)^f}{g_p(w_8^6)} \quad \text{(by (9.18))}$$

$$= (-1) g_p(w_8)^{f-6}(J_p(w_8,w_8))^3\; J_p(w_8^2,w_8^2) J_p(w_8^2,w_8^4) \quad \text{(by (1.18))}$$

$$= i^{\mathrm{ind}_g(2)}\, p^{\frac{f-6}{8}}\, (A+Bi)^{\frac{f+2}{4}} (C+D\sqrt{-2})^{f/2},$$

by (1.17), (3.1), (5.5), (9.5) and (9.17).

**Subcase (viii):** $f \equiv 7 \pmod 8$. We have

$$E_q(w_8) = \frac{g_p(w_8)^f}{g_p(w_8^7)} \quad \text{(by (9.18))}$$

$$= \frac{g_p(w_8)^{f+1}}{(-1)^{\frac{p-1}{8}}\, p} \quad \text{(by (1.17))}$$

$$= (-1)^{\frac{p-1}{8}}\, p^{\frac{f-7}{8}}\, (A+Bi)^{\frac{f+1}{4}} (C+D\sqrt{-2})^{\frac{f+1}{2}} \quad \text{(by (9.17))}.$$

**Case (b):** $p \equiv 3 \pmod 8$, $f \equiv 0 \pmod 2$. As $p \equiv 3 \pmod 8$ there are integers $C$ and $D$ such that

$$p = C^2 + 2D^2. \tag{9.19}$$

If $(C,D)$ is a solution of (9.19), all (four) solutions are given by $(\pm C, \pm D)$. We distinguish between $C$ and $-C$ by requiring

$$C \equiv 1 \pmod 4. \tag{9.20}$$

Next we determine a unique solution $K \pmod p$ of

$$K^2 \equiv -2 \pmod p \tag{9.21}$$

by means of

$$K \equiv \gamma^{\frac{q-1}{8}} + \gamma^{3(\frac{q-1}{8})} \pmod p. \tag{9.22}$$

As

$$D^2 \equiv -C^2/2 \equiv (KC/2)^2 \pmod p,$$

we can distinguish between $D$ and $-D$ by choosing

$$D \equiv KC/2 \pmod p. \tag{9.23}$$

The pair of integers $(C,D)$ is uniquely determined by (9.19), (9.20) and (9.23).

As $p \equiv 3 \pmod 8$ the least positive integer $\ell$ such that $p^\ell \equiv 1 \pmod 8$ is $\ell = 2$. We first determine $E_{p^2}(w_8) = E_{p^2}(w_8, \gamma^{p^{\frac{q-1}{2}-1}})$. Appealing to Theorem A(a), we have

$$\begin{cases} E_{p^2}(w_8) = E_{p^2}(w_8^3), \\ E_{p^2}(w_8^5) = E_{p^2}(w_8^7), \end{cases} \tag{9.24}$$

showing that $E_{p^2}(w_8) \in Q(w_8 + w_8^3) = Q(\sqrt{-2})$. As $E_{p^2}(w_8)$ is an algebraic integer, we have $E_{p^2}(w_8) \in Z + Z\sqrt{-2}$ (the ring of integers of $Q(\sqrt{-2})$). $Z + Z\sqrt{-2}$ is a unique factorization domain. By Theorem A(d), for some prime $\pi \in Z + Z\sqrt{-2}$ dividing $p$, we have

$$E_{p^2}(w_8) \equiv p \frac{\left[\frac{p}{8}\right]! \left[\frac{3p}{8}\right]!}{\left[\frac{p}{2}\right]!} \pmod{\pi^2}. \tag{9.25}$$

As $p = \pi\bar{\pi}$, from (9.11) we see that $\pi = \pm C \pm D\sqrt{-2}$. Replacing $\pi$ by $-\pi$, if necessary, we may suppose that

$$\pi = C + D_1\sqrt{-2}, \quad \text{where } D_1 = \pm D.$$

Now, as $p \equiv 3 \pmod 8$, we have

$$\begin{aligned} \frac{\left[\frac{p}{8}\right]! \left[\frac{3p}{8}\right]!}{\left[\frac{p}{2}\right]!} &\equiv \frac{\left(\frac{p-3}{8}\right)! \left(\frac{3p-1}{8}\right)!}{\left(\frac{p-1}{2}\right)!} \pmod p \\ &\equiv \begin{bmatrix} \frac{p-1}{2} \\ \frac{p-3}{8} \end{bmatrix}^{-1} \pmod p \\ &\equiv \frac{1}{2(-1)^{\frac{p+5}{8}} C} \pmod p, \end{aligned}$$

as

$$\begin{bmatrix} \frac{p-1}{2} \\ \frac{p-3}{8} \end{bmatrix} \equiv 2(-1)^{\frac{p+5}{8}} C \pmod p$$

[7: pp. 111–112].

Hence we have

$$E_{p^2}(w_8) \equiv \pi\bar{\pi}(-1)^{\frac{p+5}{8}}(2C)^{-1} \pmod{\pi^2}. \tag{9.26}$$

From (9.26) we see that

$$E_{p^2}(w_8) = \lambda\pi, \tag{9.27}$$

where $\lambda(\epsilon Z + Z\sqrt{-2})$ is not divisible by $\pi$. Mapping $w_8 \longrightarrow w_8^7$ in (9.27), we obtain

$$E_{p^2}(w_8^7) = \bar{\lambda}\,\bar{\pi}. \tag{9.28}$$

Now, by Theorem A(c), we have

$$E_{p^2}(w_8)E_{p^2}(w_8^7) = p, \tag{9.29}$$

so that (using (9.27) and (9.28)) $\lambda\bar{\lambda} = 1$. This shows that $\lambda$ is a unit of $Z + Z\sqrt{-2}$, thus $\lambda = \pm 1$. From (9.26) and (9.27), we obtain

$$\begin{aligned} \lambda &\equiv \bar{\pi}(-1)^{\frac{p+5}{8}}(2C)^{-1} \pmod{\pi} \\ &\equiv (\pi+\bar{\pi})(-1)^{\frac{p+5}{8}}(2C)^{-1} \pmod{\pi} \\ &\equiv (2C)(-1)^{\frac{p+5}{8}}(2C)^{-1} \pmod{\pi} \\ &\equiv (-1)^{\frac{p+5}{8}} \pmod{\pi}, \end{aligned}$$

proving that $\lambda = (-1)^{\frac{p+5}{8}}$, that is

$$E_{p^2}(w_8) = (-1)^{\frac{p+5}{8}}\,\pi = (-1)^{\frac{p+5}{8}}(C + D_1\sqrt{-2}). \tag{9.30}$$

The next step is to show that $D_1 = D$. As $p \equiv 3 \pmod 8$ the cyclotomic polynomial

$$\phi_8(x) = \frac{x^8 - 1}{x^4 - 1} = x^4 + 1$$

is congruent to the product of two distinct irreducible quadratic polynomials modulo $p$, namely

$$\phi_8(x) \equiv \phi_8^-(x)\phi_8^+(x),$$

where

$$\phi_8^-(x) = x^2 - Kx - 1, \quad \phi_8^+(x) = x^2 + Kx - 1.$$

Hence, by Kummer's theorem, the principal ideal $\langle p \rangle$ of the ring $R$ of integers of $Q(w_8)$ factors into the product of two prime ideals namely

$$\langle p \rangle = P_1 P_2,$$

where

$$P_1 = \langle p, \phi_8^-(w_8)\rangle, \quad P_2 = \langle p, \phi_8^+(w_8)\rangle, \quad N(P_1) = N(P_2) = p^2.$$

Thus $R/P_1$ is a finite field with $p^2$ elements. Hence there exists an isomorphism $\theta : R/P_1 \longrightarrow F_{p^2}$. Let $\lambda : R \longrightarrow R/P_1$ be the canonical homomorphism defined by

$$\lambda(\alpha) = \alpha + P_1 \quad (\alpha \in R).$$

Set $\tau = \theta \circ \lambda$ so that $\tau$ is a homomorphism such that

$$\tau : R \xrightarrow{\text{onto}} F_{p^2}.$$

Clearly we have $\tau(w_8) \neq 0$ otherwise $\tau(R) = \{0\}$. Also we have $\tau(w_8) \neq 1$, otherwise $\tau(R) \subseteq F_p$. Hence $\tau(w_8) \in F_{p^2}^*\backslash\{1\}$ and so, as $\gamma^{\frac{p^f-1}{p^2-1}}$ generates $F_{p^2}^*$, there exists an integer $k_1 (1 \leq k_1 \leq p^2-2)$ such that

$$\tau(w_8) = \gamma^{\left(\frac{p^f-1}{p^2-1}\right)k_1}.$$

Then we have

$$\gamma^{8\left(\frac{p^f-1}{p^2-1}\right)k_1} = (\tau(w_8))^8 = \tau(1) = \theta(\lambda(1)) = \theta(1 + P_1) \equiv 1 \pmod{p},$$

so that $p^f - 1 \mid 8\left(\frac{p^f-1}{p^2-1}\right)k_1$, that is $\frac{p^2-1}{8} \mid k_1$, say

$$k_1 = \left(\frac{p^2-1}{8}\right)k, \quad 1 \le k \le 7,$$

and so

$$\tau(w_8) = \gamma^{\left(\frac{p^f-1}{8}\right)k}, \quad 1 \le k \le 7.$$

Next we have (recall (9.22))

$$\sqrt{-2} - K = \left(\frac{(K^2+2)}{p}w_8\right)p + (K + w_8)\phi_8^-(w_8) \in P_1 ,$$

and so

$$\lambda(\sqrt{-2}) = \sqrt{-2} + P_1 = K + P_1 ,$$

giving

$$\tau(\sqrt{-2}) = \theta(K + P_1) \equiv K \pmod{p}.$$

But

$$\tau(\sqrt{-2}) = \tau(w_8 + w_8^3)$$

$$= \gamma^{\left(\frac{q-1}{8}\right)k} + \gamma^{3\left(\frac{q-1}{8}\right)k}$$

$$\equiv \begin{cases} K \pmod{p}, & \text{if } k = 1,3, \\ -K \pmod{p}, & \text{if } k = 5,7, \end{cases}$$

proving that $k = 1$ or $3$. Hence

$$\tau(w_8) = \gamma^{\left(\frac{q-1}{8}\right)k}, \quad k = 1 \text{ or } 3.$$

Applying the homomorphism $\tau$ to (9.30), we obtain

$$\sum_{\substack{\alpha \in F^*_{p^2} \\ tr(\alpha)=1}} \gamma^{k\left(\frac{q-1}{8}\right)\mathrm{ind}_{\gamma'}(\alpha)} \equiv (-1)^{\frac{p+5}{8}}(C + D_1K) \pmod{p},$$

where $\gamma' = \gamma^{(q-1)/(p^2-1)}$, that is

$$\sum_{\substack{\alpha \in F^*_{p^2} \\ tr(\alpha)=1}} \alpha^{k\left(\frac{p^2-1}{8}\right)} \equiv (-1)^{\frac{p+5}{8}}(C + D_1K) \pmod{p}. \tag{9.31}$$

By Lemma 1, the left hand side of (9.31) is congruent to 0 (mod $p$) and by (9.23) the right hand side is congruent to $(-1)^{\frac{p+5}{8}}(C+2DD_1/C)$ (mod $p$). Hence we have $0 \equiv C^2 + 2DD_1$ (mod $p$), that is (as $C^2 \equiv -2D^2$ (mod $p$)) $D_1 \equiv D$ (mod $p$), and so $D_1 = D$, as claimed. Hence we have from (9.30)

$$E_{p^2}(w_8) = (-1)^{\frac{p+5}{8}}(C + D\sqrt{-2}). \tag{9.32}$$

Next from Theorem B, we deduce

$$E_q(w_8) = \begin{cases} \dfrac{\left(g_{p^2}(w_8)\right)^{f/2}}{g_{p^2}(w_8^{f/2})} (E_{p^2}(w_8))^{f/2}, & \text{if } f \equiv 2 \pmod 4, \\ \dfrac{\left(g_{p^2}(w_8)\right)^{f/2}}{p} (E_{p^2}(w_8))^{f/2}, & \text{if } f \equiv 0 \pmod 4. \end{cases} \tag{9.33}$$

Finally, with $\gamma' = \gamma^{\frac{q-1}{p^2-1}}$, we have for $n$ odd

$$\begin{aligned} g_{p^2}(w_8^n) &= \sum_{k \in F_p^*} w_8^{n\,\mathrm{ind}_{\gamma'}(k)} \exp(2\pi ik/p) \\ &= \sum_{k \in F_p^*} w_8^{n(p+1)\mathrm{ind}_g(k)} \exp(2\pi ik/p) \\ &= \sum_{k=1}^{p-1} (-1)^{\mathrm{ind}_g(k)} \exp(2\pi ik/p) \\ &= \sum_{k=1}^{p-1} \left(\frac{k}{p}\right) \exp(2\pi ik/p), \end{aligned}$$

that is, by (3.1) as $p \equiv 3$ (mod 4),

$$g_{p^2}(w_8^n) = ip^{1/2}. \tag{9.34}$$

From (9.32), (9.33) and (9.34), we deduce

$$E_q(w_8) = \begin{cases} (-1)^{f/4}\, p^{f/4-1}(C + D\sqrt{-2})^{f/2}, & \text{if } f \equiv 0 \pmod 4, \\ (-1)^{\frac{f-2}{4}+\frac{p+5}{8}}\, p^{\frac{f-2}{4}}(C + D\sqrt{-2})^{f/2}, & \text{if } f \equiv 2 \pmod 4. \end{cases}$$

(c) $p \equiv 5 \pmod 8$, $f \equiv 0 \pmod 2$. As $p \equiv 5 \pmod 8$ we determine integers $A$ and $B$ as in Theorem 3, that is, by

$$\begin{cases} p = A^2 + B^2, & A \equiv 1 \pmod 4, \\ B \equiv g^{\frac{p-1}{4}} A \equiv \gamma^{\frac{q-1}{4}} A \pmod p. \end{cases} \tag{9.35}$$

Set $m = \mathrm{ind}_g(2)$. As $p \equiv 5 \pmod 8$ we have $m \equiv 1 \pmod 2$. Thus, as $(B/A)^2 \equiv -1 \pmod p$, we have

$$2^{\frac{p-1}{4}} \equiv (g^m)^{\frac{p-1}{4}} \equiv \left(g^{\frac{p-1}{4}}\right)^m \equiv (B/A)^m \pmod p,$$

that is

$$2^{\frac{p-1}{4}} \equiv (-1)^{\frac{m-1}{2}} B/A \pmod p. \tag{9.36}$$

Hence, from the work of Gauss [8] (see also [6], [11], [15], [20]), it follows that

$$B \equiv 2(-1)^{(m-1)/2} \pmod 8. \tag{9.37}$$

Before proceeding we prove a lemma we will need.

**Lemma 3.** $$\begin{bmatrix} \frac{3p-7}{8} \\ \frac{p-5}{8} \end{bmatrix} \equiv (-1)^{\frac{p+3}{8}+\frac{m-1}{2}}\, 2B \pmod p.$$

Proof. We have

$$\begin{bmatrix} \frac{3p-7}{8} \\ \frac{p-5}{8} \end{bmatrix} = \frac{\left(\frac{3p-7}{8}\right)!}{\left(\frac{p-5}{8}\right)!\left(\frac{p-1}{4}\right)!}$$

$$= \frac{\left(\frac{3p+1}{8}\right)!}{\left(\frac{p-5}{8}\right)!\left(\frac{p-1}{4}\right)!\left(\frac{3p+1}{8}\right)}$$

$$\equiv \frac{(-1)^{\frac{p-5}{8}} 2^{\frac{p-5}{4}} \left(\frac{p-1}{2}\right)!}{\left(\frac{p-5}{4}\right)! \left(\frac{p-1}{4}\right)! \left(\frac{3p+1}{8}\right)} \pmod{p} \text{ (by Lemma 2)}$$

$$\equiv (-1)^{\frac{p-5}{8}} 2^{\frac{p-1}{4}} \begin{bmatrix} \frac{p-1}{2} \\ \frac{p-1}{4} \end{bmatrix} \frac{\frac{p-1}{4}}{\frac{3p+1}{4}} \pmod{p}$$

$$\equiv (-1)^{\frac{p-5}{8}} (-1)^{\frac{m-1}{2}} \frac{B}{A} \cdot (2A) \cdot (-1) \pmod{p}$$

$$\equiv (-1)^{\frac{p+3}{8}+\frac{m-1}{2}} 2B \pmod{p},$$

where we have used Gauss's result $2A \equiv \binom{\frac{p-1}{2}}{\frac{p-1}{4}} \pmod{p}$. This completes the proof of Lemma 3.

The least positive integer $\ell$ such that $p^{\ell} \equiv 1 \pmod 8$ is $\ell = 2$, so we first determine $E_{p^2}(w_8) = E_{p^2}(w_8, \gamma^{\frac{q-1}{p-1}})$. Appealing to Theorem A(a) we have

$$E_{p^2}(w_8) = E_{p^2}(w_8^5), \quad E_{p^2}(w_8^3) = E_{p^2}(w_8^7).$$

Thus $E_{p^2}(w_8)$ is fixed under the automorphism $\sigma_5 : w_8 \longrightarrow w_8^5$. Hence $E_{p^2}(w_8)$ belongs to $Q(w_8^2) = Q(i)$. As $E_{p^2}(w_8)$ is an algebraic integer, we must have $E_{p^2}(w_8) \in Z + Zi$ (the ring of integers of $Q(i)$). $Z + Zi$ is a unique factorization domain. By Theorem A(c), we have

$$E_{p^2}(w_8) E_{p^2}(w_8^7) = p, \tag{9.38}$$

so that in view of (9.35) we must have

$$E_{p^2}(w_8) = \theta(A + B_1 i), \tag{9.39}$$

where $\theta$ is a unit of $Z + Zi$ (that is $\theta = \pm 1, \pm i$), and $B_1 = \pm B$.

We show first that $\theta = \pm i$. We have (setting $\gamma' = \gamma^{\frac{q-1}{p^2-1}}$)

$$\sum_{k=0}^{7} E_{p^2}(w_8^k) = \sum_{k=0}^{7} \sum_{\substack{\alpha\in F^*_{p^2} \\ tr(\alpha)=1}} w_8^{k\,\mathrm{ind}_{\gamma'}(\alpha)}$$

$$= \sum_{\substack{\alpha\in F^*_{p^2} \\ tr(\alpha)=1}} \sum_{k=0}^{7} w_8^{k\,\mathrm{ind}_{\gamma'}(\alpha)}$$

$$= 8 \sum_{\substack{\alpha\in F^*_{p^2} \\ tr(\alpha)=1 \\ \mathrm{ind}_{\gamma'}(\alpha)\equiv 0 \ (\mathrm{mod}\ 8)}} 1$$

that is,

$$\sum_{k=0}^{7} E_{p^2}(w_8^k) \equiv 0 \ (\mathrm{mod}\ 8). \tag{9.40}$$

As

$$\begin{cases} E_{p^2}(1) = p \ (\S 1),\ E_{p^2}(w_8) = 1 & \text{(Theorem 1)}, \\ E_{p^2}(w_4) = A + Bi,\ E_{p^2}(w_4^3) = A - Bi, & \text{(Theorem 3(a))}, \end{cases} \tag{9.41}$$

we obtain from (9.40) and (9.41) (noting $p \equiv 5 \ (\mathrm{mod}\ 8)$, $A \equiv 1 \ (\mathrm{mod}\ 4)$)

$$Re(E_{p^2}(w_8)) \equiv 0 \ (\mathrm{mod}\ 2),$$

proving that $\theta = \pm i$. Thus we have

$$E_{p^2}(w_8) = \rho i (A + B_1 i), \quad B_1 = \pm B, \quad \rho = \pm 1. \tag{9.42}$$

Next we show that $\rho = (-1)^{(m-1)/2}(B_1/B)$. By Theorem A(d) we have

$$E_{p^2}(w_8) \equiv p \frac{\left[\frac{p}{8}\right]!\left[\frac{5p}{8}\right]!}{\left[\frac{3p}{4}\right]!} \ (\mathrm{mod}\ \pi^2), \tag{9.43}$$

where $\pi = A + B_1 i$. As $p \equiv 5 \pmod 8$, appealing to Wilson's theorem and Lemma 3, we obtain

$$\frac{\left[\frac{p}{8}\right]!\left[\frac{5p}{8}\right]!}{\left[\frac{3p}{4}\right]!} = \frac{\left(\frac{p-5}{8}\right)!\left(\frac{5p-1}{8}\right)!}{\left(\frac{3p-3}{4}\right)!}$$

$$\equiv (-1)^{\frac{p-5}{8}} \frac{\left(\frac{p-5}{8}\right)!\left(\frac{p-1}{4}\right)!}{\left(\frac{3p-7}{8}\right)!} \pmod p$$

$$\equiv (-1)^{\frac{p-5}{8}} \begin{bmatrix} \frac{3p-7}{8} \\ \frac{p-5}{8} \end{bmatrix}^{-1} \pmod p$$

$$\equiv (-1)^{\frac{m+1}{2}}/2B \pmod p.$$

Hence we have from (9.43)

$$E_{p^2}(w_8) \equiv \pi\bar{\pi}(-1)^{\frac{m+1}{2}}(2B)^{-1} \pmod{\pi^2},$$

and so, appealing to (9.42), we obtain

$$\rho i \equiv \bar{\pi}(-1)^{\frac{m+1}{2}}(2B)^{-1} \pmod \pi$$

$$\equiv (\bar{\pi} - \pi)(-1)^{\frac{m+1}{2}}(2B)^{-1} \pmod \pi$$

$$\equiv (-2B_1 i)(-1)^{\frac{m+1}{2}}(2B)^{-1} \pmod \pi$$

$$\equiv (-1)^{\frac{m-1}{2}}(B_1/B)i \pmod \pi,$$

proving that $\rho = (-1)^{\frac{m-1}{2}}(B_1/B)$ as asserted. Thus we have shown that

$$E_{p^2}(w_8) = (-1)^{\frac{m-1}{2}}(B_1/B)\, i\,(A + B_1 i), \quad \text{where } B_1 = \pm B. \tag{9.44}$$

Next we show that $B_1 = B$. As $p \equiv 5 \pmod 8$ the cyclotomic polynomial $\phi_8(x) = \frac{x^8-1}{x^4-1} = x^4 + 1$ is congruent to the product of two distinct irreducible quadratic polynomials modulo $p$, namely

$$\phi_8(x) \equiv \phi_8^-(x)\phi_8^+(x),$$

where

$$\phi_8^-(x) = x^2 - g^{\frac{p-1}{4}}, \quad \phi_8^+(x) = x^2 + g^{\frac{p-1}{4}}.$$

Hence, by Kummer's theorem, the principal ideal $\langle p \rangle$ of the ring $R$ of integers of $Q(w_8)$ factors into the product of two prime ideals, namely

$$\langle p \rangle = P_1 P_2 ,$$

where

$$P_1 = \langle p, \phi_8^-(w_8)\rangle, \quad P_2 = \langle p, \phi_8^+(w_8)\rangle, \quad N(P_1) = N(P_2) = p^2.$$

Thus $R/P_1$ is a finite field with $p^2$ elements. Hence there exists an isomorphism $\theta : R/P_1 \longrightarrow F_{p^2}$. Let $\lambda : R \longrightarrow R/P_1$ be the canonical homomorphism defined by $\lambda(\alpha) = \alpha + P_1 (\alpha \in R)$. Set $\tau = \theta \circ \lambda$ so that $\tau$ is a homomorphism such that $\tau : R \xrightarrow{\text{onto}} F_{p^2}$. Clearly we have $\tau(w_8) \neq 0$, otherwise $\tau(R) = \{0\}$. Also we have $\tau(w_8) \neq 1$, otherwise $\tau(R) \subseteq F_p$. Hence $\tau(w_8) \in F_{p^2}^* \backslash \{1\}$ and so, as $\gamma^{\frac{q-1}{p^2-1}}$ generates $F_{p^2}^*$, there exists an integer $k_1 (1 \leq k_1 \leq p^2 - 1)$ such that

$$\tau(w_8) = \gamma^{\left(\frac{p^f-1}{p^2-1}\right)k_1}.$$

Then we have

$$\gamma^{8\left(\frac{p^f-1}{p^2-1}\right)k_1} = (\tau(w_8))^8 = \tau(1) = \theta(\lambda(1)) = \theta(1 + P_1) \equiv 1 \pmod{p},$$

so that $p^f - 1 \mid 8\left(\frac{p^f-1}{p^2-1}\right)k_1$, that is $\frac{p^2-1}{8} \mid k_1$, say $k_1 = \left(\frac{p^2-1}{8}\right)k$, $1 \leq k \leq 7$, and so

$$\tau(w_8) = \gamma^{\left[\frac{p^f-1}{8}\right]k}, \quad 1 \leq k \leq 7.$$

Next we have

$$i - g^{\frac{p-1}{4}} = w_8^2 - g^{\frac{p-1}{4}} = \phi_8^-(w_8) \in P_1 ,$$

and so

$$\lambda(i) = i + P_1 = g^{\frac{p-1}{4}} + P_1 ,$$

giving

$$\tau(i) \equiv g^{\frac{p-1}{4}} \pmod{p}.$$

But

$$\tau(i) = \tau(w_8^2) = \gamma^{(\frac{q-1}{4})k} \equiv g^{(\frac{p-1}{4})k} \pmod{p},$$

showing that $k = 1$ or $5$. Hence we have

$$\tau(w_8) = \gamma^{(\frac{q-1}{8})k} \quad (k = 1 \text{ or } 5).$$

Applying the homomorphism $\tau$ to (9.44) we obtain

$$\sum_{\substack{\alpha\in F^*_{p^2}\\ tr(\alpha)=1}} \gamma^{k(\frac{q-1}{8})\mathrm{ind}_{\gamma'}(\alpha)} \equiv (-1)^{\frac{m-1}{2}} (B_1/B) g^{(\frac{p-1}{4})k}(A + B_1\, g^{(\frac{p-1}{4})}) \pmod{p},$$

where $\gamma' = \gamma^{\frac{q-1}{p^2-1}}$, that is

$$\sum_{\substack{\alpha\in F^*_{p^2}\\ tr(\alpha)=1}} \alpha^{k(\frac{p^2-1}{8})} \equiv (-1)^{\frac{m-1}{2}} (B_1/B) g^{(\frac{p-1}{4})k}(A + B_1\, g^{(\frac{p-1}{4})}) \pmod{p}.$$

By Lemma 1 the left hand side is $0 \pmod{p}$. Hence, as $g^{\frac{p-1}{4}} \equiv B/A \pmod{p}$, we obtain $A + (B_1B/A) \equiv 0 \pmod{p}$, showing that $B_1 \equiv B \pmod{p}$ (as $A^2 \equiv -B^2 \pmod{p}$), and thus $B_1 = B$ as asserted. Hence we have

$$E_{p^2}(w_8) = (-1)^{\frac{m-1}{2}} i(A + Bi). \tag{9.45}$$

Next, appealing to Theorem B, we have

$$E_q(w_8) = \begin{cases} (-1)^{f/2-1} \dfrac{\left[g_{p^2}(w_8)\right]^{f/2}}{g_{p^2}(w_8^{f/2})} (E_{p^2}(w_8))^{f/2}, & \text{if } f \not\equiv 0 \pmod 8, \\ \dfrac{\left[g_{p^2}(w_8)\right]^{f/2}}{p} (E_{p^2}(w_8))^{f/2}, & \text{if } f \equiv 0 \pmod 8. \end{cases} \tag{9.46}$$

Finally with $\gamma' = \gamma^{(q-1)/(p^2-1)}$ we have for any integer $n$

$$\begin{aligned} g_{p^2}(w_8^n) &= \sum_{k \in F_p^*} w_8^{n \operatorname{ind}_{\gamma'}(k)} \exp(2\pi i k/p) \\ &= \sum_{k \in F_p^*} w_n^{n(p+1)\operatorname{ind}_g(k)} \exp(2\pi i k/p) \\ &= \sum_{k \in F_p^*} w_4^{3n \operatorname{ind}_g(k)} \exp(2\pi i k/p) \\ &= g_p(w_4^{3n}). \end{aligned}$$

This proves that the value of $g_{p^2}(w_8^n)$ only depends upon $n \pmod 4$ not $n \pmod 8$. Appealing to (3.1) and (5.5), we have

$$\begin{cases} (g_{p^2}(w_8^n))^2 = \begin{cases} -(A-Bi)p^{1/2}, & \text{if } n \equiv 1 \pmod 4, \\ -(A+Bi)p^{1/2}, & \text{if } n \equiv 3 \pmod 4. \end{cases} \\ g_{p^2}(w_4) = p^{1/2}, \quad \text{if } n \equiv 2 \pmod 4. \end{cases} \tag{9.47}$$

Putting (9.45), (9.46) and (9.47), together we obtain

$$E_q(w_8) = \begin{cases} p^{\frac{3f}{4}-1}(A+Bi)^{f/4}, & \text{if } f \equiv 0 \pmod 8, \\ (-1)^{\frac{\operatorname{ind}_g(2)-1}{2}}\, i p^{\frac{3f-6}{8}} (A+Bi)^{\frac{f+2}{4}}, & \text{if } f \equiv 2 \pmod 8, \\ -p^{(3f-4)/8}(A+Bi)^{f/4}, & \text{if } f \equiv 4 \pmod 8, \\ (-1)^{\frac{\operatorname{ind}_g(2)-1}{2}}\, i p^{\frac{3f-2}{8}} (A+Bi)^{(f-2)/4}, & \text{if } f \equiv 6 \pmod 8. \end{cases}$$

(d) $p \equiv 7 \pmod 8$, $f \equiv 0 \pmod 2$. By Theorem C, with $m = 8$ and $\ell = 1$, we have

$$E_q(w_8) = (-1)^{\frac{f}{2}(\frac{p-7}{8})}\ p^{f/2-1}.$$

This completes the proof of the following theorem.

**Theorem 7.** *(a) If $p \equiv 1 \pmod 8$, let $A,B,C,D$ be the unique integers given by*

$$\begin{cases} p = A^2 + B^2, \\ A \equiv 1 \pmod 4,\ B \equiv g^{\frac{p-1}{4}} A \pmod p, \end{cases}$$

*and*

$$\begin{cases} p = C^2 + 2D^2, \\ C \equiv 1 \pmod 4,\ D \equiv \left(g^{\frac{p-1}{8}} + g^{3(\frac{p-1}{8})}\right) C/2 \pmod p. \end{cases}$$

*Then we have*

$$E_q(w_8) = \epsilon p^{\alpha}(A + Bi)^{\beta}(C + D\sqrt{-2})^{\delta},$$

*where*

$$\epsilon = \begin{cases} 1, & \text{if } f \equiv 0,1,4,5 \pmod 8, \\ (-1)^{\frac{p-1}{8}}, & \text{if } f \equiv 3,7 \pmod 8, \\ i^{3\,\mathrm{ind}_g(2)}, & \text{if } f \equiv 2 \pmod 8, \\ i^{\mathrm{ind}_g(2)}, & \text{if } f \equiv 6 \pmod 8; \end{cases}$$

$$\alpha = \begin{cases} f/8 - 1, & \text{if } f \equiv 0 \pmod 8, \\ [f/8], & \text{if } f \not\equiv 0 \pmod 8; \end{cases}$$

$$\beta = \begin{cases} f/4, & \text{if } f \equiv 0 \pmod 4, \\ \frac{f-1}{4}, & \text{if } f \equiv 1 \pmod 4, \\ \frac{f-2}{4}, & \text{if } f \equiv 2 \pmod 8, \\ \frac{f+2}{4}, & \text{if } f \equiv 6 \pmod 8, \\ \frac{f+1}{4}, & \text{if } f \equiv 3 \pmod 4; \end{cases}$$

$$\delta = \begin{cases} f/2, & \text{if } f \equiv 0 \pmod 2, \\ (f-1)/2, & \text{if } f \equiv 1,3 \pmod 8, \\ (f+1)/2, & \text{if } f \equiv 5,7 \pmod 8). \end{cases}$$

*(b) If $p \equiv 3 \pmod 8$ and $f \equiv 0 \pmod 2$ define the integers $C$, $D$ uniquely by*

$$\begin{cases} p = C^2 + 2D^2, \quad C \equiv 1 \pmod 4, \\ D \equiv \left[\gamma^{\frac{q-1}{8}} + \gamma^{3\left(\frac{q-1}{8}\right)}\right] D/2 \pmod p. \end{cases}$$

*Then we have*

$$E_q(w_8) = \begin{cases} (-1)^{f/4} p^{f/4-1}(C + D\sqrt{-2})^{f/2}, & \text{if } f \equiv 0 \pmod 4, \\ (-1)^{\frac{(f-2)}{4} + \frac{p+5}{8}} p^{(f-2)/4}(C + D\sqrt{-2})^{f/2}, & \text{if } f \equiv 2 \pmod 4. \end{cases}$$

*(c) If $p \equiv 5 \pmod 8$ and $f \equiv 0 \pmod 2$ define the integers $A$, $B$ uniquely by*

$$\begin{cases} p = A^2 + B^2, \quad A \equiv 1 \pmod 4, \\ B = g^{\frac{p-1}{4}} A \pmod p. \end{cases}$$

*Set $m = \mathrm{ind}_g(2)$. Then we have*

$$E_q(w_8) = \begin{cases} p^{3f/4-1}(A + Bi)^{f/4}, & \text{if } f \equiv 0 \pmod 8, \\ (-1)^{\frac{m-1}{2}} i p^{\frac{3f-6}{8}} (A + Bi)^{\frac{f+2}{4}}, & \text{if } f \equiv 2 \pmod 8, \\ -p^{\frac{3f-4}{8}} (A + Bi)^{f/4}, & \text{if } f \equiv 4 \pmod 8, \\ (-1)^{\frac{m-1}{2}} i p^{\frac{3f-2}{8}} (A + Bi)^{\frac{f-2}{4}}, & \text{if } f \equiv 6 \pmod 8). \end{cases}$$

*(d) If $p \equiv 7 \pmod 8$ and $f \equiv 0 \pmod 2$ then we have*

$$E_q(w_8) = (-1)^{\frac{f}{2}\left(\frac{p-7}{8}\right)} p^{f/2-1}.$$

Some numerical examples illustrating Theorem 7(a), (b), (c) are given in Tables 22–31 in §11.

## 10. Acknowledgement

The authors would like to thank Nicholas Buck and Iain deMille for their valuable assistance in the computer related areas of this project.

## 11. Tables

The values of the Eisenstein sums given in the tables below were computed on a Honeywell DPS 8/47 computer at Carleton University. The programs were written in PASCAL. For each prime $p$ and integer $f \geq 2$, an irreducible polynomial $x^f + a_{f-1}x^{f-1} + \dots + a_1x + a_0 \pmod{p}$ of degree $f$ was found using a modification of Berlekamp's procedure. This gave a concrete realization of $F_{p^f}$ as

$$F_{p^f} = \{b_0 + b_1x + \dots + b_{f-1}x^{f-1} \mid x^f = -a_{f-1}x^{f-1} - \dots - a_1x - a_0\}.$$

A generator $\gamma$ of $F^*_{p^f}$ was then found by checking that $\gamma^{(p^f-1)/p_i} \neq 1$, for every prime $p_i \mid p^f - 1$. The sum $E_q(w_n)$ was then calculated by means of the formula

$$E_q(w_m) = \sum_{t=0}^{m-1} w_m^t \text{ card } \{s \mid 0 \leq s < \tfrac{q-1}{p-1},\ s - \left(\tfrac{q-1}{p-1}\right) \text{ind}_g(tr(\gamma^s)) \equiv t \pmod{m}\}.$$

TABLE 1

| $m = 3 \quad : \quad f = 2 \quad : \quad p \equiv 1 \pmod 3 \qquad (\alpha = 0,\ \beta = 1)$ | | | | | | | |
|---|---|---|---|---|---|---|---|
| $p \equiv 1 \pmod 3$ | $F_{p^2}^* = \langle\gamma\rangle$ | $E_{p^2}(\omega_3)$ | $g = \gamma^{\frac{q-1}{p-1}}$ | $g^{\frac{p-1}{3}}$ | $L$ | $M$ | $\frac{1}{2}(L + 3M\sqrt{-3})$ |
| 7 | $\gamma_2 = 1 + x$, $x^2 = -2$ | $\frac{1}{2}(-1 + 3\sqrt{-3})$ | 3 | 2 | $-1$ | $+1$ | $\frac{1}{2}(-1 + 3\sqrt{-3})$ |
| 13 | $\gamma_2 = 2 + x$, $x^2 = 2$ | $\frac{1}{2}(5 + 3\sqrt{-3})$ | 2 | 3 | $+5$ | $+1$ | $\frac{1}{2}(5 + 3\sqrt{-3})$ |
| 19 | $\gamma_2 = 2 + x$, $x^2 = 2$ | $\frac{1}{2}(-7 + 3\sqrt{-3})$ | 2 | 7 | $-7$ | $+1$ | $\frac{1}{2}(-7 + 3\sqrt{-3})$ |
| 31 | $\gamma_2 = 1 + x$, $x^2 = -2$ | $\frac{1}{2}(-4 - 6\sqrt{-3})$ | 3 | 25 | $-4$ | $-2$ | $\frac{1}{2}(-4 - 6\sqrt{-3})$ |
| 37 | $\gamma_2 = 2 + x$, $x^2 = 2$ | $\frac{1}{2}(11 - 3\sqrt{-3})$ | 2 | 26 | $+11$ | $-1$ | $\frac{1}{2}(11 - 3\sqrt{-3})$ |

TABLE 2

| $m = 3 \quad : \quad f = 3 \quad : \quad p \equiv 1 \pmod 3 \qquad (\alpha = 0,\ \beta = 1)$ | | | | | | | |
|---|---|---|---|---|---|---|---|
| $p \equiv 1 \pmod 3$ | $F_{p^3}^* = \langle\gamma\rangle$ | $E_{p^3}(\omega_3)$ | $g = \gamma^{\frac{q-1}{p-1}}$ | $g^{\frac{p-1}{3}}$ | $L$ | $M$ | $\frac{1}{2}(L + 3M\sqrt{-3})$ |
| 7 | $\gamma_3 = 5 + x$, $x^3 = -1 - x$ | $\frac{1}{2}(-1 + 3\sqrt{-3})$ | 3 | 2 | $-1$ | $+1$ | $\frac{1}{2}(-1 + 3\sqrt{-3})$ |
| 13 | $\gamma_3 = 4 + x$, $x^3 = -5 - x$ | $\frac{1}{2}(5 + 3\sqrt{-3})$ | 11 | 3 | $+5$ | $+1$ | $\frac{1}{2}(5 + 3\sqrt{-3})$ |
| 19 | $\gamma_3 = 3 + x$, $x^3 = -1 - x$ | $\frac{1}{2}(-7 - 3\sqrt{-3})$ | 10 | 11 | $-7$ | $-1$ | $\frac{1}{2}(-7 - 3\sqrt{-3})$ |
| 31 | $\gamma_3 = 4 + x$, $x^3 = -3 - x$ | $\frac{1}{2}(-4 - 6\sqrt{-3})$ | 3 | 25 | $-4$ | $-2$ | $\frac{1}{2}(-4 - 6\sqrt{-3})$ |
| 37 | $\gamma_3 = 9 + x$, $x^3 = -3 - x$ | $\frac{1}{2}(11 + 3\sqrt{-3})$ | 32 | 10 | $+11$ | $+1$ | $\frac{1}{2}(11 + 3\sqrt{-3})$ |

TABLE 3

| $m=3$ : $f=4$ : $p\equiv 1 \pmod 3$ $(\alpha=\beta=1)$ | | | | | | | |
|---|---|---|---|---|---|---|---|
| $p\equiv 1 \pmod 3$ | $F_{p^4}^* = \langle\gamma\rangle$ | $E_{p^4}(\omega_3)$ | $g=\gamma^{\frac{q-1}{p-1}}$ | $g^{\frac{p-1}{3}}$ | $L\equiv -1 \pmod 3$ | $M$ | $p\left(\frac{1}{2}(L+3M\sqrt{-3})\right)$ |
| 7 | $\gamma = 3+2x$, $x^4 = -3-2x^2$ | $\frac{1}{2}(-7-21\sqrt{-3})$ | 5 | 4 | $-1$ | $-1$ | $\frac{1}{2}(-7-21\sqrt{-3})$ |
| 13 | $\gamma = 4+x$, $x^4 = -2$ | $\frac{1}{2}(65+39\sqrt{-3})$ | 11 | 3 | $+5$ | $+1$ | $\frac{1}{2}(65+39\sqrt{-3})$ |
| 19 | $\gamma = 1+2x$, $x^4 = -2-2x^2$ | $\frac{1}{2}(-133+57\sqrt{-3})$ | 3 | 7 | $-7$ | $+1$ | $\frac{1}{2}(-133+57\sqrt{-3})$ |

TABLE 4

| $m=3$ : $f=5$ : $p\equiv 1 \pmod 3$ $(\alpha=1,\ \beta=2)$ | | | | | | | |
|---|---|---|---|---|---|---|---|
| $p\equiv 1 \pmod 3$ | $F_{p^5}^* = \langle\gamma\rangle$ | $E_{p^5}(\omega_3)$ | $g=\gamma^{\frac{q-1}{p-1}}$ | $g^{\frac{p-1}{3}}$ | $L$ | $M$ | $p\left(\frac{1}{2}(L+3M\sqrt{-3})\right)^2$ |
| 7 | $\gamma = 1+x$, $x^5 = -5-x-x^2$ | $\frac{1}{2}(-91-21\sqrt{-3})$ | 3 | 2 | $-1$ | $+1$ | $\frac{1}{2}(-91-21\sqrt{-3})$ |
| 13 | $\gamma = x$, $x^5 = -6-x-x^2$ | $\frac{1}{2}(-13-195\sqrt{-3})$ | 7 | 9 | $+5$ | $-1$ | $\frac{1}{2}(-13-195\sqrt{-3})$ |

TABLE 5

| $m=3$ : $f=6$ : $p\equiv 1 \pmod 3$ $(\alpha=1,\ \beta=2)$ | | | | | | | |
|---|---|---|---|---|---|---|---|
| $p\equiv 1 \pmod 3$ | $F_{p^6}^* = \langle\gamma\rangle$ | $E_{p^6}(\omega_3)$ | $g=\gamma^{\frac{q-1}{p-1}}$ | $g^{\frac{p-1}{3}}$ | $L$ | $M$ | $p\left(\frac{1}{2}(L+3M\sqrt{-3})\right)^2$ |
| 7 | $\gamma = 2+x$, $x^6 = -2-2x-x^2$ | $\frac{1}{2}(-91-21\sqrt{-3})$ | 3 | 2 | $-1$ | $+1$ | $\frac{1}{2}(-91-21\sqrt{-3})$ |

TABLE 6

| $m=4$ : $f=2$ : $p\equiv 1 \pmod 4$ $(\alpha=0,\ \beta=1,\ \varepsilon=1)$ | | | | | | | |
|---|---|---|---|---|---|---|---|
| $p\equiv 1 \pmod 4$ | $F_{p^2}^{*}=\langle\gamma\rangle$ | $E_{p^2}(\omega_4)$ | $g=\gamma^{\frac{q-1}{p-1}}$ | $g^{\frac{p-1}{4}}$ | $A$ | $B$ | $A+Bi$ |
| 5 | $\gamma_2 = 2+x$, $x^2 = 2$ | $1+2i$ | 2 | 2 | +1 | +2 | $1+2i$ |
| 13 | $\gamma_2 = 2+x$, $x^2 = 2$ | $-3+2i$ | 2 | 8 | −3 | +2 | $-3+2i$ |
| 17 | $\gamma_2 = 2+x$, $x^2 = -3$ | $1+4i$ | 7 | 4 | +1 | +4 | $1+4i$ |
| 29 | $\gamma_2 = 4+x$, $x^2 = 2$ | $5+2i$ | 14 | 12 | +5 | +2 | $5+2i$ |

TABLE 7

| $m=4$ : $f=3$ : $p\equiv 1 \pmod 4$ $(\alpha=0,\ \beta=2,\ \varepsilon=(-1)^{(p-1)/4})$ | | | | | | | |
|---|---|---|---|---|---|---|---|
| $p\equiv 1 \pmod 4$ | $F_{p^3}^{*}=\langle\gamma\rangle$ | $E_{p^3}(\omega_4)$ | $g=\gamma^{\frac{q-1}{p-1}}$ | $g^{\frac{p-1}{4}}$ | $A$ | $B$ | $(-1)^{(p-1)/4}(A+Bi)^2$ |
| 5 | $\gamma_3 = 4+x$, $x^3 = -1-x$ | $3-4i$ | 2 | 2 | +1 | +2 | $3-4i$ |
| 13 | $\gamma_3 = 4+x$, $x^3 = -5-x$ | $-5-12i$ | 11 | 5 | −3 | −2 | $-5-12i$ |
| 17 | $\gamma_3 = x$, $x^3 = -3-x$ | $-15-8i$ | 14 | 13 | +1 | −4 | $-15-8i$ |
| 29 | $\gamma_3 = 1+x$, $x^3 = -4-x$ | $-21+20i$ | 27 | 17 | +5 | −2 | $-21+20i$ |

**TABLE 8**

| $m=4 \;:\; f=4 \;:\; p\equiv 1 \pmod 4 \quad (\alpha=0,\ \beta=2,\ \varepsilon=1)$ | | | | | | | |
|---|---|---|---|---|---|---|---|
| $p\equiv 1$ (mod 4) | $F^*_{p^4}=\langle\gamma\rangle$ | $E_{p^4}(\omega_4)$ | $g=\gamma^{\frac{q-1}{p-1}}$ | $g^{\frac{p-1}{4}}$ | $A$ | $B$ | $(A+Bi)^2$ |
| 5 | $\gamma_4 = 1+x$<br>$x^4=-2$ | $-3-4i$ | 3 | 3 | +1 | −2 | $-3-4i$ |
| 13 | $\gamma_4 = 4+x$<br>$x^4=-2$ | $5+12i$ | 11 | 5 | −3 | −2 | $5+12i$ |
| 17 | $\gamma_4 = 1+x+x^2$<br>$x^4=-3$ | $-15+8i$ | 7 | 4 | +1 | +4 | $-15+8i$ |

**TABLE 9**

| $m=4 \;:\; f=5 \;:\; p\equiv 1 \pmod 4 \quad (\alpha=1,\ \beta=2,\ \varepsilon=1)$ | | | | | | | |
|---|---|---|---|---|---|---|---|
| $p\equiv 1$ (mod 4) | $F^*_{p^5}=\langle\gamma\rangle$ | $E_{p^5}(\omega_4)$ | $g=\gamma^{\frac{q-1}{p-1}}$ | $g^{\frac{p-1}{4}}$ | $A$ | $B$ | $p(A+Bi)^2$ |
| 5 | $\gamma_5 = 1+x$<br>$x^5=-4-x-x^2$ | $-15+20i$ | 2 | 2 | +1 | +2 | $-15+20i$ |
| 13 | $\gamma_5 = x$<br>$x^5=-6-x-x^2$ | $65+156i$ | 7 | 5 | −3 | −2 | $65+156i$ |

**TABLE 10**

| $m=4 \;:\; f=6 \;:\; p\equiv 1 \pmod 4 \quad (\alpha=1,\ \beta=3,\ \varepsilon=1)$ | | | | | | | |
|---|---|---|---|---|---|---|---|
| $p\equiv 1$ (mod 4) | $F^*_{p^6}=\langle\gamma\rangle$ | $E_{p^6}(\omega_4)$ | $g=\gamma^{\frac{q-1}{p-1}}$ | $g^{\frac{p-1}{4}}$ | $A$ | $B$ | $p(A+Bi)^3$ |
| 5 | $\gamma_6 = 2+x$<br>$x^6=-1-x-x^2$ | $-55-10i$ | 2 | 2 | +1 | +2 | $-55-10i$ |

TABLE 11

| $m = 5 \; : \; f = 2 \; : \; p \equiv 1 \pmod 5 \quad (\alpha = 0, \beta = 1, \delta = 0, \varepsilon = -1)$ | | | | | | |
|---|---|---|---|---|---|---|
| $p \equiv 1 \pmod 5$ | $F_{p^2}^* = \langle \gamma \rangle$ | $E_{p^2}(\omega_5)$ | $g = \gamma^{\frac{q-1}{p-1}}$ | $g^{\frac{p-1}{5}}$ | $x, u, v, w$ | $-\tau(x, u, v, w)$ |
| 11 | $\gamma_2 = 2 + x$, $x^2 = 2$ | $\frac{1}{4}(-1 - 2i\sqrt{10+2\sqrt{5}} + i\sqrt{10-2\sqrt{5}} - 5\sqrt{5})$ | 2 | 4 | 1, 0, 1, 1 | $-\frac{1}{4}(1 + 2i\sqrt{10+2\sqrt{5}} - i\sqrt{10-2\sqrt{5}} + 5\sqrt{5})$ |
| 31 | $\gamma_2 = 1 + x$, $x^2 = -2$ | $\frac{1}{4}(-11 + 4i\sqrt{10+2\sqrt{5}} + 3i\sqrt{10-2\sqrt{5}} + 5\sqrt{5})$ | 3 | 16 | 11, −2, −1, −1 | $-\frac{1}{4}(11 - 4i\sqrt{10+2\sqrt{5}} - 3i\sqrt{10-2\sqrt{5}} - 5\sqrt{5})$ |
| 41 | $\gamma_2 = 2 + x$, $x^2 = -3$ | $\frac{1}{4}(9 + 6i\sqrt{10+2\sqrt{5}} - 3i\sqrt{10-2\sqrt{5}} + 5\sqrt{5})$ | 7 | 37 | −9, 0, −3, −1 | $-\frac{1}{4}(-9 - 6i\sqrt{10+2\sqrt{5}} + 3i\sqrt{10-2\sqrt{5}} - 5\sqrt{5})$ |
| 61 | $\gamma_2 = 2 + x$, $x^2 = 2$ | $\frac{1}{4}(-1 + 2i\sqrt{10+2\sqrt{5}} + 9i\sqrt{10-2\sqrt{5}} - 5\sqrt{5})$ | 2 | 9 | 1, −4, 1, 1 | $-\frac{1}{4}(1 - 2i\sqrt{10+2\sqrt{5}} - 9i\sqrt{10-2\sqrt{5}} + 5\sqrt{5})$ |

TABLE 12

| $m = 5 \; : \; f = 3 \; : \; p \equiv 1 \pmod 5 \quad (\alpha = 0, \beta = 1, \delta = 1, \varepsilon = 1)$ | | | | | | |
|---|---|---|---|---|---|---|
| $p \equiv 1 \pmod 5$ | $F_{p^3}^* = \langle \gamma \rangle$ | $E_{p^3}(\omega_5)$ | $g = \gamma^{\frac{q-1}{p-1}}$ | $g^{\frac{p-1}{5}}$ | $x, u, v, w$ | $+\tau(x, u, v, w)\tau(x, v, -u, -w)$ |
| 11 | $\gamma_3 = x$, $x^3 = -4 - x$ | $\frac{1}{4}(-31 - 6i\sqrt{10+2\sqrt{5}} - 7i\sqrt{10-2\sqrt{5}} + 5\sqrt{5})$ | 7 | 5 | 1, −1, 0, −1 | $\frac{1}{4}(-31 - 6i\sqrt{10+2\sqrt{5}} - 7i\sqrt{10-2\sqrt{5}} + 5\sqrt{5})$ |
| 31 | $\gamma_3 = 4 + x$, $x^3 = -3 - x$ | $\frac{1}{4}(-1 - 9i\sqrt{10+2\sqrt{5}} - 38i\sqrt{10-2\sqrt{5}} + 5\sqrt{5})$ | 3 | 16 | 11, −2, −1, −1 | $\frac{1}{4}(-1 - 9i\sqrt{10+2\sqrt{5}} - 38i\sqrt{10-2\sqrt{5}} + 5\sqrt{5})$ |
| 41 | $\gamma_3 = 3 + x$, $x^3 = -1 - x$ | $\frac{1}{4}(-11 + 12i\sqrt{10+2\sqrt{5}} + 39i\sqrt{10-2\sqrt{5}} + 45\sqrt{5})$ | 29 | 18 | −9, −3, 0, 1 | $\frac{1}{4}(-11 + 12i\sqrt{10+2\sqrt{5}} + 39i\sqrt{10-2\sqrt{5}} + 45\sqrt{5})$ |

TABLE 13

| $m=5 \;:\; f=4 \;:\; p\equiv 1 \pmod 5 \quad (\alpha=0,\ \beta=2,\ \delta=1,\ \varepsilon=-1)$ | | | | | | |
|---|---|---|---|---|---|---|
| $p\equiv 1$ (mod 5) | $F_{p^4}^{*}=\langle\gamma\rangle$ | $E_{p^4}(\omega_5)$ | $g=\gamma^{\frac{q-1}{p-1}}$ | $g^{\frac{p-1}{5}}$ | $x,u,v,w$ | $-\tau(x,u,v,w)^2\tau(x,v,-u,-w)$ |
| 11 | $\gamma = 5+x$<br>$x^4=-2-2x^2$ | $\frac{1}{4}(89-20i\sqrt{10+2\sqrt5}+25i\sqrt{10-2\sqrt5}+25\sqrt5)$ | 6 | 3 | $1,0,-1,1$ | $\frac{1}{4}(89-20i\sqrt{10+2\sqrt5}+25i\sqrt{10-2\sqrt5}+25\sqrt5)$ |
| 31 | $\gamma = 7+x$<br>$x^4=-3-2x^2$ | $\frac{1}{4}(409+135i\sqrt{10+2\sqrt5}+70i\sqrt{10-2\sqrt5}-125\sqrt5)$ | 22 | 8 | $11,1,-2,1$ | $\frac{1}{4}(409+135i\sqrt{10+2\sqrt5}+70i\sqrt{10-2\sqrt5}-125\sqrt5)$ |
| 41 | $\gamma = 2+x$<br>$x^4=-3$ | $\frac{1}{4}(981+90i\sqrt{10+2\sqrt5}-75i\sqrt{10-2\sqrt5}-25\sqrt5)$ | 19 | 37 | $-9,0,-3,-1$ | $\frac{1}{4}(981+90i\sqrt{10+2\sqrt5}-75i\sqrt{10-2\sqrt5}-25\sqrt5)$ |

TABLE 14

| $m=5 \;:\; f=5 \;:\; p\equiv 1 \pmod 5 \quad (\alpha=0,\ \beta=2,\ \delta=1,\ \varepsilon=-1)$ | | | | | | |
|---|---|---|---|---|---|---|
| $p\equiv 1$ (mod 5) | $F_{p^5}^{*}=\langle\gamma\rangle$ | $E_{p^5}(\omega_5)$ | $g=\gamma^{\frac{q-1}{p-1}}$ | $g^{\frac{p-1}{5}}$ | $x,u,v,w$ | $-\tau(x,u,v,w)^2\tau(x,v,-u,-w)$ |
| 11 | $\gamma = x$<br>$x^5=-4-x-x^2$ | $\frac{1}{4}(89-25i\sqrt{10+2\sqrt5}-20i\sqrt{10-2\sqrt5}-25\sqrt5)$ | 7 | 5 | $1,-1,0,-1$ | $\frac{1}{4}(89-25i\sqrt{10+2\sqrt5}-20i\sqrt{10-2\sqrt5}-25\sqrt5)$ |

TABLE 15

| $m=5 \;:\; f=6 \;:\; p\equiv 1 \pmod 5 \quad (\alpha=1,\ \beta=2,\ \delta=1,\ \varepsilon=-1)$ | | | | | | |
|---|---|---|---|---|---|---|
| $p\equiv 1$ (mod 5) | $F_{p^6}^{*}=\langle\gamma\rangle$ | $E_{p^6}(\omega_5)$ | $g=\gamma^{\frac{q-1}{p-1}}$ | $g^{\frac{p-1}{5}}$ | $x,u,v,w$ | $-p\tau(x,u,v,w)^2\tau(x,v,-u,-w)$ |
| 11 | $\gamma = 4+x$<br>$x^6=-1-x-x^2$ | $\frac{1}{4}(979-220i\sqrt{10+2\sqrt5}+275i\sqrt{10-2\sqrt5}+275\sqrt5)$ | 6 | 3 | $1,0,-1,1$ | $\frac{1}{4}(979-220i\sqrt{10+2\sqrt5}+275i\sqrt{10-2\sqrt5}+275\sqrt5)$ |

TABLE 16

| $m=6$ : $f=2$ : $p\equiv 1 \pmod 6$ $(\alpha=0,\ \beta=1,\ \varepsilon=(-1)^{\frac{p-1}{2}}\omega_3^{\mathrm{ind}_g(2)})$ | | | | | | | | |
|---|---|---|---|---|---|---|---|---|
| $p\equiv 1$ (mod 6) | $F_{p^2}^{*}=\langle\gamma\rangle$ | $E_{p^2}(\omega_6)$ | $g=\gamma^{\frac{q-1}{p-1}}$ | $g^{\frac{p-1}{3}}$ | $L$ | $M$ | $\mathrm{ind}_g(2)$ | $(-1)^{\frac{p-1}{2}}\omega_3^{\mathrm{ind}_g(2)}\frac{1}{2}(L+3M\sqrt{-3})$ |
| 7 | $\gamma_2 = 1+x$, $x^2=-2$ | $\frac{1}{2}(-5+\sqrt{-3})$ | 3 | 2 | −1 | +1 | 2 | $\frac{1}{2}(-5+\sqrt{-3})$ |
| 13 | $\gamma_2 = 2+x$, $x^2 = 2$ | $\frac{1}{2}(-7+\sqrt{-3})$ | 2 | 3 | +5 | +1 | 1 | $\frac{1}{2}(-7+\sqrt{-3})$ |
| 19 | $\gamma_2 = 2+x$, $x^2 = 2$ | $\frac{1}{2}(1+5\sqrt{-3})$ | 2 | 7 | −7 | +1 | 1 | $\frac{1}{2}(1+5\sqrt{-3})$ |
| 31 | $\gamma_2 = 1+x$, $x^2=-2$ | $2+3\sqrt{-3}$ | 3 | 25 | −4 | −2 | 24 | $2+3\sqrt{-3}$ |
| 37 | $\gamma_2 = 2+x$, $x^2 = 2$ | $\frac{1}{2}(-1+7\sqrt{-3})$ | 2 | 26 | +11 | −1 | 1 | $\frac{1}{2}(-1+7\sqrt{-3})$ |

TABLE 17

| $m=6$ : $f=3$ : $p\equiv 1 \pmod 6$ $(\alpha=0,\ \beta=2,\ \epsilon=(-1)^{(p-1)/2})$ | | | | | | | |
|---|---|---|---|---|---|---|---|
| $p\equiv 1$ (mod 6) | $F^*_{p^3}=\langle\gamma\rangle$ | $E_{p^3}(\omega_6)$ | $g=\gamma^{\frac{q-1}{p-1}}$ | $g^{\frac{p-1}{3}}$ | $L$ | $M$ | $(-1)^{\frac{p-1}{2}}\left(\frac{1}{2}(L+3M\sqrt{-3})\right)^2$ |
| 7 | $\gamma_3=5+x$<br>$x^3=-1-x$ | $\frac{1}{2}(13+3\sqrt{-3})$ | 3 | 2 | −1 | +1 | $\frac{1}{2}(13+3\sqrt{-3})$ |
| 13 | $\gamma_3=4+x$<br>$x^3=-5-x$ | $\frac{1}{2}(-1+15\sqrt{-3})$ | 11 | 3 | +5 | +1 | $\frac{1}{2}(-1+15\sqrt{-3})$ |
| 19 | $\gamma_3=3+x$<br>$x^3=-1-x$ | $\frac{1}{2}(-11-21\sqrt{-3})$ | 10 | 11 | −7 | −1 | $\frac{1}{2}(-11-21\sqrt{-3})$ |
| 31 | $\gamma_3=4+x$<br>$x^3=-3-x$ | $23-12\sqrt{-3}$ | 3 | 25 | −4 | −2 | $23-12\sqrt{-3}$ |
| 37 | $\gamma_3=9+x$<br>$x^3=-3-x$ | $\frac{1}{2}(47+33\sqrt{-3})$ | 32 | 10 | +11 | +1 | $\frac{1}{2}(47+33\sqrt{-3})$ |

TABLE 18

| $m=6$ : $f=4$ : $p\equiv 1 \pmod 6$ $(\alpha=0,\ \beta=3,\ \epsilon=\omega_3^{2\mathrm{ind}_g(2)})$ | | | | | | | | |
|---|---|---|---|---|---|---|---|---|
| $p\equiv 1$ (mod 6) | $F^*_{p^4}=\langle\gamma\rangle$ | $E_{p^4}(\omega_6)$ | $g=\gamma^{\frac{q-1}{p-1}}$ | $g^{\frac{p-1}{3}}$ | $L$ | $M$ | $\mathrm{ind}_g(2)$ | $\omega_3^{2\mathrm{ind}_g(2)}\left(\frac{1}{2}(L+3M\sqrt{-3})\right)^3$ |
| 7 | $\gamma_4=3+2x$<br>$x^4=-3-2x^2$ | $\frac{1}{2}(17-19\sqrt{-3})$ | 5 | 4 | −1 | −1 | 4 | $\frac{1}{2}(17-19\sqrt{-3})$ |
| 13 | $\gamma_4=4+x$<br>$x^4=-2$ | $\frac{1}{2}(89+17\sqrt{-3})$ | 11 | 3 | +5 | +1 | 7 | $\frac{1}{2}(89+17\sqrt{-3})$ |
| 19 | $\gamma_4=1+2x$<br>$x^4=-2-2x^2$ | $\frac{1}{2}(107-73\sqrt{-3})$ | 3 | 7 | −7 | +1 | 7 | $\frac{1}{2}(107-73\sqrt{-3})$ |
| 31 | $\gamma_4=7+x$<br>$x^4=-3-2x^2$ | $154-45\sqrt{-3}$ | 22 | 5 | −4 | +2 | 12 | $154-45\sqrt{-3}$ |

TABLE 19

| $m=6$ : $f=5$ : $p\equiv 1 \pmod 6$ $(\alpha=0,\ \beta=4,\ \epsilon=+1)$ | | | | | | | |
|---|---|---|---|---|---|---|---|
| $p\equiv 1$ (mod 6) | $F^*_{p^5}=\langle\gamma\rangle$ | $E_{p^5}(\omega_6)$ | $g=\gamma^{\frac{q-1}{p-1}}$ | $g^{\frac{p-1}{3}}$ | $L$ | $M$ | $\left(\frac{1}{2}(L+3M\sqrt{-3})\right)^4$ |
| 7 | $\gamma_5=1+x$<br>$x^5=-5-x-x^2$ | $\frac{1}{2}(71+39\sqrt{-3})$ | 3 | 2 | −1 | +1 | $\frac{1}{2}(71+39\sqrt{-3})$ |
| 13 | $\gamma_5=x$<br>$x^5=-6-x-x^2$ | $\frac{1}{2}(-337+15\sqrt{-3})$ | 7 | 9 | +5 | −1 | $\frac{1}{2}(-337+15\sqrt{-3})$ |
| 19 | $\gamma_5=4+x$<br>$x^5=-2-x-x^2$ | $\frac{1}{2}(-601-231\sqrt{-3})$ | 3 | 7 | −7 | +1 | $\frac{1}{2}(-601-231\sqrt{-3})$ |

TABLE 20

| $m=7 \;:\; f=3 \;:\; p\equiv 2 \text{ or } 4 \pmod 7$ | | | | | | |
|---|---|---|---|---|---|---|
| $p\equiv 2$ or 4 (mod 7) | $F_{p^3}^* = \langle\gamma\rangle$ | $E_{p^3}(\omega_7)$ | $S\equiv\sum_{k=1}^{6}\left(\frac{k}{7}\right)\gamma^{k\left(\frac{q-1}{7}\right)}$ (mod $p$) | $G$ | $H$ | $G+H\sqrt{-7}$ |
| 11 | $\gamma = x$<br>$x^3 = -4-x$ | $2-\sqrt{-7}$ | 2 | +2 | −1 | $2-\sqrt{-7}$ |
| 23 | $\gamma = x$<br>$x^3 = -3-x$ | $4-\sqrt{-7}$ | 4 | +4 | −1 | $4-\sqrt{-7}$ |
| 37 | $\gamma = 9+x$<br>$x^3 = -3-x$ | $-3-2\sqrt{-7}$ | 17 | −3 | −2 | $-3-2\sqrt{-7}$ |
| 53 | $\gamma = x$<br>$x^3 = -5-x$ | $-5-2\sqrt{-7}$ | 24 | −5 | −2 | $-5-2\sqrt{-7}$ |
| 67 | $\gamma = 2+x$<br>$x^3 = -3-x$ | $2+3\sqrt{-7}$ | 44 | +2 | +3 | $2+3\sqrt{-7}$ |
| 79 | $\gamma = 1+x$<br>$x^3 = -6-x$ | $4+3\sqrt{-7}$ | 25 | +4 | +3 | $4+3\sqrt{-7}$ |

TABLE 21

| $m=7 \;:\; f=6 \;:\; p\equiv 2 \text{ or } 4 \pmod 7$ | | | | | | |
|---|---|---|---|---|---|---|
| $p\equiv 2$ or 4 (mod 7) | $F_{p^6}^* = \langle\gamma\rangle$ | $E_{p^6}(\omega_7)$ | $S\equiv\sum_{k=1}^{6}\left(\frac{k}{7}\right)\gamma^{k\left(\frac{q-1}{7}\right)}$ (mod $p$) | $G$ | $H$ | $p\left(G+H\sqrt{-7}\right)^2$ |
| 11 | $\gamma = 4+x$<br>$x^6 = -1-x-x^2$ | $-33+44\sqrt{-7}$ | 9 | +2 | +1 | $-33+44\sqrt{-7}$ |

TABLE 22

| $m = 8 \; : \; f = 2 \; : \; p \equiv 1 \pmod 8 \quad (\alpha = 0, \beta = 0, \delta = 1, \varepsilon = i^{3\,\mathrm{ind}_g(2)})$ | | | | | | | | | |
|---|---|---|---|---|---|---|---|---|---|
| $p \equiv 1 \pmod 8$ | $F_{p^2}^{*} = \langle\gamma\rangle$ | $E_{p^2}(\omega_8)$ | $g = \gamma^{\frac{q-1}{p-1}}$ | $g^{\frac{p-1}{8}}$ | $g^{\frac{p-1}{8}} + g^{3\left(\frac{p-1}{8}\right)}$ | $C$ | $D$ | $\mathrm{ind}_g(2)$ | $i^{3\,\mathrm{ind}_g(2)}(C + D\sqrt{-2})$ |
| 17 | $\gamma_2 = 2 + x$<br>$x^2 = -3$ | $3 + 2\sqrt{-2}$ | 7 | 15 | 7 | −3 | −2 | 10 | $3 + 2\sqrt{-2}$ |
| 41 | $\gamma_2 = 2 + x$<br>$x^2 = -3$ | $3 - 4\sqrt{-2}$ | 7 | 38 | 11 | −3 | +4 | 14 | $3 - 4\sqrt{-2}$ |
| 73 | $\gamma_2 = 3 + x$<br>$x^2 = -5$ | $1 - 6\sqrt{-2}$ | 14 | 10 | 61 | +1 | −6 | 16 | $1 - 6\sqrt{-2}$ |

TABLE 23

| $m = 8 \; : \; f = 3 \; : \; p \equiv 1 \pmod 8 \quad (\alpha = 0, \beta = 1, \delta = 1, \varepsilon = (-1)^{(p-1)/8})$ | | | | | | | | | | | |
|---|---|---|---|---|---|---|---|---|---|---|---|
| $p \equiv 1 \pmod 8$ | $F_{p^3}^{*} = \langle\gamma\rangle$ | $E_{p^3}(\omega_8)$ | $g = \gamma^{\frac{q-1}{p-1}}$ | $g^{\frac{p-1}{8}}$ | $g^{\frac{p-1}{4}}$ | $g^{\frac{p-1}{8}} + g^{3\left(\frac{p-1}{8}\right)}$ | $A$ | $B$ | $C$ | $D$ | $(-1)^{(p-1)/8}(A + Bi)(C + D\sqrt{-2})$ |
| 17 | $\gamma_3 = x$<br>$x^3 = -3 - x$ | $-3 - 8\sqrt{2}$<br>$+12i - 2i\sqrt{2}$ | 14 | 9 | 13 | 7 | +1 | −4 | −3 | −2 | $-3 - 8\sqrt{2}$<br>$+12i - 2i\sqrt{2}$ |
| 41 | $\gamma_3 = 3 + x$<br>$x^3 = -1 - x$ | $15 + 16\sqrt{2}$<br>$+12i - 20i\sqrt{2}$ | 29 | 38 | 9 | 11 | +5 | +4 | −3 | +4 | $15 + 16\sqrt{2}$<br>$+12i - 20i\sqrt{2}$ |
| 73 | $\gamma_3 = 3 + x$<br>$x^3 = -4 - x$ | $3 - 48\sqrt{2}$<br>$-8i - 18i\sqrt{2}$ | 26 | 51 | 46 | 61 | −3 | +8 | +1 | −6 | $3 - 48\sqrt{2}$<br>$-8i - 18i\sqrt{2}$ |

**TABLE 24**

| $m = 8 \;:\; f = 4 \;:\; p \equiv 1 \pmod 8 \quad (\alpha = 0, \beta = 1, \delta = 2, \varepsilon = 1)$ | | | | | | | | | | |
|---|---|---|---|---|---|---|---|---|---|---|
| $p \equiv 1 \pmod 8$ | $F_{p^4}^{*} = \langle\gamma\rangle$ | $E_{p^4}(\omega_8)$ | $g = \gamma^{\frac{q-1}{p-1}}$ | $g^{\frac{p-1}{4}}$ | $g^{\frac{p-1}{8}} + g^{3\left(\frac{p-1}{8}\right)}$ | $A$ | $B$ | $C$ | $D$ | $(A+Bi)(C+D\sqrt{-2})^2$ |
| 17 | $\gamma_4 = 1 + x + x^2$<br>$x^4 = -3$ | $1 - 48\sqrt{2}$<br>$+4i + 12i\sqrt{2}$ | 7 | 4 | 7 | +1 | +4 | −3 | −2 | $1 - 48\sqrt{2}$<br>$+4i + 12i\sqrt{2}$ |
| 41 | $\gamma_4 = 2 + x$<br>$x^4 = -3$ | $-115 + 96\sqrt{2}$<br>$+92i + 120i\sqrt{2}$ | 19 | 32 | 30 | +5 | −4 | −3 | −4 | $-115 + 96\sqrt{2}$<br>$+92i + 120i\sqrt{2}$ |
| 73 | $\gamma_4 = 3 + x$<br>$x^4 = -5$ | $213 + 96\sqrt{2}$<br>$-568i + 36i\sqrt{2}$ | 13 | 46 | 61 | −3 | +8 | +1 | −6 | $213 + 96\sqrt{2}$<br>$-568i + 36i\sqrt{2}$ |

**TABLE 25**

| $m = 8 \;:\; f = 5 \;:\; p \equiv 1 \pmod 8 \quad (\alpha = 0, \beta = 1, \delta = 3, \varepsilon = 1)$ | | | | | | | | | | |
|---|---|---|---|---|---|---|---|---|---|---|
| $p \equiv 1 \pmod 8$ | $F_{p^5}^{*} = \langle\gamma\rangle$ | $E_{p^5}(\omega_8)$ | $g = \gamma^{\frac{q-1}{p-1}}$ | $g^{\frac{p-1}{4}}$ | $g^{\frac{p-1}{8}} + g^{3\left(\frac{p-1}{8}\right)}$ | $A$ | $B$ | $C$ | $D$ | $(A+Bi)(C+D\sqrt{-2})^3$ |
| 17 | $\gamma_5 = x$<br>$x^5 = -6 - x - x^2$ | $45 - 152\sqrt{2}$<br>$+180i + 38i\sqrt{2}$ | 11 | 4 | 10 | +1 | +4 | −3 | +2 | $45 - 152\sqrt{2}$<br>$+180i + 38\sqrt{2}$ |

TABLE 26

| $m = 8$ : $f = 2$ : $p \equiv 3 \pmod 8$ | | | | | | |
|---|---|---|---|---|---|---|
| $p \equiv 3 \pmod 8$ | $F_{p^2}^* = \langle\gamma\rangle$ | $E_{p^2}(\omega_8)$ | $K \equiv \gamma^{\frac{q-1}{8}} + \gamma^{3\left(\frac{q-1}{8}\right)} \pmod p$ | $C$ | $D$ | $(-1)^{\frac{p+5}{8}}(C + D\sqrt{-2})$ |
| 3 | $\gamma_2 = 1 + x$, $x^2 = 2$ | $-1 - \sqrt{-2}$ | 2 | +1 | +1 | $-1 - \sqrt{-2}$ |
| 11 | $\gamma_2 = 2 + x$, $x^2 = 2$ | $-3 + \sqrt{-2}$ | 3 | −3 | +1 | $-3 + \sqrt{-2}$ |
| 19 | $\gamma_2 = 2 + x$, $x^2 = 2$ | $-1 + 3\sqrt{-2}$ | 13 | +1 | −3 | $-1 + 3\sqrt{-2}$ |
| 43 | $\gamma_2 = 6 + x$, $x^2 = 2$ | $5 + 3\sqrt{-2}$ | 27 | +5 | +3 | $5 + 3\sqrt{-2}$ |

TABLE 27

| $m = 8$ : $f = 4$ : $p \equiv 3 \pmod 8$ | | | | | | |
|---|---|---|---|---|---|---|
| $p \equiv 3 \pmod 8$ | $F_{p^4}^* = \langle\gamma\rangle$ | $E_{p^4}(\omega_8)$ | $K \equiv \gamma^{\frac{q-1}{8}} + \gamma^{3\left(\frac{q-1}{8}\right)} \pmod p$ | $C$ | $D$ | $-(C + D\sqrt{-2})^2$ |
| 3 | $\gamma_4 = 1 + x$, $x^4 = -2 - 2x^2$ | $1 + 2\sqrt{-2}$ | 1 | +1 | −1 | $1 + 2\sqrt{-2}$ |
| 11 | $\gamma_4 = 5 + x$, $x^4 = -2 - 2x^2$ | $-7 - 6\sqrt{-2}$ | 8 | −3 | −1 | $-7 - 6\sqrt{-2}$ |
| 19 | $\gamma_4 = 1 + 2x$, $x^4 = -2 - 2x^2$ | $17 - 6\sqrt{-2}$ | 6 | +1 | +3 | $17 - 6\sqrt{-2}$ |
| 43 | $\gamma_4 = 1 + x$, $x^4 = -2 - 2x^2$ | $-7 - 30\sqrt{-2}$ | 27 | +5 | +3 | $-7 - 30\sqrt{-2}$ |

TABLE 28

| $m = 8$ : $f = 6$ : $p \equiv 3 \pmod 8$ | | | | | | |
|---|---|---|---|---|---|---|
| $p \equiv 3 \pmod 8$ | $F^*_{p^6} = \langle\gamma\rangle$ | $E_{p^6}(\omega_8)$ | $K \equiv \gamma^{\frac{q-1}{8}} + \gamma^{3\left(\frac{q-1}{8}\right)} \pmod p$ | $C$ | $D$ | $(-1)^{\frac{p-3}{8}} p(C + D\sqrt{-2})^3$ |
| 3 | $\gamma_6 = 2 + x^2$, $x^6 = -1 - x - x^2$ | $-15 - 3\sqrt{-2}$ | 1 | $+1$ | $-1$ | $-15 - 3\sqrt{-2}$ |
| 11 | $\gamma_6 = 4 + x$, $x^6 = -1 - x - x^2$ | $99 + 275\sqrt{-2}$ | 8 | $-3$ | $-1$ | $99 + 275\sqrt{-2}$ |

TABLE 29

| $m = 8$ : $f = 2$ : $p \equiv 5 \pmod 8$ | | | | | | | | |
|---|---|---|---|---|---|---|---|---|
| $p \equiv 5 \pmod 8$ | $F^*_{p^2} = \langle\gamma\rangle$ | $E_{p^2}(\omega_8)$ | $g$ | $g^{\frac{p-1}{4}}$ | $A$ | $B$ | $\mathrm{ind}_g 2$ | $(-1)^{(\mathrm{ind}_g 2-1)/2}\, i(A + Bi)$ |
| 5 | $\gamma_2 = 2 + x$, $x^2 = 2$ | $-2 + i$ | 2 | 2 | $+1$ | $+2$ | 1 | $-2 + i$ |
| 13 | $\gamma_2 = 2 + x$, $x^2 = 2$ | $-2 - 3i$ | 2 | 8 | $-3$ | $+2$ | 1 | $-2 - 3i$ |
| 29 | $\gamma_2 = 4 + x$, $x^2 = 2$ | $-2 + 5i$ | 14 | 12 | $+5$ | $+2$ | 13 | $-2 + 5i$ |
| 37 | $\gamma_2 = 2 + x$, $x^2 = 2$ | $6 + i$ | 2 | 31 | $+1$ | $-6$ | 1 | $6 + i$ |
| 53 | $\gamma_2 = 2 + x$, $x^2 = 2$ | $-2 - 7i$ | 2 | 30 | $-7$ | $+2$ | 1 | $-2 - 7i$ |

## References


[1] *B. C Berndt* and *R.J. Evans*, Sums of Gauss, Jacobi, and Jacobsthal. J. Number Theory, **11** (1979), 349–398.

[2] *B.C. Berndt* and *R.J. Evans*, Sums of Gauss, Eisenstein, Jacobi, Jacobsthal, and Brewer. Illinois J. Math. **23** (1979), 374–437.

[3] *A. Cauchy*, Mémoire sur la théorie des nombres. Mém. Inst. France, **17** (1840), 249–768. (Oeuvres Completes (I) Vol. **3**, 1911, pp. 5–83.)

[4] *H. Davenport* and *H. Hasse*, Die Nullstellen der Kongruenzzetafunktionen in gewissen zyklishchen Fällen. J. Reine Angew. Math. **172** (1934), 151–182.

[5] *L.E. Dickson*, Cyclotomy, higher congruences, and Waring's problem. Amer. J. Math. **57** (1935), 391–424.

[6] *P.G.L. Dirichlet*, Ueber den biquadratischen Charakter der Zahl "Zwei". J. Reine Angew. Math. **57** (1860), 187–188.

[7] *G. Eisenstein*, Zur Theorie der quadratischen Zerfällung der Primzahlen $8n + 3$, $7n + 2$ und $7n + 4$. J. Reine Angew. Math. **37** (1848), 97–126.

[8] *C.F. Gauss*, Untersuchungen über Höhere Arithmetrik. Chelsea Publishing Company, Bronx, New York (reprinted 1965), 511–586.

[9] *A. Genocchi*, Solution de la question 293 (J.A. Serret) voir t. XIII, p. 314. Nouvelles Annales de Mathematiques, **14** (1855), 241–243.

[10] *H. Hasse*, Vorlesungen über Zahlentheorie. Springer–Verlag, Berlin–Heidelberg, New York, 1950.

[11] *R.H. Hudson* and *K.S. Williams*, Extensions of Theorems of Cunningham–Aigner and Hasse–Evans. Pacific J. Math. **104** (1983), 111–132.

[12] *K. Ireland* and *M. Rosen*, A Classical Introduction to Modern Number Theory, Graduate Texts in Mathematics No. 84. Springer–Verlag, New York (1982).

[13] *C.G.J. Jacobi*, Über die Kreistheilung und ihre Anwendung auf die Zahlentheorie. J. Reine Angew. Math. **30** (1846), 166–182.

[14] *S.A. Katre* and *A.R. Rajwade*, Unique determination of cyclotomic numbers of order five. Manuscripta Math. **53** (1985), 65–75.

[15] *E. Lehmer*, On Euler's criterion. J. Austral. Math. Soc. **1** (1959), 64–70.

[16] *P.A. Leonard* and *K.S. Williams*, The cyclotomic numbers of order seven. Proc. Amer. Math. Soc. **51** (1975), 295–300.

[17] *R. Lidl* and *H. Niederreiter*, Finite Fields. Addison–Wesley Publishing Co., Reading. Mass., U.S.A. (1983).

[18] *M.A. Stern*, Eine Bermerkung zur Zahlentheorie. J. Reine Angew. Math. **32** (1846), 89–90.

[19] *L. Stickelberger*, Ueber eine Verallgemeinerung der Kreistheilung. Math. Ann. **37** (1890), 321–367.

[20] *A.L. Whiteman*, The sixteenth power residue character of 2. Canad. J. Math. **6** (1954), 364–373.

---

Department of Mathematics and Statistics, Carleton University,
Ottawa, Ontario, CANADA K1S 5B6

Department of Mathematics, Okanagan College, Vernon, British Columbia
CANADA V1B 2N5

# Digital Sum Sets

*Samuel Yates*

Digital sum sets have been studied for centuries [10], and interest in particular sets still arises from time to time. The answers to dormant questions are suddenly found and expanded upon, and new ones spring up and are investigated. A shorter article dealing with this topic is being prepared for the British publication THETA [13], but this paper contains additional new original results and details, and also discusses another significant digit sum set not mentioned there.

The digit sum of a natural number $n$ is the sum of the digits of $n$, and is designated by $S(n)$. The digit sum of 157 is $S(157) = 1 + 5 + 7 = 13$. Here we limit ourselves to the decimal natural numbers and to four sets. The first set was cited in a 1987 newsletter by Michael Ecker [2]. It uses digit sums of all the divisors of a number. The second set was developed as a result of a lecture at a mathematics conference in 1977 by Ivan Niven [3]. A Niven number is an integer that is divisible by its digit sum. The third set is Harshad numbers, much the same as Niven numbers but with a different emphasis. The fourth set, Smith numbers, was first described by Albert Wilansky [12]. His brother–in–law, Harold Smith, observed that the digit sum of his seven–digit telephone number had the peculiar property of being equal to the sum of the digit sums of its prime factors.

Are these sets finite or infinite? How are they distributed among the natural numbers? Can generating their numbers be facilitated by special number theoretical theorems and procedures? Do they possess subsets with noteworthy properties? These are questions of the type that are usually asked about newly–described mathematical entities, and they are certainly apropos here.

## 1. Black Hole Numbers

To obtain a member of the first set, write all the divisors of any natural number $n$, including 1 and $n$ but not including repetitions. Add the digits of the divisors, obtaining a new number. Repeat these two steps as often as necessary until a pattern emerges, such as having two equal successive sums, for example. In early trials, it appears that one may expect ultimate consecutive divisor digit sum equality with a digit sum of 15. The digits of the divisors—15, 5, 3, and 1—of 15 have a combined digit sum of 15. Ecker calls numbers with ultimate equal divisor digit sums "black hole" numbers. As an example, the factors of 19 are 19 and 1. The divisor digit sum is 11, whose factors are 1 and 11, which in turn gives 3, which then yields 4, then 7, then 8, and finally 15. Then 19 is a "black hole" number.

Ecker wondered if all integers fell into the "black hole" of 15, found no immediate counterexamples, and conjectured that they do. We note that all integers up through 48 eventually yield 15. In the next few hundred numbers, the first divisor digit sum of $n$ is less than $n$ itself; and since each smaller number's divisor digit sum has already been shown to yield 15, $n$ must also do so. It is evident that if the initial divisor digit sum of every $n$ greater than 48 is less than $n$, all numbers are "black hole" numbers.

Carl Pomerance proved [11] that when $n$ is greater than 4538, the initial divisor digit sum must be less than $n$. Samuel Yates wrote and ran a short computer program that confirmed that 48 is the largest number less that 4538 whose initial digit sum exceeds $n$. Then all natural numbers are "black hole" numbers.

The program also reveals a trend toward smaller initial sums relative to the size of $n$. In the following table, $n$ is the number with the largest initial divisor digit sum $Si(n)$ in the given interval.

Table 1

| Interval | $n$ | $Si(n)$ | $\frac{Si(n)}{n}$ |
|---|---|---|---|
| 0 to 500 | 432 | 142 | .329 |
| 500 to 1000 | 864 | 198 | .229 |
| 1000 to 1500 | 1440 | 252 | .175 |
| 1500 to 2000 | 1980 | 288 | .145 |
| 2000 to 2500 | 2376 | 315 | .133 |
| 2500 to 3000 | 2772 | 330 | .119 |
| | 2880 | 330 | .115 |
| 3000 to 3500 | 3168 | 360 | .114 |
| 3500 to 4000 | 3960 | 405 | .102 |
| 4000 to 4500 | 4032 | 374 | .093 |
| 4500 to 5000 | 4752 | 417 | .088 |

## 2. Niven Numbers

The testing of "black hole" numbers and Smith numbers requires knowledge of their prime factorizations. It is not so with Niven and Harshad numbers, which are divisible by their digit sums. All one–digit numbers and the products of such numbers by any powers of 10 are Nivens, but 11, 13, 14, 15, 16, 17 and 19 are not. Robert Kennedy showed that there cannot be more than 21 consecutive integers each of which is a Niven number [5]. Kennedy and Curtis Cooper proved that, although the set of Nivens is infinite, its natural density is zero [7].

They also investigated Niven repunits [6]. A repunit $Rn$ is a number written as $n$ 1's; for example, $R4$ is 1111. If a prime $p$ divides some $Rn$ but does not divide any smaller repunit, the shortest repeating decimal that corresponds to any proper rational fraction $a/p$ has a period (or repetend) that is $n$ digits long [15]. The prime $p$ is said to have $n$ as its period length. Kennedy and Cooper showed that an infinite set of repunits that are Nivens can be constructed by using powers of 3 and primes whose period lengths are powers of 3. It might be noted here that the book Prime Period Lengths, which contains a table of the period lengths of the first 105,000 primes, is a useful tool for generating Niven repunits [14].

The product of a Niven repunit and any integer $k$ less than 11 is a Niven. If $k = 9$, the resulting repdigit $9Rn$ (string of $n$ 9's) has the additional property that multiplying it by any number no greater than itself produces a Niven. For example, $R\,9 =$ 111111111 is a Niven. So is $9R\,9$, because $S\,(9R\,9) = 9 \times 9$, which obviously divides $9R\,9$. It has been shown that $S\,(9mRn) = S\,(9Rn) = 9n$ for all $m \leq 9Rn$. (For instance, the sum of the digits in 83461653, the product of 8347 and 9999, is the same as the sum of the digits in 9999, which is 36.) Since 81 divides 999999999, it will also divide the product of 999999999 and a smaller number. Because the digit sum of the product is 81, the product is a Niven.

Kennedy and others tested the first 100 factorials and found that, in each case, $n!$ is a Niven [8]. They anticipated the possibility that there exist larger $n!$'s that are not Nivens. $S\,(n!)$ is a multiple of 9 when $n > 5$. If $S\,(n!)$ is the product of 9 and any prime larger than $n$, then $S\,(n!)$ cannot divide $n!$, and $n!$ is not a Niven. In March 1988, Pomerance wrote to me explaining why he expected that non–Nivens should surface at about 500!, and possibly somewhat earlier, and that such non–Nivens should be more common as $n$ grows larger. I then wrote and ran a short computer program to confirm his assertion, and found that the smallest non–Niven factorial is 432!, whose digit sum is $3897 = 9 \times 433$. Every integer up to and including 600 was tested. The factorials of all of them were Nivens except for 432, 444, 453, 458, 474, 476, 485, 489, 498, 507, 509, 532, 539, 541, 548, 550, 552, 554, 555, 556, 560, 565, 567, 576, 593, and 597.

## 3. Harshad Numbers

Both Niven and Kaprekar described the set of numbers each of which is divisible by its digit sum. Their approach and emphasis were independent and different. Because he felt that "there is great joy in creating these numbers," Kaprekar called them Harshads. Harshad is a Sanskrit word meaning "giving joy" [4].

Let $d = S\,(n)$. If $d$ divides $n$, $n$ is said to be "Harshad for $d$." The sequence of Harshads for a given $d$ is an irregular progression. The difference between any two of them is a multiple of 9 and a multiple of $d$.

Kaprekar stated that there is a Harshad for any given $d$. It follows that there is an infinity of Harshads for any $d$ because new Harshads can be formed by multiplying a Harshad by powers of 10. Kaprekar was also interested in finding the largest Harshad for any $d$, such that none of the digits of the Harshad is zero. For a given $d$, there is a finite number of such Harshads.

The least Harshad for a one–digit number is the number itself. We showed above that the repdigit $9R\ 9 = 999999999$ is a Niven (or Harshad). It is also the least Harshad for a relatively small number, 81. The repdigit $9R\,k$ is the least Harshad for $3^j$ when $k = 3^{j-2}$. A convenient technique for obtaining members of a set of smallest Harshads for a given $d$ employs the period length table [14].

## 4. Smith Numbers

The study of Smith numbers and their ramifications began and grew in the 1980's. Its possibilities have not yet been exhausted, by any means. There are questions already raised that have yet to be answered.

A Smith is a composite number $n$ whose digit sum $S(n)$ is equal to the sum of the digits $Sp(n)$ of its prime factors, including repetitions. For instance, let $n$ be Dr. Smith's phone number, 4937775. Then $S(n) = 4 + 9 + 3 + 7 + 7 + 7 + 5 = 42$ and $Sp(n) = Sp(3 \times 5 \times 5 \times 65837) = 42$, and the phone number is a Smith. There are infinitely many Smiths in every congruence class (mod 9), although some congruence classes always have more than others in a given range. In particular, most Smiths are congruent to 4 (mod 9) [16]. There are 29,928 Smiths among the first 1,000,000 natural numbers. Several people have conjectured that Smiths have zero density.

Properties of repunits play a significant role in the study of Smiths. McDaniel's proof of the existence of an infinity of Smiths [9] uses the 9's repdigit property mentioned above. He shows that multiplying a repunit, whose prime factors are known, by an easily obtained suitable number, produces a Smith. Whereas there is an infinity of repunits; and every repunit is a multiple of at least one prime that does not divide any smaller repunit, all primes except 2 and 5 being thus provided for; and 2 and 5 are easily shown to be also factors of an infinity of Smiths; and every repunit is a factor of an infinity of repunits; it follows that every prime is a factor of an infinity of Smiths.

A related pursuit is the generation of very large Smiths. Patrick Costello has obtained Smiths which are tens of thousands of digits long and which have as their respective nuclei each of the largest known Mersenne primes [1]. At present, the largest known Smith is over ten million digits long. It was determined by Yates, and it is described below. Its computation is based on pertinent theorems not requiring finding every digit of the Smith. Such developments are accelerated by employing tables of factors of repunits, and retarded by the difficulty of obtaining their factorizations. Although such a tabulation was of interest to Gauss two centuries ago, it was not until 1988 that the set of the first hundred repunits was completely factored!

## 5. Smith With More Than Ten Million Digits

Here we show that $N = (10^{1031} - 1)(10^{4594} + 3 \times 10^{2297} + 1)^{1476} \times 10^{3913210}$ is a Smith.

In 1985, the repunit $R\,1031$ was shown to be prime by Hugh Williams and Harvey Dubner. Any repdigit $9Rk$ has a digit sum of $9k$. As noted earlier, the product of $9Rk$ by any natural number less than or equal to $9Rk$ also has a digit sum equal to $9k$. When $10^{1031} - 1 = 9R\,1031$ is multiplied by any smaller number, the product has a digit sum of $9 \times 1031 = 9279$.

Let $M = 10^{4594} + 3 \times 10^{2297} + 1$, a palindromic prime discovered in 1987 by Dubner. $S(M) = Sp(M) = 1 + 3 + 1 + 5$. The expansion of $M$ is written as a 2953–term expression (trinomial expansions to the $n$th power produce $2n + 1$ terms) which is ordinarily arranged in terms of powers of 10 with descending exponents in steps of 2297, going from $2952 \times 2297$ to zero. The largest coefficient of any term in this expression (the middle term has the largest) has 1030 digits. It is not necessary to know what these digits are; we need only know their number. The number of digits in the coefficient of such an expansion of $(10^{2k} + 3 \times 10^{k} + 1)^{t}$ has very close to, but not more than, $0.7t$ digits. We therefore choose an exponent for $M$ which produces a largest coefficient that is almost as large as, but less than, $9R\,1031$. When $M^{1476}$ is multiplied by $9R\,1031$, each coefficient of the product has no more than $2 \times 1031 = 2062$ digits and has a digit sum of 9279. Another criterion for the exponent of $M$ is that the number of digits in this product must be less than 2297.

Because the exponent of 10 in each term is 2297 more than that of the following term, its non–zero digits do not overlap those of any adjacent term even after multiplication by $9R$ 1031, so that the final expression is merely a concatenation without any "carries."

Because each of the 2953 terms has a digit sum of 9279, $S((10^{1031} - 1) \times M^{1476})$ $= 2953 \times 9279 = 27400887$. Multiplication by $10^{3913210}$ appends 3,913,210 zeros and does not alter the digit sum, but it does increase the prime factorization digit sum by $7 \times 3913210$. Then $S(N) = 27400887$; and because $Sp(abc) = Sp(a) + Sp(b) + Sp(c)$, $Sp(N) = 1037 + 5 \times 1476 + 7 \times 3913210 = 27400887$. Because $S(N) = Sp(N)$, $N$ **is a Smith number.** It is written with $1031 + 1476 \times 4594 + 3913210 = 10{,}694{,}985$ digits.

## References


[1] *P. Costello*, Unpublished paper presented at the West Coast Number Theory Conference, Asilomar, California, 1984.

[2] *M.W. Ecker* , Mathemagical black holes. REC Newsletter **2** (1987), 6.

[3] Fifth Annual Miami University Conference on Number Theory, Oxford, Ohio.

[4] *D.R. Kaprekar* , On Kaprekar's Harshad Numbers. Science Today, (1979).

[5] *R.E. Kennedy* , Digital sums, Niven numbers, and natural density. Crux Mathematicorum **8** (1982), 129—133.

[6] *R.E. Kennedy* and *C.N. Cooper* , Niven repunits. Fibonacci Quart.

[7] *R.E. Kennedy* , On the natural density of the Niven numbers. College Math. J. **15** (1984), 309—312.

[8] *R.E. Kennedy* , *T.A. Goodman* and *C.H. Best* , Mathematical discovery and Niven numbers. MATYC J. **14** (1980), 21—25.

[9] *W.L. McDaniel* , The existence of infinitely many k–Smith numbers. Fibonacci Quart. **25** (1987), 76—80.

[10] *W.L. McDaniel* and *S. Yates* , The sum of digits function and its application to a new problem. (To be published.)

[11] *C. Pomerance* , Note to S. Yates at the Florida section meeting of the Math. Assoc. of America, March 1988.

[12] *A. Wilansky* , Smith numbers. Two Year College Math. J. **13** (1982), 21.

[13] *S. Yates* , Digit sums. THETA **2** (1988).

[14] *S. Yates* , Prime Period Lengths. New Jersey (1975).

[15] *S. Yates* , Repunits and Repetends. Florida (1982).

[16] *S. Yates* , Smith numbers congruent to 4 (mod 9). J. of Recreational Math. **19** (1987), 139—141.

---

157 Capri–D, Kings Point, Delray Beach, Florida 33484, USA.

# New Invariants of Real Quadratic Fields

*Hideo Yokoi*

The purpose of this paper is to define some new invariants and to discover many more relationships among these new invariants and already known invariants in the case of real quadratic fields $\mathbb{Q}(\sqrt{p})$ with prime $p$.

For the sake of simplicity, we restrict $p$ to the primes congruent to 1 mod 4. Let $P$ denote the set of all primes congruent to 1 mod 4, and let $\mathbb{N}_0$ denote the set of all non–negative integers. Then, "*p–invariant* " is defined as a mapping from $P$ to $\mathbb{N}_0$. For instance, class numbers $h_{\pm p}, H_p$ of $\mathbb{Q}(\sqrt{\pm p})$, $\mathbb{Q}(\zeta_p)$ and $t_p, u_p$ in the formula $\varepsilon_p = (t_p + u_p\sqrt{p})/2$ of the fundamental unit $\varepsilon_p > 1$ of $\mathbb{Q}(\sqrt{p})$ are already known $p$–invariants.

Now, we define new $p$–invariants. First of all, we define an essential $p$–invariant $n_p$ by the inequality

$$\left| t_p/u_p^2 - n_p \right| < \frac{1}{2},$$

and put

$$t_p = u_p^2 \cdot n_p \pm a_p \qquad (0 \leq a_p < \frac{u_p^2}{2}).$$

Then, $a_p$ is also $p$–invariant and satisfies the congruence relation

$$a_p^2 \equiv -4 \pmod{u_p^2}$$

(*cf.* Proof of Theorem 1, (ii)). Hence, if we put

$$b_p = (a_p^2 + 4)/u_p^2,$$

then $b_p$ is also $p$–invariant.

For the properties of these $p$–invariants, we can prove the following:

**Theorem 1.** (i) *For any prime* $p$ *in* $P$, *the fundamental unit* $\varepsilon_p > 1$ *of the real quadratic field* $\mathbb{Q}(\sqrt{p})$ *is expressed in the form*

$$\varepsilon_p = (u_p^2 \cdot n_p \pm a_p + u_p\sqrt{p})/2$$

(ii) *Any prime* $p$ *in* $P$ *is expressed in the form*

$$p = u_p^2 \cdot n_p^2 \pm 2a_p \cdot n_p + b_p,$$

*and there exists a natural number* $n_0$ *such that this is uniquely expressed in any* $n_p$ *satisfying* $n_p \geq n_0$.

**Proof.** (i) is trivial because of $t_p = u_p^2 \cdot n_p \pm a_p$.

(ii) Since $t_p^2 - pu_p^2 = -4$, we get easily

$$\begin{aligned} pu_p^2 &= t_p^2 + 4 = (u_p^2 \cdot n_p \pm a_p)^2 + 4 \\ &= (u_p^2 \cdot n_p^2 \pm 2a_p \cdot n_p)u_p^2 + (a_p^2 + 4). \end{aligned}$$

Hence, we obtain first

$$a_p^2 + 4 \equiv 0 \pmod{u_p^2},$$

and next

$$p = u_p^2 \cdot n_p^2 \pm 2a_p \cdot n_p + b_p$$

since $a_p^2 + 4 = b_p \cdot u_p^2$.

The uniqueness follows from the following lemma:

**Lemma 1.** *(Yokoi–Nakahara). Let*

$$U = \left\{ u = 2^\delta \prod_{i=1}^{r} p_i^{e_i} : \delta = 1 \text{ or } 2,\ e_i \geq 1,\ p_i \in P \right\}$$

and let

$$A_u = \left\{ \pm a_j : 0 \leq a_j < u^2/2,\ j = 1, \ldots, 2^{\delta + r - 1} \right\},$$

*which is the system of representatives of residue classes of the solutions of*

$$x^2 \equiv -4 \pmod{u^2}.$$

*Then, if rational prime* $p$ *congruent to* 1 mod 4 *is expressed in the form*

$$p = u^2 \cdot n^2 \pm 2an + b, \; u \in U, \; a \in A_u, \; n \in N_0, \; b = (a^2 + 4)/u^2,$$

*there exists a natural number* $n_0$ *such that* $(u^2 n \pm a + u\sqrt{p})/2$ *is the fundamental unit of* $\mathbb{Q}(\sqrt{p})$ *provided* $n \geq n_0$.

This lemma follows from results of Yokoi–Nakahara (*cf.* Yokoi [2] and Nakahara [1]).

For any square–free integer $D > 1$ and a natural number $m > 1$, we say that an integral solution $(u, v)$ of the diophantine equation $x^2 - Dy^2 = \pm 4m$ is trivial if and only if $m = n^2$ is a square and $u \equiv v \equiv 0 \pmod{n}$.

Then we can prove the following:

**Proposition 1.** *For any prime* $p$ *in* $P$ *and a natural number* $m > 1$, *if the diophantine equation* $x^2 - py^2 = \pm 4m$ *has at least one non–trivial integral solution, then* $m \geq n_p$ *holds.*

**Theorem 2.** *For the class number* $h_p$ *of the real quadratic field* $\mathbb{Q}(\sqrt{p})$ $(p \in P)$ *and the least odd prime* $q_p$ *splitting completely in* $\mathbb{Q}(\sqrt{p})$, *which is also* $p$*–invariant,* $h_p \geq \log n_p / \log q_p$ *holds provided* $n_p \neq 0$.

For proofs of Proposition 1 and Theorem 2, see Yokoi [5].

**Remark.** It follows from Theorem 2 that $h_p = 1$ implies $q_p \geq n_p$. In the special case $p \equiv 1 \pmod{12}$, we get $q_p = 3$, and hence in this case $h_p = 1$ implies $n_p = 1, 2,$ or $3$. On the other hand, in the case $a_p = 0$ (i.e., $u_p = 1$ or $2$), it is known that $h_p = 1$ if and only if $n_p = q_p$. (*Cf.* for example, Yokoi [3], Theorem 2.)

**Theorem 3.** *If* $n_p \neq 0$ *for a prime* $p$ *in* $P$, *then* $u_p \not\equiv 0 \pmod{p}$, *i.e., Artin's conjecture on the fundamental unit of real quadratic fields is true in the case* $n_p \neq 0$.

To prove Theorem 3 we need the following lemma:

**Lemma 2.** *Any two of the following conditions are equivalent to each provided* $u_p > 2$:

(i) $n_p = 0$

(ii) $t_p > 2p$

(iii) $u_p^2 > 4p$.

**Proof of Lemma 2.** First of all, in the case $u_p > 2$, we obtain

$$p > t_p \cdot \frac{t_p}{u_p^2} > p - \frac{1}{2} \tag{1}$$

For, since $t_p^2 - pu_p^2 = -4$ implies $t_p^2/u_p^2 - p = -4/u_p^2 < 0$, in the case $u_p > 2$ we get at once

$$p > t_p \cdot t_p/u_p^2 = p - 4/u_p^2 > p - 1/2.$$

(i) $\Leftrightarrow$ (ii) Since $n_p = 0$ is equivalent to $t_p/u_p^2 < 1/2$ by the definition of $n_p$, if we assume $n_p = 0$, then we get at once

$$t_p/2 > p - 1/2 \qquad \text{i.e., } t_p > 2p - 1,$$

which implies $n_p = 0$ at once.

(ii) $\Leftrightarrow$ (iii) From $t_p^2 - pu_p^2 = -4$, we know first $t_p > 2p$ if and only if $4p^2 - pu_p^2 < -4$, i.e., $4p + 4/p < u_p^2$, which is equivalent to $u_p^2 > 4p$ because $p \geq 5$.

**Proof of Theorem 3.** In the case $u_p = 1$ or $2$, Theorem 3 is true trivially. In the case $u_p > 2$, by Lemma 2, $n_p = 0$ if and only if $u_p^2 > 4p$. On the other hand, if we assume $u_p \equiv 0 \pmod{p}$, then $u_p \geq p$, and hence $u_p^2 \geq p^2 > 4p$. Therefore, if $n_p \neq 0$, then $u_p \not\equiv 0 \pmod{p}$.

**Theorem 4.** *For any given and fixed prime* $p_0$ *in* $P$, *there exists only a finite number of primes* $p$ *in* $P$ *such that* $u_p = u_{p_0}$ *and* $h_p = 1$.

**Proof.** Since $t_p^2 \equiv -4 \pmod{u_p^2}$, if we put

$$U = \left\{ 2^{\delta} \prod_i p_i^{e_i} : \delta = 0 \text{ or } 1,\ e_i \geq 1,\ \text{prime } p_i \equiv 1 \pmod{4} \right\},$$

then $u_p$ is clearly an element of $U$. Therefore, Theorem 4 follows easily from the following lemma:

**Lemma 3.** *For any fixed element* $u_0$ *in* $U$, *there exists only a finite number of real quadratic fields* $\mathbb{Q}(\sqrt{p})$ *of class number 1 such that* $p$ *is a prime congruent to* 1 mod 4 *and* $u_p = u_0$ *for the fundamental unit of* $\mathbb{Q}(\sqrt{p})$

$$\varepsilon_p = \frac{t_p + u_p\sqrt{p}}{2} > 1.$$

On the other hand, Lemma 3 follows by using the class number formula of Dirichlet from the following lemma:

**Lemma 4.** *(Tatuzawa) For any positive number* $c$ *satisfying* $1/2 > c > 0$, *let* $d$ *be any positive integer such that* $d \geq \max(e^{1/c}, e^{11.2})$. *Moreover, let* $\chi$ *be any non–principal primitive real character to modulus* $d$, *and let* $L(x, \chi)$ *be the corresponding* $L-$series. *Then*

$$L(1,\chi) > 0.655(c/d^{c})$$

*holds with one possible exception.*

(For the proof of Lemma 3 and Lemma 4, see Yokoi [4], Theorem 1.)

## References


[1] *T. Nakahara,* On the determination of the fundamental units of certain quadratic fields. Mem. Fac. Sci. Kyushu University **24** (1970), 300—304.

[2] *H. Yokoi,* On real quadratic fields containing units with norm –1. Nagoya Math. J. **33** (1968), 139—152.

[3] *H. Yokoi,* Class–number one problem for certain kinds of real quadratic fields. Proc. Int. Conf. Class Numbers and Fundamental Units of Algebraic Number Fields, 24–28 June, Katata, Japan, 125—137 (1986).

[4] *H. Yokoi,* Class number one problem for real quadratic fields (The conjecture of Gauss). Proc. Japan Acad. **64–2** (1988), 53—55.

[5] *H. Yokoi,* Some relations among new invariants of prime number $p$ congruent to 1 mod 4. Advances in Pure Math. **13** (1988), 493—501.


---

Department of Mathematics, College of General Education, Nagoya University,
Chikusa–ku, Nagoya 464–01, Japan

# A Limit Formula for the Canonical Height of an Elliptic Curve and its Application to Height Computations

*Horst G. Zimmer*

## 1. Basic Facts about Heights

Let $E$ be an *elliptic curve* defined over a local or global field $K$ by a *generalized Weierstrass equation*.

$$Y^2 + a_1XY + a_3Y = X^3 + a_2X^2 + a_4X + a_6 \quad (a_i \in K) \tag{1}$$

A *global field* is either an algebraic number field $K$ of finite degree over the rational number field $\mathbb{Q}$ or an algebraic function field $K$ in one indeterminate over a field of constants $k$. By a *local field,* we understand a field $K$ which is complete with respect to an absolute value. For the sake of simplicity, we shall restrict ourselves mainly to algebraic number fields $K$ as global fields and their completions $K_{\mathfrak{p}}$ with respect to a place $\mathfrak{p}$ of $K$ as local fields. However, the basic facts we wish to report about are valid also in the function field case, and we shall briefly mention the modifications to be made for function fields. For a unified treatment of both cases, in the function field case the curve $E$ will be assumed *non–constant,* which means that $E$ is not birationally isomorphic to an elliptic curve $E'$ already defined over the field of constants $k$ of the function field $K$.

In the sequel we shall need Tate's entities (see [3], [10], [20])

$$\begin{aligned} b_2 &= a_1^2 + 4a_2, & b_4 &= a_1a_3 + 2a_4, \\ b_6 &= a_3^2 + 4a_6, & 4b_8 &= b_2b_6 - b_4^2, \\ c_4 &= b_2^2 - 24b_4, & c_6 &= -b_2^3 + 36b_2b_4 - 216b_6. \end{aligned} \tag{2}$$

The curve $E$ over $K$ has *discriminant*

$$\Delta = -b_2^2 b_8 - 8b_4^3 - 27b_6^2 + 9b_2 b_4 b_6 \neq 0$$

and *absolute invariant*

$$j = \frac{c_4^3}{\Delta}.$$

*Birational isomorphisms* of $E$ over $K$ are of the form (see [3], [10], [20])

$$X = u^2 X' + \rho, \quad Y = u^3 Y' + u^2 \sigma X' + \tau \tag{3}$$

for constants $u, \rho, \sigma, \tau \in K$ such that $u \neq 0$. They transform $E$ into an elliptic curve $E'$ in the generalized Weierstrass form analogous to (1) but with coefficients $a'_1, a'_2, a'_3, a'_4, a'_6 \in K$ constituting the discriminant

$$\Delta' = u^{-12} \Delta$$

and the absolute invariant

$$j' = j$$

of $E'$ over $K$.

By the Theorem of Mordell and Weil (*cf.* [3], [10]), the group $E(K)$ of rational points of $E$ over a global field $K$ is finitely generated and hence isomorphic to the direct sum of the finite *torsion group* $E(K)_{tor}$ and a free group $\mathbb{Z}^r$ of rank $r$, the *rank* of $E$ over $K$:

$$E(K) \cong E(K)_{tor} \oplus \mathbb{Z}^r.$$

On the group $E(K)$, there exists a unique real-valued positive-semidefinite function

$$\hat{h} : E(K) \to \mathbb{R}.$$

This *canonical height* on $E$ over $K$ plays a crucial role in the theory of elliptic curves. For instance, $\hat{h}$ is used

- to characterize torsion points:

$$P \in E(K)_{tor} \Leftrightarrow \hat{h}(P) = 0,$$

- to derive the strong Mordell–Weil Theorem from the weak Mordell–Weil Theorem by infinite descent (see [3], [4], [10]),

- to determine the rank $r$ and a basis of the free part of $E(K)$ by Manin's "conditional algorithm" (see [6], [17]),
- to prove a quantitative version of Siegel's Theorem on $S$-integral rational points in $E(K)$ (see [11]),
- to establish the analogue of the Riemann Hypothesis for a curve $E$ over a finite field $\mathbb{F}_q$ of $q = p^f$ elements (see [15]):

$$\# E(\mathbb{F}_q) \le (1 + \sqrt{q})^2 .$$

Indeed Néron [7] and Tate (*cf.* [2]) proved the following global theorem.

**Theorem 1.** *For an elliptic curve $E$ over a global field $K$, there exists a unique positive-semidefinite function*

$$\hat{h} : E(K) \to \mathbb{R}$$

*such that*

(1) $\hat{h}$ *is equivalent to the ordinary global height $h$ on $E(K)$:*

$$\hat{h} \sim h;$$

(2) $\hat{h}$ *is a quadratic form on $E(K)$, i.e.,*

(i) $\hat{h}(P+Q) + \hat{h}(P-Q) = 2\hat{h}(P) + 2\hat{h}(Q)$ *for* $P, Q \in E(K)$,

(ii) $\hat{h}(mP) = m^2 \hat{h}(P)$ *for* $P \in E(K)$ *and* $m \in \mathbb{N}$;

(3) $\hat{h}$ *is invariant under birational isomorphisms of $E$ over $K$.*

This global theorem has a local analogue. Any proof of the local analogue gives rise to a proof of the global theorem.

Let $K_\mathfrak{p}$ be a local field with a place $\mathfrak{p}$ and corresponding additive absolute value $v_\mathfrak{p}$. Then the local theorem is the following (*cf.* [3], [7], [18], [19]).

**Theorem 2.** *For an elliptic curve $E$ over a local field $K_\mathfrak{p}$ with additive absolute value $v_\mathfrak{p}$, there exists a unique function*

$$\hat{h}_\mathfrak{p} : E(K_\mathfrak{p}) \setminus \{O\} \to \mathbb{R}$$

*such that*

(1) $\hat{h}_\mathfrak{p}$ *is equivalent to the ordinary local height $h_\mathfrak{p}$ on $E(K_\mathfrak{p})$:*

$$\hat{h}_\mathfrak{p} \sim h_\mathfrak{p};$$

(2) $\hat{h}_{\mathfrak{p}}$ *is "almost" a quadratic form on* $E(K_{\mathfrak{p}})$, *i.e.*,

(i) $\hat{h}_{\mathfrak{p}}(P+Q)+\hat{h}_{\mathfrak{p}}(P-Q)=2\hat{h}_{\mathfrak{p}}(P)+2\hat{h}_{\mathfrak{p}}(Q)$

$$+ v_{\mathfrak{p}}(x_P - x_Q) - \frac{1}{6}v_{\mathfrak{p}}(\Delta)$$

*for* $P=(x_P, y_P)$, $Q=(x_Q, y_Q)\in E(K_{\mathfrak{p}})$ *such that* $P, Q$, $P \pm Q \neq 0$,

(ii) $\hat{h}_{\mathfrak{p}}(mP) = m^2\hat{h}_{\mathfrak{p}}(P) + \frac{1}{2}v_{\mathfrak{p}}(\Psi_m^2(P)) - \frac{m^2-1}{12}v_{\mathfrak{p}}(\Delta)$

*for* $P=(x_P, y_P)\in E(K\ )$ *and* $m\in\mathbb{N}$ *such that* $mP\neq 0$;

(3) $\hat{h}_{\mathfrak{p}}$ *is invariant under birational isomorphisms of* $E$ *over* $K_{\mathfrak{p}}$.

Here the polynomials $\Psi_m(P)\in\mathbb{Z}[b_2, b_4, b_6, b_8][x_P, y_P]$ arise from taking $m$-fold points $mP=(x_{mP}, y_{mP})\in E(K_{\mathfrak{p}})$ in accordance with the multiplication formula

$$mP = \left(\frac{\Phi_m(P)}{\Psi_m^2(P)}, \frac{\Omega_m(P)}{\Psi_m^3(P)}\right). \tag{4}$$

Referring to the generalized Weierstrass equation (1) they satisfy the recursion formulas (see [3], [10], [14])

$$\left.\begin{aligned}
&\Psi_1(P)=1,\quad \Psi_2(P)=2y_P+a_1x_P+a_3,\\
&\Psi_3(P)=3x_P^4+b_2x_P^3+3b_4x_P^2+3b_6x_P+b_8,\\
&\Psi_4(P)=\Psi_2(P)[2x_P^6+b_2x_P^5+5b_4x_P^4+10b_6x_P^3\\
&\qquad\qquad +10b_8x_P^2+(b_2b_8-b_4b_6)x_P+(b_4b_8-b_6^2)],\\
&\text{and for } m\geq 2,\\
&\Phi_m = x_P\Psi_m^2-\Psi_{m-1}\Psi_{m+1},\\
&\Psi_2\Psi_{2m}=\Psi_m(\Psi_{m-1}^2\Psi_{m+2}-\Psi_{m-2}\Psi_{m+1}^2),\\
&\Psi_{2m+1}=\Psi_m^3\Psi_{m+2}-\Psi_{m-1}\Psi_{m+1}^3.
\end{aligned}\right\} \tag{5}$$

$$\hat{h}(P) = \lim_{m \to \infty} \frac{h(g^m P)}{g^{2m}} = \lim_{m \to \infty} \frac{d(g^m P)}{g^{2m}}, \tag{15}$$

where $g > 1$ is any fixed positive integer.

If $K$ is an algebraic function field with field of constants $k$ and $E$ is a non–constant elliptic curve over $K$, the above remains true with only minor modifications. The changes to be made arise from the fact that the sum formula (7') for the valuations of $K$ holds with the $v_{\mathfrak{p}}$ replaced by the normalized additive exponential valuations $w_{\mathfrak{p}}$ and the $n_{\mathfrak{p}}$ replaced by the residue degrees $f_{\mathfrak{p}}$ (see [16]). Also it should be noted that there are no archimedean places of $K/k$; that is, we have $M_K^0 = M_K$.

How to define the canonical local height $\hat{h}_{\mathfrak{p}}$ figuring in Theorem 2 in the case of an elliptic curve $E$ over a local field $K_{\mathfrak{p}}$? This is achieved by a suitable modification of the global limit formula (15) on the basis of the relation (2) (ii) in Theorem 2, where the positive integers $m \in \mathbb{N}$ are to be replaced by the powers $g^m$ of a fixed positive integer $g > 1$. We now present the fundamental local limit formula.

**Theorem 3.** *For an elliptic curve $E$ over a local field $K_{\mathfrak{p}}$ with absolute value $v_{\mathfrak{p}}$, the canonical local height $\hat{h}_{\mathfrak{p}}$ of a rational point $P \in E(K_{\mathfrak{p}})$ such that $g^m P \neq 0$ for all $m \in \mathbb{N}$ is given by*

$$\hat{h}_{\mathfrak{p}}(P) = \lim_{m \to \infty} \left\{ \frac{d_{\mathfrak{p}}(g^m P) - \frac{1}{2} v_{\mathfrak{p}}(\Psi_{g^m}^2(P))}{g^{2m}} \right\} + \frac{1}{12} v_{\mathfrak{p}}(\Delta), \tag{16}$$

*here $g > 1$ is a fixed positive integer.*

e proof of this theorem will be published in a forthcoming paper (but *cf.* [14], [18], ]). It consists in establishing the existence of the limit (16) and verifying that $\hat{h}_{\mathfrak{p}}$ is d the unique function appearing in Theorem 2. The method of proof is to make precise the equivalence statement (1) in Theorem 2 by replacing the ordinary local $h_{\mathfrak{p}}$ by the modified local height $d_{\mathfrak{p}}$ and providing strong estimates for the e $\hat{h}_{\mathfrak{p}} - d_{\mathfrak{p}}$ on $E(K_{\mathfrak{p}})$.

e proves the following theorem which is important in its own right. Put (*cf.*



## 2. A Limit Formula for the Canonical Local Height

How does one define these local and global height functions?

Let us consider an algebraic number field $K$ as the basic field of the elliptic curve $E$. The ring of integers in K will be designated by $R$. We denote by $M_K$ the set of all places $\mathfrak{p}$ of $K$, by $M_K^\infty$ the (finite) subset of archimedean places of $K$ and by $M_K^0 = M_K \setminus M_K^\infty$ the subset of nonarchimedean places of $K$. For $\mathfrak{p} \in M_K$,

$$n_{\mathfrak{p}} = [K_{\mathfrak{p}} : \mathbb{Q}_p]$$

is the *local degree* of $\mathfrak{p}$ with respect to the underlying place $p \in M_{\mathbb{Q}}$ such that $\mathfrak{p} | p$. Furthermore, for $\mathfrak{p} \in M_K^0$,

$$\tilde{k_{\mathfrak{p}}} = R_{\mathfrak{p}} / (\mathfrak{p})$$

is the *residue field* of $K$ with respect to $\mathfrak{p}$, where $R_{\mathfrak{p}} \leq K_{\mathfrak{p}}$ stands for the ring of integers in the completion $K_{\mathfrak{p}}$ of $K$ at $\mathfrak{p}$ and $(\mathfrak{p}) \trianglelefteq R_{\mathfrak{p}}$ for its maximal ideal. The cardinality of $\tilde{k_{\mathfrak{p}}}$ is

$$|\tilde{k_{\mathfrak{p}}}| = q = p^{f_{\mathfrak{p}}}, \quad \text{where } \mathfrak{p} | p,$$

the exponent $f_{\mathfrak{p}}$ being the *residue degree* of $\mathfrak{p}$.

The normalized additive exponential valuation of $K$ or $K_{\mathfrak{p}}$ corresponding to a place $\mathfrak{p} \in M_K^0$ will be denoted by $w_{\mathfrak{p}}$. Hence we have

$$w_{\mathfrak{p}}(K^\times) = \mathbb{Z} \quad \text{and} \quad w_{\mathfrak{p}}(K_{\mathfrak{p}}^\times) = \mathbb{Z}.$$

For $\mathfrak{p} \in M_K^\infty$, we shall refer to the associated complex embedding

$$\sigma_{\mathfrak{p}} : K \to \mathbb{C}.$$

We introduce the additive absolute values $v_{\mathfrak{p}}$ associated to the places $\mathfrak{p} \in M_K$ by putting

$$v_{\mathfrak{p}}(x) = \begin{cases} -\log|\sigma_{\mathfrak{p}}(x)| & \text{if } \mathfrak{p} \in M_K^\infty \\ -\log|x|_{\mathfrak{p}} & \text{if } \mathfrak{p} \in M_K^0 \end{cases} \tag{6}$$

for $x \in K$, where $|\ |$ stands for ordinary absolute value on $\mathbb{C}$ and

$$|x|_{\mathfrak{p}} = q^{-\frac{w_{\mathfrak{p}}(x)}{n_{\mathfrak{p}}}}$$

denotes the multiplicative valuation corresponding to a place $\mathfrak{p} \in M_K^0$ such that $\mathfrak{p}|p$ and $q = p^{f_{\mathfrak{p}}}$.

The *product formula* for the absolute values on $K$

$$\prod_{\mathfrak{p} \in M_K} |x|_{\mathfrak{p}}^{n_{\mathfrak{p}}} = 1 \qquad (x \in K^{\times}) \tag{7}$$

will be supplemented by the *sum formula*

$$\sum_{\mathfrak{p} \in M_K} n_{\mathfrak{p}} v_{\mathfrak{p}}(x) = 0 \qquad (x \in K^{\times}) \tag{7'}$$

with the local degrees $n_{\mathfrak{p}}$ as multiplicities.

We define the *ordinary global height* $h$ of a non–zero rational point $P = (x, y) \in E(K)$ with $X$–coordinate in arbitrary fraction representation

$$x = \frac{\xi}{\eta} \quad \text{for } \xi, \eta \in R,\ \eta \neq 0,$$

by setting

$$h(P) = \frac{1}{2} \log \prod_{\mathfrak{p} \in M_K} \max\{|\xi|_{\mathfrak{p}}, |\eta|_{\mathfrak{p}}\}^{n_{\mathfrak{p}}} \tag{8}$$

$$= \frac{1}{2} \log \prod_{\mathfrak{p} \in M_K} \max\{1, |x|_{\mathfrak{p}}\}^{n_{\mathfrak{p}}},$$

observing the product formula (7). Taking the logarithms yields the corresponding additive expressions



$$h(P) = -\frac{1}{2} \sum_{\mathfrak{p} \in M_K} n_{\mathfrak{p}} \min\{v_{\mathfrak{p}}(\xi), v_{\mathfrak{p}}(\eta)\} \tag{9}$$

$$= -\frac{1}{2} \sum_{\mathfrak{p} \in M_K} n_{\mathfrak{p}} \min\{0, v_{\mathfrak{p}}(x)\}$$

On the basis of the formula (9) we define the *ordinary local height* $h_{\mathfrak{p}}$ of a rational point $P = (x, y) \in E(K)$ with respect to a place $\mathfrak{p} \in M_K$ by setting $h_{\mathfrak{p}}(P) = 0$ if $P = O$ and

$$h_{\mathfrak{p}}(P) = -\frac{1}{2} \min\{0, v_{\mathfrak{p}}(x)\} \quad \text{if } P \neq O. \tag{10}$$

The ordinary global height $h$ of a point $P \in E(K)$ is then obtained as the sum with multiplicities $n_{\mathfrak{p}}$ over the ordinary local heights $h_{\mathfrak{p}}$ of $P$ formed with respect to all places $\mathfrak{p} \in M_K$:

$$h(P) = \sum_{\mathfrak{p} \in M_K} n_{\mathfrak{p}} h_{\mathfrak{p}}(P). \tag{1'}$$

We shall modify the ordinary local and global heights in order to derive sharp es' in connection with property (1) in Theorems 1 and 2. To this end, we intro each place $\mathfrak{p} \in M_K$, the quantities

$$\mu_{\mathfrak{p}} = \min\{v_{\mathfrak{p}}(b_2), \frac{1}{2} v_{\mathfrak{p}}(b_4), \frac{1}{3} v_{\mathfrak{p}}(b_6), \frac{1}{4} v_{\mathfrak{p}}(b_8)\}$$

in terms of the entities displayed in (2). The *modified local hei* $P = (x, y) \in E(K)$ with respect to a place $\mathfrak{p} \in M_K$ i $d_{\mathfrak{p}}(p) = -\frac{1}{2}\mu_{\mathfrak{p}}$ if $P = O$ and

$$d_{\mathfrak{p}}(P) = -\frac{1}{2} \min\{\mu_{\mathfrak{p}}, v_{\mathfrak{p}}(x)\} \quad \text{if } P \neq$$

This leads to the definition of the *modified global heigh* the sum with multiplicities $n_{\mathfrak{p}}$ over the modified loca

$$d(P) = \sum_{\mathfrak{p} \in M_K} n_{\mathfrak{p}} d_{\mathfrak{p}}($$

The *canonical global height* $\hat{h}$ on E(K) can r

$$\alpha_{\mathfrak{p}} = \begin{Bmatrix} -\log 2 & \text{if } \mathfrak{p} \text{ is archimedean} \\ 0 & \text{if } \mathfrak{p} \text{ is nonarchimedean} \end{Bmatrix}. \tag{17}$$

**Theorem 4.** *For an elliptic curve $E$ over a local field $K_{\mathfrak{p}}$ with additive absolute value $v_{\mathfrak{p}}$, the canonical local height $\hat{h}_{\mathfrak{p}}$ differs from the modified local height $d_{\mathfrak{p}}$ on $E(K_{\mathfrak{p}})$ at most within the bounds*

$$\frac{1}{6}(6\mu_{\mathfrak{p}} - v_{\mathfrak{p}}(\Delta)) + \frac{4}{3}\alpha_{\mathfrak{p}} \leq \hat{h}_{\mathfrak{p}}(P) - \{d_{\mathfrak{p}}(P) + \frac{1}{12}v_{\mathfrak{p}}(\Delta)\} \leq -\frac{1}{2}\alpha_{\mathfrak{p}} \tag{18}$$

*valid for any point $0 \neq P \in E(K_{\mathfrak{p}})$.*

These estimates are generalizations to the long Weierstrass form (1) of $E$ as well as improvements of estimates the author had obtained in previous papers ([14], [16], [18]). They will also be verified in a forthcoming paper (but see [13]).

Having constructed the canonical local height $\hat{h}_{\mathfrak{p}}$, one obtains the canonical global height $\hat{h}$ on the group $E(K)$ of rational points of the elliptic curve $E$ over a global field $K$ as the sum with multiplicities over the canonical local heights $\hat{h}_{\mathfrak{p}}$. This is in complete analogy to the way in which the ordinary or modified global height $h$ or $d$ on $E(K)$ is obtained as the sum with multiplicities over the ordinary or modified local heights $h_{\mathfrak{p}}$ or $d_{\mathfrak{p}}$ in accordance with the formulas (11) or (14), respectively. Indeed we have for any point $0 \neq P \in E(K)$:

$$\hat{h}(P) = \sum_{\mathfrak{p} \in M_K} n_{\mathfrak{p}} \hat{h}_{\mathfrak{p}}(P). \tag{19}$$

By virtue of this formula, the local estimates (18) in Theorem 4 can be turned into global estimates, taking into account the sum formula (7') for the absolute values $v_{\mathfrak{p}}$ on the global field $K$. To this end, referring to (12) and (17), we introduce the real numbers (see [16], [18])

$$\mu = -\sum_{\mathfrak{p} \in M_K} n_{\mathfrak{p}} \mu_{\mathfrak{p}} \geq 0$$

and

$$\alpha = -\sum_{\mathfrak{p} \in M_K} n_{\mathfrak{p}} \alpha_{\mathfrak{p}} \geq 0.$$

By (17), the latter number is simply

$$\alpha = \begin{cases} n \log 2 & \text{if } K \text{ is a number field} \\ 0 & \text{if } K \text{ is a function field} \end{cases},$$

where in the number field case

$$n = [K : \mathbb{Q}]$$

is the degree of $K$ over $\mathbb{Q}$. Observe that we have the sum formula

$$n = \sum_{\substack{\mathfrak{p} \in M_K^0 \\ \mathfrak{p} | p}} n_{\mathfrak{p}}.$$

Now the global estimates arising from Theorem 4 are the following.

**Theorem 5.** *For an elliptic curve $E$ over a global field $K$, the canonical global height $\hat{h}$ differs from the modified global height $d$ on $E(K)$ at most within the bounds*

$$-(\mu + \frac{4}{3}\alpha) \leq \hat{h}(P) - d(P) \leq \frac{1}{2}\alpha$$

*valid for any point $P \in E(K)$.*

These estimates are quite important in the theory of elliptic curves (*cf.* [1], [3], [6], [8], [9], [17]). They are stronger than those obtained in [16].

## 3. Computing the Canonical Global Height

The height formula (19) also serves as a basis for explicit calculations of the canonical global height $\hat{h}$ on the rational point group $E(K)$ of an elliptic curve $E$ over a global field $K$. For this purpose one first computes the local heights $\hat{h}_{\mathfrak{p}}$ by means of certain explicit expressions for $\hat{h}_{\mathfrak{p}}$ due to Tate and then combines them into the global height $\hat{h}$ by virtue of (19). Tate's explicit expressions for the $\hat{h}_{\mathfrak{p}}$ (*cf.* [3]) can be derived from the fundamental limit formula (16) in Theorem 3 (see [18], [19]). They depend on the reduction behaviour of $E$ modulo $\mathfrak{p}$.

To explain that derivation, let us start form the implication

$$K \leq K_{\mathfrak{p}} \Rightarrow E(K) \leq E(K_{\mathfrak{p}})$$

by considering $E$ as an elliptic curve over the completion $K_{\mathfrak{p}}$ of $K$ with respect to a place $\mathfrak{p} \in M_K^0$. From the set of birational isomorphism classes of $E$ over $K_{\mathfrak{p}}$, we choose a $\mathfrak{p}$*–minimal model* given by a generalized Weierstrass equation (1) which satisfies the condition that

$$w_{\mathfrak{p}}(\Delta) \geq 0 \quad \text{is minimal subject to } a_1, a_2, a_3, a_4, a_6 \in R_{\mathfrak{p}}.$$

Reduction modulo $\mathfrak{p}$ of the coefficients of such a $\mathfrak{p}$–minimal model yields a cubic curve $\tilde{E}$ defined over the residue field $\tilde{k}_{\mathfrak{p}}$ of $K_{\mathfrak{p}}$ by a Weierstrass equation

$$\tilde{Y}^2 + \tilde{a}_1\tilde{X}\tilde{Y} + \tilde{a}_3\tilde{Y} = \tilde{X}^3 + \tilde{a}_2\tilde{X}^2 + \tilde{a}_4\tilde{X} + \tilde{a}_6 \quad (\tilde{a}_i \in \tilde{k}_{\mathfrak{p}}).$$

The curve $E$ over $K_{\mathfrak{p}}$ is said to have *good, (split) multiplicative* or *additive reduction* modulo $\mathfrak{p}$ according as the reduced curve $\tilde{E}$ over $\tilde{k}_{\mathfrak{p}}$ is non–singular, singular with a node having two distinct (rational) tangents or singular with a cusp having a double tangent, respectively. Let $E_0(K_{\mathfrak{p}})$ be the inverse image of the group $\tilde{E}_{ns}(\tilde{k}_{\mathfrak{p}})$ of non–singular points of the reduced curve $\tilde{E}$ over $\tilde{k}_{\mathfrak{p}}$ under the reduction map. Then the three reduction types of $E$ over $K_{\mathfrak{p}}$ are characterized by the isomorphisms $\tilde{E}_{ns}(\tilde{k}_{\mathfrak{p}}) \cong \tilde{E}(\tilde{k}_{\mathfrak{p}}), \cong \tilde{k}_{\mathfrak{p}}^{\times}$ or $\cong \tilde{k}_{\mathfrak{p}}^{+}$, respectively. In the sequel we shall need the following result due to Kodaira and Néron (see [10]).

**Theorem 6.** *For an elliptic curve* $E$ *over a local field* $K_{\mathfrak{p}}$ *with a nonarchimedean place* $\mathfrak{p}$, *the set* $E_0(K_{\mathfrak{p}})$ *is a subgroup of finite index of the group of rational points* $E(K_{\mathfrak{p}})$:

$$E_0(K_{\mathfrak{p}}) \leq E(K_{\mathfrak{p}}),$$

*and the quotient group*

$$E(K_{\mathfrak{p}})/E_0(K_{\mathfrak{p}})$$

*is*

$$\begin{Bmatrix} \textit{a cyclic group of order } w_{\mathfrak{p}}(\Delta) = -w_{\mathfrak{p}}(j) \\ \textit{a finite group of order } \leq 4 \end{Bmatrix}$$

*according as* $E$ *has*

$$\begin{Bmatrix} \textit{split multiplicative} \\ \textit{good or additive} \end{Bmatrix} \textit{reduction modulo } \mathfrak{p}.$$

Of course, in the case of good reduction, we have equality

$$E_0(K_{\mathfrak{p}}) = E(K_{\mathfrak{p}}).$$

If the absolute invariant $j$ of $E$ over $K_{\mathfrak{p}}$ has value

$$v_{\mathfrak{p}}(j) \geq 0,$$

one speaks of *potentially good reduction* of $E$ modulo $\mathfrak{p}$. If $E$ has split multiplicative reduction modulo $\mathfrak{p}$, so that in particular

$$v_{\mathfrak{p}}(j) < 0,$$

there exists a unique element $q \in K_{\mathfrak{p}}^{\times}$ for which

$$v_{\mathfrak{p}}(j) = -v_{\mathfrak{p}}(q), \tag{20}$$

and the group of rational points of $E$ over $K_{\mathfrak{p}}$ is isomorphic to the quotient group of the multiplicative group $K_{\mathfrak{p}}^{\times}$ by $q$:

$$K_{\mathfrak{p}}^{\times}/q \xrightarrow{\sim} E(K_{\mathfrak{p}})$$

$$t \,(\mathrm{mod}^{\times}\, q) \longmapsto P = (\wp(t), \wp'(t)).$$

This parametrization of $E$ over $K_{\mathfrak{p}}$ is accomplished by the analogue of Weierstrass' $\wp$–function $\wp(t)$ and its derivative $\wp'(t)$. The corresponding theory is due to Tate (*cf.* [3], [10]). The curve $E$ over $K_{\mathfrak{p}}$ is called a *Tate curve.*

Now we are in a position to write down Tate's explicit expressions for the canonical local height $\hat{h}_{\mathfrak{p}}$ which will be used for calculating $\hat{h}_{\mathfrak{p}}$ and hence for calculating the canonical global height $\hat{h}$ itself. In the subsequent theorem,

$$B_2(X) = X^2 - X + \frac{1}{6}$$

will denote the second *Bernoulli polynomial.*

**Theorem 7.** *Let $E$ be an elliptic curve over a local field $K_{\mathfrak{p}}$ with a nonarchimedean place $\mathfrak{p}$.*

(a) *If $E$ has potentially good reduction modulo $\mathfrak{p}$,*

$$\hat{h}_{\mathfrak{p}}(P) = d_{\mathfrak{p}}(P) + \frac{1}{12} v_{\mathfrak{p}}(\Delta) \quad \textit{for} \quad O \neq P \in E(K_{\mathfrak{p}}).$$

## 2. A Limit Formula for the Canonical Local Height

How does one define these local and global height functions?

Let us consider an algebraic number field $K$ as the basic field of the elliptic curve $E$. The ring of integers in K will be designated by $R$. We denote by $M_K$ the set of all places $\mathfrak{p}$ of $K$, by $M_K^\infty$ the (finite) subset of archimedean places of $K$ and by $M_K^0 = M_K \setminus M_K^\infty$ the subset of nonarchimedean places of $K$. For $\mathfrak{p} \in M_K$,

$$n_\mathfrak{p} = [K_\mathfrak{p} : \mathbb{Q}_p]$$

is the *local degree* of $\mathfrak{p}$ with respect to the underlying place $p \in M_\mathbb{Q}$ such that $\mathfrak{p} | p$. Furthermore, for $\mathfrak{p} \in M_K^0$,

$$\tilde{k}_\mathfrak{p} = R_\mathfrak{p} / (\mathfrak{p})$$

is the *residue field* of $K$ with respect to $\mathfrak{p}$, where $R_\mathfrak{p} \leq K_\mathfrak{p}$ stands for the ring of integers in the completion $K_\mathfrak{p}$ of $K$ at $\mathfrak{p}$ and $(\mathfrak{p}) \trianglelefteq R_\mathfrak{p}$ for its maximal ideal. The cardinality of $\tilde{k}_\mathfrak{p}$ is

$$|\tilde{k}_\mathfrak{p}| = q = p^{f_\mathfrak{p}}, \text{ where } \mathfrak{p} | p,$$

the exponent $f_\mathfrak{p}$ being the *residue degree* of $\mathfrak{p}$.

The normalized additive exponential valuation of $K$ or $K_\mathfrak{p}$ corresponding to a place $\mathfrak{p} \in M_K^0$ will be denoted by $w_\mathfrak{p}$. Hence we have

$$w_\mathfrak{p}(K^\times) = \mathbb{Z} \text{ and } w_\mathfrak{p}(K_\mathfrak{p}^\times) = \mathbb{Z}.$$

For $\mathfrak{p} \in M_K^\infty$, we shall refer to the associated complex embedding

$$\sigma_\mathfrak{p} : K \to \mathbb{C}.$$

We introduce the additive absolute values $v_\mathfrak{p}$ associated to the places $\mathfrak{p} \in M_K$ by putting

$$v_\mathfrak{p}(x) = \begin{cases} -\log|\sigma_\mathfrak{p}(x)| & \text{if } \mathfrak{p} \in M_K^\infty \\ -\log|x|_\mathfrak{p} & \text{if } \mathfrak{p} \in M_K^0 \end{cases} \tag{6}$$

for $x \in K$, where $|\ |$ stands for ordinary absolute value on $\mathbb{C}$ and

$$|x|_{\mathfrak{p}} = q^{-\frac{w_{\mathfrak{p}}(x)}{n_{\mathfrak{p}}}}$$

denotes the multiplicative valuation corresponding to a place $\mathfrak{p} \in M_K^0$ such that $\mathfrak{p}|p$ and $q = p^{f_{\mathfrak{p}}}$.

The *product formula* for the absolute values on $K$

$$\prod_{\mathfrak{p} \in M_K} |x|_{\mathfrak{p}}^{n_{\mathfrak{p}}} = 1 \qquad (x \in K^{\times}) \tag{7}$$

will be supplemented by the *sum formula*

$$\sum_{\mathfrak{p} \in M_K} n_{\mathfrak{p}} v_{\mathfrak{p}}(x) = 0 \qquad (x \in K^{\times}) \tag{7'}$$

with the local degrees $n_{\mathfrak{p}}$ as multiplicities.

We define the *ordinary global height* $h$ of a non-zero rational point $P = (x, y) \in E(K)$ with $X$-coordinate in arbitrary fraction representation

$$x = \frac{\xi}{\eta} \quad \text{for } \xi, \eta \in R,\ \eta \neq 0,$$

by setting

$$h(P) = \frac{1}{2} \log \prod_{\mathfrak{p} \in M_K} \max\{|\xi|_{\mathfrak{p}}, |\eta|_{\mathfrak{p}}\}^{n_{\mathfrak{p}}} \tag{8}$$

$$= \frac{1}{2} \log \prod_{\mathfrak{p} \in M_K} \max\{1, |x|_{\mathfrak{p}}\}^{n_{\mathfrak{p}}},$$

observing the product formula (7). Taking the logarithms yields the corresponding additive expressions

$$h(P) = \frac{1}{2} \sum_{\mathfrak{p} \in M_K} n_{\mathfrak{p}} \min \{v_{\mathfrak{p}}(\xi), v_{\mathfrak{p}}(\eta)\} \tag{9}$$
$$= -\frac{1}{2} \sum_{\mathfrak{p} \in M_K} n_{\mathfrak{p}} \min \{0, v_{\mathfrak{p}}(x)\}$$

On the basis of the formula (9) we define the *ordinary local height* $h_{\mathfrak{p}}$ of a rational point $P = (x, y) \in E(K)$ with respect to a place $\mathfrak{p} \in M_K$ by setting $h_{\mathfrak{p}}(P) = 0$ if $P = O$ and

$$h_{\mathfrak{p}}(P) = -\frac{1}{2} \min \{0, v_{\mathfrak{p}}(x)\} \quad \text{if } P \neq O. \tag{10}$$

The ordinary global height $h$ of a point $P \in E(K)$ is then obtained as the sum with multiplicities $n_{\mathfrak{p}}$ over the ordinary local heights $h_{\mathfrak{p}}$ of $P$ formed with respect to all places $\mathfrak{p} \in M_K$:

$$h(P) = \sum_{\mathfrak{p} \in M_K} n_{\mathfrak{p}} h_{\mathfrak{p}}(P). \tag{11}$$

We shall modify the ordinary local and global heights in order to derive sharp estimates in connection with property (1) in Theorems 1 and 2. To this end, we introduce, for each place $\mathfrak{p} \in M_K$, the quantities

$$\mu_{\mathfrak{p}} = \min \{v_{\mathfrak{p}}(b_2), \frac{1}{2} v_{\mathfrak{p}}(b_4), \frac{1}{3} v_{\mathfrak{p}}(b_6), \frac{1}{4} v_{\mathfrak{p}}(b_8)\} \tag{12}$$

in terms of the entities displayed in (2). The *modified local height* $d_{\mathfrak{p}}$ of a point $P = (x, y) \in E(K)$ with respect to a place $\mathfrak{p} \in M_K$ is now defined as $d_{\mathfrak{p}}(p) = -\frac{1}{2} \mu_{\mathfrak{p}}$ if $P = O$ and

$$d_{\mathfrak{p}}(P) = -\frac{1}{2} \min \{\mu_{\mathfrak{p}}, v_{\mathfrak{p}}(x)\} \quad \text{if } P \neq O. \tag{13}$$

This leads to the definition of the *modified global height* $d$ of a point $P \in E(K)$ as the sum with multiplicities $n_{\mathfrak{p}}$ over the modified local heights $d_{\mathfrak{p}}$ of $P$:

$$d(P) = \sum_{\mathfrak{p} \in M_{\mathrm{K}}} n_{\mathfrak{p}} d_{\mathfrak{p}}(P). \tag{14}$$

The *canonical global height* $\hat{h}$ on E(K) can now be defined by the limit formulas

$$\hat{h}(P) = \lim_{m \to \infty} \frac{h(g^m P)}{g^{2m}} = \lim_{m \to \infty} \frac{d(g^m P)}{g^{2m}}, \tag{15}$$

where $g > 1$ is any fixed positive integer.

If $K$ is an algebraic function field with field of constants $k$ and $E$ is a non–constant elliptic curve over $K$, the above remains true with only minor modifications. The changes to be made arise from the fact that the sum formula (7') for the valuations of $K$ holds with the $v_{\mathfrak{p}}$ replaced by the normalized additive exponential valuations $w_{\mathfrak{p}}$ and the $n_{\mathfrak{p}}$ replaced by the residue degrees $f_{\mathfrak{p}}$ (see [16]). Also it should be noted that there are no archimedean places of $K/k$; that is, we have $M_K^0 = M_K$.

How to define the canonical local height $\hat{h}_{\mathfrak{p}}$ figuring in Theorem 2 in the case of an elliptic curve $E$ over a local field $K_{\mathfrak{p}}$? This is achieved by a suitable modification of the global limit formula (15) on the basis of the relation (2) (ii) in Theorem 2, where the positive integers $m \in \mathbb{N}$ are to be replaced by the powers $g^m$ of a fixed positive integer $g > 1$. We now present the fundamental local limit formula.

**Theorem 3.** *For an elliptic curve $E$ over a local field $K_{\mathfrak{p}}$ with absolute value $v_{\mathfrak{p}}$, the canonical local height $\hat{h}_{\mathfrak{p}}$ of a rational point $P \in E(K_{\mathfrak{p}})$ such that $g^m P \neq 0$ for all $m \in \mathbb{N}$ is given by*

$$\hat{h}_{\mathfrak{p}}(P) = \lim_{m \to \infty} \left\{ \frac{d_{\mathfrak{p}}(g^m P) - \frac{1}{2} v_{\mathfrak{p}}(\Psi^2_{g^m}(P))}{g^{2m}} \right\} + \frac{1}{12} v_{\mathfrak{p}}(\Delta), \tag{16}$$

*where $g > 1$ is a fixed positive integer.*

The proof of this theorem will be published in a forthcoming paper (but *cf.* [14], [18], [19]). It consists in establishing the existence of the limit (16) and verifying that $\hat{h}_{\mathfrak{p}}$ is indeed the unique function appearing in Theorem 2. The method of proof is to make more precise the equivalence statement (1) in Theorem 2 by replacing the ordinary local height $h_{\mathfrak{p}}$ by the modified local height $d_{\mathfrak{p}}$ and providing strong estimates for the difference $\hat{h}_{\mathfrak{p}} - d_{\mathfrak{p}}$ on $E(K_{\mathfrak{p}})$.

In fact one proves the following theorem which is important in its own right. Put (*cf.* [16])

(b) *If $E$ has split multiplicative reduction modulo $\mathfrak{p}$ (and hence is a Tate curve),*

$$\hat{h}_{\mathfrak{p}}(P) = v_{\mathfrak{p}}(1-t) + \frac{1}{2}B_2(\alpha(t))v_{\mathfrak{p}}(q)$$

*for* $0 \neq P = (\wp(t), \wp'(t)) \in E(K_{\mathfrak{p}})$,

*where*

$$\alpha(t) = \frac{v_{\mathfrak{p}}(t)}{v_{\mathfrak{p}}(q)}. \tag{21}$$

We remark that an expression similar to the one in (b) is obtained for $\hat{h}_{\mathfrak{p}}$ if $E$ is defined over the completion $K_{\mathfrak{p}} \cong \mathbb{C}$ of $K$ with respect to an archimedean place $\mathfrak{p}$ (see [3], [18], [19]).

The calculation of the canonical global height $\hat{h}$ requires another theorem due to Silverman ([8], *cf.* also [14]).

**Theorem 8.** *Let the elliptic curve $E$ over a local field $K_{\mathfrak{p}}$ with a nonarchimedean place $\mathfrak{p}$ be given by a $\mathfrak{p}$-minimal equation* (1). *Then a rational point $0 \neq P \in E_0(K_{\mathfrak{p}})$ has canonical local height*

$$\hat{h}_{\mathfrak{p}}(P) = d_{\mathfrak{p}}(P) + \frac{1}{12}v_{\mathfrak{p}}(\Delta).$$

We shall now describe a procedure for calculating the canonical global height $\hat{h}$ on the rational point group $E(K)$ of an elliptic curve $E$ over an algebraic number field $K$. The procedure carries over to the function field case too. As pointed out above, we proceed by first computing the canonical local heights $\hat{h}_{\mathfrak{p}}$ and then summing up the $\hat{h}_{\mathfrak{p}}$ in accordance with the sum formula (19). We have to distinguish between four cases.

Case 1: If $E$ over $K$ has potentially good reduction modulo $\mathfrak{p}$ for $\mathfrak{p} \in M_K^0$, we employ part (a) of Theorem 7 to compute $\hat{h}_{\mathfrak{p}}$ by means of the formula

$$\hat{h}_{\mathfrak{p}}(P) = d_{\mathfrak{p}}(P) + \frac{1}{12}v_{\mathfrak{p}}(\Delta) \quad \text{for } 0 \neq P \in E(K).$$

Case 2: If $E$ over $K$ has split multiplicative reduction modulo $\mathfrak{p}$ for $\mathfrak{p} \in M_K^0$, we employ part (b) of Theorem 7 to compute $\hat{h}_\mathfrak{p}$ by means of the formula

$$\hat{h}_\mathfrak{p}(P) = \frac{1}{2} B_2(\alpha(t)) v_\mathfrak{p}(q) \quad \text{for } O \neq P \in E(K)$$

which holds provided that the parameter $t \in K^\times$ of the given point $P = (\wp(t), \wp'(t)) \in E(K)$ is normalized to

$$0 \leq v_\mathfrak{p}(t) < v_\mathfrak{p}(q).$$

In that formula, the expression $\alpha(t)$ defined by (21) is calculated by virtue of the relation (see [12], [20])

$$\alpha(t) = \frac{v_\mathfrak{p}(\Psi_2(P))}{v_\mathfrak{p}(q)}$$

involving the polynomial $\Psi_2$ defined by (5) and the value $v_\mathfrak{p}(q)$ known from (20).

Case 3: If $E$ over $K$ has additive reduction modulo $\mathfrak{p}$ for $\mathfrak{p} \in M_K^0$ while

$$v_\mathfrak{p}(j) < 0,$$

we utilize the fact implied by Theorem 6 that, for any point $P \in E(K)$,

$$mP \in E_0(K) \quad \text{for } m \in \mathbb{N},\ m \leq 4.$$

Then Theorem 8 gives

$$\hat{h}_\mathfrak{p}(mP) = d_\mathfrak{p}(mP) + \frac{1}{12} v_\mathfrak{p}(\Delta) \quad \text{for } O \neq P \in E(K).$$

From this formula, the value $\hat{h}(P)$ can be calculated on the basis of the relation (2) (ii) in Theorem 2.

Case 4. If $E$ over $K$ is considered as a curve over the completion $K_\mathfrak{p}$ of $K$ at a place $\mathfrak{p} \in M_K^\infty$, we proceed as follows. We have

$$K_\mathfrak{p} \cong \mathbb{R} \quad \text{or} \quad K_\mathfrak{p} \cong \mathbb{C}.$$

In the case in which $K_{\mathfrak{p}} \cong \mathbb{C}$, Silverman [12] has suggested a modification of a fast–converging series proposed by Tate for calculating the canonical local height $\hat{h}_{\mathfrak{p}}$. Tate's original series applies to the case in which $K_{\mathfrak{p}} \cong \mathbb{R}$ and was used in [14] for computing $\hat{h}_{\mathfrak{p}}$.

We restrict ourselves here to giving an account of the method that was employed in [14] in the case of $K_{\mathfrak{p}} \cong \mathbb{R}$. Tate's series is described in the theorem below. In this theorem,

$$\Gamma = \{P = (x, y) \in E(\mathbb{R});\ x \neq 0\} \leq E(\mathbb{R})$$

is the open subgroup of $E(\mathbb{R})$ consisting of all points with non–zero $X$–coordinate. Furthermore, $v_{\mathfrak{p}}$ is the additive absolute value on $K_{\mathfrak{p}} \cong \mathbb{R}$ defined by (6).

**Theorem 9.** *For an elliptic curve $E$ over the field of real numbers $K_{\mathfrak{p}} \cong \mathbb{R}$, the canonical local height $\hat{h}_{\mathfrak{p}}$ on $E(\mathbb{R})$ is calculated as follows. For a point $P = (x, y) \in \Gamma$ such that $2^{\nu} P \neq 0$ for all $\nu \in \mathbb{N}$, put*

$$T_0 = \frac{1}{x}, \quad T_{\nu+1} = \frac{W_{\nu}}{Z_{\nu}} \quad \text{for } \nu \in \mathbb{N}_0,$$

*where*

$$W_{\nu} = 4T_{\nu} + b_2 T_{\nu}^2 + 2b_4 T_{\nu}^3 + b_6 T_{\nu}^4,$$
$$Z_{\nu} = 1 - b_4 T_{\nu}^2 - 2b_6 T_{\nu}^3 - b_8 T_{\nu}^4.$$

*Then*

$$\hat{h}_{\mathfrak{p}}(P) = -\frac{1}{2} v_{\mathfrak{p}}(x) - \frac{1}{8} \sum_{\nu=0}^{\infty} \frac{v_{\mathfrak{p}}(Z_{\nu})}{2^{2\nu}} + \frac{1}{12} v_{\mathfrak{p}}(\Delta). \tag{22}$$

This is Tate's fast–converging series.

**Proof** (see [13]). We first remark that the restriction to points $P \in \Gamma$ imposed in Theorem 9 does not mean a loss in generality. For by a suitable birational isomorphism (3) with $\rho \neq 0$, the curve can be transformed into an isomorphic curve $E$ for which

$$\Gamma = E(\mathbb{R}).$$

In fact, by applying to $E$ another birational isomorphism, one can even ensure that $x > 1$ holds for all points $P = (x, y) \in E(\mathbb{R})$. In the sequel we shall assume this to be the case.

In the convergence proof regarding the infinite series for $\hat{h}_{\mathfrak{p}}$ in (22) and in the verification of properties (2) (i), (ii) in Theorem 2 for $\hat{h}_{\mathfrak{p}}$, the following estimates are required (see [13], [18]).

**Lemma.** *For an elliptic curve $E$ over any local field $K_{\mathfrak{p}}$, the modified local height $d_{\mathfrak{p}}$ on $E(K_{\mathfrak{p}})$ satisfies the inequalities*

$$\frac{1}{2}(6\mu_{\mathfrak{p}} - v_{\mathfrak{p}}(\Delta)) + 5\alpha_{\mathfrak{p}}$$
$$\leq d_{\mathfrak{p}}(P+Q) + d_{\mathfrak{p}}(P-Q) - 2d_{\mathfrak{p}}(P) - 2d_{\mathfrak{p}}(Q) - v_{\mathfrak{p}}(x_P - x_Q) \leq -2\alpha_{\mathfrak{p}} \quad (23)$$

*for $P = (x_P, y_P)$, $Q = (x_Q, y_Q) \in E(K_{\mathfrak{p}})$ such that $P, Q, P \pm Q \neq O$;*

$$\frac{1}{2}(6\mu_{\mathfrak{p}} - v_{\mathfrak{p}}(\Delta)) + 4\alpha_{\mathfrak{p}}$$
$$\leq d_{\mathfrak{p}}(2P) - 4d_{\mathfrak{p}}(P) - v_{\mathfrak{p}}(\Psi_2(P)) \leq -\frac{3}{2}\alpha_{\mathfrak{p}} \quad (24)$$

*for $P \in E(K_{\mathfrak{p}})$ such that $2P \neq O$.*

The proof of this lemma will be provided in the above mentioned forthcoming paper.

For proving Theorem 9, we shall choose $g = 2$ in the limit formula (16) of Theorem 3. Let $P = (x, y) \in E(K_{\mathfrak{p}})$ be any point such that $2^m P \neq O$ for all $m \in \mathbb{N}$. We introduce the notation

$$s_m(P) = \frac{d_{\mathfrak{p}}(2^m P) - \frac{1}{2}v_{\mathfrak{p}}(\Psi_{2^m}^2(P))}{2^{2m}} + \frac{1}{12}v_{\mathfrak{p}}(\Delta)$$

for the $m$-th term in the sequence (16) of Theorem 3 and

$$t_m(P) = -\frac{1}{2}v_{\mathfrak{p}}(x) - \frac{1}{8}\sum_{\nu=0}^{m-1} \frac{v_{\mathfrak{p}}(Z_\nu)}{2^{2\nu}} + \frac{1}{12}v_{\mathfrak{p}}(\Delta)$$

for the $m$-th partial sum of the series (22) in Theorem 9.

Then our task is to provide convergence proofs for the limit relations

$$\hat{h}_{\mathfrak{p}}(P) = \lim_{m \to \infty} s_m(P), \tag{25}$$

corresponding to (16) in Theorem 3, and

$$\hat{h}_{\mathfrak{p}}(P) = \lim_{m \to \infty} t_m(P), \tag{26}$$

corresponding to (22) in Theorem 9.

To prove (25), one uses the Lemma to derive the estimate

$$\left|\hat{h}_{\mathfrak{p}}(P) - s_m(P)\right| \le \frac{1}{2^{2m}} \max\left\{\left|\tfrac{1}{6}(6\mu_{\mathfrak{p}} - v_{\mathfrak{p}}(\Delta)) + \tfrac{4}{3}\alpha_{\mathfrak{p}}\right|, -\tfrac{1}{2}\alpha_{\mathfrak{p}}\right\}. \tag{27}$$

From (27) we shall derive a corresponding estimate for $\left|\hat{h}_{\mathfrak{p}}(P) - t_m(P)\right|$, which will prove the relation (26), by finding bounds for the difference $t_m(P) - s_m(P)$. This is accomplished by means of the multiplication polynomials

$$\Phi_m, \Psi_m \in \mathbb{Z}[b_2, b_4, b_6, b_8][X, Y]$$

from (4) in the following manner. Writing the coordinates of the $2^{\nu}$–multiples of $P \in E(K_{\mathfrak{p}})$ in the form $2^{\nu}P = (x_{2^{\nu}P}, y_{2^{\nu}P})$, we find for the auxiliary quantities in Theorem 9 the following formulas (see [14]):

$$T_{\nu} = \frac{1}{x_{2^{\nu}P}}, \qquad W_{\nu} = \frac{\Psi_2^2(2^{\nu}P)}{x_{2^{\nu}P}^4}, \qquad Z_{\nu} = \frac{\Phi_2(2^{\nu}P)}{x_{2^{\nu}P}^4}.$$

Employing these formulas and taking into account the recursion formulas (5), we end up with the estimate (see [13])

$$\left|t_m(P) - s_m(P)\right| \le \frac{\left|\mu_{\mathfrak{p}}\right|}{2^{2m}}. \tag{28}$$

The inequalities (27) and (28) can be combined into

$$\left|\hat{h}_{\mathfrak{p}}(P) - t_m(P)\right| \le \frac{1}{2^{2m}}\left[\max\left\{\left|\tfrac{1}{6}(6\mu_{\mathfrak{p}} - v_{\mathfrak{p}}(\Delta)) + \tfrac{4}{3}\alpha_{\mathfrak{p}}\right|, -\tfrac{1}{2}\alpha_{\mathfrak{p}}\right\} + \left|\mu_{\mathfrak{p}}\right|\right] \tag{29}$$

This proves (26) and at the same time unravels the speed of convergence of Tate's series (22) in Theorem 9 which is used in the calculation of the canonical local height $\hat{h}_{\mathfrak{p}}$ at the (archimedean) places $\mathfrak{p} \in M_K^{\infty}$ such that $K_{\mathfrak{p}} = \mathbb{R}$.[1]

In conclusion, we mention that in [14] a somewhat less efficient procedure was employed for calculating the canonical global height $\hat{h}$ on the rational point group $E(K)$ of the curve $E$ over the field $K = \mathbb{Q}$. The procedure was efficient enough however, to be applied to estimating the constants in the conjecture of Lang (see [5], [20]) about upper bounds for the values of $\hat{h}$ at the points $P_1, \ldots, P_r$ of a basis of the free part of $E(\mathbb{Q})$, where the basis had to be chosen according to ascending height values.

## References


[1] *M. Hindry* and *J.H. Silverman,* The canonical height and integral points on elliptic curves. Invent. math. **93** (1988), 419—450.

[2] *S. Lang,* Les formes bilinéaires de Néron et Tate. Sém. Bourbaki no. **274** / 1—11, mai 1964.

[3] *S. Lang,* Elliptic Curves: Diophantine Analysis. Springer–Verlag, Berlin and New York 1978.

[4] *S. Lang,* Fundamentals of Diophantine Geometry. Springer–Verlag, Berlin and New York 1983.

[5] *S. Lang,* Conjectured Diophantine estimates on elliptic curves. Progr. in Math. **35**, 155—171, Birkhäuser–Verlag, Basel 1983.

[6] *Ju.I. Manin,* Cyclotomic fields and modular curves. Russian Math. Surveys **26** (1971), 7–78.

[7] *A. Néron,* Quasi–fonctions et hauteurs sur les variétés abéliennes. Ann. of Math. **82** (1965), 249—331.

[8] *J.H. Silverman,* Lower bound for the canonical height on elliptic curves. Duke Math. J. **48** (1981), 633—648.

[9] *J.H. Silverman,* Lower bounds for height functions. Duke Math. J. **51** (1984), 395—403.


---

[1] In MR 87m : 14025, the reviewer of [14] points out that no error estimates had been provided for the $t_m$. However, it is exactly (29) that we had used in [14].

[10] *J.H. Silverman,* The Arithmetic of Elliptic Curves. Springer–Verlag, Berlin and New York[2] 1986.

[11] *J.H. Silverman,* A quantitative version of Siegel's theorem: integral points on elliptic curves and Catalan curves. J. reine angew. Math. **378** (1987), 60—100.

[12] *J.H. Silverman,* Computing heights on elliptic curves. Math. Comp. **51** (1988), 359—373.

[13] *H.M. Tschöpe,* Berechnung der lokalen Néron–Tate Höhe von Punkten auf elliptischen Kurven. Diploma thesis, Saarbrücken 1985.

[14] *H.M. Tschöpe* and *H.G. Zimmer,* Computation of the Néron–Tate height on elliptic curves. Math. Comp. **48** (1987), 351–370.

[15] *H.G. Zimmer,* An elementary proof of the Riemann hypothesis for an elliptic curve over a finite field. Pacific J. Math. **36** (1971), 267—278.

[16] *H.G. Zimmer,* On the difference of the Weil height and the Néron–Tate height. Math. Z. **147** (1976), 35—51.

[17] *H.G.Zimmer,* Generalization of Manin's conditional algorithm. Proc. 1976 ACM Sympos. on Symb. Alg. Comp., Yorktown Heights, N.Y., 1976, 285—299.

[18] *H.G. Zimmer,* Quasifunctions on elliptic curves over local fields. J. reine angew. Math. **307/308** (1979), 221—246.

[19] *H.G. Zimmer,* Corrections and remarks concerning quasi–functions on elliptic curves. J. reine angew. Math. **343** (1983), 203—211.

[20] *H.G. Zimmer,* Computational aspects of the theory of elliptic curves. To appear in Proc. NATO–ASI Conf. Number Theory and Applications, Banff, Canada 1988.

---

Fachbereich 9.1 Mathematik, Universität des Saarlandes, D–6600 Saarbrücken.